MW01136905

IRON FIST

The Lives of Carl Kiekhaefer

ELMER CARL KIEKHAEFER

1906-1983

IRON FIST

The Lives of Carl Kiekhaefer

JEFF RODENGEN

Write Stuff Syndicate, Inc.

1108 Citrus Isle
Fort Lauderdale, Florida 33315
(305) 462-6657

Library of Congress Catalog Card Number: 91-67461

ISBN 0-945903-04-9

Produced in the United States of America

10 9 8 7 6 5 4 3 2

Pour Karine

Acknowledgments

Very special gratitude is extended to the Kiekhaefer family: Freda Kiekhaefer, Frederic Carl Kiekhaefer, Dr. Carol Kiekhaefer, Anita Kiekhaefer, Helen Kiekhaefer Wimberly, and the sisters of the late Elmer Carl Kiekhaefer, Florence Kiekhaefer Brandt; Isabelle Kiekhaefer Gottinger, Palma Kiekhaefer Moerschel, Ruth Kiekhaefer Pike, Marion Kiekhaefer Scheuneman and Almira Kiekhaefer Westendorf.

Without question, the most critical resources for this work have been the unselfish time and candor of Rose Smiljanic, Charles D. Strang and the late James R. Wynne. Combined, these individuals gave hundreds of hours of their time to the memory of Carl Kiekhaefer, and without whose help this book would not have been possible. The author would also like to acknowledge the overwhelming support given to the preparation of this work by his wife Karine, who gave many thousands of hours of her time to assist in the retrieval and organization of the Rodengen/Kiekhaefer Research Library assembled for the project. The author would also like to acknowledge the generous support of the Dorgambide family, Frederique and Jean, for their many individual favors and assistance throughout the life of the project. Special thanks is also extended to Grace Silvis, transcriber, who struggled through endless hours of recorded interviews to accurately memorialize the oral histories critical to the life of Carl Kiekhaefer. Special thanks are also extended to Clement J. Koehler, director of public relations for Mercury Marine, who suggested to the author to write a book on the life of Carl Kiekhaefer, and whose initial advice, "Look behind the public icon," has served the author well.

The following individuals donated generously of their time for in-depth interviews, in some cases over several days, and in others repeated interviews over many years. Their candor, honesty, trust and courtesy is greatly appreciated. Among these are: Alexander, Charles F.; Anderegg, Robert C.; Anderson, Joseph C.; Anthony, Joseph; Backus, Ruth Olive; Bakos, Johnny; Baurle, Paul M.; Behnisch, Ernest; Brandt, Florence

Kiekhaefer; Callaghan, Brian; Castle, Donald E.; Chapman, James; Coll, Harry H.; Conover, Clay; Cotter, Jack; Craig, David; Crouse, John; Dumbolton, Joseph; Edgarton, Allan L.; Edwards, Gen. Mabry; Garbrecht, Gary; Geer, Abbot; Genth, Richard; Gilchrist, George; Gores, Bill; Gores, Stan; Gottinger, Isabelle Kiekhaefer; Hamberger, Sylvan L. "Ham"; Hauser, Armand; Hull, John C.; Jacobs, Irwin; Jones, C.W. "Doc"; Jones, Ted O.; Jordan, Richard; Jost, James H.; Karls, Ted; Kiekhaefer, Anita; Kiekhaefer, Dr. Carol; Kiekhaefer, Freda; Kiekhaefer, Frederic Carl; Klumb, Vernon; Koehler, Clement J.; Kowalski, Joseph; Kraus, Gerald W.; Lambrecht, Ralph; Lechnir, Thomas P.; McClean, Eugene; Meyer, Mildred; Meyer, Wayne; Moerschel, Palma Kiekhaefer; Morris, Andrew L. "Chick"; Mundy, Frank; Nabb, Edward; Napier, Jeff; Newberg, William C.; Oleson, Niel F.; Page, Roger William; Pike, Ruth Kiekhaefer; Poberezny, Paul; Powers, Allison; Powers, Richie; Predmore, Wayne Herbert; Reichert, Jack F.; Rose, Edgar; Rudeen, Luisa; Scalpone, Francis J. "Frank"; Shoenfeldt, Fred "Fritz"; Smiljanic, Rose; Snyder, Richard; Snyder, Wilson; Sommerville, Vinton H. "Slim"; Steele, Owen L. "Billy"; Stevenson, John C.; Stieg, Herman; Strang, Charles D.; Stuth, Robert; Swift, Joseph; Thornton, Ray; Wagner, Gene; Weigt, Thomas; Westendorf, Almira K. Kiekhaefer; Whiting, John; Wiklund, Harald; Wimberly, Helen Kiekhaefer; Wynne, James R.

During the nearly six years of research for this book, a great many individuals and organizations assisted the author generously with their time, materials, occasional and informal interviews and services. The author regrets that some may inadvertently not be listed, and to all the author is exceedingly grateful for their support. Among these are: American Power Boat Association (APBA); Antique & Classic Boat Society; Antique Outboard Club; *Air Progress* magazine; *Atlanta Journal 500* magazine; Baldiotti, Regina; Berens, Abram, M.D.; Bishop, Jack; Bishop, Kay; *Boating* magazine; *Boating Business* magazine; *Boating Industry* magazine; *Boston Sunday Globe*; Bonno, Paula K.; Brunswick Corporation; Buchholz, Harley; Butler, Denise; Butler, Mike; *Cedarburg News*; *Cedarburg News Graphic*; Chapus, Jean-Pierre; Chapus, Jeremie; Chapus, Marie-Francoise; Chapus, Mathieu; *Chicago Journal of Commerce*; *Cincinnati Post*; Coigney, Joel; Collinson, Robert; Conover, Phyllis; Craig, Jeanne; D'arcy, Kelly; Dedgwick, Alice; Diamandis Communications, Inc.; *Dock Lines*, newsletter of Evinrude

Motors; Dorgambide, Frederique; Dorgambide, Jean; Evinrude Motors; *Forbes* magazine; *Fortune* magazine; Fisk, Judy; Fisk, William; Fond du Lac *Commonwealth Reporter*; Fond du Lac *Reporter*; Fond du Lac Library; Fond du Lac Tourist & Convention Bureau; Fort Lauderdale Public Library; Franzen, Lynn; *Globe and Mail*; *Go Boating* magazine; Graphic Dynamics, Inc.; Gypsy the Research Vehicle; Hanna, Vic; Hart, Dex; Hasselbalch; Kurt; Henkel, Jean; Henkel, Peter; Holmes, Deona; Holmes, Jack; Huck, Edward; Jaeger, Mark; Jansen, Phil; Janssen, Peter A.; Johnson Outboards; Kahn, Sharon; Keim, John; Kogan, Rick, author of *Brunswick, The Story of an American Company from 1845 to 1985*; Kiekhaefer Aeromarine; Kozyak, John; Kuykendall, Ron; *Lakeland Boating* magazine; Larson, Barbara; Lehner, Lucy; *Little Rock Gazette*; *Marian*, newsletter of Marian College; *Mirror News*; Marchello, Joseph; McClean, Eugene; *MercComment*, employee newsletter of Mercury Marine; Mercury Marine; *Mercury News*, Mercury Marine employees newsletter; *Mercury Outboard Motor News*, employee newsletter of the Kiekhaefer Corporation; Mequon Library Mequon Men's Club; *Miami Herald*; *Miami Pictorial*; Michalowski, Carla; Miller, Fred; *Milwaukee Journal*; *Milwaukee Sentinal*; Mitcheltree, Sandy; Modas, Dan; Modas, Lynn; *Mopar* magazine; Mossberg, Kate; *MoToR BoatinG* magazine; *Motorboating & Sailing* magazine; *Motor Guide* magazine; Mueller, Tom; National Marine Manufacturer's Association (NMMA); Naugle, Cindy; *New York Herald Tribune*; *New York Times*; *Old Cars* magazine; Outboard Marine Corporation; Owens, John; Ozaukee County Historical Society; *Ozaukee County-Wee Wonder Cedarburg News*; *Ozaukee Press*; *Peterson's Circle Track* magazine; Peterson, Judy Rodengen; *Popular Mechanics*; *Powerboat* magazine; *Profile*, Brunswick employee magazine; *Reader's Digest*; Rodengen, James M.; Rubinson, David; *Saga* magazine; Sass, Karen; Schall, Donna; Schlump, John; Schryver, Doug; Shattuck School; Siebold, Douglas; Silvis, Edward; *Speed Age* magazine; Silvis, Grace; Spohnholz, Diane; *Sports Illustrated* magazine; Skiff, Gail; Stein, Mark; Stemer, Ross; *Stock Car Racing* magazine; Thousand Islands Shipyard Museum (Antique Boat Museum); *Time* magazine; Toriello, Larry; Titcombe, Lloyd M.; *Toronto Star*; *True*, The Man's Magazine; Van Vleet, et al, authors of *The Four Men From Terre Haute*; Vesper, Mary; *Wall Street Journal*; Ward, Joanne; Wendt, Alice; *Yachting* magazine; Zipps, Robert.

CONTENTS

Foreword

"It must be awfully nice to be a writer and be privileged to point to anyone from on high and never be pointed back at. How would you like to compare notes, if it were possible, on respective accomplishments on our day of reckoning -- as well as our wrongs?"[1]

Carl Kiekhaefer

"Tell Charlie to forgive me," are acknowledged to be the last words of E.C. "Carl" Kiekhaefer. 'Charlie' is Charles D. Strang, the brilliant M.I.T. professor turned marine engineering patriarch. He was Carl's closest friend, shared an office with the eccentric industrialist for twelve years, and is today chairman of the board of Outboard Marine Corporation, the billion dollar manufacturer of Johnson and Evinrude outboards. It is among only a scant few times during a stormy and controversial life that Kiekhaefer would acknowledge a mistake in judgment. Carl had precious little time for excuses and apologies, and considered those who extended them weak and prone to fault. For him, action was everything. In the face of obstacles, the only thing to do was press on, work around, over, or simply crash through the problem with absolute stamina.

Carl Kiekhaefer was proud of his celebrated capacity for work. To achieve his goal of making Mercury, the company he founded, the largest manufacturer of marine propulsion in the world, he would sacrifice anything except quality. He worked evenings, Saturdays, Sundays, holidays, Christmas Day, and wouldn't even stop work while his wife was giving birth to his three children. He once promoted his chief engineer to vice president because "he was the only guy who could stay awake working as late as I could." His managers and executives cringed when the telephone rattled at all hours of the evening and while they were on vacation. He would boast, "I pay my men twice what they're worth, and then I make 'em earn it!" He *exploded* with fury when his world record-breaking racing teams in stock cars, hydroplanes, offshore ocean and outboard motor powered boats came in second. He fired his key managers and top executives over and over again, only to call them at home days later and howl at them for not being on the job. Dissatisfied with workmanship, he threw a man's entire tool chest through his factory's plate glass windows in the middle of a sub-freezing Wisconsin winter, followed by the worker himself.

Every fiber of Kiekhaefer's business life vibrated with passion. To stimulate his national dealers to fever pitch he would hang brand new models of Johnson and Evinrude outboard motors over enormous bonfires ablaze in the woods near his factories in Fond du Lac Wisconsin, exhorting the nearly frenzied participants to "kill the enemy" as the engines melted to globs of charred aluminum in the inferno. He modeled his secret proving grounds after military installations, complete with armed patrols around the property perimeters and kept his men "confined to base" during periods of secret engine testing at the mysterious and private *Lake X*.

Though the demands Kiekhaefer placed on his employees were extraordinary, he generated the deepest measure of loyalty and awe from them, most admitting that he had absolutely, unquestionably changed their lives. A club was formed of men who had been fired by Kiekhaefer, this time never to return, known as the A.O.K., or Alumni of Kiekhaefer. Among the alumni are the top leaders of today's $18 billion marine industry. To poke fun at the prestigious group, Carl once named his champion offshore race boat *"A.O.K."*, competing and winning against his incensed rivals. His legendary zest for racing competition on both land and sea was even made the focus of an entire comic book modeled after his total commitment and frequently bizarre antics in pursuit of absolute victory. Through his enormous instincts for publicity, his Mercury outboard motors became known throughout the world for speed and durability. He bought an entire lake in Florida, including all the land surrounding it, and transformed it into the most prestigious marine high-performance proving grounds in the world. He ran his engines with teams of drivers day and night for 50,000 miles to prove the reliability of his products. He pulled elephants on water skis to demonstrate the raw power of his designs. When ground was broken for his new *mile long* factory it was Carl Kiekhaefer, hand-rolled Cuban cigar jammed into the corner of his grinning mouth, completely in his element, driving the enormous earthmover to signal the start of construction at the world's largest outboard motor plant. Carl was big, he was tough, he was brilliant, he was vulnerable.

IRON FIST: *The Lives of Carl Kiekhaefer* carefully traces the humble origins of this legendary industrialist, from the homesteading heritage of his German ancestry, through his lonely and awkward youth and stormy adolescence, his years of dues-paying struggle in the factories of Milwaukee, the perilous and exasperating beginnings of outboard production, the marketing conquests of the postwar years, his unprecedented victories and heart-pounding strategies in stock car and boat racing, the spectacular rise to the top of a worldwide marine empire, and his tragic decline and excruciating death in 1983. During Carl's extraordinary career, he was granted over 200 patents for his innovations in products ranging from outboard motors to snowmobiles to skylights. His blustering, bellowing and intimidating style made him the curious target of a worldwide press corps, eager to be included in one of the many classic Kiekhaefer stories that circulated wildly throughout his realm. Carl was upset and buried a brand new Cadillac with a bulldozer at his *Lake X* proving grounds in Florida, a favorite one begins. Carl, upon seeing a worker sitting on a crate drinking a soft drink, fired the man on the spot, paid him two weeks wages and threw him out, only to discover later that he was the delivery man bringing Coca-Cola to one of his factories, another Kiekhaefer tale promises. The one-time farmhand from Wisconsin became friends with the elite leaders of American industry, Governors and Senators. His piquant and vocal observations about himself, his intriguing industry and the American promise that he had realized, are on a par with Jonathon Swift, Benjamin Franklin and Will Rogers. "Wherever I am going in the future," he wrote ten years before his death, "heaven or hell, I do hope they have an engineering department because I might have some fun being reduced back to a plain, old engineer. The simplest and sweetest things in life are complex engineering; it's the people problem that's responsible for today's ills. I hope some day I can still get a job as a draftsman or as an engineer with a boss who can hand me the work on a platter -- with no other responsibilities except that it works."[2]

Recognizing the unusual scope of his career and accomplishments, Carl occasionally considered the possibility of writing a book about his life. Of his prospective place in history, Carl was certain.

"A good many of the ideas that brought innovation ... originated with small entrepreneur inventors such as Henry Ford I. ... Thomas Edison was certainly another as was Alexander Graham Bell. In a very modest, more modern version, I would like to mention Carl Kiekhaefer, who with his innovations and inventions, made the smallest company the largest in marine propulsion in a matter of 20 years. ..."[3]

"As to my book," Carl wrote, "I think I shall wait until I will have some time to reflect on my life and the past and then record what has happened in my own words. This, of course, if the Good Lord will allow me the time. ..."[4] He never would find time apart from his perpetual work, and passed away on October 15, 1983 without having commenced an autobiography. Fortunately, he left behind a largely intact archive of his life's work, including over a million documents, and nearly 100,000 pages of correspondence. Remarkably, he recorded many of his crucial telephone conversations, with friend and foe alike, which were transcribed and filed away for posterity.

Among the smorgasbord of Carl's life-long paranoia, was the fear that an author would someday write a biography of his life that was incomplete, or that focused on only a single facet or two of his complex life. "When it comes to Elmer Carl," he said of himself, "I am afraid this would fill a pretty good sized volume in itself, not because of any inflated ego, but the life intensity that developed since leaving the farm amazes even myself -- what a human being can accomplish when he finds his niche in life."[5]

He also feared that a biographer might devote undue emphasis to his weaknesses, and allow them to overshadow the great breadth of his accomplishments.

"The writers usually take the privilege of casting one in one light or another -- sometimes in a sarcastic light and sometimes in a humorous or condescending light. Once the writers get going on you, there seem to be a lot of additional writers who pick up the note and continue along the same vein. ... I feel that some dignity should remain for the work accomplished and the success maintained. ... When it gets time to write a book, I would like to do it myself and 'tell it like it is' from where I sit."[6]

Carl feared that his flaws and indiscretions might be blown out of proportion, explaining, "Many biographies are boring if the individual is not an internationally-known figure or it is not handled in a risque and sensational manner."[7]

Over 300 individual interviews were conducted during research for IRON FIST: *The Lives of Carl Kiekhaefer*, resulting in nearly 10,000 pages of transcribed interviews. In the nearly six years of research dedicated to the work, over a million documents were read, and over 100,000 were painstakingly photocopied, catalogued, indexed and studied in-depth to insure the most accurate possible assessment and portrayal to separate the man from the myth. It is interesting to note that of the hundreds of individuals that "knew Carl well" or "were friends" or "close associates," their descriptions often read as if they were each describing some different person. Carl's impact on others was seldom subtle, and those who interacted with him most closely were often riveted by his powerful personality, and consequently perceived him within the confines of their own special relationship. A mosaic began to take form from these many perspectives, which allowed a very elusive and unique subject to reveal himself. Wherever possible, Carl himself has been allowed to tell his story, to "tell it like it is," from the many documents which were preserved throughout his long life.

His life is instructive. His bold approach to business and his tragic family life are intertwined throughout his career in unique ways. His deep-rooted principles of success, derived from his humble origins, combined to make him among the most interesting, and the most lonely, figures in modern industry. Of this unavoidable paradox, Carl was defensive to the very end.

> "I don't know what image people have in their minds when they think of a company president, but in most large companies, I have found no one puts in more hours or more days per week than the president. ...
>
> "Very few people understand this and it is one of the paradoxes of success that the farther up the ladder you go, the more lonesome the world becomes. But the greater the success, the sweeter the satisfaction and the greater the peace of mind eventually; peace of mind that can only come from from accomplishment, not money, not popularity but the accomplishment of contribution and the recognition that goes with it.
>
> "I find dreams of this type to be one of the wonders of the world and the human race. It is a miracle what can be built from a dream with nothing but two hands, an honest heart and ambition."[8]

To those who worked with Carl Kiekhaefer, even briefly, no matter how great their future accomplishments, nothing seems to compare to those glory- and agony-filled days of industrial drama and pure excitement that swirled in his wake wherever he ventured. He is today revered as an icon of invention and tireless energy, an industrial Caesar of a marine industry empire, and competitive gladiator in perpetual combat. He was, though, a perplexingly complex man with a dizzying array of emotional impediments and deeply rooted fears of rejection, failure and love. The clash of his oversized abilities and fragile emotions combined to produce one of the most unforgettable individuals in 20th century industry.

CHAPTER ONE

Origins

"To be German means to carry on a matter for its own sake."

Richard Wagner[1]

It wasn't always spelled 'Kiekhaefer', or pronounced 'Key Kay Fur'. Three centuries before the birth of Elmer Heinrich Carl Kiekhaefer the family name was Kieckhofel, 'Key Koh Fell', meaning "look inside the home" in low German. In fact, a young Elmer spoke low German until his grade school years, well enough to flawlessly recite religious catechisms in the stern dialect. His rugged Teutonic ancestors plowed and planted the fields of Trieglaff, State of Pomerania in northeastern Europe, in a section of historical Germany which would be ceded to Poland following the Second World War. Lying on the Baltic coastal plains between the Oder and Vistula rivers, the region also known as West Prussia was in the grasp of an unjust Prussian aristocracy, when one of the oppressed clan fled to America. Thirty-five year-old Carl Gottfried Kieckhofel, along with his wife Friedericka and two children, arrived by ship to American shores in 1850, settling near Mequon, Wisconsin, 15 miles north of a small village known as Milwaukee.

The brisk and robust climate of easternmost Wisconsin, on the shores of massive Lake Michigan, were comparable to the Baltic shores of mother Prussia, and held the added allure of thickly wooded, beautifully rolling lands for $1.25 an acre. Carl Kieckhofel began to clear the lands, shipping raw timber to Milwaukee from his wharf on the great lake, and constructed a modest homestead that was to be the birthplace and legacy to three more generations. Within ten years, he had two more children, 4 "milch" cows, two plowing oxen, a few cattle, sheep and swine, and had improved thirty acres sufficiently to produce wheat, rye, oats, peas, beans, potatoes, barley and Indian corn beyond the requirements of his family's own needs.

It was a solemn, meagre and back straining existence for both Carl and his wife Friedericka. Long hours of manhandling the plow behind the plodding wake of the oxen for him, and even longer hours of cooking, cleaning, tending to the children and baking bread in outdoor ovens for her. A typical fare of delights for the Kieckhofel table included "suelze or blood sausage and gruetz wurst, a sausage made from buckwheat groats ground up with pork and pork blood; Schwarzauer, a stew made of goose and duck wings, livers, gizzards and hearts cooked with apples, plums, blood, and dumplings; and Spickgans - breast and shanks of goose [soaked] in brine for 9 days and then smoked."[2]

Carl Kieckhofel continued to acquire land and harvest the rich hardwood timber until the family holdings swelled to two-hundred acres. By 1870 the growing brood were spelling their name 'Kiekhaefer', which is the present form. When Carl and Friedericka retired in 1876, the homestead was taken over by their youngest son, Heinrich (Henry), who had to stipulate to thirteen stringent conditions for the privilege of obtaining title to the homestead. Designed to guarantee an uncomplicated and secure retirement for his mother and father, Heinrich agreed to pay $1500 outright for the two-hundred acre homestead, reserve for their exclusive use the east wing of the house, and provide:

2. "Sufficient fire wood, fit for the stove, consisting of good, sound & dry timber and delivered at their dwelling.
3. One Barrel of good Rye flour, four Barrels of good wheat flour & Ten Bushels of Potatoes annually.
4. One fat hog weighing at least 280 pounds annually.
5. Thirty Dollars in cash, lawful currency of the United States of America, annually.
6. Eight pounds of good wool, annually.
7. One quart of milk, daily. Two pounds of butter, weekly, & one half a dozen eggs weekly.
8. One Barrel of salt annually.
9. The free use of one horse team, at any time required, & also the privilege to keep one cow & sufficient fodder. ...
10. Good and careful attendance and medical aid in case [Carl and Friedericka] are getting old, sick or helpless.
11. One half of all the fruit raised in the orchard of the farm ... and also the free use of one eight [sic] of an acre for a garden. ...
12. To pay the sum of Fifty Dollars at the time of marriage of Anna Kiekhaefer.
13. To do the washing, cleaning in case of sickness of said Carl or his wife.[3]

Heinrich has been described as "strict but just", a "humble, self-educated man." According to family tradition, he used to tell the story of how, when he was a boy, strong rumors of an imminent Indian attack caused great panic and preparations throughout the homestead, but the attack never materialized. Though Heinrich never attended high school, he sought out the instruction of a nearby retired teacher, and studied diligently enough to impress his family with recitations of mathematical square roots and other advanced subjects. He would often extract from the newspaper new words, and challenge the family to reveal their meaning. He also owned a massive, 2,200 pound bull named Hans that "had the misfortune of falling into a well in the barn. The hired man found him standing on his hind legs with his front feet braced against the side to keep his head above water. The water was pumped out and the bull was coaxed onto hay thrown in until he was able to climb from side to side to the top."[4]

The deed to the homestead in his pocket, Heinrich, along with his wife Augusta, cleared the stumps from the fields left by his father's logging operations, and slowly transformed the rough landscape into a dairy farm. Also during his career, he became an insurance agent for the Cedarburg Mutual Fire Insurance Co. and served as secretary of St. John's Lutheran Church in Mequon for fifty years.

In 1882, Arnold Carl Kiekhaefer was born to Heinrich and Augusta on the homestead, and it would be Arnold that would eventually complete the land development, modernize and mechanize the operation, and acquire a herd of pure-bred dairy stock. On his twenty-third birthday in 1905, Arnold married Clara Wessel, and six months later she gave birth to Elmer Carl Kiekhaefer on June 4, 1906. Elmer, as he would be known for the next thirty-five years, was born in the same room as his father, and his grandfather before him.[5]

Life should have been far from lonely for young Elmer, who would grow to take on the features of his mother, rather than his father. The tiny Kiekhaefer farmhouse was bulging with family and workers. Arnold and Clara provided Elmer with six sturdy and vigorous sisters, Palma, Almira, Isabelle, Ruth, Florence and Marion, ranging in age from five years to sixteen years younger than the solitary boy in the house. Elmer's mother, Clara, was a very strong woman, and unaccustomed to pampering. Two weeks from giving birth she would always be back in the fields.[6] She, along with the six girls, worked shoulder to shoulder with the men, accomplishing the same work that the men were expected to perform. When Elmer was three years old, a younger brother was born, Arnold Frederick, but died before reaching his fifth month of life. Also living in the teeming household were his grandparents Heinrich and Augusta, a spinster Aunt Martha, and two hired hands for a whopping total of fourteen.

There was only one heated room in the house, the room in which Elmer, his father and grandfather had been born. The room was heated by a kitchen stove which needed to be stoked frequently, and required wood to be continually hauled in during the bitter sub-zero winters of eastern Wisconsin. Kerosene lamps provided lighting for the family, and later the household was jubilant when asbestos mantled lamps were installed, considered the most modern of improvements. The single outhouse was a brisk jaunt from the house, and a line fourteen deep could form on cold mornings, with each ready to bolt for the half-moon silhouette on the door.

For Elmer and his sisters, a refuge of relative comfort in the winter months was the barn, warmed by some eighty livestock and their steaming hides. The Kiekhaefer barn itself was a curious sight to behold, and became a landmark for miles around. It was an octagonal barn, seventy-four feet in diameter and towering sixty-feet over the countryside. The massive structure was designed by Dutch inspired local farmer Ernest Clausing and built by Elmer's grandfather Heinrich in 1895. It was a clever design which made use of ponderous five-sided hand-hewn timbers of up to a foot across. Half a century later, the young boy who played in the hay lofts, praised the unique qualities of the configuration.

> "This type of structure ... required less material per cubic content. There was an additional space advantage in that there was no truss work, as in the case with rectangular barns, and thus it had more usable floor area, both in the upper part of the structure which was used for hay and feed storage (not machinery) and the lower structure, or basement structure, which was built of field stones, which in turn housed the livestock, making a real comfortable abode for the cattle in even the severest sub-zero weather. Due to the animal heat and the excellent insulation provided by hay, straw and feed in the

upper structure, it was well insulated. Also, with reduced peripheral square footage, it had less radiating surface or cold absorbing surface than the rectangular barns."[7]

Perhaps the earliest impressions of Elmer came from an Aunt who has a vivid recollection of the young boy being fascinated - almost mesmerized by anything that moved, or that he could persuade to move. He would grasp an object in his pudgy little hands and push it along, all the while making a "brrrrrrrrr" sound with his lips, imitating the gasoline engines of the day, or the threshing machines in the fields.[8] His mother also would recant memories of young Elmer constantly striving to produce a machine-like motion from the objects around him, like looping strings over the dresser knobs, and pulling on the ends to make the knobs rotate back and forth.

The farm exposed Elmer to a veritable wonderland of practical machinery. In the first decades of the twentieth century, the function of a device was often quite apparent at a glance, lacking the shields, cosmetic panels and hidden parts that characterize modern farm equipment. The gears were large and heavy, solid iron and steel that oozed oil and squished grease as they turned and meshed, and required constant vigilance and maintenance. The implements worked hard for the family: plowing, tilling, cutting, raking, planting, fertilizing and hauling. Elmer's father and grandfather were forward-thinking farmers, and realized the importance of owning the most up-to-date equipment they could afford. As Elmer reached seven or eight years of age, he began to help with the equipment, and to work in the fields alongside his family, three generations of Kiekhaefers working the rich Wisconsin soil together.

By Elmer's turn in the fields, his grandfather's oxen had been replaced by big, bony draft horses, like Kate, the one-eyed work horse that became a favorite of the girls. These powerful, even tempered horses were bred for the yoke and harnesses of leather and iron that bound them to the heavy tools that they dragged back and forth across the land. Elmer learned to differentiate between the many sounds generated by all of the components in operation. The sharper the plows and blades, the better lubricated the linkage, chains and gears, the less fatigue and steam showed on the back of the horse. The rhythmic squeak of the yoke and leather harness rubbing against the foaming back of the beast, the clang of the chains, was different from the soft rumble of the gears, the whoosh of the blades, or the subdued crunch of the steel wheels across the uneven soil. Each sound had a purpose, and Elmer's job was to orchestrate them into a symphony of efficiency. When the horse slowed down, it was usually something to do with the implement, rather than the soil. Elmer learned to adjust and improve the devices around him. It was a basic lesson in horsepower that would stay with him for the rest of his life. Elmer would estimate that he and his family walked for several thousands of miles "behind the plow and other implements, steering a horse."[9]

In 1917, when Elmer's father was the first farmer in his area to secure a powered tractor, whole new mechanical devices arrived to challenge the lad's eager and curious mind. Hydraulics, pneumatics, internal combustion, valves, crankshafts, power takeoffs, brakes, transmissions, ignition and electrical systems succumbed to his disassembly and inspection to yield their secrets. The arrival of the tractor was not without penalty, however, as Elmer would describe some sixty years later.

> "My father, wishing to be up-to-date and modern, was the first to have a tractor, one of the more noteworthy instances in my farm life. I was at first delighted with the honor of operating the tractor, even at the age of 11, but I began to suspect it was an honor in disguise since to steer the bloody thing took 30 turns of steering from lock to lock, which was an operation done at each end of the field with every turn of a furrow. This is what developed the upper parts of my body, which completed the complete barrel shape physique. But for my stature, I would have made a pretty good football halfback."[10]

Later, Elmer would be exposed to a wide variety of two-cycle engines that were used throughout the homestead. His chore-boy work gave him early experience with the small engines that ran the milk separator, milking machine, corn sheller and water pumps. He soon learned the value of a spark plug brush and the proper gas and oil mixture. The loud and cantankerous iron engines billowed blue smoke and consumed his best knots at the end of many starting ropes.

Perhaps the strongest and most lasting lesson drilled into Elmer and his sisters was an uncompromising, inflexible and absolute work ethic: the unspoken, unwritten code of the homestead. In the fields, everyone divided up responsibility for the seasonal task at hand. Harvesting was an especially demanding time, when so many different operations were required to complete the yearly cycle. One of the girls would specialize in mowing, another in raking, another in bundling shocks of grain, and yet another in operating the threshing machine. Elmer, father Arnold and grandpa Heinrich would prepare the heaviest of machinery, setting it up for the appropriate task, and crank the engines over for the girls.

After morning chores and a bracing breakfast, work began in the fields. As the sun rose higher in the sky, mother Clara, Aunt Martha or one of the sisters would bring cool water to quench the parched throats of the small Kiekhaefer army in the fields. At noontime, great baskets of food were driven out by buckwagon from the homestead kitchens, loaded with great mounds of beef and pork sandwiches, gallons of fresh soups and lemonade, hot potatoes, bowls of fruit and sweet pastries of pie or cake.

Each afternoon, the long 3:30 freight train would blow a deep throated whistle as it rumbled and rattled through Ozaukee County towards Milwaukee, signaling all hands in the fields to a well earned break of water, lemonade and coffee. By nightfall, the long day left the family and crew caked with a mud of dust and sweat, and their hair filled with the shrapnel of chopped straw and earth. Exhausted, they would rinse their red hands and faces, and shuffle slow-footed towards the dining room table.

Six days a week the routine was nearly identical. Six days of rolling out of bed in the dark, six days of hard labor in the fields, six days of short breaks, catered lunches and afternoon train whistles. Asking for a day off was unthinkable. The thought of everyone else working shoulder to shoulder while the missing pair of hands engaged in some idle or selfish amusement, was enough to dismiss the idea. "Everybody else worked, and you had no business taking a day off," Elmer's sister Almira recalled. "We were afraid to ask because everybody else worked, and you didn't dare be the one that didn't."[11] It was an atmosphere of absolute responsibility and dedication, a sacred family tradition of labor and commitment. As hard as those in the fields worked, the ladies in the kitchen struggled equally. It was the code. No one had the opportunity to whisper or grumble that someone had worked less arduously that day, for each had made sure that their contribution was equally large and equally obvious to each of the others.

Sundays and holidays were an entirely different matter. They were very special days for the Kiekhaefers, and except for a single, solitary emergency with a pea cannery, it was a day of absolute rest from the fields. Naturally, the cows still needed milking, and the animals couldn't be neglected, but even these chores took on a special cheer on Sunday. The entire household would don their very best suits and dresses and squeeze into Arnold's Model T for the short ride to St. John's Lutheran Church in east Mequon. Young Elmer was particularly welcome in the pews of St. John, for he had a most wonderful singing voice, ranging from bass through baritone to tenor.

Often there was a special church social or picnic, luncheon or other event associated with Sunday, and the whole family would attend. Witness to the forbidden nature of work on Sundays, it was the only day that father Arnold could be seen dressed up and wearing his very best clothing *all day long*, there never being the temptation nor reason to change. Following church, the family either visited other families, or were hosts to visitors. Frequent guests were Arnold's youngest sister, Flora, and her husband, John Blank, and Elmer's favorite cousin, Willis Blank.

Holidays were always observed by the Kiekhaefers, and were dates of great anticipation and preparation throughout the homestead. Thanksgiving and Christmas were the annual highlights, and the family would take the week off between Christmas and New Years without fail each year. Family baptisms and confirmations were also treated as holidays, and were always commemorated with a formal, elaborate dinner at home.[12] In later years, Elmer would completely abandon this tradition of rest on Sundays and holidays during the growth of his industrial empire. His later intolerance of rest, along with an almost complete disregard for "family," is in absolute contrast with his own early life.

The average day in the life of young Elmer at the Kiekhaefer homestead could have been the inspiration for an endless series of Norman Rockwell portraits for the cover of the Saturday Evening Post. It was a wholesome, solid, visceral environment of hard work, hearty appetites, fresh air and strict moral values. The big oval oak table in the dining room was always spread open by five well oiled extension leaves to seat fourteen hungry farmers for daily meals, and could accept four more extensions on Sundays when company would inevitably arrive. Preparing nourishing menus for three square meals a day for fourteen was nearly a full time endeavor for Grandma and Aunt Martha, Arnold's unwed sister. As Elmer's six sisters grew older, they would in turn help out with setting and clearing the table, and help to clean what could amount to over a hundred-fifty individual items following the meal, desert and coffee. Any month of the year meant the serving of over twelve-hundred individual meals.

Grandfather Heinrich reigned at the head of the Kiekhaefer table, with father Arnold to his immediate right, and grandmother Augusta followed by Aunt Martha to his left. Next to father sat Elmer, followed by the hired men, the girls, and finally mother Clara commandeered the foot where she would tend to the youngest, from high-chair age onward. It was a strict and status ranking seating order which was never modified by whimsy or arrival time at the table. The hot casserole dishes and serving bowls would arrive from the kitchen in long, steaming procession, and would be passed, family-style, from the head of the table to the foot and up the other side.

Thickly marbled beef of all descriptions and preparations dominated the evening meal, alternating with rich pork dishes drowning in thick brown gravies, or lean country chicken. Mountains of mashed potatoes were leveled atop the Kiekhaefer table, dug away in giant plopping scoops and topped with fresh creamery butter and dark green garden chives. Peas and carrots, string beans and cauliflower, corn and tomatoes poured from the gardens and fields onto the big table and were devoured by the often exhausted and ravenous household.

Breakfast was also an eagerly attended event, the wondrous aroma of which would overcome all hands, following nearly two hours of muscle burning pre-dawn chores. Dozens of fresh brown speckled farm eggs would be fried and scrambled. Yards of fresh pork and beef sausages would appear, steaming in great piles, with crackling thick strips of salted barnyard bacon. Fried potatoes with big brown onions were served with thick, palm-sized slices of homestead ham. Skyscrapers of grandma's dark wheat toast were prepared, and kettles of black, billowing coffee would all disappear before the ritual of the day's work began.[13]

Horse-play was strictly verboten around the Kiekhaefer table. Even the occasional giggling of the girls was not tolerated. One of the girls remembered an instance where the infectious, meaningless and irrepressible giggles which often afflict young ladies at inopportune and awkward moments seized the sisters at the table. Arnold, assuming that the girls were laughing at him, sternly admonished them, "My children don't laugh at me." The discipline and control of the household was so explicit and conditioned into the children, that the impact of his brief words would jolt the girls from the frothing, uncontrollable spasms of adolescent laughter to near tears.

Each morning, a herd of thirty cows had to be milked. Arnold, considered quite progressive, had mechanized the milking process as early as practical in his farming career. Elmer would clean the udders of the cows, fasten the suction cups, wait for the level of yield to fall to a certain extent, and then complete the milking by hand. He spent many pre-dawn hours sitting on the three-legged milking stool, his cheek pressed up against the warm hides, pulling and squeezing the last few cups of milk from each of the thirty cows. When the milking was completed, all of the implements and pails were washed thoroughly, and after the cows were released the stalls were shoveled and fresh straw scattered about.

On more than one occasion, Elmer would demonstrate his agility by running over the backs of the yoke-shackled cows from one end of the barn to the other to show-off to his cousin Roland Lederer, who would then take up the challenge. Had they been caught by either Arnold or Heinrich they would have been punished severely.[14]

Like clockwork, grandpa Heinrich would come to the barn each morning to cool the milk, so that it was the proper temperature when the milk hauler came to the Kiekhaefer farm on his route. The Kiekhaefers made blocks of ice in the winter which were put in cooling tanks, and the milk was plunged in shiny stainless steel cans beneath the ice-cold water until the mercury on the milk thermometer fell to the green zone. It was then poured into shipping containers for the milk truck. During the winter, ice was made and placed in the ice-house, where it was covered with a thick blanket of saw dust. The homestead would have plenty of ice all summer, even when the temperature climbed into the high 90s, when all hands turned out to help crank the home made ice cream machine. Next to the annual check for the Kiekhaefer barley crop from the breweries of Milwaukee, the monthly milk check was the most anticipated and important revenue for the homestead.

The Kiekhaefer sisters were able to invent fun for almost any occasion as they worked around the farm. They enjoyed singing and harmonizing together, and would often raise a chorus while milking, using the rhythmic chuga chuga chuga of the milking machine or the rasping squirt of the milk in the bucket for cadence. There were times when Arnold, passing by and hearing one of the slower, more romantic melodies by these milking maidens, would stick his head into the barn and request a faster tune.

In fact, the family was quite well represented musically. Arnold was an accomplished sousaphone (tuba) player, and played regularly at church and social functions with the Cedarburg Legion Band. He would perform with the band at the Fireman's Picnic, and even the County Fair, playing the ever popular German ooom pah pah dancing songs, like the Beer Barrel Polka, old country waltzes and other crowd pleasing favorites. Unfortunately, whenever Dad played in the band, only two of the children could go along for the occasion. Arnold drove, Clara held two of the kids on her lap in the front seat, and the enormous sousaphone case took up the entire back seat of the aging family Maxwell.

Since the sousaphone was the only instrument in the house, Elmer attempted to master it on his own. For a time, when Elmer was in his early teens, he would practice a tuba in the barn, an instrument which closely matched his stocky and widening physique, and gave hours of solo performances to his captive bovine audience who would bellow and roar in agonizing disapproval. Not the easiest instrument to carry a complicated melody, it could certainly exasperate and derange the sensibilities of a busy and crowded household. As Elmer entered his teens, he began to sense that his image would not be enhanced by being associated with an instrument that so closely matched his own profile and physique. So, he switched to the violin.

It was a wondrous sight, watching Elmer, five-foot-ten, two-hundred thirty pounds, with the barrel chest of an olympic weight lifter and the arms of a blacksmith, gently drawing the bow of his violin across the face of the fragile instrument, his wide, pink chin trapping the innocent device against his chest. The image of this great contradiction was somewhat mesmerizing. Eventually, Elmer could sustain a modest melody, recognizable to all who listened - carefully. While Elmer practiced the unforgiving instrument, a dog from the neighboring farm would run across the stubble fields, frothing and panting, and press its black, wet nose up against the screen door and howl in concert with the screeching and squealing emanating from within. Elmer became sufficiently accomplished to win a chair in the Cedarburg High Orchestra, and on occasion would play with his father and the Cedarburg Legion Band. People, however, seldom begged Elmer to play.[15]

Sometimes, Elmer accompanied his singing sisters with the violin as they milked the cows, especially as they were keen to learn the musical phrases of the latest tunes on the RCA radio. The girls enjoyed the impromptu concerts thoroughly, and these rare occasions have endured as among the few truly fond memories that they have of their brother.

CHAPTER TWO

Farmboy

"Did I not work 15 out of the first twenty years of my life on that farm too?

"Did I not do the work better than any hired man or the girls? Did I not operate tractors, threshing rigs and do all heavy work besides? I can remember the terrible summer 'vacations' of World War I when, as a child, I could not rise in the morning with fatigue -- broken out continually with skin rash that prevented sleep. Who repaired the machinery? I think I did my share. ... I, for one, would never want to re-live my childhood in that old homestead."[1]

Elmer Kiekhaefer

Even though Elmer might have had the benefit of thirteen companions under one roof, he remained somewhat withdrawn and aloof, especially from his six sisters. The sisters developed an unusually strong camaraderie with one another, given the differences in their ages and the sheer odds of so many girls getting along in a small house. Young Elmer's own insecurities, being the oldest and being the only boy, acted as a barrier between he and his sisters, and blocked his admission to this seemingly private girls club.

Further complicating the relationship between Elmer and his sisters, when the adults were not at the farm he was transformed into *Elmer The Great*, the all powerful domestic disciplinarian and ruler of his often horrified subjects. "He would just sort of be a little mean to us," one sister explained, "he'd *make* us behave when we weren't even doing anything wrong. We sort of stayed away from him."[2] The girls developed a common bond of unity against Elmer's forays into authority, learning the value of "strength in numbers." Frustrated by some advantage enjoyed by the girls, he one day blurted out to his mother in exasperation, "I've just got *too many sisters*!" Clara, who possessed great maternal instincts and often demonstrated a real penchant for a folksy, down-home brand of philosophy, quickly responded, "Oh really? Which one is too many?" Elmer, ashamed to imagine that he would actually have to choose among the six for some imaginary elimination, could not answer, and quietly withdrew.

For both Elmer and the girls, the sprawling lands of the homestead provided many diversions when the work and chores were done. Walks down to Lake Michigan in the summer were popular, where they would swim and wade along the rocky shoreline, skipping stones in the bracing, clear water, and imagine the lives of the people who lived on the opposite shore. A well used cow path was the most popular route, worn smooth by generations of Kiekhaefer cattle and horses, plodding and bobbing along to partake of the ever-cool waters. The sight was most beautiful in the morning, as the sun slowly rose above the watery horizon, and made the great lake shimmer like a giant bowl of wet diamonds. In the distance, huge freighters could be seen, bearing inbound cargos of coal and oil to Milwaukee, and timber, grain and goods of local manufacture outbound across the lake.

In the winter, a frozen pond on the property became a skating rink, and Elmer especially enjoyed gliding across its glassy surface, briefly free from the mounting responsibilities and frustrations that accompanied his transition to adolescence. Elmer's sisters remember how he would sneak out ahead of them when they planned an outing of snow sledding, camouflaging great holes and trenches which he would dig, and then lie in ambush. He would explode with laughter when the girls, shrieking with both delight and terror, would disappear into his bone-jarring traps. Except for these isolated moments of amuse-ment, his was for the most part a very lonely and unhappy childhood.

Around his eighth Christmas, Elmer received a modest erector-set, complete with a tiny steam powered engine which was fueled with an alcohol burner. He would fill the small boiler with water from the well pump, scratch a kitchen match along the floor and light the small alcohol wick. Before long the little engine would puff and chug and begin to spin the miniature flywheel, and Elmer was transported to another dimension. He took great pride in lecturing to his sisters how the tiny engine worked, but would lash out if they tried to touch it.[3] Later in life, his sisters would reveal that they felt as if they never really knew their brother at all.

Contrary to his often expressed feeling, Elmer was the one who consistently received preferential treatment. Being the oldest, and being the only boy, had many advantages, especially

Top left: Clara Kiekhaefer, Carl's mother with the A.C. Kiekhaefer Memorial Trophy established by Carl to honor Carl's father, the first president of the Kiekhaefer Corp.

Top right: Arnold C. Kiekhaefer, Carl's father.

Below: Standing, second from right, Heinrich Kiekhaefer (grandfather) and far right, Arnold Kiekhaefer (father). Seated, center is Augusta Kiekhaefer (grandmother).

Top left: Carl holding oars on his honeymoon in June, 1932.

Top right: First company portrait of Carl, 1940.

Below: One of Carl's first inventions, a design for a new magnetic separator, drawn at Stearns Magnetic in Milwaukee in 1930.

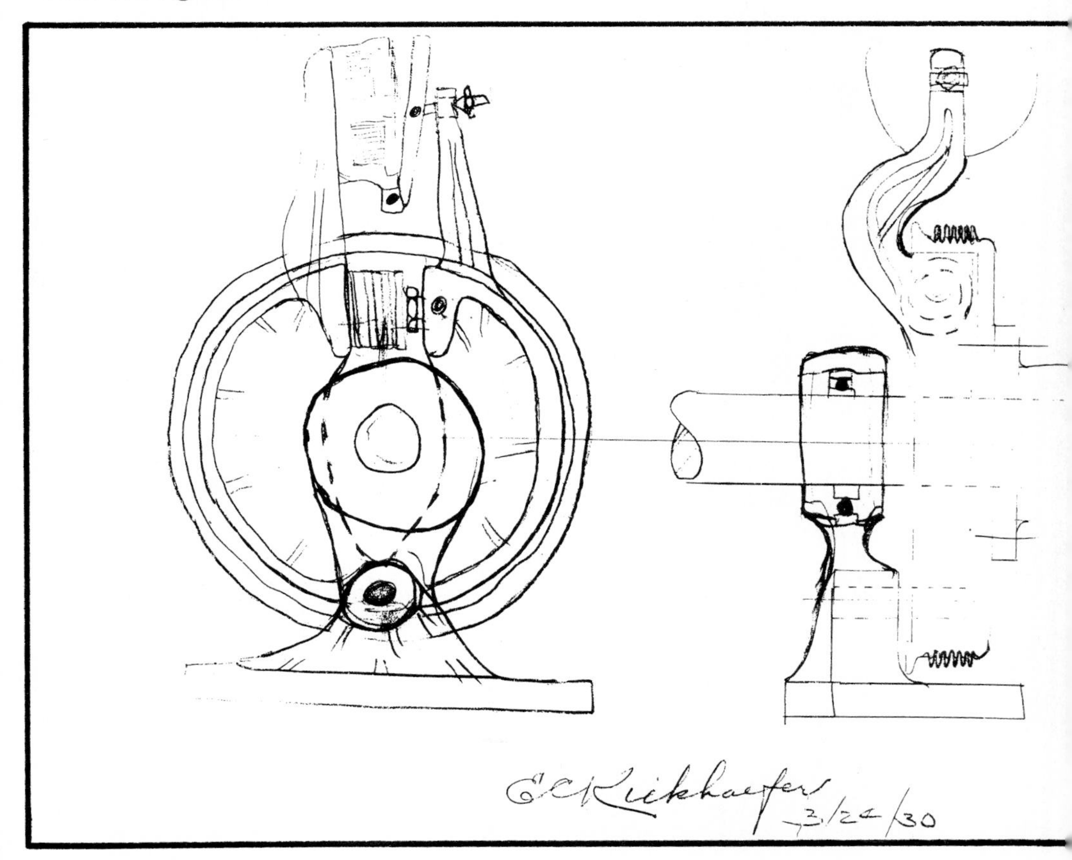

Left: The Cedarburg, Wisconsin, plant following a heavy snowfall in 1939.

Below: The first shipment of Montgomery-Ward *Sea King* outboard motors is boxed in 1939.

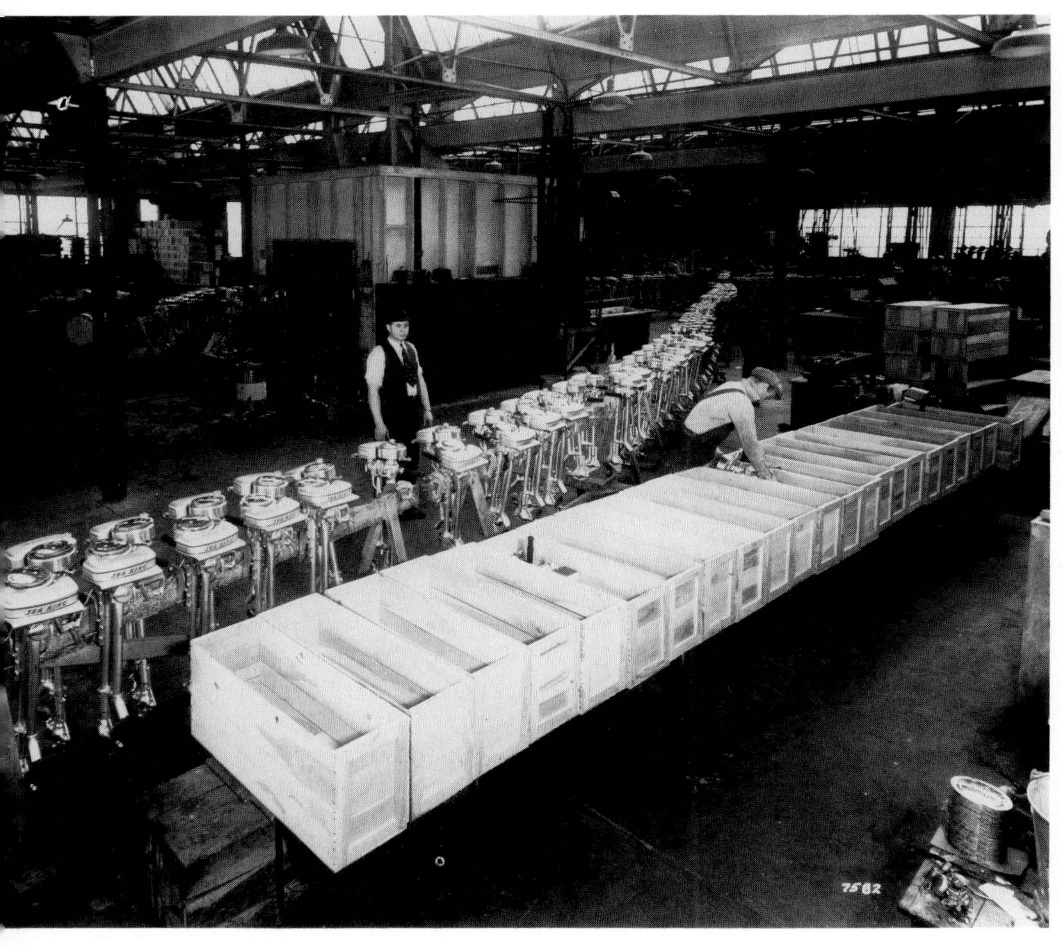

Top left: Thorwald Hansen's first ad for the *Thor*, 1935.

Top right: Carl Kiekhaefer's first ad, for the new *Thor Streamliner* in 1939.

Below left: The first *Mercury*, in the first Mercury brochure, 1940.

Below Right: The 1941 Mercury brochure and new twin-cylinder Mercury.

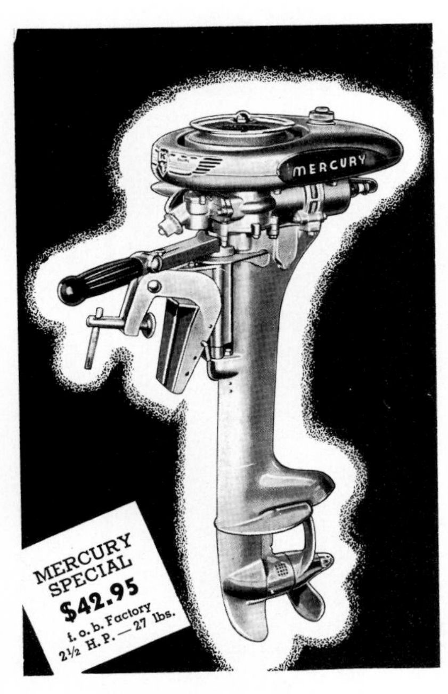

If it's a full sized quality motor you want PLUS LOW PRICE here's the motor to buy! It has the power, stamina and quality to deliver the kind of performance you would expect from a higher priced motor . . . it has exclusive features that mark it the "best buy of the year," bar none.

Above: A batch of twin-cylinder Mercury outboards roaring in the test tank, 1940. Technician is checking rpm at various throttle settings.

Below: The Montgomery-Ward *Sea King* assembly lines, 1940, in Cedarburg, WI.

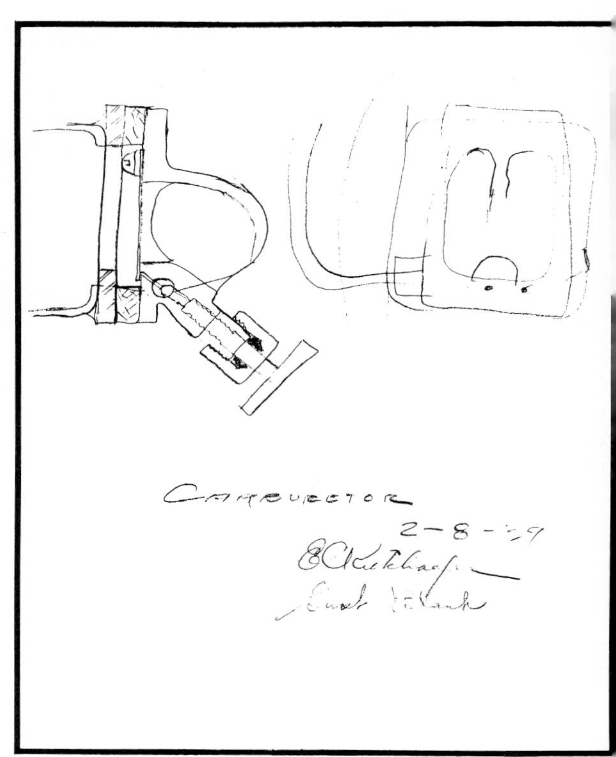

Above: Kiekhaefer testing a *Thor* single-cylinder outboards through a weed patch in 1939.

Right: The first of hundreds of improvements he would make to outboard engines, drawn exactly two weeks after starting business.

Below: The first exhibit, at the Milwaukee Sportsman's Show on April 22, 1939.

Right and below: Carl always took the heavy end of his two-man chain-saw, the first to be manufactured in America, and gave hundreds of demonstrations throughout the timber country.

Above: During World War II, Carl toured timber camps in a specially designed trailer to demonstrate portable chain-saws and portable generator sets.

Below: Following the war, Kiekhaefer Mercury personnel crisscrossed the country in search of dealers.

on the farms of the early 1900s. Elmer had his own attic room, which he shared with the hired men most of the year. Though both Elmer and his sisters worked equally hard in the fields and in the barn, the added burden of tending to younger sisters always fell to the girls. The girls acquired the well worn clothing of their older sisters, while Elmer would periodically receive new clothes. Elmer complained about the agony of receiving new shoes on the farm, though, and would later write, "... when the head of the household was buying shoes, he guessed at the sizes and when delivery came after a month or two, you were stuck with it -- small matter though since the shoes formed themselves to your feet in those mushy spring thaws and after a short time, fit like a glove."[4]

A delicate and largely unspoken tradition also separated Elmer from his sisters. The eldest male of the line generally inherited the land and carried on the homestead. From Carl Kieckhofel to Heinrich to Arnold to Elmer. There was always a subtle difference in the way grandfather Heinrich or father Arnold expressed sentiments about the land and family to Elmer than to the sisters. The underlying assumption that Elmer would carry forth the family legacy, be the next link in the hereditary chain between the Kiekhaefers and the soil, created a certain aura about Elmer that could never be shared by the girls. The historic homestead and hearthside of the sisters was somehow transient and temporary, but if Elmer, the scion of the Kiekhaefer line so wished, it would be his forever.

Elmer and his six sisters attended the same one-room, Jefferson elementary schoolhouse, which was a mile and a half hike from the homestead. A mile and a half can seem like a marathon arctic expedition for children as young as six in the dead of winter. Many mornings the huge cast-iron, wood burning stove that dominated the tiny schoolroom was covered with steaming, sizzling mittens, and the children would linger by its pink-hot warmth until the teacher sent them to their wooden bench seats. Though the Kiekhaefer family's spread in ages precluded more than a few attending at the same time, Elmer shared the classroom with his oldest sisters, and a few dozen children from the local farming community. It was a traditional education, of "Readin' and Ritin' and 'Rithmetic," taught by a solitary teacher from grades one through eight, and an advantage for the inquisitive Elmer to be continually exposed to information in advance of his designated grade. In fact, Elmer graduated from the eighth and final grade of the little schoolhouse when he was eleven. But, considered too young to attend the high school in nearby Cedarburg, he was held back, completing a frustrating two more years of generalized study.

Though in a few short years Elmer would swell to imposing proportions, barrel-chested with thick, broad shoulders and the intimidating arm strength of rural life, in his younger years he lagged somewhat behind the other children in stature and bearing. When he finally reached Cedarburg High School at the age of thirteen, his body began to grow rapidly, and he began to assume the profile that would one day be recognized around the world.

The high school was a full seven and a half miles from the homestead. A nettlesome distance, uncomfortably far for walking, but much too close to consider year-round boarding in the town of Cedarburg, even if the family could have afforded the double luxury of losing a valuable farm hand and paying for room and board. Elmer walked the fifteen mile round trip daily, except when he could catch a ride with a neighboring farmer who delivered milk to Cedarburg, or when the weather was impossibly inclement. The daily marathon continued to strengthen his already muscular, but stocky physique, and led to his only limited forays into organized athletics. Fifty years later, he would describe his athletic experiences with a certain nostalgia in response to a comment made about the sport of fencing from a business associate.

> "Fencing has always been a fascinating sport to me, possibly because I was too heavy and too slow under foot to be anything but a setup, but I, too, do like sports and with my gorilla-like build, I was always able to pass the ball a good many yards farther than I could punt it so I never made anything but the second team. This was before quarterback football, obviously, so my specialties were of no appeal to the coach in 1922.
>
> "I had worse luck in basketball because I had to walk fifteen miles each day to get to high school from Dad's farm and when I got to school, I steamed like a horse, and perhaps smelled like one, and being old enough to notice girls, a new frustration set in which I tried to erase by playing basketball before starting my seven and a half-mile trek back to the farm. Since this was in the years 1921 through 1924, the family Model T was reserved for church-going and unavailable, and except on rare occasions when Father would come to town to

do some shopping, I seldom had a ride in it. However, one evening while I was playing basketball, you can imagine my horror when Dad walked in through the door and viewed the proceedings with amazed disapproval. He thought that exercise was strictly for city people and while he didn't interrupt the game, I was, however, immediately collared thereafter and lectured on the way home. From that time on, my weekends were designed to utilize the energy of youth (rather than wasting it on ball playing) to the removal and spreading of all the manure generated by some seventy head of cattle and ten horses and I forget how many heifers, calves, pigs and hens. That was the end of my athletic career. My grandfather, of course, fully [agreed with father] since he lectured, 'Never make a move without having it count,' feeling that when your energy supply had been used up, you were ready to meet your Maker."[5]

Elmer's relationship with his father was in stark contrast with the closeness and warmth that characterized the balance of the family. He began to view himself as more and more of an outsider, and developed in his youth an emotional buffer between himself, his father, and the concept of family. Though he truly loved his mother, and would rarely argue with her, his growing aloofness and detachment became obvious. Elmer rebelled against his dominating father, and felt that somehow his spirit and initiative was being smothered by the complex relationships between members of the large family.

The one exception was his Uncle John Blank, the husband of Arnold's sister Flora. He was the only member of Elmer's immediate family who had a real education, and Elmer truly admired and loved him. When the bitter Wisconsin winter would send temperatures to twenty and thirty degrees below zero, Elmer would stay with Uncle John and Flora in Cedarburg rather than attempt the long hike to the homestead. It was John Blank who encouraged Elmer's interest in mechanics, and became his greatest booster. Uncle John was the only person who would ever talk to Elmer about his future, or about life and the profound topics that young men are still sorting out. In many ways, John became a surrogate father for Elmer, and from their many fireside conversations sprang the earliest and deepest friendship of Elmer's life. One of Blank's sons, Willis, became Elmer's favorite cousin, and in later years would enjoy a bond of friendship and loyalty in business life shared with no other member of the Kiekhaefer family.

Though Elmer earned only average grades at Cedarburg High School, he was singled out by the principal to take the daily cash revenues from the school to the local bank. It irritated Elmer because he assumed he was chosen because of his farming background, and the traditional guaranty of honesty that accompanied farm life.

It was during his sophomore year at Cedarburg High School, in 1920, that Elmer first heard the radio. Radio Corporation of America (RCA) was formed early in the year, and by August, the first scheduled radio program aired over station 8MK in Detroit. It was the memory of listening to his first static-laden radio program that prompted Elmer, over sixty years later, to say, "I was born right. The only thing that happened before I was born was the printing press and gun powder."[6]

When Elmer began to notice girls, he became quite fastidious about his appearance and clothing, always keeping his shoes polished, and attempted to obtain what the girls of 1923 Cedarburg might consider the most fashionable clothing. He enlisted the assistance of his sisters to press his Sunday best, admonishing them to "really push down hard", and make the crease in his rather short pants razor sharp. He wanted them to be able to withstand the torment of his big knees, flattening the creases out as he sat in the car on the way to pick up his date. And to his dates, his sisters remember, he was always true, never playing the field, and was a gentleman whom their parents were happy to see at the door.

It was obvious to the family that Elmer had a genuine capacity for mechanics, and an unusual affinity for the many devices so common to agriculture. Elmer made many small improvements in the machinery he operated, either in performance, strength or decreased maintenance. It wasn't a complete surprise to his family, then, when he announced in early 1924 that he wished to go to engineering school following graduation from Cedarburg High.

The announcement did little to assuage the growing conflict between Elmer and his father. The cost of sending Elmer to the Milwaukee School of Engineering was compounded by the labor lost to the homestead while he attended classes or required study at home. But more to the point, it was the open, undisguised declaration that he would not be available to perpetuate the family homestead, that ruffled the sensibilities of his father and grandfather.

Elmer began attending the Milwaukee School of Engineering in the fall of 1924, a little over three months following his graduation from Cedarburg High School. The school was located only around thirty miles round-trip from the homestead, so Elmer purchased a motorcycle to make the daily commute, driving it well into early winter, until the ice on the roads forced him to secure a car. Unfortunately, Elmer then acquired a well-used convertible, and the exposure to the ice cold air of winter nearly cost him his life. Driving with the top down in the dead of winter, flying along the frozen two lane highway that linked Mequon with Milwaukee, Elmer contracted a nearly fatal case of pneumonia. He was eighteen, and experienced perhaps the greatest pain of his life. Concern was so high in the household that his bed was moved from the attic to the main floor so that his mother could attend to him more carefully. He would scream and gasp for breath, bringing his mother running to his bedside. One night the pain got so bad that Elmer asked his sister Almira to run and bring his mother from the barn. His fever had escalated to a dangerous degree, and the doctor was beckoned at once to the homestead. It would be four more years before Alexander Fleming would discover the miracle antibiotic of penicillin, and with only the common chest plaster remedies available, young Elmer was forced to endure the excruciating pain for a full six weeks. When at last his fever subsided, though unsteady and weak, he dragged himself out of bed and returned to school, this time with the convertible top securely up.

Elmer attended the Milwaukee School of Engineering for only one year, withdrawing from classes the following June of 1925. Some evidence suggests that his prolonged absence from classes due to his severe illness destroyed his willingness to continue.[7] The more plausible explanation is that, though he enjoyed the curriculum of basic electrical engineering and mechanical drafting, he was frustrated and anxious to find work. He had been commuting either to high school or the engineering school for five years, and was impatient with the slow, plodding regimen of education.

Tensions continued to grow between father and son, and in 1925, the summer of his nineteenth year, one final, explosive argument occurred which was to change Elmer's life forever. No one, even Elmer, would remember how it began, but the two were having so many more frequent and angry exchanges that it hardly required a special incident to bloom into another seething, vitriolic shouting and door slamming match. Elmer had become increasingly independent, feeling detached from the tightly knit family circle since finishing his year of studies in Milwaukee. He came to resent the incursions of the incessant and unremitting farm work into his growing academic and increasingly private life, and most likely the argument was born of conflicting opinions over future work on the homestead.

Elmer roared at his father that he was leaving the farm. Arnold, matching his son's Teutonic indignation and intractable head-strong style, delivered to Elmer the challenge and ultimatum that the son would never forget: "If you're leaving, really leaving, then leave for good. Don't ever come back." Events had gone too far for either man to retract his statements. Elmer's mother and some of his sisters overheard the deafening exchange, and sensed at once that something dreadful had just occurred in the household. Mother Clara went to comfort her shaking and crying son, asking him, "Don't you want to see any of your sister's again, either?"[8] Elmer seemed to imagine that the whole of the Kiekhaefer family was against him, and leaving the home of his birth, the legacy of his ancestors and all that he knew and loved, fled to Milwaukee. There would be no turning back, he knew, but he was eager - almost desperate - to apply his classroom knowledge to a bona fide engineering work-place, and prove he could make the long anticipated rite of passage from farm to city.

CHAPTER THREE

Dues

"When the rest of the family celebrated the reunions, I spent years at night school. When the boys played cards and drank beer Sunday afternoons, I designed and invented, and did my college lessons 'till two and three o'clock in the morning -- and yet was at work ... the same time as the others. I worried about earning my future. I did not wait for someone to lay it on a platter. ... I had ambitions for bigger things."[1]

Elmer Kiekhaefer

Elmer was broke and without work. He was fortunate to have an aunt, his mother's sister, living in Milwaukee who allowed him to stay until he could manage on his own. Casting about for any sort of income which would save him from the humiliation of returning to face his father, Elmer accepted a position as a heavy equipment maintenance mechanic in a Milwaukee construction yard. It was a natural opportunity for Elmer, so accustomed to the regular maintenance and overhauls that were required on the tractors and implements of the farm. Elmer always enjoyed driving the tractor and heavy machines which farming required, so the opportunity to try his hand at even larger equipment, like graders and bull dozers, offset the mere subsistence wages he received. He loved the feel of the hydraulic controls, and the power that he could control with just the tips of his strong fingers and the balls of his feet. Like a little boy moving loads of sand around a backyard pile, Elmer enjoyed the power that became an extension of his arms and legs, harnessing the might and muscle of the giant machines. The job also exposed him to the fundamental principles of construction, and the basic steps required to level land, pour foundations, frame structures, construct walls and landscape completed projects.

Knowing that the construction yard offered little promise in terms of career development, Elmer put aside his work boots, oily rags and tool box within six months, and in 1926 started work as an office boy at the Nash Motors Body Division in

Milwaukee. It would be Elmer's first true change in avocation, for since he was old enough to hold a wrench he had been servicing machinery. It was his first job that required clean hands and shoes, a pressed shirt and tie, and an eight to five schedule. It was a significant and genuine turn of events in his development, and at twenty years of age, was his first opportunity to apply the knowledge and skills learned at the Milwaukee School of Engineering. Elmer began to work with his head, rather than with his hands.

Nash Motors was celebrating its tenth year of manufacturing since taking over the automotive concern of Thomas B. Jeffery Company, the pioneering enterprise that produced the first Rambler automobile in 1902. The body division, also known as the Seaman Body Company, was responsible for the styling and design of Nash body components, and had a busy drafting and blueprinting section which would be the proving grounds for young Elmer over the next two years.

He quickly moved up the ranks, from delivering materials and specifications to the various drafting benches as an office boy, to blueprint boy, assisting the blueprint section in converting the many hundreds of drawings required for each automobile into the familiar white-line, Prussian blue background duplicates which could be sent to tool makers, outside vendors, plant engineers, prototype builders and engineering departments. He would check the accuracy of the blueprints made against the original drawings before they were sent out to their destinations. His keen eye for detail, combined with his basic engineering skills, earned an early promotion to drawing detailer. He was then allowed to actually contribute his marks to drawings, detailing various minor areas of body panels, fenders, hinge mechanisms, as well as illustrating rudimentary electrical system routing and termination. Eventually, as a layout draftsman, he was permitted to execute complete drawings, based on the stringent specifications established by the engineering department.

Even though Elmer had attained his goal of producing his own drawings from a fresh sheet of Nash onionskin, he was still frustrated by the creative atmosphere that rewarded accuracy but not innovation. But even more frustrating to Elmer was the annual layoff. In the late 1920s, it was still common practice to layoff the majority of production workers for three, and even four months to change over to the next model year. It was a confusing interruption that he would bear for two changeover seasons, forcing him to work odd jobs until he was hopefully called back. And at Nash, Elmer felt anonymous and unappreciated, just another time-card which he had dutifully punched for two years. "I took to wandering around the plant," Elmer would write in later years, "getting to meet the foremen, getting to know the executives, learning what machines did what ... in short, satisfying my curiosity about what made things tick - until finally I was fired for spending too much time talking to a foreman."[2]

In late 1927, when a classified listing appeared for a draftsman at Evinrude Motors in Milwaukee, Elmer eagerly interviewed for the position and was hired. Evinrude Motors, formed in 1909 by Ole and Bess Evinrude, was sold to Chris Meyer in 1914. By the time twenty-one year old Elmer was put on the payroll thirteen years later, the company was making many thousands of engines a year, spread out among five models, and was among the most respected names in the recreational marine industry. Ole Evinrude re-entered the outboard business in 1920 as the ELTO Company, which stood for Evinrude Light Twin Outboard, and putting Ole in the strange position of being in heated competition against products which he had also designed, and still bore his family name.

Elmer was fired within three months by Evinrude's chief engineer for the young draftsman's frequent, disquieting and brazenly insubordinate arguments concerning both design and product development. For an entry level draftsman, who had just reached the age of majority, to debate such fundamental issues with the busy head of engineering could not be tolerated. Elmer, as far as can be gleaned from his communications in these years, had never even *used* an outboard motor before starting at Evinrude, much less had any knowledge of their development, operation or production. But, here he was, boldly proselytizing for his point of view, with only the faintest blush of experience with the subject at hand. His final day, no doubt concluding with an explosive exchange of oaths and razor edged expletives with the chief engineer, would be regarded as so ghastly by his employer, that Elmer would prudently avoid listing Evinrude on his resume for future employment, and would not even discuss the incident with other than close friends for the next twelve years. To be hurled from his job, and humiliated before his fellow workers for what he considered to be jealousy, was a demeaning reversal in his short career.

Out of work for the third time in three short years, Elmer was interviewed by Rosswell H. Stearns of the Magnetic Manufacturing Company of Milwaukee in early 1928, and, based on his successful work at Nash, began work as a layout draftsman. The company, nicknamed MagMan by employees to match their catchy cable address, was celebrating its tenth anniversary, and by the time of Elmer's arrival, occupied a one and one-half story brick building, squeezed between an interior woodworking factory and a machine shop near downtown Milwaukee. The firm was solidly into the magnetic separation, magnetic clutch and magnetic brake business, with around thirty-five employees and two salesmen.[3]

Magnetic separators performed many automated tasks for a wide variety of industry. In the animal feed, grain and flour industries, it was a rather common nuisance to discover nails, tacks, baling wire, barbed wire, springs, nuts and bolts, chunks of stray iron and other metallic debris inadvertently mixed in with feed which could accidentally be fed to animals with fatal results. Many lawsuits resulted from livestock which were killed by this so-called "tramp" iron, and many granary explosions occurred because of stray metal causing a spark in the fine dust atmosphere. A magnetic separator, combining a conveyer belt with a powerful electromagnet, would separate these bits of scrap from a stream of grain moving along the belt. A number of variations of this basic scheme provided for vibrating the target material to loosen the foreign matter, causing the material to fall over a revolving "magnetic drum", or through a "magnetic trap-door", or through a zigzag shaped spout. Other products in the Stearns line, like their magnetic clutches and brakes, operated on the principle that a strong electromagnetic field can cause sufficient drag to quickly stop a spinning motor or tool, as in a factory machining environment, or enough power to engage or disengage rotating shafts with tools or gears.

It was the perfect job for Elmer, combining the basic electrical engineering skills he had learned at the Milwaukee School of Engineering, with the mechanical principles he had mastered on the Kiekhaefer homestead. Even better, virtually every order received by Magnetic Manufacturing required a unique, custom-designed configuration, so there was a continuous flow of lively engineering discussion and creative brainstorming. Because the products manufactured by Stearns were constantly changing, creativity was encouraged and rewarded by management, rather than discouraged and punished. It was a gratifying change in atmosphere from his previous jobs at Nash and Evinrude.

Encouraged by the charged atmosphere at Stearns, Elmer enrolled in the University of Wisconsin Extension Division program of night courses, with his heart set on a degree in electrical engineering. His classes concentrated on engineering related mathematics, such as algebra and electricity equations, and electrical theory. Though he worked long days at Stearns, he would attend classes late into the evening, feeling that a degree would be the key to his ultimate success at work, and in industry generally.

On October 24, 1929, a wave of panic rolled over a sea of dazed brokers, investors and bankers on Wall Street, as the stock market plummeted to an unprecedented and spectacular loss. Within an hour during the frantic day, blue chip certificates of companies like General Electric, Johns-Manville and Montgomery Ward tumbled, in some cases losing twenty-five percent of their value. Fortunately for Elmer, Stearns Magnetic Manufacturing Company served a critical need for companies which served even more critical needs, supplying the feeds and grains, flour and foodstuffs of a hungry human and animal population.

Within two years of starting at Stearns, which then promoted itself as *Stearns - The House of Magnetic Magic*, Elmer was contributing new ideas and designs to a growing product line. Sketches bearing the signature of E.C. Kiekhaefer were starting to make the rounds of the engineering department by the spring of 1930, and his evening courses at the University were giving him the confidence and ability to design more sophisticated devices and configurations virtually every month. A chain reaction of responsibilities and design-related activity began to transform Elmer's job description. When one of Elmer's new concepts matched the specifications that a prospective customer required, he was asked by Stearns to assist in sales, and extol the virtues and benefits of his ideas to existing and potential industrial accounts. As his designs slowly worked their way from his drafting table into production, he helped to build them in the shop. Once his equipment was installed in the field, Elmer would be sent out to adjust, repair or expand his designs for Stearns clients. One responsibility led to another, and soon Elmer was proving himself to be an employer's dream, with equally powerful abilities in sales, design, manufacturing and service.

After four years of night classes at the University, Elmer abruptly quit going to classes in 1932. He was within a year of completing his requirements for the BSEE degree he had so coveted. His grades and grasp of the subject matter were good, and the close relationship between his studies and job benefited both activities. But Elmer was becoming consumed with his new responsibilities, some of which required travel to distant installations, forcing him to miss his evening classes for many nights on end. America was in the throes of a viciously downward spiraling economy, and a secure job had become the most important possession a young man could have. If Elmer had to make the choice between his commitment to his employer and his commitment to his education, one look at the headlines or bread lines were enough to convince him that his education could wait.

Some facets of his education couldn't be postponed, however, and Elmer sought out very specialized training to help him with his expanded responsibilities. He intermittently attended the Milwaukee Vocational School, taking various shop courses, including electrical arc welding and gas welding techniques. The manufacture of Stearns devices required a lot of welding, and Elmer became so proficient that he taught others back at the shop. He also took the I.C.S. (International Correspondence School) course on alternating currents. These courses would prove to be Elmer's final reckoning with formal education.

One Saturday evening, in the humid Wisconsin summer of 1931, twenty-five year old Elmer went to the Eagle Club in Milwaukee. It was, of course, a dry dance hall that served only iced tea and sarsaparilla, in conformance with the Prohibition of alcohol in effect since the passage of the 18th Amendment in 1920. Elmer noticed two young girls who were enjoying themselves, and asked one, a pretty girl named Freda (pronounced Free-dah) Greenfield, then twenty-two, for a dance. From all appearances, he seemed to really enjoy dancing with her, the girl would later recall, but when the song was finished he took her back to her seat and asked her friend if she'd like to dance. Feeling somewhat dejected at being passed over by the handsome, well dressed, barrel chested stranger with the odd sounding last name, she felt even worse when Elmer proceeded to take her friend out that evening. She was much relieved when her friend told her that Elmer was merely on a reconnaissance mission, and spent most of his time finding out all he could about Freda - whether she had a boyfriend, where she came from, what she was doing, anything he could uncover.

Freda recalled that he had a long, very unusual name that she found impossible to remember. When Elmer picked her up for their first date, she noticed that he had a temporary license certificate in the back window of his car which spelled out the name Elmer Carl Kiekhaefer. When Elmer wasn't looking, she stealthily copied it down, and practiced pronouncing and spelling the awkward name until it seemed almost natural.[4]

Freda was from Blue River, Wisconsin, population one-hundred-ninety-seven, situated in southeastern Wisconsin, twenty-five miles from the Iowa state line in Grant County, between Boscobel and Muscoda, near the banks of the Wisconsin River. She was a beautiful and petit young lady of English, Dutch, Scotch and Irish descent. Her grandparents, named Garner, had come to the region by covered wagon from Philadelphia, establishing themselves first in Iowa, and then moved to begin farming in Sun Ferry, Wisconsin. Her father sold furniture in the tiny town, and was also the town barber in the days when a shave and a hair cut was two-bits, so if there was any gossip to hear, the Greenfields generally heard it first. Freda remembered that she used to be able to lie in her hammock in front of her house and name everybody in town who walked by. It was a peaceful, quiet childhood, and one that emphasized the basic values of honesty and modesty that were so essential in a small rural setting. Her mother had a police whistle, and when it was time for meals or for homework, the shrill message was broadcast loudly throughout the community. If Freda or her only brother, Wally, didn't hear the call, someone else in town was bound to tell them that Mrs. Greenfield was whistling, and they had better scoot home quick.

When Freda met Elmer, she had just graduated from nursing school, obtaining the highest marks in her class. She was in her first year of nursing, dispensing comfort and medicine to patients at the Milwaukee County Hospital, along with maintaining some private nursing clients.

Elmer drove to Freda's home in Blue River many times that summer, and on the occasions when he would arrive to find that Freda was out with someone else, he would nervously wait at her house, learning to play bridge with her mother. After a few unannounced visits like this from Elmer, Freda began to realize that he was quite serious about his feelings for her. She knew, though, that Elmer was much different from the other boys she had met. He was so completely serious, such an idealist and perfectionist. Freda had come from a background of strict moral values, combined with trusting parents and a healthy attitude about the world and its inhabitants. Elmer continually cautioned her against the evils of the world, and how *other men* could take advantage of her naive confidence in strangers, and her wholesome optimism even during the bleakest days of the Great Depression. He became increasingly protective and jealous of her association with others, and would let her know that only he could protect her morals from the insidious temptations of others.[5]

The police interrupted their budding romance on more than one occasion. One evening, while Elmer was amorously and passionately kissing his young date in the front seat of his car, steaming the windows and oblivious to the world, a policeman rapped his nightstick on the windshield. "I suggest you take her home and continue on the back porch," he barked. Only too willing to obey the law, Elmer did.

Elmer was supposed to pick Freda up one morning, when she was getting off the exhausting night shift at Milwaukee County Hospital. She waited and waited, and when it looked like he wasn't going to show at all, she began to walk along 27th street, towards downtown Milwaukee, a distance of several miles, and with each step getting more and more irritated at Elmer. When she was almost all the way to Milwaukee, Elmer came screeching up in his car. Freda completely ignored him and continued to huff and puff and stomp along, elevating her nose ever so slightly and looking straight ahead. He pulled up beside her again, rolled down the window, and began to plead for mercy, testifying how it wasn't his fault at all, that his landlady had forgotten to set the alarm. The story didn't sound too convincing to Freda, who continued to ignore him, and with her arms swinging to an even faster pace, accelerated along the sidewalk towards the tall brownstone buildings of Milwaukee. Elmer, too, increased his efforts, and weaving along the road, leaning towards the open window, began to shout at her to get in the car. She kept on moving. Observing this suspicious behavior was a Milwaukee City Policeman on patrol, who switched on his blinking red light, bringing this shouting and pouting procession to a halt. "Is this man bothering you, Miss?" the policeman demanded, looking straight into Elmer's sheepish face. Freda was in the kind of mood that would have enjoyed seeing Elmer arrested for anything from reckless driving to indecent proposition. Her carefully measured delay in answering the officer was almost too much for Elmer's imagination, and his countenance took on the most pathetic and pitiful expression she had ever seen. Both Elmer and the officer were prepared to hear her say that she'd never seen this man before, and that yes, he was bothering her quite a lot. But at the last moment she told the officer, "No, he's O.K.," and feeling victorious in her punishment of Elmer, finally got into the car.

Elmer took his sweetheart to meet the Kiekhaefer family on the homestead, having negotiated at least a temporary truce with his father. The family simply adored her, and many of Elmer's sisters couldn't get over how well she seemed to fit with the family, and even commented how her coloring seemed to match their own family shade. But her visits only served to convince Freda how fundamentally different the philosophies of two families could be. The Kiekhaefer homestead was organized somewhat like a regimented factory, where the constant topic of conversation was what work needed to be performed next. When Arnold would come in from the fields, any of the girls that weren't working, and working hard at some basic chore around the house, made sure they were real busy by the time he hit the doorway. She could see how hard it was to organize recreation at the Kiekhaefers. At her home in Blue River, she and her lone brother were free, by and large, to do what they pleased with their time.

Elmer and Freda fell in love that summer of '31, and the passionate courtship that continued into the fall and winter gave pause to Freda's parents. The Greenfields, wishing to be certain that this brash young engineer was truly the man for their daughter, and wanting to be sure that she wasn't merely confused by and infatuated with his strong personality, whisked her off to Florida for the winter, separating the two star-crossed lovers by nearly two thousand miles. Their romance continued to blossom, as Elmer and Freda wrote to one another nearly every day. He would write of his loneliness and sincerity, of his ambition and ability. She would describe the wondrous blanket of stars that seemed so bright and close that you could "just reach up and touch them". In more playful exchanges, she would squash the large Florida mosquitoes onto a sheet of paper to show him how even this tropical paradise had its drawbacks. She even tempted him to come to visit her, though the lean times, heavy work load and night classes he was still attending were too much of an obstacle for Elmer to even consider the long trip.

Of all the correspondence during that lonely winter, the letter that Freda remembered most vividly was a three-page work of classic prose that Elmer had composed. The ambitious work rhymed in sections, and worried her more than just a little. In it Elmer extolled the virtues of their love in strong, forceful language, and spoke of how hard he would work to make life comfortable for her, and how great he considered his, and their future to be. The tone of the letter was so intense and esoteric, that she felt slightly panicked that the relationship was becoming so serious, so quickly. The last lines of his long, heart-wrenching love-poem were these:

"Fritzie, Sweetheart, say it won't be long,
Ere we can kiss and glory in life's most beauteous song,
Fritzie, Sweetheart, don't tell me I'm all wrong,
When those eyes tell me sweetly 'It cannot be so long,'
Lord, may conscience guide us, Pray, that I do her no wrong"

Sincerely, Elmer

Though she was quite enamored of Elmer, on occasions his unrelenting perfectionism and obsessive competitiveness disturbed her in a way she couldn't exactly define. When he asked her to be his wife, Freda told him she would have to think about it for a while. The long separation, though, from January to April of 1932, only served to fan the rising flames of their romance, and on June 25th, three weeks after Elmer's twenty-sixth birthday, they were married in a simple ceremony in the Greenfield family home in Blue River.

On their wedding day, Elmer was engaged in a heated lawn tennis match with his best man, and both of them were so unyielding and unrelenting in their efforts that they completely forgot about the pending marriage. As the ceremony was about to begin, someone was dispatched to retrieve the soaked and steaming groom and best man to the house. The Greenfield home was tailor made for the intimate family wedding, complete with long staircase for the slowly descending couple, their halting steps accompanied by the harmonious strains of the family piano, issuing forth the traditional and familiar melody. An uncomfortable moment occurred during the ceremony, at the point in time when the parson asked whether there was anyone present that had cause to object to the union, as Freda's young niece yelled out amidst the expected silence, giving goose bumps to all those in attendance.

Their honeymoon didn't exactly go as planned, but their plans were dictated by the austere times of the nation's widening depression. They traveled to northern Wisconsin, and set up camp in a small tent. When they awoke at dawn, they discovered that their tent had been invaded by hundreds upon hundreds of harvestman or daddy-longlegs spiders which had literally covered the walls and bedding, giving them both a jolt as they opened their eyes. They quickly rolled up their wriggling and twitching honeymoon suite, and though they had only a few dollars between them, moved into a small motel for the remainder of the week, which Elmer dedicated to hunting. It was a landmark occasion for the newlyweds, for, except for one aborted attempt, it would be the last vacation they would share together in the next over half-century of marriage.

Elmer continued to advance in responsibility at Magnetic Manufacturing, which had changed its name to Stearns Magnetic. Three of Elmer's designs resulted in patents, and though he had assigned ownership of them to Stearns and received no particular benefit from them, he found the experience of invention exhilarating, and before long he was modifying nearly every configuration that went out the door. Elmer's formidable design and fabrication talents resulted in his promotion to Superintendent of the Plant in 1933, which placed him in the third highest position of responsibility for the entire operation. It was the first time in his young career that Elmer would have men reporting directly to him.

Unfortunately, Stearns was barely surviving by the mid-thirties, and were lucky if they posted a few thousand dollars profit for the year. But even though Elmer assumed greater and greater responsibilities at Stearns, and his young bride was pregnant, his salary failed to keep pace with his position and self-estimate of value. But even more frustrating was Elmer's dim view of his personal long-term future with the company. He would later reflect on the trap he felt he was falling into.

"My contributions to the company were substantial, but I was hemmed in. Mr. Stearns was a frugal Scotsman, and he thought he had me tied up. He wouldn't raise my salary, and he wouldn't sell me any stock in the company, so that I couldn't participate in it's growth. In short, he thought he had me anchored there for the rest of my life.'[6]

On November 8, 1933, Freda gave birth to a daughter, Helen Jean, and soon discovered that wholesale changes began to occur in her relationship with Elmer. He insisted that her nursing career was over, and that her place would be in the home. Though at first it sounded somewhat inflexible to Freda, it was a traditional concept, and perfectly in tune with the morays and experience of most households of the thirties. And when she began to worry about her growing isolation, she imagined that it must be normal for a busy mother and housewife.

By 1935, Elmer calculated that he was responsible for a full seventy percent of Stearns engineering designs. He began to feel that his career had become stagnant, without sufficient reward in either position or salary.[7] He was still earning less than $4,000 a year, and the only job that interested him at Stearns was that of Chief Engineer. Because of the hard times, Stearns wasn't making sufficient profit to give Elmer an adequate raise in salary, and Elmer refused to take on additional responsibilities without more money. He began to develop some of his concepts on the side, in order to receive direct remuneration for his efforts. He produced successful designs in motor disc-brakes, air handling equipment and air pump mechanisms, which he built in his spare time to supplement his small income.[8]

In October 1935, Elmer began a systematic search for other employment that could mean an advancement in his career. Learning of the death of a key engineer at Dings Magnetic Separator Co., also of Milwaukee, Elmer quickly wrote them to see if they would be interested in his talents. The year 1935 is widely considered the very depth of the Depression, and Elmer was careful to hide his search from Stearns, using his home address for all correspondence. When Dings Magnetic Separator wrote Elmer saying that they had a great many applicants for the position, he responded with a letter extolling his abilities, in hopes that his application might be reconsidered. It read in part:

"My relations with the Magnetic Mfg. Co., while of the most faithful, and cordial, are for the present at least, somewhat out of line with my ability. My position as shop Supt. does not permit as much time for designing and inventing, as I should like. ... With practical as well as theoretical background, I can be of real service to you as a designer and field engineer. I have not only designed separators, clutches, brakes, apron conveyer magnets; I have personally repaired them, rebuilt them and sold them. Since designing is my particular calling, I am planning to leave the Magnetic Mfg. Co., in search of greater opportunity, if not responsibility."[9]

In another letter to a prospective employer, Elmer managed to sum up his estimation of his worth in four sentences:

"My theoretical training plus real experience in the various phases of manufacturing has given me the perspective so necessary to a good engineer.

"The fact that I was born and brought up on a large farm, fortunately has given me very good health and a genuine love for the practical in design.

"The early habits of hard work has done me good stead and has helped me to carry on long after the other 'chap' has quit--be it design, sales or trouble repair."[10]

Though Elmer corresponded with many organizations, and was interviewed by several, like the Line Material Company of Milwaukee, a manufacturer of street lighting equipment, he was unable to secure any position which he could consider an improvement over Stearns. Though his needs were changing, and the added expense of his young daughter forced him to even suspend the $34 annual premium on Freda's life insurance policy, he abandoned his search for a change, and remained at Stearns Magnetic.

It was during these dark days of the Depression that Elmer began to read books about the lives of prominent industrialists in various fields of endeavor. Eschewing fiction, the reality of biographies gave Elmer an education into the progression of events in famous lives, and he began to search for clues to their success. He would later reflect:

"Among my idols when I was a young man, were Henry Ford, Walter Chrysler, Andrew Carnegie, Thomas Edison, Wilbur and Orville Wright, Charlie Nash and many others of the

> old timers. I used to read such books as I could find on their respective lives, and I used to ask myself, "What did they have that I didn't?". I felt that I had everything that they had except I had to pay income taxes. They were able to plow their earnings back into the corporation they founded, whereas I [couldn't]."[11]

He also perfected a cost-saving technique during these lean times, of ordering mail-order books, reading them quickly, and returning them before the free examination period had ended. Among the books that Elmer devoured in this fashion were *Achievement* by Roberts, *Life Begins at Forty* by Pitkin, and *Technical Man* by Hurst.[12] He also continued to look for books to assist him in his growing work in alternating current circuits and polyphase motors, and searched unsuccessfully for English translations of famous German electrical engineering works such as Arnold's *Gleichstrommaschine*, and *Wechselstromtechnik*.

After ten years frustration, Elmer was finally appointed Chief Engineer of Stearns Magnetic in 1938. The press release which announced his appointment to the industrial society of Milwaukee stated in part, "His new position calls for supervision of research and development work. ..." It was an understatement, since Elmer had been supervising the research and development work at Stearns for over six years by the time his new title caught up with his efforts. His long struggle for position and prestige over, Elmer was absolutely convinced that his career lay in the design and manufacture of magnetic separation equipment, and began to consider the possibility of striking out on his own, in direct competition with Stearns. "I decided it was time to move. ...," he would later reflect, "Life is not static; it is dynamic, always in a state of flux. I decided then and there that if I was as good as I thought I was I'd better start proving it to the world, organize my own company and build equipment the way it ought to be built."[13]

CHAPTER FOUR

Cedarburg

"What I didn't know then about running a business of my own would have filled many a volume. [But] once you are an engineer, you're one of God's chosen people; and you can do no wrong."[1]

Elmer Kiekhaefer

John Blank, Elmer's favorite uncle, had taught school for a number of years before Elmer's grandfather Heinrich appointed him president of the Cedarburg Mutual Fire Insurance Company, a position the elder Kiekhaefer had held in addition to his homestead responsibilities. Blank was a natural for the job, for he was well liked, widely known and had a reputation for involvement in community affairs. It was at Uncle John's and Aunt Flora's house in Cedarburg that Elmer boarded during the icy winter months, attending Cedarburg High School. When John Blank was elected mayor of the small town, no one was surprised, especially not Elmer, who had always looked up to John as his confidant, and, as Elmer would later write, "He seemed to be the only educated person on my side of the family tree."[2]

Uncle John kept abreast of events in the small town of Cedarburg and, as the leading civic authority and pre-eminent booster of the municipality, he maintained an active dialogue with fellow citizens in local enterprises, from farming and banking to manufacturing. He had closely observed the changing fortunes of Cedarburg commerce, and on occasion, because of his position and influence, was able to put people together who could assist each other in cooperative business ventures.

Because of Cedarburg's diminutive size, it was easy to count the industrial employers in the area on one hand. Always an interesting topic of conversation was the fate of the old manufacturing building on the outskirts of town, that just couldn't seem to keep its operators in the black. Built in the 20s, the building had suffered a succession of owners and industries, each flawed in product or management, and each in turn failing. The Schmidtdorf Electric Corporation had entertained high hopes of squeezing themselves into the highly competitive spark-plug business, and when this began to look hopeless, started building small electrical equipment. When Schmidtdorf ultimately abandoned the facilities, the American Electric Motors Company began operations with a flourish. They designed and built the first fully-enclosed polyphase electric induction motor, and were soon swamped with orders. However, success can often be unkind to the ill-prepared, and as Elmer would later write, "the business was so successful that the three partners drank themselves into disaster and ruined the business."[3]

And so, in 1934, when the American Electric Motor Company joined the avalanche of business failures during the Great Depression and went bankrupt, it was with a great deal of civic pride that John Blank welcomed Thorwald Hansen to town. Thor Hansen had been chief production engineer for the A.O. Smith Company, manufacturers of stamped steel automobile frames. He was a gifted engineer, and with his considerable experience in metal stamping, was determined to fill what he considered to be a need for an inexpensive entry in the outboard motor industry. Taking over the abandoned building, Thor and his optimistic band of seventeen minor investors, including his son Royal Hansen, named their enterprise the Cedarburg Manufacturing Company.

It was Thorwald Hansen's son Royal "Roy" Hansen that prompted Thorwald's interested in outboard engines. Roy raced Evinrude outboards in the twenties and early thirties, and became quite knowledgeable about their design, manufacture and performance. Between the two of them, they decided that aluminum castings, which everyone in the industry was using for the manufacture of outboards, were not durable enough, and hit upon the idea of manufacturing them out of steel stampings, Thorwald's specialty.

Thor Hansen designed a small, two-horsepower, two-cycle engine, using as many inexpensive stamped metal components as possible, to keep his manufacturing costs to a bare minimum. Weighing in at only 35 pounds, the single iron cylinder engine had an aluminum piston just two-inches across (bore), with a stroke of 1 5/8 inches. To further conserve expenses, the *Thor*, as Hansen called his diminutive engine, used the least costly metals and alloys for components, like bronze connecting rod and bearings, nickel alloy gears, a low grade steel drive shaft and an aluminum propeller. When conditions were ideal, all these parts would spin, twist and pump at about 3,000 r.p.m. In his first promotional flyer, Hansen proclaimed, "You are going to appreciate THOR's utter simplicity. Look at the motor. Notice how absolutely clear it is of gadgets of any sort. The THOR is a motor reduced to the simplest form."[4]

Thor Hansen's decision to enter the most highly competitive segment of the marine industry at the worst of possible economic times would turn out to be a critical mistake. The peak year of production for the outboard motor industry had been 1929, with the sale of 58,456 engines, sold at an average retail price of $134. But, by 1933, the season just prior to Hansen's entry, sales had fallen drastically to fewer than 18,000 units, accompanied by an equally sharp drop in the average retail price of outboard motors, to around $70 per unit. When even the basic necessities of food, clothing and shelter were beyond the means of countless millions of American households, purchases of luxuries, like the trappings of leisure pursuits, were unavoidably suspended. The entire industry, which contained four principal manufacturers, producing nearly forty models of engines, *collectively* would gross only slightly over a million dollars for the year in which Thor Hansen decided to compete.

Actually, some pretty fuzzy thinking had taken place. Among the minor shareholders of the Cedarburg Manufacturing Company was a wide-eyed and ambitious optimist who was to become Thor Hansen's sales manager. After all, to someone with a name like E.C. Freshwaters, the outboard motor business seemed to be an ideal avocation. Thor Hansen had built a few prototypes of his new engine in 1934, while Freshwaters made a survey of the potential market. In a formal, four page forecast of Thor sales, Freshwaters concluded that, even though the *entire industry* had sold less than 18,000 engines in the previous year, Thor would sell 18,000 units by themselves as a start-up, totally unknown organization, competing against the legendary names in the business like Evinrude, Johnson, Elto and Muncie.

To prove his point, Freshwaters first determined that his forecast for 1935 should be based on the average number of motors sold over the previous eight years, including the peak years of 1927-1929, ignoring the fact that America was in the throes of the worst depression in its history. Next, he determined that the market for used and reconditioned motors, ranging in price from $35 to $50, would be virtually wiped out by the entry of the Thor at $39.75. And finally, an entirely new market, unbeknownst and totally overlooked by the best minds in the industry, would be established because of the irresistibly low-priced Thor. In his report, Freshwaters stated that "I have talked to a great many of these men and they all tell me the same story, that they are willing to pay $40. to $50. for an outboard motor to take the place of Oars, but they refuse to pay $80 to $100.00 for a new motor and also that they would not have a reconditioned motor at any price."[5]

In the early winter of 1934, Freshwaters embarked on a mad dash throughout the southern states on a whirlwind sales trip which was to cover 6,000 miles in thirty days. Offering the Thor at a discount of 25%, or $29.75, Freshwaters established 67 dealers and collected signed orders for 708 Thor engines. The concept of the inexpensive little engine had met with a favorable response and glowing estimates of future sales wherever Freshwaters and his Thor prototype could ambush an audience. In his closing remarks to Hansen, Freshwaters would leave little room for doubt about his enthusiasm, and one can only imagine the manufacturer's rejoice in his sales manager's conclusion:

> "In concluding, I wish to stress the fact that as long as there is no competition in price and performance for the THOR Outboard Motor, we have sufficient evidence in the form of actual orders, letters received from Dealers throughout the States and verbal opinions of dealers from all parts of the country to convince nearly anyone that an enormous market exists for our motor and that we should have no trouble in selling 18,000 THOR Outboard Motors per year."

Though Hansen would suggest that two spins of the rope on the flywheel were sufficient to get the Thor purring - one spin to clear the cylinder of moisture and old vapors, and one spin to draw a fresh charge of gasoline mixed with air into the motor and fire it - sometimes owners would work up less sweat rowing after they abandoned their efforts with the engine. "The Thor requires no priming to start," his first brochure suggested. The fact was, there was no way to prime the engine. Thor Hansen had eliminated some of the best features of competitive engines, and used well phrased advertising expressions to make his lack of application of recent advances in the industry sound like an advantage. "It operates with a new kind of carburetion, which is direct flow from Gas Tank to Engine," he would boast, but the truth was, some of the adjustments and refinements consumers had long since taken for granted were missing in his design. The mild steel he used rusted quickly in fresh water, and corroded rapidly in salt water. To save even more money, he eliminated the use of seals in the lower unit gear box, instead employing a thin cork washer. But unfortunately, the grease for the nickel alloy gears would run out, allowing water to seep in, hastening the demise of these critical components. "Look at the motor. Notice how absolutely clear it is of gadgets of any sort." Indeed. Prominent authorities in the outboard motor business had tried to steer Hansen out of potential problems, but Hansen wouldn't listen. When W.J. Webb, then with Ole Evinrude's ELTO outboard motor company, saw Hansen's first Thor at the 1935 New York Motor Boat Show, he was surprised by the reaction he received to his advice.

> "Thor had a very square head and no one could tell him anything about the problems that his motor would face. [When] I saw his motor ... I asked him what steel he was using and when he told me, I tried to tell him he was headed for trouble in salt water. Others in the industry tried to tell him the same thing, but Thor knew better. Also, no one could tell Thor that selling outboards was quite a different thing from selling stampings to auto makers. Had Thor listened to the good advice which was offered, we might have steel stamped outboards around today."[6]

By the end of the year, Thor Hansen was making a variety of minor improvements to his engines, most notably cosmetic changes to the lower unit. His first Thor looked somewhat like a plumber's nightmare, with three gangly tubes protruding downward from the little single cylinder engine. One tube, the drive shaft, led to a spindly gear box brandishing ten large stove bolts to keep it clamped together with the propeller shaft. A second tube, about the diameter of an index finger, led downwards from the engine to a small ball and plunger water pump located between the propeller and the gear box. The third tube, the largest of the three, was for exhaust, leading from the engine to just above the propeller. Over the next several years, the most noticeable changes to his original design would be the angle, diameter and placement of these elements.

By the fall of 1935, Hansen had designed and built a twin-cylinder version of his engine, which, except for the addition of a second cylinder, piston, different crankshaft, and a few other component changes, was identical to the single engine version. One cylinder was simply stacked on top of the other, and fired alternately, so that pistons were always moving in opposite directions. This new engine weighed just eight pounds more than the single engine model, and, at 4.8-horsepower, packed exactly double the power of the single engine model.[7] Thor Hansen's twin-cylinder engine was modestly successful with his dealers, but the single-cylinder version was by far the more popular. Thorwald proudly sent W.C. Clausen, general sales manager of the giant Outboard Motors Corporation, manufacturers of Evinrude and Elto outboards, a copy of his modest single-page flyer on his new line of Thors. OMC responded quickly with an order for one of the engines, naturally to keep an eye on what Hansen was up to. But when the engine still hadn't arrived a month later, Clausen pushed for delivery and enclosed a copy of OMC's new 16-page, 1935 Elto and Evinrude catalog, displaying nine new models from 1.5-horsepower singles to the 31.2-horsepower, four-cylinder *Speediquad Imperial*.[8]

In early 1936, Hansen received his first of many orders from the giant retail chain of the Montgomery Ward Company. Purchasing the Thor single cylinder engine in lots of 500 engines, Montgomery Ward marketed the bargain priced models under their private *Sea King* brand name. For the Cedarburg Manufacturing Company, the steady stream of orders and dollars from the far-flung chain of stores meant unusual security amidst generally uncertain economic times. Except for decalomania, labeling and literature, there were absolutely no differences between the Thor and Sea King engines. Customers of Montgomery Ward would have no idea that the Cedarburg Manufacturing Company had produced the engines, for the origins of the engine were nowhere to be found in the operating handbook or parts list. In 1936, Montgomery Ward would offer Sea Kings in 2-hp., 4-hp. and 8.5-hp. models only, down from a high of six engine sizes in 1932. Except for the 2-hp. Thor, all other models of Sea King engines were manufactured by Evinrude Motors in Milwaukee, a division of OMC.

Hansen had offered a one year guarantee against any defects in workmanship or materials for the Thor. The guarantee was surprisingly generous considering the quality of the materials used. Fortunately, most customers found the "gadget-free" engine simple enough to trouble-shoot themselves. But Montgomery Ward, who offered a 90-day warranty for the Sea King version of the Thor, had always been proud of the exceptional service which they afforded their customers, and every Sea King that customers returned to Montgomery Ward for anything ranging from a fouled spark plug to a completely demolished engine was promptly returned to the Cedarburg Manufacturing Company for warranty service. Warranty work began to pile up.

As a result of their large orders, Hansen, along with Evinrude, had to sell engines to Montgomery Ward at a higher discount than even their own best dealers. For Evinrude, this was less of a problem, for they were manufacturing nineteen models in 1936, and had a distribution system with thousands of dealers. For Thor Hansen, though, the orders from Montgomery Ward were a two-edged sword. On the one hand, the orders kept his Cedarburg Manufacturing Company working at peak capacity, building as many as 2,000 Sea King engines a year. But his profit margin was so small, only a few dollars per engine, that the more engines Montgomery Ward ordered, the harder Hansen had to fight to keep costs down. Because Montgomery Ward paid upon delivery, Hansen had to constantly scramble to finance his supply of parts and labor to keep the assembly line moving to meet contract shipping schedules.

Hansen also knew that to be able to compete with Elto, Evinrude, Johnson and Muncie, he had to develop a broader range of offerings, and give his dealers a full line of engines. In late 1937, Hansen designed the world's first three-cylinder outboard engine, which developed 6.2-hp. at 3100 r.p.m. Like the development of the twin-engine Thor, the new *Pyramid 3* was simply another expanded version of the original single-engine model. Like the twin, it was an alternating firing engine, but with three pistons and cylinders, a larger crankshaft and exhaust manifold, it topped the scales at 56 pounds and retailed for a $110.00. The only problem was, the over six-horsepower of the Pyramid 3 was delivered through the same steel stampings and nickel gears designed originally for the two-horsepower model, giving the engine only modest performance and a high rate of failure. Montgomery Ward decided not to order Thor's *Pyramid 2* or *Pyramid 3*, and the Thor dealer network couldn't sell them either. Debts began to mount.

By 1938, Hansen was in trouble. He had managed to sell off more of his stock in the Cedarburg Manufacturing Company, first offering it to his current investors, and then to new sources. At $100 per share, he was only able to raise slightly more than $3,000. He had run out of money, and was in debt nearly $40,000 to his suppliers and lenders. Though he had orders from Montgomery Ward, some of his critical suppliers refused to deliver parts and materials until he reduced the large balances due on their accounts. Thor Hansen turned to Elmer's Uncle John Blank for help.

Among John Blank's network of friends and business associates was Edgar H. Roth, vice president of the Cedarburg State Bank. After analyzing the financial condition of the Cedarburg Manufacturing Company, the bank considered Hansen's predicament as too risky for the small community bank, which was only capitalized for $40,000. The discussion began to swing toward the possibility of liquidating the company in some fashion, discharging Hansen of his enormous debt, and allowing a group of investors to use the facilities to embark on a new business. It would be the third business failure in a row for the unlucky manufacturing facilities, and so the conversations must have seemed like business as usual for the concerned mayor and banker.

John Blank knew that the building occupied by Hansen was sound, well heated, and among the assets were machinery and tooling from the American Electric Motor Company which went bankrupt before Hansen arrived. Knowing that his nephew, Elmer, as Chief Engineer of Stearns Magnetic Company, might be capable of operating a new electrical company at the location, Blank began to formulate a plan to acquire Hansen's operation. Blank first spoke with Freda, Elmer's wife. He carefully quizzed her about Elmer's current situation and ambitions, and whether she considered him as capable as Blank himself assumed he was. Freda answered, without hesitation, that Elmer was like a steamroller, and that once he set his mind on something there was no holding him back. After contacting Elmer to see if he would be interested in being involved in the venture, Blank contacted a handful of leading citizens which included Edgar Roth of the Cedarburg State Bank, Adlai Horn, the editor of the *Cedarburg News*, Dr. William H. Wiesler (a local dentist), Matthew P. Becker and Palmer J. Wirth (one of which was the local butcher) and Arnold Kiekhaefer, Elmer's father.

John Blank's plan was contingent on several key factors. First, he estimated that approximately $25,000 was needed to complete a complex transaction. Secondly he needed Elmer's full support and involvement to administer the operation, as general manager of a new company. Thirdly and most critically, he estimated that he needed to persuade at least 80% of Hansen's 108 creditors to liquidate their debts to the new organization for 30% of their value. For example, if Hansen owed $1000 to a supplier, they would have to be convinced to sell their debt to the new company for $300, which would then be paid immediately. The theory was, if they didn't sell their debt for the greatly discounted value, the possibility could exist that the new company wouldn't take over Hansen's debts, and the creditors might not receive a dime.

On November 8, 1938 a letter was drafted to be sent to each of Thor Hansen's creditors, signed by Milwaukee attorney Lester H. Gunsburg, representing John Blank. It would be the only opportunity and notice that Thor Hansen's suppliers would receive, and it read in part:

> "I represent a group of clients who are interested in purchasing your claim, due and owing you from the [Cedarburg Manufacturing Company], at 30% of its face value. None of my clients are interested as an officer or stockholder of the debtor. Advise me forthwith if you are interested in the assignment of your claim.
>
> The forgoing offer is conditioned upon your ability to prove your claim and validly assign the same to my clients and upon my clients obtaining consents to so assign to them at least 80% of all claims outstanding against said debtor, within the current month."[9]

Several factors made the proposed acquisition seem like a sure bet investment. Royal Hansen, who had assumed the presidency of the Cedarburg Manufacturing Company, was in receipt of an order from Montgomery Ward for 500 single-engine Thor engines, which could only be built if he could raise an additional $6,000 to meet his payroll and cost of materials. The order was worth around $12,000, so Blank decided that while their efforts to buy the discounted debts continued, they should try to lock in the order from Ward's.

Meanwhile, Montgomery Ward had learned that the Cedarburg Manufacturing Company was in dire straits, and the order for 500 engines might not be delivered. In order to assuage the retail giant's fears, a letter was sent from Hansen's attorneys in Milwaukee to Ward's headquarters in Chicago to describe how this new group of investors would guarantee the production of the engines. It stated, in part:

> "We understand that you are about to go to press with your catalogue and that you wish some written assurance that your order will be fulfilled at the time specified, and that you are concerned about this matter particularly because of the company's present financial difficulties.
>
> A group of persons not identified with the company have, within the past two days, made a written offer to all of the company's creditors to purchase its accounts payable at a discount. We expect substantial progress will be made by this group within the next week. This group has assured us that if it acquires these accounts and takes over this business, it will fulfill your contract. However, at the present time, we cannot be certain that this group will acquire the company. The company has gone as far as it can in the direction of completing your order. We enclose a copy of an agreement made today with Mr. John G. Blank of Cedarburg, Wisconsin, whereby the company will acquire sufficient funds to meet pay roll and material requirements. However, if the unforeseen should happen, that is, if a creditor should succeed in getting an judgment against the company and forcing a liquidation, we could not be certain that your order would be filled. ...
>
> Under the circumstances, we feel that you are reasonably safe in including this item in your catalogue and trust that you will do so. ..."[10]

With John Blank's guarantee in the form of a check for $6,000, Montgomery Ward was satisfied.[11] The two-horsepower Thor made the 1939 catalog, and production began before Christmas of 1938. The response of creditors to the letter proved to be nearly unanimous in favor of accepting the discounted payment, and Blank made plans to close a deal with Hansen for the company by the end of January 1939.

Because three different companies had failed in the same facilities, a large inventory of valuable equipment had collected over the years. The replacement value of the machine tools alone was nearly $234,000. Even if an emergency occurred and the investors had to bail out, they all knew that with any kind of luck they could sell off the equipment and, at the very least, get their money back. Even the land, valued at $1200 in 1935, along with the building which was valued in excess of $44,000, would guarantee their $25,000 investment. The combination of the Montgomery Ward order, the machinery, the land and the buildings, made the venture seem a pretty safe investment.

As soon as Hansen completed the Montgomery Ward order in January, the new organization was set to move in. Under Elmer Kiekhaefer's direction, the new company would start a magnetic separation, magnetic clutch and magnetic brake enterprise, competing in markets that Elmer knew very well.

During the month of December, 1938, Arnold Kiekhaefer mortgaged the farm. It wasn't a decision that he took lightly, for the homestead was all that he and his family owned. Mortgage money wasn't easy to come by during the Depression, as hundreds of thousands of homes, farms and businesses had mortgaged themselves to the hilt. When people couldn't make the loan payments, their lands and buildings were turned over to the banks. The banks in many areas had more land than cash, and so it was through the efforts of Edgar Roth, the vice president of the Cedarburg State Bank, that a mortgage was granted for the Kiekhaefer homestead. When the papers were signed, Arnold told his daughter Almira, "If this doesn't work, I'm broke. You're going to have to take care of mother."[12] The mortgage money became the backbone of the funds to acquire the company, and Arnold Kiekhaefer became the largest single stockholder with a full forty percent interest. The butcher, the dentist and the newspaper editor each invested $1,000, and John Blank invested $6,000.

Elmer was the last to put money into the acquisition fund. He knew that to ask his father for funds to contribute to the enterprise would be fruitless, because Arnold had already put the entire homestead on the line. So, reluctantly he approached his grandfather. But Heinrich was still upset with him for leaving the farm. In fact, on several occasions when Elmer had seen Heinrich during the preceding eleven years, his grandfather had completely ignored him. At a family picnic, after his grandfather had managed to avoid him all afternoon, Elmer confronted him and demanded to know what was wrong. Heinrich looked him straight in the eyes and said, "When are you coming home and get back to work?" Heinrich finally relented, however, and consented to loan Elmer $500. So, of the initial $25,000 that was raised to begin the new company, only $500, or a scant two percent, was contributed by Elmer. On December 30, 1938, Elmer delivered his $500 to John Blank and Edgar Roth at the Cedarburg State Bank, and signed an agreement that if they weren't able to issue stock for some unforeseen reason in the venture, Elmer would be able to redeem his proportionate share of any accounts receivable due the Cedarburg Manufacturing Company.[13]

Even though Elmer was by far the smallest cash contributor in the group, his contribution in knowledge and leadership would be the largest, and as a consequence, his stake in the new company was raised to eleven percent. The shares of Arnold and Elmer Kiekhaefer would together represent a fifty-one percent majority of holdings. Arnold Kiekhaefer, though he pursued other business interests, was a full-time farmer. Edgar Roth was a full-time banker, Adlai Horn a full-time newspaper editor, and neither the butcher nor the dentist were interested in new careers. Elmer's true investment in the new venture was to be himself. The others knew that he would give himself tirelessly to the effort, and if anyone could break the string of failures in the humble facility along highway 27 in Cedarburg, it would be Elmer Carl Kiekhaefer.

CHAPTER FIVE

New Directions

"Nothing could get in his way. If he had been an army general he would have commanded the dead to stand up and do their job."[1]

Bob Stuth
Early Cedarburg employee

All was proceeding according to plan, when, in January of 1939, disaster struck the Cedarburg Manufacturing Company. The previous order of 500 single-cylinder Thor engines which Thorwald Hansen had delivered to Montgomery Ward were almost completely rejected and returned. Three-hundred eighty-four engines failed to either pass Montgomery Ward's strict inspection for quality, reliability and performance, or were returned to Wards by consumers who were fit to be tied. Not only were this latest batch of engines difficult, if not impossible to start, but if Herculean efforts managed to get them started, they would promptly quit. If they even got wet they would quit. It seemed to Montgomery Ward's buyers that if you even looked at the little engines sideways they would quit. The Cedarburg Manufacturing Company was not in a financial position to absorb the cost of new shipping labels, much less a huge return of engines for alteration and warranty repair. On top of everything else, Montgomery Ward canceled the current order for 500 engines that Hansen had already begun to assemble. It was the final defeat for the small company.

On January 22, 1939, Elmer Carl Kiekhaefer walked through the door of the Cedarburg Manufacturing Company and took possession. It was a bitterly cold day, with a thick sheen of ice covering all roads to the plant. Years later, Elmer recalled his impressions.

"I walked down to that plant through all the snow. I couldn't get within 300 feet of it, there was 12 foot snow drifts around that building. ... It was 22 degrees below zero inside that plant too. There was no coal in the bins or no steam in the boiler. When I walked through that plant, I saw a bunch of broken-down lathes and other old-fashioned broken-down machinery. It was probably the most depressed day I have ever had in my life. I had no product. I had no employees and no money, but I did have resourcefulness."[2]

It was Sunday, and Elmer wasn't legally supposed to take over until Wednesday the 25th, but the thought of doing nothing for three whole days was driving him crazy. The new company would be known as the Kiekhaefer Corporation, but not entirely for the reason Elmer would have preferred. The principal investor and president of the corporation was his father, Arnold Kiekhaefer, and Elmer was vice president and general manager.

On that first day, Elmer brought along his sister, Isabelle, to prepare announcements that would herald the opening of the new enterprise. Isabelle reminded him that he didn't have a mailing list, and he really had no idea what he was going to manufacture. She sensed that her brother was just lonely, and didn't want to go down to the plant alone.[3]

It would be a most memorable day in the life of Elmer, for at almost exactly the same time as he opened the doors to the plant, his wife Freda gave birth in Milwaukee to his second daughter, Anita Rae Kiekhaefer, joining Helen, who was then six years old. It is important to understand that even though it was Sunday, and Elmer knew that his wife was in labor, that instead of being by her side chose to tour his empty and freezing plant. It was an emotional syndrome that would afflict Elmer throughout his entire professional career: factory first, family second.

The plant was just as the Cedarburg Manufacturing plant had left it, with an abandoned look about the piles of parts, scattered islands of machine tools, tall stacks of wooden crates containing rejected Thor engines, and a row of partially assembled engines clamped to wooden saw horses that curved in a gentle arc across the middle of the silent plant. But, it was the electrical tooling equipment that interested Elmer the most, for he was convinced that he would design a line of magnetic separators, clutches, brakes and electric motors, and immerse himself in the line of business that he understood best.

The initial balance sheet of the Kiekhaefer Corporation was drawn up on the first official day of business, January 25, 1939. Of the original $25,000 raised to acquire the Cedarburg Manufacturing Company, only $5,146 remained, and less than $2,000 was due in highly questionable accounts receivable. It looked as if the Kiekhaefer Corporation had about ninety days to prove itself, or suffer the same grievous fate as the previous three firms to have occupied the plant. Elmer knew from his eleven years at Stearns Magnetic that it takes time to design new products, market and manufacture them. Reluctantly, Elmer conceded that he needed to raise additional working capital to properly develop the business. Looking around, it was obvious that by far the most liquid assets, with the exception of shop tools and machinery, were the abandoned Thor engines.

Standing between Elmer and the liquidation of the Thor engines, though, were two prickly obstacles: the reluctance of the existing Thor dealer network to accept additional engines, and the literal rejection of the Thor by its biggest customer, Montgomery Ward. The Thor dealer network had shrunk dramatically since 1934. The first cancellations had come from coastal dealerships that supplied engines for use in salt water. To accurately stamp the steel components, Thor Hansen had used a soft alloy steel which was extremely vulnerable to rust.

Working from dawn to often after midnight, Elmer began to labor as a man possessed. He retained a number of men who had worked for Hansen and understood Thor assembly and testing. Two of the men, Ed Dehling and George Sekas, recalled that Elmer positively bubbled with confidence, and would tell them, "Stick with me, fellas." Their wages were thirty cents an hour, when cigarettes like Marvel and Cavalier sold for a dime a pack. Some of the machinery was so old that it was powered by long leather belts driven by overhead shafts, a technique that had been obsolete for decades.[4]

Work was accomplished at the factory with dizzying speed. Within the first week, Elmer had re-designed the carburetor of the single cylinder model, to provide a better mixture to the cylinders, and allow the fitting of a streamlined cowling. He directed the roughing out of a new Thor brochure, stressing inland water fishing, scrupulously avoiding any mention of salt water use.

Elmer Kiekhaefer's powerful personality began to emerge through the literature copy he prepared: "A man's motor, built for fisherman." Beginning a tradition that would last throughout his long career, Elmer discarded the misleading copy prepared by an advertising agency for the company's first brochure. Suggestions such as, "The owner of a Thor always recommends it enthusiastically to his friends!", and "Thousands of fishermen enthusiastically recommend Thor Motors, ask them why," were rejected, and replaced with solid facts and honest logic such as, "A motor heavy enough to drive your boat regardless of wind, weeds, waves and weight." Kiekhaefer's penchant for honesty and modest claims started with his very first promotional pamphlet, where he pointed out why he didn't make high-speed claims for his new motors.

> "So many things govern the speed of a boat such as design, weight, load, head wind, rough water, skill of operation, that it is impossible to list accurately what your boat will do. THOR Motors will probably give you the same and probably more speed per H.P. than any other motor of similar rating."[5]

One of the hardest things Elmer would ever have to do in his career, was to face the Montgomery Ward buyer who had rejected the previous batch of engines, and canceled all future orders. After refusing to return Elmer's phone calls, the buyer, Earnway Edwards, Director of Purchases for the Sporting Goods Division, finally relented.[6] When Elmer announced himself at Ward's Chicago headquarters, he was "met with a cold reception from a very disinterested sporting goods buyer."[7] The nervous exchange was recalled by Elmer twenty years later, and was remembered as a crucial turning point in his career.

> "When I announced myself - after he finally did see me - as the new owner of the Thor outboard motor company, he never looked up from his desk but said, "Yes, what's that supposed to mean?" I replied, "Frankly, we'd like to reinstate the order for Thor outboard motors that you have canceled."
>
> "His next question posed a problem. He asked, "Did you ever work for an outboard motor company before?" Here's where I had to tell a little white lie. I couldn't very well tell him that I had been fired by Evinrude, so I decided that Evinrude was not really an outboard motor company and answered, 'No.'
>
> "He countered by saying, 'Pray tell me, what makes you think you could make that piece of junk run?'
>
> "'Will you try them?' I questioned.
>
> "He looked up at me for a long time with a steady gaze. I could read his thoughts, and believe me if I ever looked serious and intent I did then. Something must have convinced him because he said, 'Well, all right. Send me a half a dozen that will run.'
>
> "'When?' I queried.
>
> "'Just as soon as you can get them here,' was his reply.

"'Thank you, I'll be in touch with you,' I said and left."[8]

Elmer was elated as he turned his car north onto the highway towards Milwaukee. The three hour drive must have gone by in a blur, as his mind exploded with the many tasks ahead, posing and discarding the many options to get the Thor engines running again. The successful modification and sale of the 384 engines to Montgomery Ward would double his cash reserves in one transaction. But there were so many obstacles. Ignition, carburetion, timing, even the crankshafts would have to be reworked. How would he ever be able to solve enough problems to stabilize this querulous engine that seemed to have a mind of its own?

Elmer decided to invoke the ancient Norse legend of Thor as part of his cure for the rejected engines. Thor, a deity common to all of the early Germanic peoples of Europe, was the mythological god of thunder, war and strength. Thor, the son of the God Odin, as it turns out, actually had a great deal in common with Elmer. The great warrior God was revered as a middle-aged man of enormous strength, and as a God of order who was an unrelenting foe against a race of giants. Among Thor's greatest weapons was the thunderbolt of lightning, and in ancient times whenever the distant rumble of thunder was heard, his name was invoked with trepidation.

Elmer employed the concept of the thunderbolt to overcome one of the inherent weaknesses of the Thor engine. Among the assets of the Cedarburg Manufacturing company was a carload of nearly obsolete Bosch magnetos which Thor Hansen had used on the Pyramid Two. These were of the opposed twin-engine variety, and had two coils which were designed to fire two cylinders simultaneously. The inability of the Thor single-cylinder engine to start quickly, or to run smoothly when the wiring got wet, was in part due to the insufficient spark being delivered to the cylinder. Elmer rewired the surplus twin-coil magnetos to deliver the entire voltage to one cylinder, instead of two, thus doubling the strength of the spark. He installed these twin magnetos in the six prototype engines for Montgomery Ward.

The increased spark was helpful, but it still couldn't overcome the lack of an adequate gas and air mixture. The Thor lacked a traditional carburetor, which blends the air and gas in a proper mixture to explode inside the cylinder. Instead, it had a mixing valve arrangement, which was nearly impossible to fine tune. On February 8, 1939, exactly two weeks after taking over the building, Elmer drew his first sketch of improvement for outboard motors. It was an enlarged intake manifold for the Thor that incorporated a finely adjustable thumb-screw jet that would allow a better fuel mixture, and take advantage of the hotter spark he had already created. Again, the engine performed better, but still sounded too rough to risk a Montgomery Ward inspection.

In a last ditch effort, Elmer tore the engines down again, and decided that what Montgomery Ward didn't know wouldn't hurt them. In a masterful exercise in skullduggery, he pulled out the crude and poorly balanced crankshaft of the Thor, and replaced it with a nicely forged one from a 2-1/2 H.P. Elgin Waterwitch outboard engine, sold by his client's largest rival, Sears Roebuck and Company. With some minor alterations, it worked beautifully. The engines started quickly as a result of the hot spark, idled nicely as a result of the more accurate air/fuel mixture, and delivered smoother power as a result of the secret crankshaft.

Elmer quickly built up six modified engines, and sent them to Chicago for inspection. The buyer called Kiekhaefer and said, "All right. Fix up the rejects, and we'll see how they go."[9] Elmer, knowing that he couldn't very well get away with installing Sears-Waterwitch crankshafts in his Montgomery Ward *Sea King* engines, reworked his inventory of Thor crankshafts with the Waterwitch shafts as a guide. The remaining rejected engines were modified and sent back to the very same stores they had come from, where they sold, and sold well.

Montgomery Ward was impressed with the modifications which Kiekhaefer incorporated in the *Sea King*, and reinstated the canceled order for 500 engines. He was told that if his first solo production run of 500 engines were acceptable, that he would receive additional orders for more. Elmer was in business. Within his self-imposed 90 day period to succeed, he had sold 884 outboard engines, and was determined that his first production of *Sea Kings*, assembled by the Kiekhaefer Corporation's own lines, would warrant the future orders.

With the Montgomery Ward order in his pocket, Elmer went to work to strengthen Thor's anemic distribution network. By the third week in February, as the Kiekhaefer Corporation's neared its first full month of business, he was ready to introduce an "all new" product. The greatest improvement to the Thor single was the addition of a real carburetor, manufactured by Tillotson of Toledo. The Thor mixing valve, which would still be used on Elmer's *Sea King* models in 1939, had forced the engine to be regulated by the "spark lever", or magneto advance, so that when you reduced speed, the engine still had a full gas charge with a retarded magneto. This caused fouling of the spark plug, poor gas mileage, and rough performance. The new carburetor allowed Elmer to boast that his "motor will accelerate smoothly, and without complaint, from idling to full throttle. Speed is controlled by throttle or 'Hand Accelerator' with a power response comparable to a fine automobile."[10] A simple linkage had been devised so that the movement of one lever would coordinate both the throttle and the spark advance, a concept employed in other outboard motors for many years.

More importantly, Elmer had accomplished a complete styling overhaul of the Thor Single, the final design of which was carefully hand shaped in clay for his approval. He added a sleek, compact looking clam-shell style engine cowling, which was made possible by integrating the old U-shaped, boxy-looking fuel tank into the upper half of the design. It transformed the character of the little engine almost completely, eliminating the fragile, exposed look of the original Thor power head, and giving it a new sensation of strength, compactness and value. It was a bold and imaginative design stroke, and when Elmer sent out releases to the Thor dealer network and outdoor magazines, he included an "electro" image of the new Thor *Streamliner* Single, made by the *New Thor* Outboard Motors of Kiekhaefer Corporation. It looked for all appearances as if a completely new engine had been designed in just two weeks flat.

On Saturday, April 22, 1939, less than two months after starting business, the Kiekhaefer Corporation displayed its new line of Thor engines at the Milwaukee Sentinel's Sportsman Show, held in the Milwaukee Auditorium. Elmer had designed the little booth himself, which consisted of a backdrop sign bearing the stern countenance of the thunder God Thor shooting lightening bolts that were labeled *stamina, power and economy.* His booth was set up next to the Poynette State Experimental Game and Fur Farm exhibit which proudly displayed a collection of blackneck, ring necked, mongolian and mutant pheasants. On display was the new Thor *Streamliner* along with the all but discontinued two- and three-cylinder Thor engines. Elmer spread key components for the engines over a table cloth of black felt, and encouraged hecklers accustomed to the poor performance of Thor engines to see for themselves. The hit of the show was the new Thor Streamliner, which, because of Elmer's close fitting, stylized engine housing, looked years ahead of its time.

Elmer hired "Red" Parkhurst as his first sales manager, a well-known automobile and motorcycle racer of midwest renown, who was for many years a test rider for the Harley-Davidson Company of Milwaukee. "Red" was quite a well-built and agile man who, like Elmer, was hardened by the realities of rural life in the twenties and thirties. As a consequence of his various exploits and motorcycle mishaps, "Red" allegedly had a silver plate surgically implanted into his skull, a consequence of a severe accident. But, "Red" was considered tough as nails, and no one considered him the worse for wear, except one day when it seemed quite convenient.

Elmer and "Red" had gotten into a heated discussion that quickly escalated into a vitriolic shouting match. Elmer always had a reputation for excitability, but as the pressure of the first few months of 16 and 18 hour days built up, it seemed he blew up more and more regularly, and with louder and louder results. This time, "Red" Parkhurst wouldn't take it. As both men bumped broad, inflated chests like an enraged baseball manager and umpire, the language flew with breakneck speed from "You're Fired!!" to stinging personal affronts and violent oaths sputtered in a shower of saliva. "Red" had heard enough, and took a wide and mighty swing at Kiekhaefer. But Elmer, wide-eyed at the boldness and strength of Parkhurst, was already moving, and exploded through the side door of the plant, with a cursing and screaming blur of red hair in hot pursuit. Elmer kicked up a blizzard of snow as he raced through the stubble field adjoining the plant, his breath looking like the billowing white clouds of a steam locomotive in the icy afternoon air. Each of the twenty-five noses inside the plant were pressed against the factory's windows as Elmer, in a flood of adrenalin, quickly outdistanced his tiring pursuer. With sweat streaming down his face and shirt, he jogged back to the plant, where to an assembly of mesmerized onlookers he casually announced: "I didn't want to kill him, you know, *he's got a silver plate in his head!!*[11]

Elmer had organized the young company into two divisions, Automotive and Electric. The Automotive Division was envisioned as manufacturers of automotive brakes and superchargers, along with the Thor outboard engine line. The Electric Division, still Elmer's lingering emotional target for the business, would manufacture magnetic brakes, clutches, separators, electric motors and any other custom electrical apparatus the market would bear.

For the better part of 1939, Elmer, Freda, six year-old Helen and newborn baby Anita lived in a miserable clapboard house located in a gravel pit not far from the factory. The house was nearly beyond repair, and in places you could see between the boards in the walls to the pit outside, and between the floorboards underfoot. There was a modest pond at the bottom of the pit, and Elmer, traveling in tight circles, tested Thor engines there, clamped to the sterns of tiny rowboats as soon as the ice thawed.

Freda recalled those early days in Cedarburg with memories of rare glimpses of Elmer. He would be exhausted when he finally dragged himself home from the plant, often after midnight. His baggy clothes would be streaked with oil, perspiration and grease, and he would quickly collapse into bed. Sometimes he would bolt upright in their bed in the middle of the night, having dreamed of some solution to a vexing problem at the plant, and scribble it down on a list that, during the day, he kept stuffed in his shirt pocket.

Elmer very rarely took time to even acknowledge his growing children. He was so often in moods of shortened patience and self-imposed agony that his six-year old daughter Helen began to hide in her room when she heard the unmistakable approach of her father's souped-up Plymouth grinding and crunching the gravel on the winding road to the house. It was a habit of shrinking fear and retreat that she would practice in one form or another for many years to come. It was a habit that many who worked for Elmer began to practice as well.

Bob Stuth was working as a mechanic at a filling station and shop near his Milwaukee home when Elmer, still working at Stearns, drove up one day. He was blustering about not being able to keep cylinder head gaskets intact on the big Chrysler engine which he had jammed under the hood of his jaunty Plymouth sedan. The engine was so much larger than the original, Stuth remembered, that the radiator had to be moved quite a ways forward. Stuth replaced the cylinder head gasket for Elmer, and must have impressed him, because he was invited to stop by Stearns and see a super charger that Elmer had designed for the car.

Stuth, an accomplished motorcycle racer, ultimately took on a second job during the late depression, in the machine shop of Allis Chalmers. He was falling asleep at the breakfast table, working two full time jobs, each requiring 40 to 48 hours a week. He would wrap up the night shift at Allis Chalmers, from 11 at night to 8 in the morning, and then rush to the filling station to start another full shift. So, the following year, when Stuth saw an announcement in the Milwaukee paper that Elmer was looking for men to work with small engines in Cedarburg, he quickly wrote and offered his services.

A handwritten reply from Elmer brought Stuth to Cedarburg, where he was hired for 45 cents an hour. Testing had begun on the reconditioned *Sea King* engines for Montgomery Ward, and the test booths weren't ventilated. Stuth went home his first three nights with banging, excruciating headaches from the barely survivable levels of carbon monoxide that spewed from the little engines all day long. Finally, he ripped out the test booths and dragged them outside in the open to keep from suffocating on the job.

Elmer insisted on being personally responsible for every facet of the new operation. From his many years as chief engineer at Stearns Magnetic, there wasn't a shop tool made that Elmer couldn't operate. He would personally demonstrate drill press, lathe, grinder and boring equipment to every new employee until they could operate them to his high standards. Wherever he went in the plant, he speeded up assembly through some suggestion, usually accompanied by a demonstration. But he rarely commented on good work, seldom praised an excellent performance. His reward, rather, was to spare the worker his fury. A quiet glance without comment could be dividend enough to those who had experienced his wrath in the past.

Arnold Kiekhaefer seldom came to the plant though he was the president of the corporation. His full-time job farming the Kiekhaefer homestead, along with a number of other small business pursuits, made his visits to the plant a rare sight for the employees. In stark contrast to Elmer, Arnold would stroll through the plant, praising workers for their efforts and efficiency, and giving short inspirational speeches wherever he went.

It was said that you could hear the son's teeth gnashing and grinding all the way to the other end of the factory.

As the quality of the Thor *Single/Sea King* engine steadily improved, Montgomery Ward awarded new production contracts. By the end of his first year in business, the Kiekhaefer Corporation had shipped over 2,300 engines to Ward's. The Thor *Streamliner*, had been only marginally successful, with only 473 engines shipped to Elmer's own dealers. Only 90 Thor *Alternate 2*, were sold, and only 12 Thor *Alternate 3* engines were shipped for the entire year, sounding the death knell for these awkward engines.

Montgomery Ward was impressed with the miracle that Elmer Kiekhaefer had accomplished by getting the ill-designed Thor engines to run, and by substantially increasing the performance of the engines without increasing their wholesale cost. Earnway Edwards, whose faith in Elmer had gotten the Ward's business back on track, contacted him and asked, "If you can get results out of a piece of junk like this, how about designing us a new outboard."

"No sweat," Elmer answered cockily.[12] No one at the plant had designed an entire engine before, but Elmer was flush with confidence and accepted the assignment with almost childlike enthusiasm. He sought out and hired the best engineers he could find, recruiting one from his old nemesis, Evinrude. Elmer and Montgomery Ward agreed on a set of design and cost specifications for a new three-horsepower, single-cylinder *Sea King* outboard. When Elmer delivered the prototype to them in Chicago less than five weeks later, he walked away with an order for twenty-thousand engines.

The dust hadn't settled from the frantic activity of preparing the *Sea King* prototype, when Western Auto Stores, a major chain of automotive parts and accessories, contacted Elmer. Western Auto had been buying engines manufactured by Muncie Gear, and selling them under Western's brand name of *Wizard* Outboard Motors. George W. Derse, the buyer for Western Auto, said, "Kiekhaefer, we've been watching you. Would you consider selling us an engine? We would like to have our own style outboard."

Swollen with success, Kiekhaefer replied, "Why not? But will Western Auto pay for the tooling?"

"Yes," Derse replied, and after settling on a design, gave Kiekhaefer an order for ten-thousand single-cylinder engines.

Before the ink was dry on the order, Western Auto asked if Elmer would build them a twin-cylinder model as well.

"Sure," he replied, "No problem."[13] Elmer promptly welded together two single-cylinder engines back in the plant. Twenty-seven years later, Elmer described what happened next.

> "Believe me, those were the days of fast moving nimble engineers. The engine was finished and ready for testing just one day ahead of our appointment to demonstrate it to Western Auto. Merlyn Culver was our sales manager by this time. He and I were out testing the new twin on a little lake near Cedarburg. It was a bit of a mess because it was snowing that November day, and we were testing a new engine that was a hodgepodge of parts welded and brazed together.
>
> "I remember that it had a Champion outboard propeller and probably parts from a half dozen kinds of outboards. The weight of the snow was continually slowing down our boat, and we had to turn it over occasionally and get the ice out of it so that we could continue our testing.
>
> "It held together and did the job. We loaded an automobile and drove all night to keep an 8 o'clock appointment the following morning to demonstrate the engine to the Western Auto people.
>
> "Wouldn't you know ... it was a beautiful day in Kansas City. On the dock to greet us were ten people who took turns all day long driving this poor little old welded-together twin to see if it would hold up.
>
> "Let me tell you that little old twin got the job done because we ended up with an order for 2500 new Wizard twin outboard motors.
>
> "Along about this time, we remembered full well the fate of Thor - our predecessor - in dealing with one mail-order house. Here we were dealing with two. If one was bad, two could be worse. It was decided then and there that if these people could sell outboard motors we could, too; and Mercury outboard motors were born that day."[14]

CHAPTER SIX

Messenger of The Gods

"Things have changed a great deal since the old days. In those far-distant days, a former chief engineer, an ambitious young man wanted to become a millionaire. This is no longer true today. The ambitious young man is satisfied just to *live* like a millionaire."[1]

E.C. Kiekhaefer

Elmer Kiekhaefer's choice of *Mercury* as the name for his new line of engines is loaded with coincidence with Elmer's own life, and was perhaps the most appropriate name he could have invoked to bless his venture. In the religious beliefs of ancient Rome, the god Mercury was revered as the god of merchandise and merchants, a symbol of gain and financial success. Even better, Mercury was considered the protector of the traveler, whose grand influence would shield and safeguard paths, roads, and even the unmarked routes across the waters. Mercury was most often portrayed as a young athlete, nude and beardless, with a winged helmet and winged sandals, carrying a herald's staff or caduceus, and was widely revered as the fleet-footed messenger of the gods. By coincidence, the sacred number of Mercury was four, and the fourth day of the month was his birthday, just as it was for Elmer.

Elmer and the Kiekhaefer Corporation had a terrific head start in designing a new line of Mercury outboard engines. The combined contracts from Montgomery Ward and Western Auto Supply would total thirty-two thousand engines by January 1940. Elmer could boast that Evinrude production was only slightly higher. Targeting the introduction of his new line for the January 1940 New York Motor Boat Show in Grand Central Palace, he began a feverish prototype development program in the last 6 weeks of 1939. Satisfied with the new single and twin-cylinder engine designs he had finalized with Wards and

Western Auto, he concentrated his design efforts on new lower units.

The lower unit of outboard engines, with its ungainly and clumsy looking combination of plumbing, bothered Elmer. This array of exposed tubing created a great deal of hydrodynamic drag as the motor moved through the water, robbing the engine of efficiency and horsepower. The exposed components were prone to damage, and could be bent or broken by floating debris, or while in transit, rattling and bumping around inside the trunk of a car. And lastly, it just *looked* wrong to him.

Elmer had good instincts for practical engineering. He could tell at a glance whether or not a component or structure could handle the load for which it was designed. He would cruise the drafting tables of his designers, recommending that this "be beefed up here", or "this needs to be thicker there", and "make this stronger for Chrissake, what are you guys, a bunch of ladyfinger engineers or what?" His rigid rules for over-engineering were carry-overs from his days on the farm and at Stearns, where equipment was brutalized in use.

Elmer established a design goal of making the slimmest and most compact lower unit possible. To determine the right shape, he molded a thin, streamlined housing from clay to form the nearly knife-edged sections he desired. The results were dramatic, and all who saw the final result were amazed at both the beauty and the utility of the innovation. It looked almost like modern sculpture, a smooth flowing shape of metal that made other outboard engines appear crude and outdated.

In order to cool the engine of most outboard motors of the day, it was necessary to pump water through a water jacket surrounding the cylinders, and then exhaust the heated water. Several varieties of mechanical pumps were being used by manufacturers, including schemes using a plunger pump, siphon assisted pump, vacuum pressure pump, and centrifugal pumps. Each had advantages, but they shared a common fault. They operated adequately under ideal conditions, but when they ingested sand, weeds or other debris, they failed. If the engine were allowed to continue operating without water to cool the cylinders, the pistons could seize and cause great damage.

Elmer and his small band of designers set out to develop a new, more reliable pump. After experimenting with a number of circular impeller style pumps, they combined the best features of a round, multi-bladed rubber impeller, turning inside a round housing with enough tolerance to admit some debris without plugging up the works. Even at idling speeds, Kiekhaefer's new *Rotex Positive Water Pump* could double the water capacity of competitive engines. It is among the first innovations on Kiekhaefer's list of "Industry Firsts" that would eventually exceed over a hundred pioneering improvements to marine propulsion.

Another significant Kiekhaefer innovation was the use of reed valves between the carburetor and the crankcase. Reed valves, like the little wafer reeds in a harmonica, flow outward against the pressure of air in one direction, and lie flat against a seat in the other direction, preventing the return of air. For an outboard motor, it is a combination of air and gasoline that is drawn into the crank case of the engine, before it is compressed in the cylinder and ignited by the spark plug. The idea is to prevent this explosive mixture from being pushed back out of the crank case, through the carburetor and into the air, where it could explode in a "backfire." The reed must be able to withstand millions of bends without cracking or deforming. The idea wasn't new, in fact Evinrude had used reed valves in four models of engines as early as 1934. Evinrude called their design an "automatic air valve," and claimed that it would "insure fuel economy and prevent 'blow-back' - helps keep the boat clean."[2] They were considered too temperamental, however, could flutter wildly, pieces could break off and damage the piston or cylinder, and they were rather quickly abandoned. In fact, the "automatic air valves" worked so poorly, that Evinrude had to repurchase many of the engines equipped with them that year. Elmer, though, stubbornly appreciated the simple, self-adjusting, self-timing idea of reed valves. He experimented relentlessly, and finally discovered the right material, shape, thickness, pressure path and seat combination to make them work properly.

Five models of Mercury were introduced in January 1940: three singles and two twins. Elmer less than imaginatively named these the K1, K2, K3, K4 and K5. The K1 Special Single was identical to the new 2.5-hp. *Sea King* designed for Montgomery Wards, while the K2 Standard Single and K3 Deluxe Single were the same 3-hp. single-cylinder engine designed for Western Auto Supply, with the K3 offering a more streamlined tank, lower unit and mounting bracket. The twins, K4 Standard and K5 Deluxe, were the same alternate firing twin *Wizard* which had been commissioned by Western Auto Supply, the Deluxe version having a chrome tank and the throttle control located on the steering handle. The retail prices he established for the first crop of Mercury engines ranged from $42.95 to $98.50. Among other innovations on these engines were his *Twin-Flux* magneto, which, true to his long career with electromagnets, were heavy duty and generated a 20,000 volt spark that would snap over a quarter-inch even at idling speeds. For these new engines he also pioneered a multiple friction-disk steering system that allowed hands-free tracking while underway.

With these new features, and five new models to prepare, Elmer mercilessly drove himself and his crew of 25 workers, 18 hours a day, seven days a week. No one dared to leave while Elmer was still in the plant, for he made it clear that everybody would have to give 100% of their time and effort to complete the work in time for the introduction at the New York Motor Boat Show in early January. Weekends and holidays became meaningless to Elmer, as the steady stream of work continued through Thanksgiving, Christmas and New Year's Day, with workers given barely enough time to have holiday dinner with their families.

Elmer himself spent precious little time at home with his wife and daughters. What Freda had assumed were the demanding labors and exigencies of a new business was beginning to look more and more like a permanent pattern of business first and family second. Whenever Elmer did arrive at home, he was exhausted and short tempered, refusing to discuss business with Freda, further isolating her from his life. It was all the more stressful for her when they would dine with another couple from the plant. The conversation quickly turned to business, leaving her with no clear idea about what to say. She felt it was obvious to their few friends that Elmer never confided in her, and she was the only person that he didn't talk shop with.

When the new Mercury prototypes were completed, Elmer, father Arnold and sales manager Merlyn Culver headed for New York. They jammed the back seat and trunk of Elmer's souped-up Plymouth Business Coupe with the new engines and boat show booth, and the three of them sat cheek to jowl in the front seat.

Elmer resisted taking his father along to New York. In fact, Elmer may have wished for almost anyone else. Arnold was indeed the president of the Kiekhaefer Corporation, but he had not taken an active part in the day to day affairs of the business.

The lingering anger from the day when Elmer left the homestead had continued to be a provocative barrier between them. "If you're really leaving, don't ever come back," still rang in Elmer's ears when he thought of his dad, and now here was Arnold, still outranking him, even though Elmer had managed to move heaven and earth in the new business. Elmer was anticipating a warm and excited reception to the new engines, and now he was forced to step back and introduce his father as the president of the company.

Elmer had traveled rather extensively while he was chief engineer for Stearns Magnetic, on occasion staying in the field for months at a time. Arnold, because of the constraints of farming the homestead, and the lack of reason or opportunity, had always been pretty well cemented to home. The trip to New York was for Arnold the opportunity to see the fabled city, to stand with his back arched, marvel at the tops of the tallest buildings in the world, and experience the sounds and smells of the busiest place in America. For all of his father's intelligence, Elmer still considered his father just a farmer. Though Arnold was a successful businessman in his own right, with a long history of small and profitable ventures in the community, Elmer looked on him as a liability, an embarrassing hayseed that could scare off what Elmer imagined to be the smooth and sophisticated wheeler-dealers of the sporting goods world.

Nothing could have been further from the truth, for Arnold had a capacity for easy-going conversation that attracted even the most jaded businessman. He exuded trust, for his straightforward manner and speech were wholesome and unadorned, creating a distinct air of confidence. His honesty was self-evident, through his humble body language and simple gestures. And when Arnold donned his brown, three piece "banker's suit", complete with silver watch fob and steel rimmed glasses, he was transformed, creating a handsome, dignified appearance worthy of a governor or senator. It would make no difference to Elmer, however, for Elmer was reluctant to share the credit or the accolades for the accomplishments of the Kiekhaefer Corporation with anyone, and following the trip to New York, he began to actively push his father even further into the background of business affairs.

Elmer faced his first genuine crisis in business during the New York Motor Boat Show of January, 1940. He was working hard to establish the very first of his new Mercury outboard dealer groups, and convincing the old Thor dealers that there was a whole new opportunity for them in the new line. Just as his momentum was starting to build, and it began to look as if success was in his grasp, he received a tip from someone back at the plant that somebody was trying to sell him out. The banker, Edgar Roth, had received an offer to sell the Kiekhaefer Corporation to the Flambeau outboard motor company for fifty percent more than the investors had originally advanced. So, for $37,500, Roth and the others were actively preparing to sell the company out from under Elmer while he was out of town and couldn't block the sale.

Elmer boarded a Ford Tri-Motor and raced back to Cedarburg in an effort to save the company. The group of investors aligned with Edgar Roth of the Cedarburg Bank had decided that "you can never take a loss taking a profit,"[3] and here was a real opportunity to not only get their money back, but to make $12,500, all within eleven months of their original investment. To them it sounded just too good to be true, and they weren't about to pass up the golden opportunity.

For two days Elmer held forth, extolling the progress that the company had made in such a short time, describing the future that he saw for the organization, and briefing the doubters on the great promise shown in New York for the engines. "The future of Kiekhaefer hung in the balance," Elmer would later write, "but believe me I did a selling job to save our company."[4] He reminded the eager profit takers that in the eleven months of 1939, they had managed to do $300,000 in sales, and promised them that if they just held on tight that he could double it during 1940, so confident was he of his products and marketplace. One by one he turned the group around, and when they finally agreed to decline the Flambeau offer, he vowed to somehow eliminate the influence of the non-believers, personally get controlling interest of the corporation, and avoid having to look over his shoulder every time someone wanted to cash in.

Orders were received at the New York Motor Boat Show for 16,000 Mercury engines, spread out over two and three manufacturing seasons. The combination of *Sea-King*, *Wizard* and Mercury orders was taxing the relatively inexperienced Kiekhaefer organization to their limits. The lack of any critical components, like magnetos, could quickly spell disaster for the young company. Elmer calculated he could build 12,000 engines between January 1st of 1940 and July 15, the end of the outboard selling season. As early as October 1939, Elmer began to put pressure on the Eisemann Magneto Corporation to supply 10,000 magnetos during the coming manufacturing season, 2,000 magnetos by February 1st, and then a production quantity of 200 magnetos a day. Eisemann was to supply both the single-cylinder and twin-cylinder magnetos for all three brands of Kiekhaefer engines. By March 1, after heated telephone exchanges, telegrams and many letters, the factory had still only received a paltry 109 magnetos, and none of them worked properly. Elmer was enraged.

His orders began to back up so far that he started getting cancellations. He had scheduled mass assembly to take place for the single-cylinder engines first, and then shift over to the twin-cylinder *Wizard* models later in the spring. His lack of magnetos, though, threw the entire schedule out of whack, and he was desperate. When he spoke with the Chicago representative for Eisemann Magneto, a Mr. Stanley, he was stunned to hear that they had never intended to supply such a large quantity of magnetos to the Kiekhaefer Corporation, and told him, "We did not believe it conceivable for a young organization to get into production so quickly and to build 200 motors per day, and have therefore felt all the way that you did not need magnetos at that rate."[5]

Elmer's ambitious contracts succeeded in tagging him with the moniker *Carloads* Kiekhaefer, because everything he did was on such a grand scale, and because parts for his motors began to arrive in box carloads instead of delivery vans.[6] In fact, the factory that looked so huge and cavernous less than a year ago, was getting crowded with parts inventory, assembly benches, lathes, drill presses, test cells, offices, shipping boxes, and with scores of barrels of grease and oil. The organization of the factory followed no grand scheme, and equipment and people were more or less placed wherever there was enough room. New equipment began to arrive to gear up for the rush of orders received, and new men had to be hired and trained to operate the new equipment. Even by the early months of 1940, the Kiekhaefer Corporation had become the largest employer in Cedarburg, and Elmer's reputation for no-nonsense business was beginning to spread like the growing smoke from his engine test cells.

His paranoia fueled by the attempted takeover by Flambeau, Elmer began to take steps to insure that the current stockholders, his father included, sell their stock only back to the company. To this end he instructed his combined corporate and patent attorney, Guy Conrad, to prepare documents for every stockholder to sign. With the Flambeau fiasco still ringing in his ears, he also demanded that the stockholders not be able to "hijack" the company by having to meet a price that an outside influence might wish to pay for the stock, and limited the buy-back price to the prevailing book value of the stock. An indication of his deep apprehension and insecurity appeared in a letter to Conrad, urging him to make haste in providing the document.

> "If you can get this in the very near future, we would appreciate it since prospects look good for next year and I have reason to feel that a little manipulation may be forthcoming. Our sales possibilities for next year are very, very excellent and this little company is going places."[7]

Following the execution of the document by all stockholders, Elmer must have breathed a heavy sigh of relief. In one sweep of the pen he had ingeniously guaranteed that the company would never expand the current list of shareholders, and never have to pay more than the book value to buy back any currently issued shares of stock. He might never have to fear losing control of the company again. It was the first of a long and often vicious sequence of people and stock manipulations that would ultimately place 90 percent of the stock of the Kiekhaefer Corporation in his own pocket.

Once Elmer had convinced his suppliers that his high volume orders were justified, he inaugurated two assembly lines down the middle of the Cedarburg plant. One line alternated between Montgomery Ward *Sea King* models, and Western Auto Supply *Wizards*. The other line was dedicated to Mercury outboards exclusively. In this way he could continue to meet the sporadic deadlines of the chain store orders, and at the same time assure his new Mercury dealers and distributors that they would have enough product to satisfy their customers. By September, the end of the fiscal year, he had shipped 2901 Mercury outboards, with singles outselling the twins by well over two to one. In addition, 4,000 *Sea King*, 2,000 *Wizard* singles, and 500 twins were produced for a total of 6,500 private label engines.

All together, the dedicated crew of *Carloads* Kiekhaefer had managed to build 9,401 engines by the end of 1940.

Though short of his goal of 12,000 engines, he had learned the often cruel lessons of manufacturing and assembly well, and was poised to increase his production substantially for 1941. However, in the frenzied rush to finish up orders for Western Auto Supply *Wizard* engines, Elmer had completely neglected to design a new line of engines for 1941. As late as December 19th of 1940, the production department had to beg Elmer to decide on new models to show at the New York Motor Boat Show, only a month away.

Because the outboard industry had been flourishing in the United States for over thirty-five years before the Kiekhaefer Corporation was formed, Elmer and his designers had little choice but to adopt many of the innovations which had long been proven and long demanded by consumers. Elmer was determined to add new features to his lines for 1941, even if he had only a few weeks to perform another miracle. The success of the introduction of the Mercury left little doubt that customers would respond to advanced features. He realized, though, that in many other ways his engines were years behind the competition. He made three basic improvements to his 1941 line, each of which had become standard features in the outboard industry for many years. The first change was in the fuel tanks of his models. A lingering legacy from the days of Thorwald Hansen were his stamped steel gas tanks. They had the unpopular tendency to dent easily, and to rust quickly, both resulting in an engine that looked old before its time. Elmer designed a two-piece cast aluminum tank, which looked almost identical to the steel tanks which adorned Mercurys in 1940. They were, though, as the Kiekhaefer Corporation would advertise, "rust-proof, leak-proof and dent-proof".[8]

Another improvement, the "Twinflex, Propeller-Protecting Clutch"[9], was designed to absorb some of the initial impact of a propeller and a submerged object, and protect the rather fragile shear pin which was common to all Kiekhaefer Corporation engines. But this feature had been used in models of Johnson Outboards since 1934, and by 1941, Evinrude had done away with shear pins altogether.

Elmer also introduced the "Magnapull Starter" on two models for 1941, which would eliminate the winding and pulling of a rope starter. This self-winding feature, introduced by Evinrude as the "Simplex Starter" six years earlier on two models, had been quickly accepted by consumers. But the combination of each of these improvements were sufficient to put Mercury outboards into the same performance league as Evinrude and Johnson, and pull ahead of other competitors, like Elto, Champion, Waterwitch, Hiawatha, Lausen, and Muncie (Neptune).

When Elmer arrived at the 1941 New York Motor Boat Show, his new engines sported new names like *Comet*, *Streamliner*, *Torpedo* and *Rocket*. The advertising slogan he chose for his new line was "She's got no bad habits,"[10] and prices for his five models from 2.9-hp. to six-hp. ranged from $47.95 to $114.50, which, on the average, were as much as twenty percent higher than comparable Evinrude models. Evinrude, though, could boast a line of nine engines through four-cylinders and over 33-hp., along with a thirty year advantage in market share and name recognition. Johnson displayed ten engines for 1941, known coincidentally as *Streamliners*, including twins to 22-hp., and *Ready-Pull* starters on half of their models.

Even with the evidence of his own scrambling to catch up with advancements made by others, Elmer mounted a loud and clamorous campaign to charge that the industry was busy copying *his* designs. In the very first newsletter distributed by the company, the *Mercury Outboard Motor News* in March of 1941, he first reminded his dealers of his 1940 prophecy about his new Mercury outboards, that it was "The 'Dream Motor' Of Today That Others Will Seek To Imitate Tomorrow."[11] Then, without actually *naming* the competition, he set out to prove that his lower unit housing, water pump and even reed valve features had been blatantly copied by various leaders in the industry. The silhouettes and masked outboards shown in the newsletter were obviously those of Evinrude, Johnson and Champion engines. But, fully aware of his own encroachments on the successful concepts pioneered by others, he graciously accepted their trespass, and didn't wish to risk the start of patent infringement actions, no doubt fearing retaliation in kind.

> "We're not resentful nor alarmed that other manufacturers in our field have deemed it necessary to incorporate some of our last year's features into their 1941 models. ... Remember, Imitation At Its Best Is But Imitation!"[12]

At both the January 1941 New York Motor Boat Show and in letters to his dealers early in the year, Elmer took great delight in teasing prospective customers about the imminent arrival of the Mercury *Mystery Motor*, the *Thunderbolt*. Supposedly an engine that should have been ready for the 1941 New York show, he couldn't resist the temptation to whet appetites for what he clearly considered to be a monumental engine.

> "Yes, it's on the way! The THUNDERBOLT, the new super Deluxe Mercury Motor, will be ready soon. And, it sure IS going to be something "to write home about." Marvelous! Stupendous! Colossal! Superb! These are but a few of the adjectives that you will be looking for when the THUNDERBOLT arrives."[13]

Actually, the *Mystery Motor* was a new, four-cylinder engine, which, but for a few exceptions, used most of the same components as the twin-cylinder models. It was designed to weigh a little over sixty pounds, and have a firing impulse every ninety degrees which would enable it to throttle down to low speeds smoothly. He considered the unique "overlapping" firing order to give the engine the flexibility of four-cycle engines, but with two-cycle simplicity.

By the end of January, 1941, Elmer had a near operational prototype, but events were beginning to take shape many thousands of miles from Cedarburg that would alter not only the entire business structure of the Kiekhaefer Corporation, but affect the lives of millions throughout the world.

CHAPTER SEVEN

Suspended Dreams

"These are rather disturbing times -- most of us engaged in business have plenty to think about these days. ... For the past several months we have been hoping against hope that the situation might improve. But today we are faced with the inevitable ... government restrictions on aluminum are now such that it is simply going to be impossible for us to obtain sufficient quantities of this metal to produce our motors. ..."[1]

E.C. Kiekhaefer

Even as Elmer Kiekhaefer opened the doors of the Kiekhaefer Corporation for the first time in January 1939, a terrible drama of pending world war was unfolding throughout Europe. The rising menace of conflict dominated newspaper headlines throughout the year: "Czechs Collapse," "Italy and Germany Execute 'Pact of Steel,'" "Nazis invade Poland," and by the end of September, "Britain and France Declare War On Germany." President Franklin D. Roosevelt had proclaimed America's neutrality in the growing conflict, and forbade the shipment of arms, munitions and aircraft to any of the countries for which "a state of war unhappily exists."[2] But on November 4, 1939, the embargo was lifted, and the United States began a massive build-up to supply the beleaguered nations of Great Britain and France with armaments and the many goods of war.

By the late spring of 1940, many large industries were already switching to military production, and Elmer began to consider how his small company might also obtain the high-volume government contracts being issued. In June 1940, a week after his thirty-fourth birthday, Elmer made his first overture to the government, hoping to persuade representatives to tour his facilities and consider the potential for military production.

'To our knowledge, our factory ... has not been surveyed for industrial facilities or capacities. We felt that we should report to you, since in peace time activities we are in a position to turn out a million dollar's worth of manufactured goods a year.

"For your information, we manufacture small internal combustion engines of the outboard motor type that can, of course, be readily converted into small stationary or portable engines. ...

"May we, therefore, respectfully solicit your early consideration?"[3]

During the summer of 1940, Elmer took a trip to the War Department offices at Fort Belvoir, Virginia, to formally present the capabilities of his organization, and to solicit engineering development contracts. At a time when the Kiekhaefer Corporation's most ambitious engine was the six-horsepower twin-cylinder K5, he naively accepted a challenge to build a prototype of a 25-hp. four-cylinder engine to push twenty-five ton pontoon boats. Evinrude, Johnson and Elto had for many years been manufacturing successful four-cylinder engines in this power range, and so it was almost impossible from the start that the Kiekhaefer Corporation would last in competition against the massive lead established by these major competitors. Further compounding Elmer's problems was the fact that he had to squeeze vital development time for this big engine from his engineers who were desperately behind in production on Mercury, *Sea King* and *Wizard* outboards. He became hopelessly behind schedule after five months of delays and engineering blind alleys in building the prototype, and tried to convince the War Department to give him more time.

"[We] have, as yet, not been able to prepare a sample motor for demonstration. As you can readily understand, the development of this large engine is costly and will take probably a little more time than we first anticipated. ... We do feel, however, that we are in a position to give you the very best service since the largest of our competition has upwards of thirty-eight different models to manufacture, and we believe it goes without saying, that large bodies move slowly.[4]

Unfortunately, Elmer's own production lines were in such a state of nearly daily emergency that there was hardly a quiet moment to contemplate advanced designs. By the first week of December 1940, when Elmer's production department was begging him to concentrate on designs for a 1941 Mercury outboard line, Elmer was trying to convince the National

Advisory Committee for Aeronautics at Langley Field, Virginia, that he was in fact an active manufacturer of magnetic clutches. The truth was, the Kiekhaefer Corporation had never built a magnetic clutch, but Elmer sent the engineers at Langley ten blueprints showing clutches from four inches to four feet in diameter, and had the brazen cheek to characterize them as "our standard clutches."

As military aircraft production accelerated, Elmer found his traditional sources of aluminum drying up. First there were delays and spot shortages, until finally, his production lines began to grind to a halt. In February 1941, the U.S. Government issued order M-1-a which seriously curtailed the use and casting of aluminum without special permission. Elmer was stunned. After convincing his stockholders not to sell their shares of the Kiekhaefer Corporation to Flambeau because he would double the sales of outboard engines in 1941, it looked as if they might have to close the doors. In a desperate appeal, Elmer flew to Washington, D.C. to meet with Charles Hoge of the Priorities Division, and plead his case for more aluminum.

> "It is extremely unfortunate for us that order M-1-a should strike us at the peak of our production season. ...
>
> "Since we are but a new company ... we have not been able to build up the financial reserve necessary to carry the shock of unfinished production at the end of our season.
>
> "It will mean that over 100 men, that cannot be absorbed locally, will be out of work. It will mean that we cannot continue with the special development work that is being carried on, at our expense for a special motor for the U.S. Army ... in fact it will mean financial disaster for a business enterprise that within a few years would employ several hundred men in a community that has not enjoyed much business activity.
>
> "... Since the outboard is strictly of portable application, heavy metals cannot be substituted successfully; in fact, our plant was formerly owned by the Cedarburg Mfg. Co. who got into financial difficulties through trying to market an outboard made of steel and cast iron. ...
>
> "In behalf of employment for over 100 men, and in behalf of fine, local, civic minded stockholders, we cannot face financial failure with a product that has, within two years, achieved such merit that the entire outboard industry has imitated its engineering design. ..."[5]

Within days of his return in early April 1941 to Cedarburg, Elmer was visited by Major C. Rodney Smith of the Army Corps of Engineers. During his tour of the facilities, the Major disclosed the government's hopes for development of "a lightweight gasoline motor of approximately 4 horsepower, suitable for powering a portable chain timber saw."[6] Elmer eagerly listened to the specifications established for such a saw, which had never before been built in the United States. Only Germany, now the target of America's defense planning effort, had built one. The Major showed Elmer pictures of the German "Stihl" gasoline saw, and told him that if he was interested in competing for the contract, that he would have to have an operational prototype ready for inspection in less than six weeks.

Elmer also learned that the Henry Disston & Sons company of Philadelphia was the leading candidate to make the guide-bar and moving saw tooth portion of the device. Disston was a household word in America, was among the oldest saw firms in the world, and a family business that dated back over a century. By 1941, Disston employed 3,000 workers, with factories in Philadelphia covering sixty acres. Disston had much experience with war production, having produced defense goods for the armed forces starting with the war of 1812. Besides the proposed saw mechanisms, Disston was a manufacturer of armor plate for fighter planes and tanks, as well as "treacherous knives known as machetes, first used by the Filipinos."[7] The total weight of the proposed Disston-Kiekhaefer "two-man" portable "chain-saw" was limited to less than one hundred pounds, and the target performance would meet or exceed the abilities of the German saw.

Elmer was interested only in the development and production of the gasoline engine, and indicated he would be happy to work with Disston. He traveled to Fort Belvoir, Virginia, to personally inspect the German "Stihl" engine. Elmer managed to convince the War Department to lend him the only saw which they had in their possession, which effectively placed prospective competition in the dark.

Within two weeks Elmer delivered a complete set of blueprints for his engine, which pleased the War Department and Disston, and enabled him to proceed with a prototype. His design revealed his years of practical experience on the farm, using equipment that would suffer heavy duty cycles and dirty or wet operating conditions. The War Department responded to his ideas with praise and encouragement.

> "We are very pleased with the "clean lines" of your design, the location of the starter, the absence of exterior wires and cables, and the functional simplicity of the controls. ... Placing of control wires inside the tubing [handles] as you have shown is very desirable."[8]

Foremost among the criteria given by the War Department was that the engine be very simple to operate. The Army would have precious little time to train unskilled operators. They wanted a plain, unadorned machine that could be mastered by the rawest of recruits. Major Smith wrote Elmer to drill home his point.

> "Above all, as I have stated to you repeatedly, but cannot overemphasize, dependable starting by unskilled operators is an absolute essential. It must be accepted as basic that before we can adopt a machine of this sort for military usage it must be proven to be as dependable with soldier operators as a truck, tractor, air compressor, or other similar power equipment in general military use. Part of this dependability must be based on "fool-proof" or "tinker-proof" qualities."[9]

Until the marriage of a light, portable, air cooled two-cycle engine with a moving saw chain, the options for felling trees were limited. Two-man sawing teams drawing heavy steel saw blades back and forth was the most common method, next to the traditional lumberjack and hand-axe. A number of grotesquely oversized ideas were presented to the War Department in their search for a fast way to clear trees for roads and airfields, or to make or destroy tank barricades. They ranged from modified garden tractors with reciprocal saw attachments, huge caterpillar tractors weighing dozens of tons with large circular saws attached, to devices that used a white-hot electric wire to "melt through" trees.

The German "Stihl" saw was built using very light-weight magnesium, a material that was scarce in the United States. The single cylinder engine only developed five horsepower, weighed over 105 pounds, and was only suitable for intermittent work due to excessive heat build-up, and the jamming saw teeth. Kiekhaefer designers understood that if their prototype unit was to be accepted by the military, it would have to endure the most severe environmental and endurance testing. It would need to operate in freezing arctic cold, scorching tropical heat and humidity, in unspeakably filthy conditions with little or no maintenance, and at altitudes ranging from the desert floor to tall mountain peaks. It would have to perform with contaminated fuel, improper lubrication, and, above all, be capable of operation by operators with little or no mechanical skills.

Elmer's new twin-cylinder KB-6 solved every problem the War Department could throw at him. Elmer had simplified the controls to such an extent that all that was required was to pull out the choke, snap out a clutch to disengage the saw-chain, and pull up on the starter handle. Only a very small adjustment was available on the carburetor for high mountain elevations, small enough that if it was positioned incorrectly at sea level, the engine would still start. As Elmer pointed out, "the net result is that the operator likes his machine. ..."[10]

Elmer did his own power tests on the German "Stihl" engine, and gleefully reported the results to the Major. "A dynamometer test on the German engine seemed to indicate a maximum output of 5 1/4. However, when saw speed was pulled down to about 1000 ft. per minute, or its best cutting speed, the corresponding horsepower developed was but a little over 3."[11] Elmer's design developed a full five horsepower, was smaller than the German example, had much heavier bearings, could withstand the most arduous duty cycles, included a starter and a full engine cowling, and still weighed in at less than sixty-five pounds.

Among the competition for the powered chain-saw was the Reed-Prentice Corporation of Worcester, Massachusetts. Reed-Prentice was a leading manufacturer of saw blades, and also had a small, though largely unsuitable, line of portable gasoline engines. They submitted a combination of their engine and sawing mechanism to the War Department at the same time that the Disston-Kiekhaefer combination was delivered. The two Reed-Prentice prototypes, powered by four and eight horsepower engines, developed fractures in their crank shafts during testing. The War Department wasn't too impressed with the Reed-Prentice engines, but found merit in their saw mechanism. They encouraged them to contact Kiekhaefer to discuss a possible Reed-Prentice/Kiekhaefer team effort. It began to look as if the Kiekhaefer Corporation was going to get the engine business no matter who was awarded the saw mechanism portion of the device. In a letter to Reed-Prentice, the War Department hinted that Elmer had just about wrapped up the engine contract.

> "We ... hold no particular brief for the Kiekhaefer motor over any other gasoline drive, but feel that it has shown sufficient promise over other types to make it a serious contender. ... [C]onsequently, we believe that it would be well worth our while and your while to test the Kiekhaefer motor with your sawing device. ..."[12]

Still, the War Department issued Elmer a list of changes they wanted. They wanted the engine to be even lighter, have a new transmission case, a new set of gears, different cylinders, new fan housing, new magneto support bracket, new gas tank and new controls. Elmer was flabbergasted, as he had seen the War Department's letter, praising the merits of his design. Nonetheless, the changes were made in less than three weeks, and the improved model was re-submitted to the Corps of Engineers at Fort Belvoir.

Among an ambitious list of twelve improvements was a remarkable forty-percent increase in displacement, giving the engine additional "lugging power" and power reserve. Elmer proudly explained that the new model should eliminate all competition from the field.

> "This increase in displacement [doesn't] increase engine weight by more than one pound. Incidentally, this so-called improved engine, which we shall still call 5 horse, will greatly outperform anything the Engineer Board has had under test ... The added power and performance should surprise them and should make all the competition appear ridiculous."[13]

The War Department decided that if Disston became the preferred saw mechanism supplier, they would receive the entire defense contract, relegating the Kiekhaefer Corporation to subcontractor status. Under this scheme, Disston would receive the assembled engines, attach their saw mechanisms, and deliver them to the War Department. This subordination of the Kiekhaefer Corporation to the saw manufacturer upset Elmer. He kept feeling this notion of superiority from Disston during the development phase, and it grated on him enough to quietly enter a conspiracy with Reed-Prentice to remove Disston from the picture altogether.

As early as July 1941, Elmer put his plan in motion. He briefed his Washington, D.C. engineering representative, James Allan about his feelings. "It is apparent that it is much easier to work with Reed-Prentice than with Disston," Elmer wrote. "We have also learned from our Philadelphia representative that [Disston] is inquiring around for various engine parts ... apparently doing everything to keep from using our engines."[14]

Next, he dictated a secret letter to Mr. F.W. McIntyre, the vice president and general manager of the Reed-Prentice Corporation, that started with the ominous prelude: "... we believe you are entitled to certain information that might be termed 'laying the cards on the table.'"[15]

> "The situation at hand is simply this: Major Smith is being criticized by his superiors in not finding a conclusion to the power saw project. He will not countenance any picayunish unbusinesslike relations henceforth between engine and saw manufacturer. He has practically stated that Reed-Prentice has the best saw. The fact that Disston refused to make improvements on theirs makes Reed-Prentice's changes so much more sure in obtaining the saw business, provided you have the power plant to go with it. ... As specialized engine builders, we intend to get the volume on this particular size engine to such a degree that the cost will be very low. ...
>
> "Since both of our organizations have considerable money tied up in this Army project, we believe the time has come for action if we ever want to realize anything on this business. I shall instruct our Washington representative, and I trust that it meets with your approval, to see Major Smith and to see what his reactions may be to a proposed exclusive Reed-Prentice - Mercury arrangement. In view of such a bona fide arrangement, we believe Major Smith will drop the Mercury-Disston arrangement. ..."[16]

Elmer's plan enjoyed a simple brilliance. If he were successful in negotiating a contract with Reed-Prentice that would specify his engines as the exclusive power of their saws, then he could tell the War Department he was unable to work further with Disston. Disston, without an adequate engine supplier, would be dropped from competition, and the award would go to the Reed-Prentice/Kiekhaefer combination.

Elmer, paranoid of failure, began to play both sides of the fence to make sure that he would win, regardless of who the government selected for the saw mechanism. To stall a formal commitment to Reed-Prentice, he first invited Mr. McIntyre of Reed-Prentice to visit the Kiekhaefer Corporation facilities in Cedarburg. When McIntyre toured the plant, Elmer made a point of introducing his father, Arnold Kiekhaefer, and made sure that he was identified as the president and chairman of the board of the corporation.

To anyone remotely aware of the chain of authority at the Kiekhaefer Corporation, the thought of Elmer reporting to anyone else was preposterous, for his control of the operation was absolute, and debate with anyone, especially his father Arnold, was unthinkable. Yet, having set the stage with McIntyre during his tour by introducing him to President Arnold Kiekhaefer, Elmer now wrote that a delay had occurred, and it was out of his hands.

> "Relative to an exclusive engine contract outside of the military angle, it looks very favorable. We believe that as pioneers in the saw business, you are entitled to whatever protection you can get and we are willing to help your picture along in any way that we can.
>
> "My superiors have, however, declined at the moment to take a definite stand on it. ..."[17]

But, Elmer was running out of time. For him, the first casualties of the growing world crisis would be the *Comet*, the *Streamliner*, the *Rocket Deluxe*, his *Mercury Mystery Engine* - the *Thunderbolt*, and his dream of taking on the giants in the outboard industry. For seven months he had been waiting for some sort of solid order from the War Department, and for seven months he had heard nothing but delays, changes, politics, more modifications and more delays. From a high of 110 workers, layoffs brought the working total down to only eighteen employees by November 1941. Elmer was trapped between the prospect of future defense contracts, and the lack of aluminum to produce civilian outboard engines. He begged the government's Office of Production Management for aluminum to build more outboard engines, to bring back his labor force, and to keep the company solvent.

> "... [We] are primarily interested in keeping our organization intact. It is a cruel hard fact that our number of employees had dropped from 110 to 18. It is a fact that we are making no shipments. It is a fact also that we have developed a special

device for the U.S. Army on which contracts supposedly were to have been placed several months ago. ...

"You can understand ... that all of the above development works take money. It is a fact that we have built one device for the Engineer Board now that has cost us in the neighborhood of $50,000.00, without the benefit of a development order."[18]

Elmer's desperate appeals were to fall largely on deaf ears, and as Christmas approached, he began to wonder whether he would be able to keep the plant open. He was openly bitter about his rivals, Evinrude and Johnson, because of their relative ease in obtaining aluminum for outboard motor production. The Kiekhaefer Corporation wasn't alone in feeling abandoned by the government, and feeling that some sort of collusion with big business could mean the end to small companies.

Even the United States Senate was alarmed at the inequities which threatened small business. Senator Hatch of New Mexico, who was head of a special committee to investigate defense program contracts, was "alarmed lest America have the experience of Britain, where 20,000 manufacturing plants were shut down almost overnight in the changeover from peace to a war economy."[19] When President Roosevelt appointed Floyd Odlum as director of the Office of Production Management division of contract distribution, Odlum quickly estimated that the release of just "2 percent of the supply of strategic materials would enable 30,000 to 45,000 small metal-working U.S. plants to continue during the first half of 1942."[20]

Unfortunately for Elmer and the Kiekhaefer Corporation, no relief was forthcoming. Though the corporation had shown a modest profit when they closed their books at the end of September, the company would be virtually shut down by November, victim to their unique combination of expensive development work and lack of aluminum. The year 1941 had taken the Kiekhaefer Corporation on a roller coaster of financial promise and disaster. Over 15,000 engines left the plant before they ran out of aluminum, representing over $612,000 in sales. Because of the money poured back into the company for engine development work and new tooling, the Kiekhaefer Corporation would show only a $38,000 profit. It would seem like good news compared to the events which were to follow.

In late 1941, tragedy rocked Elmer's life. He lost what he would later consider to be his closest friend and confidant, his Uncle John Blank, in a horrible accident. In his capacity as corporate secretary of the Cedarburg Mutual Fire Insurance Company, John Blank was preparing to make a call on a client ten miles west of Cedarburg. After having lunch with the family, Blank decided to take his wife, Flora, along for the ride. Some of the Blank children who had returned from work for the noon hour encouraged them to go, saying they would stay and clear the table and take care of the dishes. Elmer's Uncle John and Aunt Flora got into their sedan and headed west out of town, approaching the railroad tracks they had crossed hundreds of times before. No one is certain what had distracted them, whether they were deeply engrossed in conversation or not, but they were struck by the full force of a train moving at high speed, and were both instantly killed.

It was a terrifying catastrophe for the entire family, and was one of those moments in life that is unforgettable, as everyone would remember where they were and what they were doing when they received the news. Arnold Kiekhaefer had to shoulder the horrible burden of making a formal identification of the bodies of his sister and brother-in-law, and Elmer's sisters remember their father crying painfully for hours when he returned. Elmer had lost the one person in his life who believed in his abilities absolutely, and who had nurtured his self-image and confidence since he was a young boy. It was Uncle John that had brought the opportunity of the Cedarburg Manufacturing Company to Elmer, had designed the plan to reduce the debts of the company, and had organized the investors to purchase the assets from Thorwald Hansen.

Willis Blank inherited most of his father's interest in the Kiekhaefer Corporation, and was named secretary, the same position that his father, John Blank, had held. It is significant, that of all the original stockholders in the Kiekhaefer Corporation, Willis Blank would eventually be the only one allowed to retain a large block (10%) of shares outside of the immediate Kiekhaefer family.

If conditions weren't already bad enough for Elmer negotiating the mountains of Washington red tape and the many political obstacles of defense contracting in peacetime, when the United States was swept into the cauldron of world conflict by the surprise attack at Pearl Harbor, confusion reigned throughout the industrial-defense community.

On December 7, 1941, over 360 Japanese warplanes participated in the disastrous attack, sinking or very seriously damaging five U.S. battleships and 14 smaller ships at Pearl Harbor, Hawaii. Over 2,000 Navy personnel perished, along with 400 civilians. President Roosevelt called the attack "a brilliant feat of deception, perfectly timed and executed with great skill." Within four days of the attack, the United States would declare war on Japan and her Axis partners, Italy and Germany, and America joined the horrible fury of world war.

CHAPTER EIGHT

Machines of War

"We believe our organization and the Kiekhaefer employees deserve tremendous credit for their unselfish time and effort. There were times when these employees worked twenty-four hours a day to take "bugs" out of the design to be submitted for breakdown tests. They have worked Sundays -- they have worked evenings, or whenever they have been asked to work. They have foregone their vacations and holidays."[1]

E.C. Kiekhaefer

Even as the United States joined the growing second World War, another battle for survival continued in Cedarburg, Wisconsin. Elmer Kiekhaefer, by constant pressure and stubborn determination, managed to secure enough stray aluminum to embark on an on-again, off-again production of outboard engines. Actually, Elmer had been very carefully planning for an aggressive 1942 season, hoping to produce no less than 25,000 motors, a blend of Mercury, Sea King and Wizard product lines. During the first three months of 1942, he was off to a great start, building over 5,000 Mercury engines in 3-, 3.2- and six-hp. models before what Elmer would later describe as his second great disaster in business occurred.

On March 27, 1942, the United States Government issued Limitation Order 80, or L-80, which prohibited the manufacture of leisure products from aluminum, and ordered that all leisure industry aluminum fabrication be suspended indefinitely. The government classified slot machine manufacturers and outboard motor builders in the same category, characterizing both products as unnecessary and wasteful of strategic materials. Two days later, United States government compliance officers entered the Kiekhaefer Corporation plant unannounced to ensure absolute submission. In the absence of any substantial, finalized defense contracts, the Kiekhaefer Corporation was effectively put out of business.

Elmer had come tantalizingly close to securing the contracts he so desperately needed. It was almost as if he were being teased and unmercifully baited with the prospect of orders, only to have them withdrawn at the last moment. Two days before Christmas 1941, Elmer received news that a contract for production of 2,000 chain-saw engines had been signed by the War Department. About two-thirds of the engines were to be manufactured for use with the Disston saw mechanism, and the other third for assembly with Reed-Prentice. But, no sooner was the contract announced, than it was held up by the Office of Production Management, in particular the division of price control and investigation. For a glorious five days, and over the Christmas weekend, Elmer felt that he could see the end in sight for his troubles. Unfortunately, during a demonstration of the prototype saw, an Army General was asked how much he thought the device should cost. Many years later, Elmer would remember well his exasperation of working with the government.

> "We had it, then we didn't have it, then we had it, then General Knutsen came along and somebody in the army asked him, "What's that damn thing supposed to cost?" He looked at it and said, "Oh, $150." We had asked $300 for it. Of course, if you build them in General Motors quantities, yes, $150 would have built it. That's one example of the problems we were up against."[2]

In the first days of January 1942, Elmer's Washington representative, James Allan, proceeded to the office of Mr. De Camp, the individual who had stalled processing of the contract at the Office of Production Management. Elmer was enraged by the reaction that Allan's conciliatory efforts received. He wrote a scathing letter to the Facilities Staff of the War Production Board, criticizing their methods and policies.

> "Mr. Allan ... offered to present a price breakdown to Mr. De Camp, but [he] declined to discuss the situation with him since the matter was termed "ridiculous" and, as a matter of fact, Mr. De Camp walked out on Mr. Allan refusing to discuss the matter further. ...
>
> "Then, too, we thought that our contribution of our entire organization and all our available funds was what was expected of United States manufacturers to help in the war effort."[3]

Finally, on May 18th, 1942, Elmer received a contract for 3,300 saw engines. Remarkably, the government had arbitrarily mandated a twenty percent price reduction because of the large quantity involved. Elmer had no choice but to accept the terms, for the contract was the only thing standing between the small Cedarburg plant and certain bankruptcy. But having a contract in his hand for slightly over a million dollars was the emotional rescue that Elmer had been waiting for.

Over two months dragged by before Elmer saw a single dollar of government funds. He was forced to undergo an agonizing and whirlwind campaign to borrow money, taking him to twelve different banks and government agencies before he gave up in disgust. It seemed that no one would help, from banks in his own backyard, to Federal Reserve banks, to the government's own Reconstruction Finance Corporation, not even the War Department. Elmer was struck by the demeaning experience of being rejected by so many institutions. He thought it impossible that United States banks, so carefully regulated by the government, would deny funds to a company manufacturing critical defense materials under signed contracts. He pleaded with the War Production Board for relief.

> "Our financial condition is, of course, the result of all of [these] difficulties and the situation is getting very critical. Materials are pouring in here by the carloads and if we are to avoid disaster, we feel that W.P.B. must give us assistance NOW. ...
>
> "Many manufacturers have ridiculed us for the part we have played in this program. On the other hand, we felt that we have an organization and talent to offer, and that it was the thing to do to help win this War. ...
>
> "So many obstacles have been encountered in our War Effort that it becomes almost unbelievable."[4]

When an initial advance was received from Disston in mid-July for just under $400,000, Elmer calculated that he was nearly $150,000 short of funds needed to produce and ship the engines on time. Again, Elmer cried foul to the War Production Board, but help was still refused.

Once more, Elmer's father, Arnold Kiekhaefer, came to the rescue. Elmer tried to secure a "V"-(for Victory) Loan from the Federal Reserve Bank in Chicago. A V-Loan would have been 90% secured by the Federal Government, exposing the bank to only a 10% risk. The $100,000 loan was denied because of the $10,000 unsecured portion. Arnold promptly put up additional collateral of his homestead acreage to satisfy the bank. It was the second time that Arnold risked the family homestead for the company, having paid back the first mortgage within two years of operation. Three other banks had to combine resources to loan the additional $50,000 to complete production of the order.

The constant strain of an empty treasury, and the insecurity of war contracts, left the Kiekhaefer Corporation vulnerable to a labor union organizing effort. As early as the spring of 1940, elements of the Machinists Union, American Federation of Labor, A.F. of L., began to muscle their way into the company rank and file. Elmer responded by forming the Mercury Independent Workers Union. A dispute over which of the two groups had the authority to negotiate with Kiekhaefer management, resulted in the issue coming before the Milwaukee office of the National Labor Relations Board.

Though the Kiekhaefer Corporation employed little more than a hundred workers, it was the largest industry in Cedarburg, and so received extra attention from the union. Elmer was confused over the vocal protests of the organizers, for he had a solid reputation for paying above average wages, and was careful following layoffs to re-hire on the basis of seniority. As the war removed men from industry, a growing number of women were hired at the plant, and Elmer offered quite an enticing array of benefits to lure applicants.

> "There is a tier of showers, with private lockers, for both men and women, where you can "fresh up." You can enjoy such group activities as tennis, softball, ice hockey, musical organizations and bowling leagues. There are frequent employee parties, and a large recreation room. You have music while you work. You will like the many other extras that make it a pleasure to work at Kiekhaefer."[5]

Though the brochure made it sound as if the Kiekhaefer Corporation were the Kiekhaefer Country Club, in truth Elmer always treated his employees more than fairly, both financially and administratively. Even so, Elmer finally gave in to the union's demand for representation in July 1942. He felt he had been unfairly coerced into a relationship with the union, however, and complained loudly to the War Production Board.

> "The Kiekhaefer Corporation upon the order of the National Labor Relations Board arbitrarily had to accept the A.F. of L. Machinists Union as the sole bargaining agency in our plant without the benefit of a vote on the part of the employees. ...

> "It is unfortunate, indeed, that we are the largest industry in Cedarburg and that, as such, we have been a center of attack for the union organizations. As far as we know, there are no unions in any other industry in Cedarburg and residents here do not appear to be too much 'union minded.'"[6]

Union or no union, the discipline and assignment of personnel remained the absolute domain of Elmer Kiekhaefer. Some employees wondered if it were the exigencies of wartime production, or the uncertainties of government contracting that caused a hardening of character in Elmer. It was increasingly apparent that he ruled his organization with an iron grip, and that he insisted on levels of performance that most employees had never experienced. When he suspected greater capabilities in his labor or management force, he turned up the heat, adding to job descriptions and responsibilities until he judged an individual saturated, or at optimum performance. He was very fast to size up potential workers, and if they passed his lightning-quick intuition, he offered them a job on the spot. Anyone who impressed him as insubordinate, however, never received a second chance.

His temper grew during the war years, along with an aura of indefatigability and mechanical genius. The temper would both flash and subside quickly, like a polished sword brandished and then returned to the scabbard. Shoddy or thoughtless workmanship was never ignored, and he would loudly and brutally point out another's stupidity, bludgeoning the hapless employee with corrections and eleventh hour warnings. He still rolled up his shirt-sleeves to demonstrate, regardless of the device or conditions around the equipment. It was not uncommon to see his white shirt, half out of his baggy trousers, smeared with blotches of shiny black grease.

As the company geared up for defense production in 1942, Elmer set up two eleven hour shifts, though many bleary-eyed employees worked fourteen or more hours before leaving the plant. Elmer, though, worked harder than most, setting the example by arriving earlier and staying later than even the youngest, healthiest workers. Contracts dribbled in from various branches of the services throughout the balance of the year, ranging from a solitary engine for the Army rigged to power a 2.5 kw generator set, to orders for thousands of gas tank caps, or a few special generator sets for the Signal Corps. An order for nearly a thousand KB4 outboard motors was received from the

Navy, and a few special engines were built to power grinders and other tools for field combat applications.

Elmer's calls for operating capital were finally answered in the fall of 1942. He had lobbied long and hard for advancement of funds against defense contracts by the government, rather than loan guarantees, and had argued that he could reduce the cost of his chain-saw engines up to ten dollars by eliminating the financial hardships imposed on a small company by a large contract. But loan guarantees were better than not being able to perform the contracts at all, and on September 18, 1942, the War Department agreed to guarantee a loan in the amount of $525,000, issued through the Marshal & Ilsley Bank in Milwaukee. Though the government guaranteed the full amount, the terms of the loan were quite strict. Among the provisions of the note was the requirement that both Arnold and Elmer Kiekhaefer also personally guarantee the loan, pledge the company's land, plant and machinery, assign an existing $100,000 life insurance policy on Elmer's life, pay no other loans with the proceeds, and pay "no dividends ... on the company's outstanding stock until the loan has been fully paid."[7]

The guarantee also contained a provision that angered Elmer in a most personal way. Along with the funds came a government assigned "financial manager" from the Federal Reserve Bank, who was given complete authority over all Kiekhaefer Corporation operations. The federal watchdog, Mr. Johnson, was to be the final authority in all critical decisions in the company, but was only available for four hours a day. If this wasn't upsetting enough to Elmer, this Mr. Johnson, installed as executive vice president, along with an assistant, was to be paid the princely sum of $1,600 per month. Elmer himself made less than $1,200 per month for the average sixteen-hour-day he worked. It took only one week for Elmer to blow his stack. In an angry and provocative letter to Colonel John Seyboldt, Chief Contracting Officer of the U.S. Corps of Engineers, Elmer lashed out at the bureaucratic nonsense that endangered his company's survival.

> "It is the opinion of the Kiekhaefer Corporation that any so called financial manager cannot with four hours daily attendance at the plant manage a business that runs twenty-four hours for six days a week. Such management just is not workable [due to the] confliction of authority, responsibility and

its resulting confusion. Responsibility without authority is one of the most useless situations in the world."[8]

In a final effort to oust the government financial manager, Elmer offered to substitute a senior accountant from Arthur Anderson & Company, who had worked extensively with General Motors in industrial accounting and expense budgeting. Even though Elmer's substitute candidate, George Reynolds, was a graduate of the Harvard Business School, and was auditor of record for the Kiekhaefer Corporation, the substitution was denied.

By the first anniversary of America's entry into the war, only 348 of the 3,300 air cooled engines ordered by the government contracts were delivered. Bureaucratic red tape, financial delays and specification changes kept the small plant swimming in nearly constant confusion. The books of the corporation showed that the company suffered a stinging loss of $56,000 during fiscal 1942, almost entirely due to burgeoning developmental costs for the chain-saw engine.

For all of his troubles, Elmer could still tell the Corps of Engineers that he was proud of his accomplishments, and would do it all over again, no matter how vulnerable his situation had become. Patriotically, he said, "Such unbankable things we have done, we would probably do over, under the same circumstances."[9]

Elmer's patience finally paid off in 1943, as the Kiekhaefer Corporation boxed over 10,000 engines in six configurations, and experienced for the first time building a thousand engines in a single month. New government contracts poured in as orders jumped to over $2.5 million, more than double 1942 levels. Though the twin-cylinder chain-saw engines for Disston still dominated production, single-cylinder engines for portable grinders and twin-cylinder engines for field compressors and generators accounted for nearly a quarter of revenues.

On October 2, 1943, Elmer received a letter from Robert P. Patterson, Under Secretary of War, informing him that the coveted Army-Navy "E", or Production Award, was being conferred on the men and women of the Kiekhaefer Corporation.[10] The "E" award originated in 1906 as a Navy commendation for Excellence in gunnery. Later, the award was extended to include outstanding performance in engineering and communications. Following American entry into the war, the Army and Navy combined to cite outstanding defense production plants contributing to the war effort.

But the "E" award could also be called the "Elmer" award, because it was the catalyst that moved Elmer Kiekhaefer to forever change his name. As the products of the Kiekhaefer Corporation evolved with ever more sophistication, Elmer wanted desperately to upgrade his own image as well. Ironically, the stronger the company became, the more employees it required from the surrounding Cedarburg community, and the more people in the plant knew about Elmer Kiekhaefer and his humble farming origins from firsthand knowledge. In the summer of 1943, many of the 150 employees at the plant had attended Cedarburg High School with him.

The 37-year-old Kiekhaefer began to cringe at the sound of his own name. "Elmer" had become a name closely associated with farming, and with rural simplicity. Even worse, the name had become a derisive joke among defense plant workers nationwide, along with "Kilroy" and "Rosie the Riveter." One of the biggest jokes circulating around was "Where's Elmer?," a search for a fictitious someone to blame for a stupid error often caused by rube ignorance. For Elmer Kiekhaefer to suffer the imagined guffaws and snickers over a name he feared was associated with hayseeds, country bumpkins and backwoods ignorance was becoming intolerable. At first, he changed only his signature from Elmer Kiekhaefer to E.C. Kiekhaefer. Though his baptismal name was legally Elmer Heinrich Carl Kiekhaefer, on his birthday in 1943 he furiously scratched through the name Elmer on the nomination forms for the Army-Navy "E" awards, and penciled E.C. over the sloppy mutilation. In the next four months, he would systematically eliminate the name Elmer from any correspondence, and sign everything as E.C. Kiekhaefer. He practiced his new signature on the back of flyers and scratch pieces of paper, and one sheet survives that contains scores of his new signature, in different pens and colors, one on top of the other, with slightly different flourishes and embellishments to master the new seal of his identity.

The next step in his transformation was to choose a new first name for himself that would reflect his growing self-assurance and increasingly cosmopolitan image of himself. Of his two remaining names, the heavily Teutonic moniker of *Heinrich* certainly wouldn't have been a popular choice given the charged atmosphere of a world war where everything German was considered the enemy, especially when German soldiers were called *Heinies* by allied servicemen. But *Carl*, the third of his three given names, though certainly Germanic in origin, rang with more universal appeal and neutrality, and would be safe from insult or political whimsy.

On Saturday afternoon of October 30, 1943, his metamorphosis would become complete. The bright sunny afternoon fairly erupted with pomp and ceremony as the Kiekhaefer Corporation was heralded as the first manufacturing facility in Ozaukee County to receive the "E" award and flag. More than this, it was a day to celebrate E.C. "Carl" Kiekhaefer as a newly emerged industrial leader of the first order. The entire day was officially proclaimed *Kiekhaefer Corporation Day* by Mayor H.A. Zeurnert, who also declared that every business in Cedarburg participate in the celebration by flying the American flag and generally join in applause for the city's largest employer.[11] The manufacturing plant wasn't large enough to hold the crowds of citizens and dignitaries flocking to this major event in Cedarburg history, so the Cedarburg High School Gymnasium, the largest facility in town, was designated as the ceremonial site.[12]

At three o'clock sharp the Cedarburg High School Band, of which Elmer had once been an enthusiastic member, struck the grandiose chords of a John Philip Sousa march in the parking lot of the Kiekhaefer Corporation, and began the long, winding march down Washington Avenue to the High School Gymnasium to herald the start of ceremonies. The Superintendent of Schools, Harry E. Olson, was master of ceremonies, and introduced Colonel C. Rodney Smith to make the actual award.

Colonel Smith was especially pleased to be making the presentation, for, as then Major Smith, he had been responsible for getting Elmer started in the defense business with his visit and challenge regarding the German chain-saw engine. The tall and handsome Army officer, four years Elmer's senior, had been Chief of the Mechanical Equipment Section at Fort Belvoir's Engineer Board, and had personally led the fight for the portable chain-saw contracts for the Kiekhaefer Corporation.[13]

It was a patriotic occasion of grand proportions. Harry Olson asked all within the packed gymnasium to rise to their feet, as a full Marine Color Guard marched solemnly down the broad center isle to the straining voices of Cedarburg citizenry singing The Star Spangled Banner with wet eyes and straight backs. And when Colonel Smith extolled the virtues of American valor, both on the firing lines and assembly lines, hearts crept up the throats of nearly everyone. The sound of squeaking wooden folding chairs fell silent as he described the magnitude of the contribution made by Cedarburg's finest factory.

"I have personally had the opportunity of seeing ... the brilliant engineering, and the tremendous gamble which you of the Kiekhaefer Corporation have put into the conversion of your plant for the production of a lightweight, portable, gasoline engine driven chain-saw for the Corps of Engineers. I have had the pleasure and privilege of working with Mr. Kiekhaefer and his associates in developing the chain-saw engine from the beginning, and have seen its evolution into the most dependable and lightest weight engine of its horsepower manufactured in the world today. ...

"When it is considered that this engine was developed in an amazingly short time and without the use of magnesium or the primary grades of aluminum -- that in thirty-one instances steel or plastics have replaced critical aluminum -- the feat becomes more outstanding -- a real tribute to that American Production that Hitler once said could never match his ten-year head start but which, today, is producing three and four times the total amount of output of the Axis powers. Forced labor never could have accomplished the results you have obtained here under the American system. ...

"One of the greatest dangers this country faces today is over-optimism! The Allied nations are on the offensive! Every newspaper headline -- every radio commentator -- heralds new victories -- for our forces in Africa -- in Sicily -- in Italy -- the Pacific -- and over Germany. Those feats by the American and British troops, coupled to the steady, deadly advances of the incredibly valorous Russians -- are apt to lead us to believe that the war will end soon.

"That is fallacy!

"As General Eisenhower has just said, the road to Berlin is long -- and that to Tokyo still longer. Before the final victory is won -- and we will win it -- there will be many casualties -- many hard battles. We of the home front -- the production front -- can do our part to lessen those casualties by sticking energetically to our jobs -- turning out the utmost in equipment for the men who are fighting over there to make it possible for us to work over here. ...

"And now, it becomes my pleasure to present to the men and women of the Kiekhaefer Corporation this "E" for Excellence banner, the visible emblem of what your Government thinks of your work. Fly it proudly, for it is the gift of our uniformed men and women to their production brothers and sisters -- your Country's recognition of a job well done."[14]

The large crowd rose as one at the conclusion of Colonel Smith's address, moved by the renewed importance that their efforts had generated. The Marine Color Guard clipped the Army-Navy "E" award to one of their standards, and posted the new flag among the American and Wisconsin Flags in their detail. It was the first of three such awards which would be earned by the corporation. Harry Olson returned to the podium to introduce the man who was largely responsible for the growth and prosperity of the Kiekhaefer Corporation, the man known for his engineering prowess and leadership, "Ladies and gentlemen, to accept the award for the Kiekhaefer Corporation, Mr. Carl Kiekhaefer."[15]

"*Carl* Kiekhaefer?," Elmer's wife Freda elbowed the person sitting next to her in the front row, "Who the hell is *Carl* Kiekhaefer?" It was a startling revelation to virtually everyone in the audience, but to the Kiekhaefer family it was the most disorienting. "It was half way through that ceremony before I realized they were talking about my husband," Freda remembered the shock, "they kept saying *Carl*, and I thought it was Elmer that was supposed to be getting the award. It was almost over before I realized that they were calling *him*!"[16] All over the audience, people were asking themselves the same question, "Who's Carl Kiekhaefer?" It was, in fact, the very moment that Elmer had chosen to reveal the new Carl Kiekhaefer, the serious industrialist and manufacturer who would be laughed at no longer.

The discipline it takes to learn to call someone Carl after knowing them as Elmer for so many years was fraught with occasional failures, and many is the time that Carl Kiekhaefer had to firmly remind everyone that he was no longer to be referred to as Elmer. It was hardest on the family, of course, particularly his mother and sisters who had known him by no other name. This person with the name Carl sounded and seemed like a stranger to them, because the association of names and people is so deep. Imagine for a moment calling your husband, wife, brother or sister by another name from one day to the next. But from Halloween, the 31st of October 1943, any references to Elmer were to be removed, forgotten, covered up, obliterated, erased and never again recalled. The penalty for deviating from this universally known but unwritten axiom was assumed to be the worst. Even many years later, even accidental references to the name Elmer were dealt with swiftly and in no uncertain terms. There was a quite popular song known as *Elmer's Tune*, that was rather universally pleasing throughout the midwestern United States for many decades. Charles D. "Charlie" Strang, who is today chairman of the board of Outboard Marine Corporation, remembered an occasion when he was head of engineering for the Kiekhaefer Corporation in the sixties, working directly under Carl, when an unusually embarrassing "Elmer" incident took place.

> "We had built, for [water] ski events mainly, a truck that had an organ inside of it. Push a button and the roof of the truck would lift and come up, and there was an organ playing and a guy sitting there. We took that to our annual press meeting in Chattanooga. Everything was set up down at the water level and there was a zigzag road [leading] to the water level. It was a big, big opening night [event], with flood lights and the organ was there. Carl is late. Finally, we see head lights coming down. I'm standing next to Fritz Shoenfeldt, and Fritz says, "Wouldn't it be hell if the organist started to play *Elmer's Tune*? He's a local guy we hired. We cut the lights, and just as Carl pulls in the guy breaks into a very popular tune, *Elmer's Tune*!
>
> "Carl *leaped* out at him, so I grabbed the guy down off the organ [as Carl said,] "Get that son-of-a-bitch the hell out of here. They had to pull the organist out of the way, and the organ sat silent all night."[17]

Even though the local organist had no idea why he was attacked and dragged down from his organ after playing only a few bars of a popular tune, the incident reflects the deep-seated fear of "farmboy" ridicule that would accompany Elmer Carl Kiekhaefer throughout his entire professional life.

CHAPTER NINE

Stockholder Rebellion

"What is wrong with the community spirit of some of our leading citizens of Cedarburg? They seem to forget that three of our business predecessors on these premises have fallen by the wayside and we believe we are beginning to see some of the reasons why."[1]

E.C. Kiekhaefer

The Kiekhaefer Corporation made banner headlines twice within a three week period in the fall of 1943, first to honor, then to humiliate. First, on October 27, the Ozaukee *Cedarburg News* dedicated an entire twelve-page issue to the glory of the Kiekhaefer Corporation, upon the occasion of being awarded the Army-Navy "E" award, a celebration joined by the whole community. The special issue contained dozens of special advertising sections, all given over to the praise and congratulations to the Kiekhaefer Corporation by neighboring businesses, and by distant suppliers anxious to add their voices to the wave of tributes. "Best Wishes and Congratulations ...," from the Cedarburg Meat Market. "May they continue to progress and achieve ...," from Irene's Beauty Salon, and "Congratulations to the men and women of the Kiekhaefer Corporation for their glorious achievement," from the Cedarburg State Bank. Page after page of salutations, heartfelt praise and a shower of compliments.

Imagine then, Carl Kiekhaefer's horror to read the double banner headlines, three weeks later on November 18, 1943 which read: "Kiekhaefer Stockholders Demand Firm's High Salaries Be Lowered," and "Notice Served on Corporation to Cut Expenses."[2] He was thunderstruck. Headlines of this caustic and stinging nature displayed on the heels of an entire issue dedicated to embracing and exalting the virtues of Cedarburg's seemingly model enterprise brought Carl's hair-trigger emotional arsenal to critical mass.

A week before, he had been the unhappy recipient of an almost unbelievable letter of demand drafted by the legal firm of Schanen & Schanen of nearby Port Washington, Wisconsin. It seems that five of the minority stockholders, amounting to an aggregate of eleven percent of the Kiekhaefer Corporation, objected to the planned construction of an administration and research building addition to the crammed Kiekhaefer Corporation facilities. Carl had broken ground on the two-story, $70,000 building on October 7, with the blessing and permission of the War Production Department. The 14,000 square-foot brick and concrete structure was designed to reduce the great burden of cramped factory quarters for research and office personnel, and also to add an environmental testing chamber for measuring engine performance against extremes of temperature and humidity. When the Kiekhaefer Corporation sent notice of the proposed building to the stockholders on November 2, 1943, work had been in progress for nearly a month. The upset stockholders, fearful that Carl was never going to issue a dividend to justify their initial investment, began to panic, and hoped to pressure Kiekhaefer into either buying their stock at inflated prices, or paying the handsome dividends they were certain the company was able to pay. They lashed out at Carl for what they assumed were exorbitant executive salaries and director's fees. Their letter of demand was hand delivered by Roland C. Schaefer, the Sheriff of Ozaukee County the morning of November 13, 1943.

"Gentlemen:

On behalf of the Cedarburg Finance Co., E.H. Roth, Dr. William H. Wiesler, Palmer J. Wirth, and M.P. Becker, owners of 105 shares of the common capital stock of the Kiekhaefer Corporation, we are directed to notify your corporation that said stockholders are opposed and object to any expenditures of moneys of said corporation for the construction of buildings, or additions and contents. ...

"These stockholders maintain that such an expenditure is entirely unnecessary and unwise, and is detrimental to the best interests of said corporation, and that the corporation is not in a financial position to make such expenditures. ...

"You are further advised that in due course proper action will be taken to compel re-payment ... of the excessive sums of money paid for directors' fees and officials' salaries, and other illegal expenditures, and to restrain you in the future from paying such enormous, illegal, and excessive sums of money for directors' fees and officials' salaries, and similar expenditures."[3]

Carl wasted no time in rubbing salt in the wounds of the renegade stockholders, the same group that had tried to sell him out to the Flambeau outboard group while he was at the 1940 New York Motor Boat show. His acerbic reply, dated three days later on November 16, 1943, baited the already steaming investors, it can be assumed with certainty, into bringing their charges out in the open through the newspaper. Carl's response, penned by corporate attorney and full-time employee Guy S. Conrad, left little doubt what Carl thought of the attempted raid on his authority and treasury.

"... The inconsistency of the named stockholders charging waste and bad management in the notice ... and on the other hand asking several hundred dollars per share for stock the par value of which is $100.00 is believed apparent.

"As a further response to the allegations of the notice ... the Kiekhaefer Corporation asserts that the sole reason for the service of said notice, resides in the preconceived plan of the named stockholders to force the purchase of their stockholdings in the Kiekhaefer Corporation at an exorbitant and unwarranted price. ...

"... [T]he stockholders ... are at liberty to take any legal action provided by law ... within their discretion, and at such time as they desire. The Kiekhaefer Corporation will welcome any such action by the ... minority stockholders."[4]

Carl's message was as clear as a schoolyard taunt: Any time, any place. In their reply, the attorneys for the troublesome five shareholders, disclosed a glimpse of the true reason for the rebellion.

"... The Cedarburg Finance Co. owns 50 shares of the common capital stock of the Kiekhaefer Corporation. The Cedarburg Finance Co. at present is contemplating the building of houses in Cedarburg, and will have to raise about $100,000.00. In order to do this it will have to pledge some of its assets for that purpose. ..."[5]

The stockholders reply was received the same day that the glaring headlines shouted foul in the *Ozaukee Press*, November

18, 1943. The lengthy and highly critical article was almost certainly the result of an organized political power play brokered by the disgruntled stockholders, but Carl was never able to prove it. The loosely constructed commentary was riddled with inaccuracies, and demonstrated an acute bias rarely displayed in the rural county newspaper.

For these vitriolic accusations to be made in his home town newspaper was a huge affront to Carl, who with 250 employees, was the community's largest employer and benefactor. It is easy to imagine the destructive and giddy small-town gossip that ensued the moment the newspaper issue hit the streets, and the exquisite rage that flowed through Carl Kiekhaefer as he read the foul denouncements born of animosity and malice. The article, resplendent with errors and misleading innuendo, set off a volley of protest and further mudslinging by Carl, starting with a general statement for whoever was interested, issued the following day.

> "What prompts us to make this reply more than any other reason, is the charge of exorbitant salaries, which smears every patriotic effort we are making to help win the war by turning out tools of victory, including our recent award the Army and Navy E.
>
> "We ask any fair-minded person if a salary total for four officers of $22,500 or an average of $5600 for each officer is too high for a company with a sales volume of $2,000,000. Compare this with the list of salaries paid by Wisconsin corporations, recently released by the Treasury department, and published in the daily newspapers. Salaries of $50,000 and $100,000 for an individual officer are not uncommon.
>
> "To keep a business going requires much financial risk, and in our own case, when money was needed to finance our war effort, Mr. A.C. Kiekhaefer, president, and Mr. E.C. Kiekhaefer, vice president, signed notes in the amount of $525,000, pledging their house and home, their life insurance, everything they owned, and yet these minority stockholders raise the question of exorbitant salaries.
>
> "Mr. E.H. Roth, one of the prime instigators of these charges of exorbitant salaries, voted the present salaries when he served as director and treasurer of this company, and received the same salaries for these offices that is now being paid. It wasn't an exorbitant amount when he received it, and he got in salary a total of some $3,900 until he was kicked out as a director and officer -- and good riddance. ...

"We always believed the requisite of a good newspaper was accurate reporting, but when statements in the Ozaukee Press are so erroneous ... we suggest the Press change its name to the Ozaukee Blabberer. ...

"We regret that it is necessary to meet much of this small talk, and to take off from a work day that covers sixteen to twenty-two hours a day, and for some of us seven days a week, but our only concern is to conduct our business successfully, to make and to keep jobs for the 250 men and women in our employ, to treat them fairly, and to treat our stockholders and ourselves fairly. We want to be an asset to this town. We want to help make it grow, and we hope that we may count on the good-will of our employees and the majority of the citizens of Cedarburg."[6]

On a headstrong roll of vengeance and retribution, Carl drafted a second statement, also signed by his father, Willis Blank and Adlai Horn. Not only were Blank and Horn named in the potentially libelous article, but both were from well respected and long standing Cedarburg families, and who agreed, independent of the Kiekhaefer family, that a vicious injustice had been perpetrated by the airing of these allegations. This new statement, though, if anything, was even more provocative than Carl's first salvo. It would serve to expose the nefarious relationship between the law firm representing the disgruntled stockholders and the newspaper, leaving little doubt as to the efficacy of the feature story and brazen headlines.

"We believe, in fairness to ourselves, and to the Corporation which we serve, that the unwarranted aspersions on us as individuals and as officers responsible for the management of the Kiekhaefer Corporation as reported by the Ozaukee Press, is of sufficient interest to the community to warrant this public denial. ...

"We would prefer to have no part of a public "haggling" brought about by disgruntled minority stockholders (all employees of the Cedarburg State Bank) and aired publicly by a newspaper (published and edited by the family of the attorney for those stockholders).

"In closing, we presently are badly in need of additional factory help and respectfully suggest to the individual disgruntled stockholders (and their attorney) that we will very gladly place them with night shift jobs if they personally care to work the hours that both management and employees of the Kiekhaefer Corporation work."[7]

And like a volcano rumbling and threatening to blow its top, Carl was preparing himself to fire the final blow to the shareholding insurgents. Carl was a quite incisive writer, able to successfully build supporting points to his final arguments, and to wound adversaries with the weapons of airtight logic and common sense. In his farewell address to the diatribe begun by the avarice of some of his original investors, he wrote a letter to the editor of the Cedarburg News in an appeal to every family in Cedarburg.

> "When the writer assumed his position of General Manager, he never dreamed there was any other policy to follow than to build a solid, thriving concern in the community of Cedarburg. Our directors policies have not changed in this respect and we do not believe our employees or any man or woman in Cedarburg would like to see us stop expanding or stop creating new jobs, or stop creating new opportunities for Cedarburg residents. ...
>
> "We do not believe these are the times to pay tremendous dividends even if it were possible. With the war casualties coming so close to home, how can anyone in this community think about dividends and place them ahead of the war effort, particularly, bankers who have sons and daughters in the armed forces.
>
> "Men and women of the Kiekhaefer Corporation, what do you think? Write what you think to the Ozaukee Press and ask them to publish it.
>
> "Men an women of Cedarburg, merchants and business men, don't you believe this dry rot should be stamped out of the community spirit? Let's build up the city, not tear it down. Let's do things. Let's first get on with the war, then let's help in this community. Let's help build up schools, manual training courses, librarys [sic], public buildings, city gymnasium, etc. Let's make this the best little place in Wisconsin to live. We will do our part."[8]

After two days of maddening politics, disrupted schedules and frayed nerves, Carl was grateful to return to business as usual. He never would forget the Cedarburg State Bank's connection to the investor uprising though, and would never really forgive the entire community of Cedarburg for the unwarranted suspicions that were generated as a result of the newspaper articles. Within a few years, when Carl would consider the alternative of a dramatic expansion of his Cedarburg facilities, or move the heart of his growing operation elsewhere, he would remember well the double banner headlines and the stinging reprisals in the autumn of 1943. And as for the Cedarburg State Bank - following the embarrassing episode, in a year when Kiekhaefer Corporation sales topped $2,615,000, the bank's closing balance of deposits for the community's largest company was $23.

CHAPTER TEN

Higher Sights

"As you know, one of the rules of the Kiekhaefer Corporation is that no executive or office employee leave the plant during working hours without disclosing details to the operator. Failure to do so in the future, will definitely result in dismissal of the offender as well as the operator if not reported."[1]

E.C. Kiekhaefer
Christmas Day, 1944

Plans were announced during the closing weeks of 1943 by President Roosevelt, Britain's Prime Minister Winston Churchill, and Russian Premier Joseph Stalin to coordinate efforts for an allied invasion of Western Europe. In January, 1944, Britain's Royal Air Force bombed Berlin, following an American bombardment by 1,400 aircraft of key German aircraft assembly plants. Day after day, week after week, American newspaper headlines charted the course of growing allied victories, and buoyed optimism that the war wouldn't last forever. The encouraging developments started many American companies thinking about a post-war marketplace, including the Kiekhaefer Corporation.

On February 18, 1944, Carl Kiekhaefer received the first post-war order for outboard engines from Western Auto Supply. Though work on the requested 15,000 engines could not begin until the factory was released from war production following the war, it was a most welcome development. In a lengthy letter to the War Department attempting to justify the smallest possible contract renegotiation rebate to the government, his frustration with losing valuable time in the marine industry boiled to the surface. Short of naming OMC, Carl took the opportunity to torpedo his future rivals in a post-war market.

'The fact that at this time, some of our competitors are in civilian outboard production - simply because they chose not to take on certain war contracts - and actually, therefore, "not interfering" with war production. The fact that this Company is the largest - and has been in business twenty years - and

> that the principals of the Company operate four different engine plants not-with-standing [sic]. ...
>
> "Our post-war models are obsolete. Wages and material costs are bound to be higher. Result? Post-war motors of new design must be developed to maintain customer appeal. Post-war models must be designed to enable use of new methods and materials.
>
> "[This] calls for more, skilled engineers - which are unavailable - due to hoarding by large companies with better manpower priorities and higher wage scales at shorter hours.
>
> "What price - loyalty and patriotism? Old fashion patriotism? Are you going to leave us in a position to continue as a going concern or are we going to be drained of our working capital?"[2]

Carl received an inquiry from the War Production Board a few weeks later, indicating its willingness to rewrite Limitation Order L-80 with respect to outboard motors and parts. Having learned through experience that he would only receive a small percentage of anything he asked for from the government, he replied that he would build 78,000 outboards in 1945 if he were allowed the critical materials to produce them. In his answer he also estimated the man hour requirements for building his single cylinder engines "at three-quarters of a motor being built per man in an eight (8) hour day," and for his larger twin-cylinder models, "one-half and one-quarter motor being built per man, per eight (8) hour day."[3]

Carl had neglected to inform the War Production Board that he had indeed been working on post-war designs, though in most cases only minor improvements to pre-war designs had been made. By July 1945, Limitation Order L-80 which had prohibited the use of aluminum for recreational products was revoked. Restrictions were still strictly enforced, however, which severely limited the practical ability to produce outboards. As Carl announced in the *Mercury News*, return to peacetime production would not be as quick as he would like.

> "We are now permitted to manufacture our Mercury Outboard Motors and sell them without priority or other restriction. ...
>
> "[Certain] provisos mean that it will be some time yet before actual volume production on outboards will occur. Although a step in the right direction has been taken, the

> industry will be held back until cutbacks and cancellations of war contracts ease the material and labor supply.
>
> "Needless to say we, at Kiekhaefer, will use every facility we possess to fill our obligations to the military and be prepared to supply our Distributor-Dealer organization with more and better Mercury Outboard Motors the moment our plant and people are honorably discharged from the Army of Essential War Production."[4]

Following the announcement of the withdrawal of L-80 restrictions, Western Auto Supply more than doubled its post-war order from 15,000 to 33,000 engines, in three new models ranging from three- to 12-horsepower. The problem was, they wanted delivery by Christmas 1945, if possible, but no later than January of 1946.[5] Carl and the Kiekhaefer Corporation began to feel the effects of what was known as "post-war fever."

On the first day of 1944, work began on a new project under the direction of the Army Air Force. A new engine was needed to power small radio-controlled target aircraft drones for air-combat gunnery practice. Since the delivery of the very first air-cooled chain-saw prototype, the Kiekhaefer Corporation had earned a reputation for fast, flexible and innovative engineering. The first contract, for $40,000, was for two prototype engines named the Y-40, which would eventually become a twin-cylinder, horizontally opposed, two-cycle engine of thirty-five horsepower. The largest budgets in the War Department were for aircraft production, so when the opportunity came along, Carl was only too eager to accept.

The target drone engine project presented unique and new challenges. The target aircraft were quite small, measuring only 10-feet, 2-inches from wing tip to wing tip, and only 8-feet, six-inches from propeller to tail, with a fuselage measuring only 13.5-inches deep. The entire aircraft weighed less than 138-pounds, including engine and propeller. The little drone was launched from a small catapult, and then flown from a radio-control panel on the ground or from pursuit aircraft.

Once launched from ground or ship deck, the aircraft would typically ascend to four or five thousand feet, level off, and accelerate to its maximum speed between 175 and 200 mph. If it wasn't shot down by student gunners firing .30- or .45-caliber machine-guns from fighter aircraft or bombers, or by ship based anti-aircraft guns, it would run out of fuel in less than an hour, be flown back to a recovery area, radio instructed to deploy a parachute in the tail section, and descend back to earth at the relatively safe rate of sixteen feet per second. Hardly the typical mission for an outboard or chain-saw engine.

The drone engine ultimately developed by Carl and his designers weighed 34 pounds and developed 35-horsepower at between 4200 and 4400 rpm. It was considered a great technical achievement to produce an engine that weighed less than a pound per horsepower. It was equipped with a float-type carburetor, and used a battery ignition system. To start the engine, the nearly three foot long wooden propeller was spun either by hand or by an electric spinner that turned the propeller from the front until the engine fired. Once the engine reached launch rpm, a release pin was electrically retracted on the catapult, and the whole works was flung into the air, most likely to be blown to smithereens.

One of the obstacles to overcome was one first thought to be the easiest: engine cooling. Since the aircraft was moving through the air at nearly two-hundred miles per hour, it was initially thought that the engine would be cooled easily. Actually, because the aerodynamics of the small target drone were so critical, the engine had to be almost completely enclosed by a tight fitting cowling, with outside air ducted to the cylinder cooling fins. Early tests revealed piston seizing, poor fuel distribution, incorrect clearances and sticking piston rings. Once these problems were overcome by providing more cylinder fin area and greater piston-cylinder clearances, the Kiekhaefer engine passed flight tests with flying colors.[6] Again, the company had demonstrated their ingenuity and flexibility by having a prototype in the air in a little over three months time. It would be nearly a year, however, until the engine, then known as the 0-45-1 and 0-45-35 had been perfected and volume deliveries made.

Carl's government contracts continued to grow from nearly $4 million in 1944, to about $5 million in 1945. Mandatory government contract renegotiation would reduce the spoils of war starting with the 1944 production year, though, as over $800,000 were trimmed from profits. Carl didn't object, however, for he had anticipated making a rebate of over $1.2 million, which meant that his heart-wrenching letters to the War Production Board had managed to save the company nearly $400,000 that year alone. Ralph Evinrude reported that OMC, by contrast, had sales of nearly $33 million in 1944, and had paid out nearly $2 million in contract rebates.[7] But whereas the Kiekhaefer Corporation sales volume was up over 80 percent, sales at OMC had tumbled by nearly 80 percent during the same period.

Though the manufacture of the chain-saw engine for Disston was the backbone of all Kiekhaefer Corporation war efforts, the sales of compressor, generator and water pump engines were nearly as large for 1944, a year when only one solitary outboard engine was manufactured, a KB4 twin-cylinder engine that sold for $55.70. The following year, no outboard engines would be manufactured whatsoever, and sales of the new target drone engine would compete equally with chain-saw engine revenues.

The war was teaching lessons to Carl and his organization that might have been impossible to learn in the relative calm of peacetime. The quick reaction to engineering challenges were often prompted by a feeling of life and death struggle, and the wide array of products and assignments put flexibility into every job description. The pace of war production put a tremendous strain on the organization, and on the wives of the men who worked long hours. In February 1945, Carl issued an order to all engineers and draftsmen, making a seven day a week schedule mandatory. The over sixty-seven hour work week went from 8:00 a.m. to 9:00 p.m., Monday through Friday, with an hour off for lunch and supper, making an eleven hour day. On Saturday, hours were from 8:00 a.m. to 5:30 p.m., with a half-hour for lunch, or an eight and a half-hour day. On Sunday, workers toiled from 8:00 a.m. to noon. Carl knew the schedule was unpopular with most men and with almost every wife, so he sweetened the blow by allowing expense money for suppers for all workers on the rough schedule.

Though somewhat tongue in cheek, a column in the *Mercury News* describes the buzz of activity during a average day during these difficult war years.

> "We arrive at the plant in the gray dawn with a full day of planned and necessary activity ahead of us, firmly resolved to carry on as scheduled -- then as our frost-encrusted parka is being hung up to thaw, an excited voice over our left shoulder greets us thus -- "There is a new directive in -- Job X is rescheduled -- our supplier of gimmicks has let us down -- the express company has lost that shipment of voo-doos and the Air Force is arriving for a conference in twenty-two minutes" --

O.K. -- let's go -- and in three minutes Lohmann has the X Schedule in work -- Curly Stuth is changing the set-up on the machines -- Ray is arranging to shift the line -- Harry has the Express Company chasers on the phone -- and Conrad is telling all our other customers what this directive is doing to their schedule of shipments. O.K.! O.K.! -- Fellows, here are the boys from the Air Force for the conference. The boss will handle them until we get their stuff lined up -- what is that on the Plant broadcast -- will you take that call from the Navy? -- O.K. Commander, we realize the urgency of your job and will put every available ounce of effort into it. Yes, Bill, those fellows from the War Manpower Commission are here -- Henry, we have just been advised that a complete breakdown of costs on Contract 123 must be mailed today -- Yes, Boss, we will be right in. A new progress report is required on Development Engine A and delivery date must be established -- O.K. we will clean that up tonight. Intercom: Doc Sherwood calling to say that Pilot Engine AEF in test cell No. 4 continues O.K., pulling 25% over rated horsepower -- come one, let's go to lunch -- WHEW!!

"Los Angeles called while you were at lunch -- will you see E.C.K. right away -- there are four loggers in the lobby sent by W.P.B. to get the dope on increasing their production by using our chain-saw -- call production and see if there is any possibility of increasing B. Company's order for KB7's. The Army wants 900 more in March -- who's expediting that steel for the R Job -- what's this new Government Agency doing in here with another request for breakdown sheets -- yes, Hank, guess we will have to take a man off production to do it -- did you see this letter order for a KB4 Mercury Outboard from Kankakee and the fellow just says ship it tomorrow -- inciden-tally, there are 46 more postwar dealer requests in today's mail -- why don't these Government guys make appointments, there is a truck load of them here to make radio suppression tests -- Oh! Oh! They came armed with one of those "must" orders -- well this means a 60-hour delay on the dynamometer curves for the Navy -- W.P.B. from Washington is calling to discuss our projected schedules for next month -- tell that facilities branch that it is only 12 degrees below here this afternoon and how come they expect our gang to get along without a shed when our neighbors are building a beer warehouse -- Becky, Chicago is on the phone for information on parts orders TOP shipments -- what? It only came in Monday and the inspector says a change order is in the mail? Guy, how many more KB3's can we schedule for August -- gotta get a wire off now -- say, can't we borrow your stenographer to make a quick pick-up at Tool

& Die --- everybody else is tied up -- Intercom paging all Department Heads -- report for conference in 10 minutes -- where the ..... is my perpetual memo book -- don't forget that Navy proposal has to be mailed today -- say, has Spike's wire on that logger job been answered -- better have two extra girls in tonight so we can clean up some mail -- My Gosh! Tomorrow is the deadline on the MERCURY news copy -- meet me back here at 6:30 -- yea, my wife is a Production Front widow too -- no it isn't too late, Harry is taking in more mail and express when he goes into town at midnight -- tomorrow we will stick right to our routine work -- Good night. Oooooooooh! My car is frozen!"[8]

Two days after Carl's thirty-eighth birthday, on June 6, 1944, the long awaited allied invasion of Europe commenced on the beaches of Normandy, and the end of the war seemed inevitable. Within months, Paris was liberated by French and American forces, and the German armies were in full retreat. In April 1945, President Franklin Roosevelt succumbed to a cerebral hemorrhage, cheating him of the chance to see the conclusion of the great struggle, and within two hours, Harry S. Truman was sworn in as the nation's new President. In the same month, Adolph Hitler committed suicide in an underground bunker in Berlin and the body of executed Italian leader Benito Mussolini was hung by his heels to the cheers of crowds in Milan. On May 7, 1945, the war in Europe was officially ended, as Germany surrendered unconditionally to the victorious allied nations. Almost exactly three months later, atomic bombs devastated the Japanese cities of Hiroshima and Nagasaki, hastening the surrender of the Japanese empire, and bringing World War II to a long awaited end.

The Kiekhaefer Corporation had designed no less than fifty-one special applications of air-cooled engines for the military services, and had endured the hardships of bureaucracy and competitive political muscle. Kiekhaefer products had performed in every major theater of the war, and had earned the highest marks for performance, reliability and simple operation. A whole new war was just beginning for Carl and the Kiekhaefer Corporation, with stakes larger than even "Carloads" Kiekhaefer could imagine. The first shots in the post-war battle for control of the marine industry market were about to be fired.

CHAPTER ELEVEN

Fond du Lac

"When he was feeling in a good mood and had faith in you as a person, he'd happily give you the shirt off his back. But then if something wrong happened, he'd take that shirt back again and he wouldn't bother opening the buttons."[1]

Herman Steig

Herman Steig was twenty-three in 1936 when he started work for Carl Kiekhaefer at Stearns Magnetic in Milwaukee. Even then, Carl depended on Herman's help with special projects, like working on weekends at Stearns, greasy elbow to elbow, helping Carl pull the engine out of his car. Carl had told Stearns to hire Herman, and Carl insisted on a raise for him when Herman threatened to leave for better wages. "I was going to quit because wages weren't too high in those days," Herman recalled, "so Carl went to the office and he demanded that they give me a ten cents an hour raise, and by God I got it! That was unheard of. You used to go by a penny at a time, but he demanded it, and I got it, and I stayed." After Carl had left and started the Kiekhaefer Corporation, he called for his old friend to join him in Cedarburg. Starting in early 1941, Herman was going to be Carl's point man in developing the magnetic separator business. When it became clear that outboard and air-cooled engines would dominate the future of the company, Carl kept Herman moving around the plant doing everything from sub-assembly to fixture lay-out to tool-making to machine shop superintendent, until he was put in charge of the plant's entire night shift in November 1942.

It was Herman's habit of volunteering to go the extra mile whenever the occasion arose that cemented Carl's confidence in him. Typical of his enthusiasm was the time on Easter Sunday in 1945, after the plant crew had been working all day Saturday, and through the long night preparing the first 50 aircraft target engines, and the men could hardly stand up. "Carl got his father's farm truck in there and loaded [the engines] into it by about ten o'clock on Sunday," Herman remembered, "He looked around and said 'Well, who's going to drive this thing to Chicago now?'" The silence of the exhausted crew made it clear that nobody was interested in putting in another eight or nine hours after working 27 straight hours. "I'll do it," Steig answered, and grabbed the keys from Carl.

Carl Kiekhaefer had an ability to make people expand their potential, to grow with responsibility and to accomplish things they never would have imagined they could do on their own. If Carl had confidence that an individual could reach a new plateau in their career, then he would make sure they learned to dig deep into themselves to tap the extra resources that he was convinced they had within to rise to the occasion. One of his favorite expressions became, "I pay a man twice what he's worth, and then I make him earn it."

The moment the war ended, Carl was faced with a major dilemma. He was told by the Army Corps of Engineers to continue production of chain-saw engines under his existing contracts, and even accepted additional contracts along with more target aircraft engine orders from the Air Corps. Now that he finally had the opportunity to build outboard motors again, he had no room left to assemble them. The entire facility at Cedarburg was jammed with chain-saw and aircraft engine production lines, with hardly enough room for employees and machinery. Still stinging with the embarrassment of the newspaper headlines exposing his stockholder rebellion the year before, Carl resolved to look elsewhere rather than expand his plant in Cedarburg. His odyssey to a dozen nearby Wisconsin cities turned up a number of potential sites ranging from Warsau to Beaver Dam to Fond du Lac. High on his list of priorities were existing facilities with room to grow, and a community that would bend over backwards to attract and keep his growing business and payrolls.

He found the unique combination he was searching for in Fond du Lac, forty miles to the northwest of Cedarburg. Local folk lore suggests that by the middle of the seventeenth century, French explorers crossed the Great Lakes into Michigan and Wisconsin, and made their way down the Fox River from Green Bay into Lake Winnebago. Named for the Indian tribe that first settled the region, Wisconsin's largest inland lake was measured as early as 1670 by grizzled trappers eager to establish a foothold in the lucrative beaver and otter saturated wilderness. The southernmost point of Lake Winnebago the French named

Fond du Lac, which loosely translated means *At the Foot* or *Bottom of the Lake.* A combination of deep surrounding forests, plentiful game, rich soil and plentiful artesian water attracted settlers, and by 1836 the first permanent home had been erected. When the first artesian well was dug in 1846, water gushed forth at a rate estimated at over 1,000 gallons per hour.[2] By the turn of the century, Fond du Lac became a thriving industrial community, with eighteen lumber and shingle sawmills, and flour and grist mills situated along the area's rivers and streams, with a population of over 15,000.

In 1902, the infamous Carry Nation, the "eccentric hatchet-swinging Prohibitionist reformer from Kansas" rolled into Fond du Lac to lecture its robust saloon-loving citizens about the evils of alcohol.[3] Even her advance agent admitted that Nation was a sight to behold. "She is homely," he confessed, "She is homelier than a mud fence. But, say, she does have a most powerful arm for swinging a hatch ax."[4] While pummeling the assembled tipplers in E.J. Schmidt's crowded saloon with all the force her six-foot and pudgy frame could muster, she nearly frothed as she announced that "every German in Wisconsin should be blown up with dynamite," no doubt alluding to the Teutonic love for beer. An insulted German at the bar promptly presented her with a bottle of whiskey, purportedly to calm her frayed nerves. The newspaper reported what happened next:

> "She surprised those present by drawing a hatchet from beneath her dress and smashed the bottle. Schmidt took the hatchet away from the woman, and that ended the seance. Strong men watched Mrs. Nation after that to see that she did not attempt any of her antics for which she is famous."[5]

Forty-four years later, Carl Kiekhaefer and Herman Steig drove by the corner of Main and Division streets in Fond du Lac, and probably wondered about the plaque which still reads, "Site of the famous Carry Nation Axe Wielding Incident, 1902." It was between twenty and twenty-five degrees below zero that cold January morning in 1946, as Kiekhaefer drove Steig southwest of town. He was eager to show him what an aggressive Fond du Lac Industrial Development Corporation had offered to sell him at a most attractive price. Steig could barely contain his astonishment. Carl pulled up beside the largest barn Steig had ever seen, nearly three-hundred feet long, almost five stories high, topped by eleven giant ventilators and surrounded by four towering grain silos. "Much to my surprise," Steig remembered, " he turned around to me and he said, 'Well, how would you like the task of making a factory out of this barn here?'"[6] Steig was speechless. "I thought, *'My God!,'* what do you think I can do? This is so way beyond me. ..." But, taking his measure of Carl, standing next to him with his hands on his hips, head tilted back to take in the whole gigantic structure, a big cigar jutting nearly straight up from one side of a broad smile, he knew anything was possible. "Sure," he said, "why not?"[7]

The Corium Farms barn was built in 1917 by Fred Rueping, owner of the Rueping Leather Company that had flourished in Fond du Lac since 1854. Corium, which means "leather" in latin, grew to be the largest and most successful dairy farm in the region, with over 500 acres and a herd of over 300 thoroughbred cows descended from lines direct from the Channel Island of Guernsey off the coast of France in the English Channel. One of the first inhabitants of the barn was the celebrated bovine *Imported Prospects Rose de Hords*, the milk and butterfat record breaking queen of the herd that was bought for nearly $18,000 in 1919. For many years the Corium Farms operations were considered the very model of modern milking operations, and swept blue-ribbon honors at a succession of Wisconsin State Fairs.[8] When the dairy interests were sold off during the war in 1945, the mammoth structure was ultimately acquired by the consortium of civic-minded Fond du Lac boosters, led by Dick Mills, that persuaded Carl Kiekhaefer to take advantage of the unique situation.

The Kiekhaefer Corporation took possession of the Corium Farms barn and surrounding thirty-eight acres on February 1, 1946. Carl had managed to acquire the land and buildings for only $25,000, which worked out to roughly $2,250 for the land at $59 an acre, and $22,750 for the buildings. Now it was a race against the clock to transform the barn into a modern engine manufacturing plant. Herman Steig quickly hired a large crew of men to clear out the hundreds of bull and cow stanchions and stalls that made the interior of the cavernous building appear like a medieval maze of pipes, corrals, gutters and troughs. Carl rolled up his sleeves to join the crew to remove the over two-hundred tons of stacked hay that lay in the loft, twenty-five feet above the floor. They drove a hay baler underneath the loft, and with a small army of pitchforks, tossed the loose hay into the machine, and neatly tied bales were trucked away for sale to surrounding farms. "And he was right with us," Steig remembered of Carl, "pitching hay down into the baler. In fact, he even went down and ran the baler for a while."[9]

Once the interior of the barn had been stripped, Steig poured a new concrete floor to fill the gutters and holes that made the remaining slab look like a detonated mine field. At one end of the structure he partitioned off areas for a machine shop and tool crib, leaving the bulk of the cavernous structure for assembly operations. Each special attribute of the unique building was put to good use. One of the four huge silos was used to house a large cylinder honing machine that had to be skidded sideways into the opening, and then set upright. Another silo became home to the air compressors needed for factory operations, while yet another became the motor spray-painting department. The last silo became a test cell for the target aircraft drone engines. Since these engines were tested with actual flight propellers, the giant silo worked perfectly to provide large volumes of air for both intake and exhaust. Offices for Carl and other administrators were built in the loft, overlooking one end of the plant floor.

During the renovation, Carl had been driving back and forth the forty miles from Cedarburg to Fond du Lac, racing along the curving two-lane highway at breakneck speeds and cursing the waste of time sitting behind the wheel. Very few things irritated Carl more than losing time. He decided there was only one way to get back and forth faster: to fly. It seemed like such a natural thing for Carl to master, after all he had been operating equipment his whole life. He had the requisite touch on the often complicated controls of farm machinery and shop hardware, and most people described his automobile driving technique as nearly airborne anyway. The previous year, in March 1945, Carl sat in the front of an aircraft for the first time, and actually handled the controls of a Fairchild 24W while on a business trip to Dayton, Ohio with a pilot who was also an instructor. He enjoyed it quite a bit, and vowed to someday get his license to fly. After putting up with the laborious drive to Fond du Lac for only a month, on March 23, 1946, he strapped himself in again, determined to go the distance.

Among the many benefits that accompanied the Corium Farms purchase was a gravel airstrip situated just behind the property. Once Carl had made up his mind to get his company off the ground and into the air, he proceeded with characteristic fervor. Two lessons in the older Fairchild were enough to convince him that flying was the way of the future, and before he had his first minute of ground school he located and bought a pair of Ercoupe aircraft, model 415-C, equipped with 75-horsepower Continental engines for $10,500. He bought two aircraft because he wanted Herman Steig to learn to fly the other one. He told Steig, "When I call you from Cedarburg for a meeting, I want you to fly down. I don't want to wait an hour!"[10]

The tiny two-seat aircraft wasn't exactly designed with Carl's five-foot ten and a half-inch, 225 pound profile in mind, making it difficult for the portly pilot to squeeze into the little seat behind the yoke. But, as witnesses to Carl's training said, after being airborne for an hour with his instructor, Carl was sweating so profusely that he would slip right out the door when the lesson was over. "He'd come out of the airplane with a *big* sweat on him," Herman Steig remembered, "it was kind of rough on him. He was somewhat scared I should say, but he went through it."[11] With only five and a half hours of dual instruction, Carl launched himself into the Wisconsin skies alone for the first time on May 16, 1945, and after fifteen anxious minutes, landed again safely. Following his first solo flight, Carl began to take advantage of his new time machine, making dozens of flights between Cedarburg and Fond du Lac. Each flight was logged at thirty minutes, which saved Carl about an hour per round-trip over the car, even driving as fast as he could.

On August 5, 1946, the forty-two year old Kiekhaefer passed his flight test for his private license, just over five months and fifty hours of flight following his first lesson. That Carl was able to fit flight training into an already brutally self-imposed work schedule is testimony to the importance he attached to saving time. Except for an occasional flight to Green Bay, Madison or Chicago, every entry in Carl's logbook shows the quick hop between his two plants. He continued to fly regularly until his forty-third birthday the following June, when he scared himself sufficiently to never get behind the controls of an aircraft again for the rest of his life.

He was coming in too low for his approach for landing at the Corium Farms gravel strip when his landing gear smashed into a set of saw horses that had been placed at both ends of the crude, one-way landing strip. He came perilously close to crashing the nose of the aircraft into the rough ground and cart wheeling, but with just over ninety-two hours of flight time under his belt, he recovered just in time to avoid disaster. The incident shook him to the core, however, and from that day forward, he rarely admitted to ever having been a pilot, and never touched aircraft controls again.

Meanwhile, Carl realized that the narrow window for seasonal outboard motor production couldn't wait for conversion of the Corium Farms barn, so he began limited production in Cedarburg. Many lessons in durability and performance had been learned as a result of the demanding conditions of wartime engine use and abuse, and Carl wanted to update his nearly obsolete line of outboards as quickly as possible. The first and biggest change came to the six-horsepower, two-cylinder *Rocket*, which became the first outboard to have anti-friction ball, roller and needle bearings at all major locations in the power head and lower unit. Before the war, Carl's engines had used bronze bushings in some areas. By using very accurate bearings at each critical load-carrying juncture of the engine, a large number of benefits were realized. With less overall friction in the engines, cold engines were easier to pull over, and started more quickly. Idle speeds were reduced, because the engines didn't need to work as hard at low speeds, like for trolling. The bearings themselves didn't need to be replaced nearly as often as the old style bronze bushings. Less accumulative drag on the engine components also meant increased spark plug life and less build up of carbon deposits on piston rings, pistons, cylinder walls and exhaust ports.

Carl was beginning to make major changes to his outboard engines for the first time since he reconditioned Thor engines for Montgomery Ward. As he coaxed higher and higher performance out of his pre-war designs, flaws appeared. He replaced aluminum connecting rods with drop forged, hardened and precision ground alloy steel, the same used for drive shafts. The dynamometer readings for the *Rocket* continued to rise with each improvement, until the motor advertised as a six-horsepower engine was roaring well past seven and a half. These improvements were then incorporated in the single-cylinder, 3.2-horsepower *Comet*. Carl capitalized on his new roller, ball and needle bearing equipped engines by calling them *full-jeweled* power heads. Watchmakers had for some time referred to their precision movements as "fully-jeweled" when gear bearings had been bored from semiprecious stones, which is undoubtedly where Kiekhaefer appropriated the friction-free slogan.

As a result of continuing aluminum shortages, only the two-cylinder *Rocket* was built in 1946, but between those labeled Mercury and those labeled *Wizard* for Western Auto Supply, the Kiekhaefer Corporation managed to ship nearly $1.5 million worth of outboards by the end of September. In addition, over $3.1 million in chain-saw, generator and other industrial engines were built in the same period. Carl Kiekhaefer was back in the outboard business with both feet.

The lessons learned from building thousands of chain-saw and other engines for the government, and the increased performance of the modified *Rocket*, were enough to convince Carl to make a complete break with the *Thor*-based models, and design a totally new engine from scratch. Work began almost immediately on a totally new, ten-horsepower engine that was destined to propel Carl Kiekhaefer and the Kiekhaefer Corporation to prominence in the marine industry, and send competitors scrambling to their drawing boards. Carl wanted an engine to compete head to head with the best engines in the industry, and to firmly establish Mercury presence in the post-war marketplace. As the engine design lab worked on the essential components of the two-cylinder, alternate firing engine, Carl directed the sculpting of clay models to achieve the most modern appearance and distinctive markings possible, to set the engine apart from the pack. He requested a fully enclosed or "hooded" engine, a cosmetic concept started by Evinrude in 1929, and one which Carl had emulated with his own 1940 Thor *Streamliner*.

The cowling design that emerged from the lab was meticulously shaved and molded from modeling clay, painted green and supported by a two-by-four. Looking somewhat like an elongated sphere, or plump football, it was proudly paraded into Carl's office for his inspection. "It looks like a God Damn Watermelon on a stick," Carl bellowed at the quivering clay-smudged designers, "or a God Damn Green Pumpkin." It was a useful shape, though, that had a tendency to grow on you, and without much modification it was this somewhat chubby profile that would finally emerge from the Kiekhaefer production lines as the Mercury *Lightning*. Among the more distinctive visual features of the bright pea-green engine was an oversized letter "K" that doubled as the throttle lever handle. The 10-horsepower engine was blowing the dynamometer past the 16-horsepower mark, but Carl insisted on calling it a 10-horse so that no other 10-horse in the water could possibly touch it. More than modesty, the underrating of horsepower was to become a strategic marketing weapon for Carl through the years, a careful positioning of his products so they would be known as the fastest in their advertised horsepower class. Along with the provocative slogan of *Full Jeweled Power*, this first, totally new engine from Kiekhaefer was to quickly gain a reputation for both speed and precision engineering, and swamp the growing company with an avalanche of new orders. Carl discussed his reasoning for underrating his engines in an interview years later, and admitted the deliberate step to establish Mercury products as *high performance*.

> "First of all, we found out when we got into the military, the 16-horsepower Evinrude was rated at 9-horse by the Army Engineers, and so down the line. They really de-rated the motors and I know that the Army was only trying to -- the military was only trying to get a realistic horsepower rating so we thought about that too, and we said, 'Well, what the hell is the cost difference? We might as well go the other way.'
>
> "Instead of building a 10-horse motor, we'll put out a 14-horse motor. And we were already beyond the 16-horsepower rating that Evinrude had. Consequently we were able to outrun them. With our Mercury 10 we outran not only all their 10's but their 16's and in some cases even their 22's. ...
>
> "So we rated our engines very conservatively. Subsequently they got the reputation of being high performance. And it's just giving the customer a little more for his dollar. That was the principle of the thing and we didn't have to be so darned exact. If we rated an engine 10-horse and some of them pulled 13 and some of them pulled 15, we weren't too worried about it."[12]

Thirty years after the 10-horsepower *Lightning* was introduced, Carl was presented with a beautifully framed, chronological montage of his accomplishments, but he complained that the Mercury *Lightning* wasn't depicted.

> "Without trying to be critical of a nice piece of art, I am sorry only that the Mercury "Lightning" did not appear. This was the most important engine Mercury ever built and also what put Mercury into orbit.

"The "Lightning" was the 20-cubic inch twin, which later stacked into a 40-cubic inch four and a 60-cubic inch six, based on a high degree of commonality. So many advancements were incorporated into this engine that it went clean over the tops of the heads of some of the competition. ... Features included ... the most horsepower per pound, the most horsepower per cubic inch and the most horsepower per dollar, of any outboard motor before or since 1947. ...

"The advent of the Mercury "Lightning" set Mercury aside into something separate and apart from the rest of the mob."[13]

By July 1946, Herman Steig had performed the near-miracle of transforming the giant Corium Farms dairy barn into a full time engine manufacturing facility. While the production lines of Cedarburg were still churning out the Thor-based six-horsepower *Rocket*, Carl transferred most chain-saw engine production to Fond du Lac to make room for outboard production. While the new ten-horse *Lightning* was prepared for production, Carl ordered the preparation of a new six-horsepower twin-cylinder *Rocket* and single-cylinder 3.2-horsepower *Comet* modeled after the *Lightning*'s fully shrouded "Green Pumpkin" design to round out a new trio of 1947 offerings. Just when both factories had begun to settle down, Carl sent notice to Herman Steig that the new *Lightning* would be the first outboard motor built in the new Fond du Lac facility.

When the *Lightning* hit the water following its introduction at the New York Motor Boat Show in January 1947, it caused shock-waves in the outboard industry. It ran away from everything even remotely close to its size and "advertised" horsepower. It beat nearly every popular motor on the market, including the 16-horsepower and even 22-horsepower Johnsons. One afternoon, when the first *Lightning* prototypes were still being tested, Carl asked Bob Stuth, among the most versatile of Carl's employees, to get his boat and 33.4-horsepower Evinrude *Speedifour* and bring it to their new outboard motor proving grounds on the Milwaukee River, just north of Thiensville, Wisconsin. Carl wanted to prove that his new 10-horse could whip even the massive four-cylinder Evinrude. As the carefully staged race began, the lower unit of the *Lightning* prototype blew apart. Red faced, Carl "reeled into the lab about screwing up his day and embarrassing him."[14]

1947 was the fastest growing year in Kiekhaefer Mercury history. From building 16,908 *Rocket* outboards in 1946, the company built over 55,000 outboards during 1947, more than tripling production. Sales skyrocketed, doubling from $5.2 million to well over $10 million as a direct result of the *Lightning*.

Everything grew for Carl that year, even his family. On August 23, 1947 a son was born to Freda and Carl, and in the Teutonic Kiekhaefer family tradition, was named Frederick Carl Kiekhaefer. "Freddie", as he would be known throughout his formative years, was a clever, restless child who, the last of Carl and Freda's children, was destined to play an important and strategic role in Carl's life in years to come. The two daughters, Helen and Anita, then 14 and eight years old respectively, were growing up without benefit of their father's attention. Though Carl was clearly proud of his family, he spent infinitely more time preoccupied with his factories than he did at home, and as a consequence, it is no exaggeration that many of Carl's key employees would know him better than his own children. In later years, Carl regretted his emotional neglect of his children, but his almost total absence wounded his three children with deep emotional scars they would carry well into adulthood.

Carl formed a new corporation in 1946 when he purchased the Corium Farms property, Kiekhaefer Aeromarine Motors, which was placed on the books as an affiliate. The formation of the new corporation allowed some advantages in monitoring the productivity of each plant independently, but also allowed Carl a way to retrieve shares in the Kiekhaefer Corporation before the new engine designs could drive up prices out of reach. Using corporate secretary Willis Blank as a shield, Carl managed to trap stock owned by the unfriendly shareholders. Two days after Christmas of 1946, just as the new *Lightning* was put into production, Carl and Willis Blank signed an agreement that was the beginning of an ingenious plan that would eventually consolidate all but 10 percent of shares into Kiekhaefer ownership.

First, Blank obtained an option to purchase the eighty-five shares of Kiekhaefer Corporation stock owned by Edgar Roth, Palmer Wirth, Matthew Becker and the Cedarburg Finance Company, the conspiratorial stockholder group that had embarrassed Carl in the headlines, not to mention trying to sell him out to Flambeau when he was out of town in 1940. Next, Carl executed an option to buy any shares that Blank would acquire, along with an additional 60 shares from the Blank family under the control of Willis, including some shares owned by Willis alone. Willis Blank had received 300 shares of the new corporation, Kiekhaefer Aeromarine Motors, which would also be transferred to Carl under the terms of the agreement. The complicated plan called for Carl to pay Willis $25,000 to secure the options, place an additional $25,000 of security in escrow as a guarantee of good faith, and then pay for blocks of shares at regular intervals until 1950, when all the options would be exercised or expire. Total payments would add up to $215,000. In the presence of Carl's corporate attorney and confidant, Guy Conrad, the papers were signed. Along with the stock options, the document provided Carl with the immediate voting proxies, which remained his unless he suspended payments for the stock. For the first time since the near catastrophe of the attempted Flambeau sell-out, Carl could breathe easier, knowing that no one could wrestle control of the company away from him and his father, or sell existing stock to anyone but himself.

For a while in 1947, it looked as if the Gamble chain of specialty stores would acquire and merge with the Western Auto Supply chain of stores, and Carl was certain that it would mean the end of his days supplying the Wizard brand to Western Auto. Outboard Marine Corporation's Gale Division was supplying private label outboard motors to Gamble Stores under the Hiawatha brand name, and Carl was convinced that if the merger was made, Western Auto would be forced to follow suit, and have their *Wizard* brand engines built by OMC as well. Carl had considered suspending production of Western Auto *Wizards* if the engine started to interfere with the growth of his own Mercury brand in the marketplace. The ideal situation would be to introduce new innovations in Mercury outboards at least a full model year ahead of the private label *Wizard* models.

Carl was a master at the game of business cat and mouse. Knowing that Western Auto Supply had no other immediate source of engines once the Gamble Stores merger collapsed, he deftly manipulated them into position to guarantee production in his factories for years to come. He pretended not to be interested in the millions of dollars of contracts he would lose if Western turned elsewhere. In a transcribed telephone conversation with Frank Wilfred of Western Auto Supply on September 25, 1947, Carl made it clear that unless they would conform to his own production schedules, they could look elsewhere for their engines.

Wilfred: What is the deal, Carl? How is it shaping up?

Kiekhaefer: Well, we're just so swamped down here and short-handed engineering wise that it's awfully difficult to get new models ready for you this fall, and rather than hold you up ... maybe we should just tell them [Western Auto] that we can't furnish engines any more. ...

Wilfred: ... [W]e aren't going out of the outboard motor business, but if Kiekhaefer's aren't going to [supply us] or we can't work out a satisfactory long term program with Kiekhaefer, well, we were going to have to go somewhere else.

Kiekhaefer: ... Now, we hardly thought that you'd be willing to go along with that because Steve Briggs [Chairman of Outboard Marine Corporation] will do anything in his power to kick us in the rump and he'd give you terms. ... But, on the other hand, you wouldn't be wanting to sell the Sea King, the Sea Bee [Goodyear Tire's private label built by OMC] and a half a dozen other motors by the same name.

Wilfred: Well, of course our feeling has always been, Carl, that we liked your motors very well. ... But we don't want two sources on it, and I don't think you would. ... Either we want you to build the motors for us and build them all, or, if that can't work out that way, then we'll have to go to somebody else. I don't think you'd want part of our business, and I don't think anyone else would. ... Starting with this basis, Carl, are you interested in a program with us on a long term basis?

Kiekhaefer: Yes.

Wilfred: I wanted you to say that. ...

Kiekhaefer: We are interested, of course, Frank, because the Western Auto Supply has been very nice people to do business with, and there are moral obligations. There was a time when we needed you badly and now that you need us, we don't want to run out of this picture ... but I'm telling you right now that we would like to continue to do business with Western if there's a way to work it out -- for instance, if you can continue to take the older motors for a while, until we get out from under a little bit, and then we design your new motor. ..."[15]

Industry predictions of a post-war recreational boom underestimated the enthusiasm with which returning war veterans would embrace outboarding. As the market grew, Carl was beginning to compete against a rash of new companies. "By the time the war was over, we hoped to get some rest," Carl said, "but by that time there were some 60 outboard manufacturers in the business. Everybody that made pots and pans wanted to get in the outboard business."[16] Although Carl exaggerated the number of companies involved, the threat was real enough. Among the companies that were determined to crowd Kiekhaefer out of the industry was the Champion Motors Company of Minneapolis, Minnesota.

In 1934 Earle L. Du Monte and Stanley G. Gray formed Champion Outboard Motors Company, and licensed the Scott-Atwater Manufacturing Company to build engines from 1934 to 1943, when the government's freeze on aluminum took effect. Champion Outboards were distributed through tire stores owned by Firestone Tire & Rubber Company, which also supplied parts and service. Out of business for lack of aluminum, the patents and licenses to build the motors were sold to The Flour City Ornamental Iron Company, which waited patiently for the war to end. In February 1946, they broke ground for an impressive, 100,000 square foot factory in Minneapolis, and five months later, the first postwar Champion outboard motor rolled off the assembly line. Within twenty-four months, over 120,000 motors were built by nearly 500 men working in three shifts.

Scott-Atwater, meanwhile, had converted to production of tools for Federal Cartridge Corporation, and became a major supplier of B-17 and B-20 engine parts. Aware of the new prosperity in outboard production, C.E. Scott and H.B. Atwater also began producing their own design of outboards in Minneapolis in 1946.

Ralph Evinrude, president of Outboard Marine Corporation, reporting sales of nearly $32 million for 1948, said that "despite maximum production, demand for our products continued to exceed supply."[17] In fact, a mad scramble for position in the wide-open outboard motor field was underway, and in a few short years, dozens of manufacturers would rise and fall, each desperate to grab a piece of the enormous pent-up demand for products following years of war-time denial and frustration. On the heels of Champion and Scott-Atwater, within a short four-year period, production started on Atco, Bantam, Brooklure (Spiegel), Buccaneer (Gale), Chris-Craft, Chrysler, Corsair, Elgin (Sears, Roebuck), Saber (Fedway), Firestone (Firestone Tire &

Rubber), Flambeau, Hiawatha (Gamble-Skogmo), LeJay, Majestic, Martin (National Pressure Cooker), Milburn Cub, Motor Troller, Atlas (Distributed by Standard Oil), Sea-Bee (Goodyear Tire & Rubber by Gale Products, OMC), Sea-Flyer (B.F. Goodrich), Silvertrol, Voyager and West Bend outboards. When you add to these the already established outboard manufacturers that began production before the war, like Waterwitch (Sears, Roebuck), Elto (Evinrude), Evinrude (OMC), Flambeau, Johnson (OMC), Lausen (Hart-Carter) and Muncie, and then throw in Sea King (Montgomery Ward by OMC) and Kiekhaefer's own Wizard for Western Auto Supply, it is easy to see that the competition had become a very real nightmare, producing a sea of confusing brand names and alliances within a short time.

As consumers became more and more acquainted with the new features offered on outboard motors, manufacturers flooded newspapers and magazines with boastful and often misleading advertisements. Speed and dependability were the most sought after virtues in new engines, and the design teams at Kiekhaefer Corporation worked at a feverish pace to prepare the new 25-horsepower *Thunderbolt*, which had been promised as the "Mercury Mystery Motor" even before aluminum restrictions stopped development in 1941.

Herman Steig had barely opened the Corium Farms barn for production when Carl ordered expansion. A 25,000 square foot addition was followed by another 40,000 square foot annex the following year. With the market filling up rapidly with competition, the name of the game quickly became production. Once initial flaws and problems with production in the new facility were smoothed out, Carl gave Steig another bold assignment.

> "One day him and I got into a little spat, and I was real kind of stubborn myself. He could tell that I was very, very disturbed over that spat, although I regulated how much I talked back to him. I knew how far I could go and where I better stop and of course he knew that too, how far to go with me. He suggested that we go down and have a drink at the bar. So we went to the bar and had a drink and he said, 'How much money are you making now,' and I told him, and I said, 'and that's all.' He said, 'Well, I'll double it, how's that make you feel now?' I said, '*Great.*' That's the way he would do it, just off hand, just like that."

But in the next breath, Carl told Steig to pack his bags. Herman was to scour the country, traveling to all the great industrial centers to buy as much machinery as he could find. The government's war asset warehouses were bulging with surplus machine tools, and Carl needed to expand as quickly as possible to keep ahead of landslide competition. Surplus machinery could be bought for less than twenty-five cents on the dollar, so when Carl gave Steig a blank check to spend up to $600,000 on machinery, Steig was able to secure nearly $3 million of prime production hardware. Much of the equipment was brand new, still in the original shipping cases, loaded with accessories and still covered by warranty. "I just went out and, on my own I'd say, 'I'll take these three machines,' Steig recalled, "and they'd write up the paper work and send the bill to the company."[18]

Carl was tooling up to take on the entire industry. In less than a year, he was able to quadruple his production capacity, and was prepared to take on all comers in a race for market share. He patrolled his plants like Patton had rolled through the lowlands of Europe, barking orders and changes with every turn of the aisle. His reputation for toughness, tenacity and engineering intuition were spreading far outside the confines of his factories. The market was ready, his facilities were ready, his products were ready, and Carl was ready to explode into high gear.

CHAPTER TWELVE

Shifting Threats

"Capitalism is a steadily shrinking island of Free Enterprise surrounded by turbulent and destructive seas of Monopoly controlled by economic and political dictators."[1]

Captain Eddie Rickenbacker
In the *Mercury Messenger*, March 1949

Orders for peacetime chain-saw engines poured in from Disston, as lumber shortages hampered a massive housing boom triggered by returning servicemen. Disston and the Kiekhaefer Corporation finalized plans for a new, lightweight one-man chain-saw, along with improvements and additions to the two-man saws developed during the war. Even during the war, the government encouraged Carl and his team to tour timber areas with the new chain-saws, to demonstrate the increased timber felling and bucking efficiency that most loggers had never before witnessed. Equipped with special station wagons and vans emblazoned with the Kiekhaefer name on the side, they toured the great commercial timber operations.

Carl was an eager participant in these forays into the forest, for the operators of the lumber camps were his kind of men: tough, self-assured, rugged entrepreneurs who worked long and hard to carve out a living. He was quick to take the heavy end of the two-man saw to cut down the hardest wood and buck the trickiest logs. His audiences were constantly amazed, not only at the efficiency and power of the new saw, but at the strength, stamina and presence of its operator.

The Kiekhaefer Corporation also participated in public demonstrations of the saw at almost every opportunity. Sport shows were quite popular near the end of the war, because frustrated hunters, fisherman and outdoorsmen could visit the many booths and exhibits to see the new products that would hopefully soon be on the market. In 1945, the Kiekhaefer Corporation staged contests and exhibitions of wood cutting at sport shows in Chicago, Minneapolis and Milwaukee. Quite elaborate demonstrations were staged in front of hundreds of spectators at each performance, and the Kiekhaefer-Disston chain-saw became one of the main attractions at the show. The Mercury newsletter of June 1945 gave an account of the many attempts to beat the new powered saw with professional hand sawing.

> "Leo Wagner of Barss Corners, Nova Scotia and Watson Peck of Bear River, Nova Scotia, were a couple of boys from the wilds of Canada who learned first hand just how fast that powerful Mercury Engine can make that Disston chain cut wood.
>
> "Wagner and Peck are a team of expert log rollers and canoe tilters who have thrilled thousands with their skill and daring on a log, but at the sport shows this year they were also pressed into service as a team of hand sawyers to compete against the Mercury powered saw.
>
> "For fifty-four performances at the three shows these two boys sawed with all their might, but never once did they get close to the power saw. When last seen, they were heading for the woods in Nova Scotia trying to figure out how they could smuggle one of the machines over the border."[2]

Disston wasn't alone in working to fill this sizable market, however, and aggressive competition began emerging soon after peace was declared. One of the new chain-saw competitors was an all too familiar name to Carl, and one that would continue to needle and frustrate him for many years to come.

Robert Paxton McCulloch was five years Carl's junior, and though raised a short 40 miles apart, they might as well have been from different planets. Bob McCulloch was born in Milwaukee in 1911, to a family that might best be described as industrial aristocracy. His grandfather on his mother's side, John I. Beggs, was "a thorny Scot and Milwaukee public-utility Tycoon who left the bulk of his estate to his three grandchildren."[3] McCulloch's father was the president of United Railways Company in St. Louis. As if these weren't sufficient industrial society credentials, his father-in-law was Stephen F. Briggs of Milwaukee's Briggs & Stratton Corporation, among the nation's largest manufacturers of small gasoline engines. In 1929, when Ole Evinrude's Elto outboard firm joined forces with Evinrude Motors and Lockwood-Ash Motor Company, it was the considerable financial might of Stephen Briggs that pulled the formidable competitors together to form the Outboard Motors Company, or OMC. Seven years later, Briggs would purchase Johnson Motor

Company to make OMC the largest producer of outboards in the world.

Both similarities and contrasts abound between Bob McCulloch and Carl Kiekhaefer. One man was born to fabulous wealth, while the other to the farm. One was swept into the heady world of industrial society, while the other plowed the fields. One would attend the finest colleges in America, the other would attend night classes in welding, engineering and vocational arts. One would inherit millions on his twenty-first birthday, while the other was being fired from his $24 a week job at Evinrude in the same city. Both were big, tough men with flashing tempers, intimidating presence, and a love for unbridled competition and speed.

When McCulloch entered Princeton's ivy league campus in 1928, he was not voted most likely to succeed.

> "... [H]e had wide shoulders, a spare six-foot-two-inch frame, sandy hair, [and] a habit of rolling his eyes up when he laughed so that the whites gleamed. He had a passion for riding behind, or in front of, the fastest motors he could build or soup up. Rich, personable, and a speed nut, he looked like a bad scholastic risk."[4]

In fact, before leaving for Princeton, McCulloch had already locked horns with Carl Kiekhaefer. Both men had been mechanics working to prepare stock cars to compete in a race at a local Milwaukee track. Carl had equipped his team's entry with his own crude supercharger design, while the McCulloch team, with virtually unlimited resources, was able to purchase and install the best equipment money could buy. Carl's team lost, and heated words were exchanged something to the effect of "Damned Milwaukee blue blood,"[5] and the stage was set for a frustrating and expensive rivalry that would cross industry and product lines for many decades to follow.

Purportedly bored with the sheltered lifestyle of the East, Bob McCulloch left Princeton after two years, moved to California and enrolled as an engineering student at Stanford University. He didn't travel lightly, though, and graduated with a checkered flag along with his diploma. But then, what else could be expected of a man whose very initials were R.P.M.

> "Part of the duffle he took with him to Palo Alto was a trailer loaded with outboard motors, hulls and a competent

> mechanic. While McCulloch sat in class, the mechanic tuned up McCulloch's motors; on Friday afternoons the pair were off to the outboard races. The 1931 payoff was an engineering degree, national championships in class C and class D outboard hydroplanes, and a serious interest in motors."[6]

Only 20 years old, McCulloch returned to Milwaukee and established a machine shop, where he built special gasoline engines for midget racers, culminating in a twin-engine, four-wheel drive midget that was so volatile and so much faster than competition that it was ruled off the tracks by officials citing any rule violations they could find. By 1942, a year in which the Kiekhaefer Corporation managed to lose $50,000 for their efforts, McCulloch was manufacturing superchargers in a modern Milwaukee factory, with sales of over $3 million. The fact that only General Motors was producing more superchargers made McCulloch's company look very attractive to the giant Borg-Warner automotive products concern, who promptly purchased the company from the bored and restless owner for $1 million. McCulloch invested the proceeds from the sale of the company into blossoming Pan American Airways, of which at one time he was considered to be the second largest stockholder.

But McCulloch was back in business within six months, this time as McCulloch Aviation, Inc. Managing to establish his company in time to secure generous war-time government contracts, Bob McCulloch began to compete directly with Carl Kiekhaefer again, this time for two-cycle engines for military applications, such as lightweight, general purpose generator motors to power portable radio equipment. At war's end, McCulloch began building target drone aircraft engines, ranging from eight-horsepower and 22 pounds, to a 71-horsepower, 77 pound model for the Army Air Force, again in direct competition with Carl.

McCulloch moved his operations to Los Angeles in 1945, buying a fifteen acre parcel of premium land across from Los Angeles International Airport for $160,000 cash. Starting with temporary barrack buildings for factory space, he was in production within 60 days from taking possession. Again drawing from the well of his own vast personal wealth, a new 80,000 square foot, $1 million factory was made ready by the summer of 1946, while Carl Kiekhaefer struggled to prepare his $25,000 dairy barn in Fond du Lac, Wisconsin. Carl Kiekhaefer bought the land for his Fond du Lac operation for $59 an acre, while McCulloch paid nearly $1,100 an acre for his. Bob McCulloch built a magnificent glass and brick, sawtooth roofed monument to modern production, resplendent with wide green manicured lawns and smoothly paved parking lots, located within the most prestigious manufacturing community in the world. At the same time, Carl was baling hay in his new barn and Herman Steig was shoveling manure from the troughs of the 30 year old, abandoned and obsolete wooden temple to milk and butterfat production.

When Carl Kiekhaefer finally secured volume contracts to produce engines for the Disston Saw Company in 1942, his relationship with rival saw manufacturer Reed-Prentice Corporation of Worcester, Massachusetts was largely severed. It came as little surprise to Carl, still working on completing wartime chain-saw engine contracts in 1946, when Reed-Prentice turned to Bob McCulloch to supply an engine for their own version of the saw. McCulloch began volume production of an engine in March 1946, which Reed-Prentice packaged with their saw mechanism and distributed through the Sears Roebuck network, selling tens of thousands of saws to rural farming communities. But a large number of saws were returned to Sears for repair, and buyer disaffection began to jeopardize future production. McCulloch was particularly frustrated that he had no direct feedback with saw users, and was unable to reconfigure the whole product for better operation. As an interview with Bob McCulloch once revealed, even more frustrating was the loss of higher direct sales profit.

> "This bothered Bob McCulloch. More to the point, maybe, he was canny enough to know that while chain-saw motors carried a low profit, there was a wide profit margin in the complete job. He wondered if there was any good reason why he should not make chain-saws. ...
>
> "There were at least two good reasons. The chain-saw ... had been carefully nursed by such U.S. tool-makers as Disston and Mall. Cranky, heavy and expensive, it had failed to cut a great swath in the Canadian woods, and U.S. organized labor had fought its use. True, the combination labor and lumber shortage of World War II had made it a part of U.S. logging, and the tool had made good money for people like Disston. But to enter the chain-saw business was to compete in a market that was tightly sewed up."[7]

McCulloch and his design team quietly developed their own ultra-lightweight chain-saw using multisection magnesium die castings, and in 1948 terminated their contract with Reed-Prentice to begin production of the new design. Bad news for Carl Kiekhaefer, the new saw, known as the 5-49, developed five-horsepower, weighed only 49 pounds and cost $385. At less than two-thirds the weight of the Kiekhaefer-Disston saw, costing $50 less, and capable of more types of work, "it swept the north woods like a flash fire."[8]

The McCulloch saw ripped into sales of the Kiekhaefer-Disston version so quickly that both engine and saw maker were stunned. McCulloch had fortuitously acquired intact distribution from another manufacturer that had built up a sizeable marketplace, but couldn't deliver product. In 1948, as McCulloch sales more than doubled to $4 million, industry estimates showed that over half of all timber was still being sawed by hand, clear-cut evidence of a vast market for the new saw.

In May 1948, the Kiekhaefer Corporation was threatened with a walk-out by its union employees, joining a flurry of demands for higher industrial wages sweeping the nation. In a speech delivered to the work force to convince them to delay any wage increases, management revealed the depth of the wound inflicted by McCulloch.

> "You men who were here last year at this time will recall that we were building and shipping about 1,000 chain-saws and about 2,000 10 H.P. outboards a month. The dollar value of a chain-saw is the equivalent of about two outboards so that in dollar value it would have equaled about 4,000 outboards a month. ...
>
> "As a result, the company has lost $374,521. Repeat - the last four months has practically wiped out the surplus built up by the company since it started. It now requires all the ingenuity of the company's executives to meet present payrolls. To make a general increase now might cripple us financially to a point where it could result in a loss of all jobs here.[9]

Appeased by Carl's plea for patience, the union accepted a tiny increase for certain categories of work, ranging from two to six cents an hour. In the same month, by contrast, a 17 day strike against Chrysler ended when workers accepted an across the board 13 cents an hour raise, and workers at General Motors received an 11 cents an hour blanket raise to avoid a strike. The big difference, however, is that the auto workers received $1.63 per hour, while the Kiekhaefer Corporation increase meant that rank and file worker wages were boosted to 89 cents per hour. The disparity is largely attributable to the differences in the cost of living in the rural settings of Fond du Lac and Cedarburg, versus Detroit, one of the nation's largest and most industrialized urban centers.

Though chain-saw sales were declining, the spectacular success of the 10-horsepower Mercury *Lightning* continued to propel the Kiekhaefer Corporation into high gear. During 1948, Carl was finally able to build the prototype of the engine he had been promising since before World War II, the *Thunderbolt*. Once known as the *Mercury Mystery Motor*, the *Thunderbolt* underwent major design changes once the *Lightning* was proven in the marketplace. Because the *Thunderbolt* had been on and off the drawing boards for so long, the engine that ultimately emerged from the engineering laboratory was the product of the widest design swings in Kiekhaefer history. Eventually, Carl decided to base the new engine almost totally on the *Lightning* design. Basically, the new *Thunderbolt* prototype became two *Lightning* power heads welded together and joined by an elongated crankshaft. The world's first in-line four-cylinder outboard engine delivered well in excess of the advertised 25-horsepower, true to Carl's penchant for underrating his motors. Carl would later admit that the *Thunderbolt* actually developed a whopping 40-horsepower, making it the "first line of outboards to produce one horsepower per cubic inch."[10] With characteristic aplomb, Carl ventured a number of comparisons about the long-awaited offspring from his lab.

> "As quiet as a 3-1/2, as easy to control as a 5, as easy to crank as a ten, weighing about that of a 16, having the performance more than a 33, and the feel and response of the finest V8 Automotive engine or inboard engine - that's the Mercury 25 with the Magnificent Thunderbolt Engine - Not only the first powerful outboard in more than a decade, but the greatest news and the finest contribution to outboards in 40 years."[11]

Using the same pistons, connecting rods and reed valves from the *Lightning* produced an engine with exactly twice the displacement at 39.6 cubic inches, and weighing in at almost twice the *Lightning* at 115 pounds. With a unique squeeze-grip throttle, the bright green power head was notable for its lack of a rewind starter. It would have weighed considerably more, and been on the market many months sooner except for an announcement by Johnson Motors in Waukegan that was destined to shake up the entire marine industry.

In 1948, Johnson leaked word of the pending production of a revolutionary new outboard engine, the *QD*. The 10-horsepower creation had two new features that nobody in the business could match, and which would quickly become new standards for the entire industry: remote fuel tank and gear shift control.

The remote fuel tank for the Johnson *QD* held five and a half gallons of fuel and two-cycle engine oil, and was connected to the engine with a twelve foot flexible fuel hose. At first blush the concept of a remote fuel tank might not seem like such a remarkable idea, but the benefits to the outboard motorist were many. First, the tank allowed the weight of the gasoline to be carried separately from the engine, reducing the weight of engines by twelve to sixteen pounds. Secondly, because most outboard motor tanks would hold only two or two and a half gallons of fuel, they had to be filled up often, and worse, had to be filled while standing up, leaning out over the engine, off balance, juggling funnels and dirty fuel cans. More often than not, there was an additional can of gas in the boat anyway, so the addition of a remote fuel tank didn't take up any additional space. The remote tank had a larger capacity than integral outboard tanks, thus increasing the range of an engine considerably. If the remote tank needed to be filled, it could be done comfortably, sitting down in the boat, or by passing the empty can over to a dock side fuel station. The twelve foot fuel hose allowed operators of lightweight, small skiffs or rowboats to place the fuel tank far forward in the boat for better balance. And then, perhaps most obviously, the external tank had a rather accurate fuel gauge, something not generally available for the integral outboard fuel tank. Johnson was able to capitalize even on the increased weight of the fuel which was contained in the remote tank, by pointing out that it was easier to carry the engine in one hand and the fuel in the other, than it was to carry both with one hand. They called it the *Two-Hand-Carry*. The connector on the end of the fuel line had three holes in it. It would "plug-in" like a light cord to three prongs on the motor. One was a guide pin, one the fuel line, and the other was the air-pressure line. It was a perfect combination of engineering and marketing which has remained virtually unchanged in modern outboards of this size.

The other revolutionary feature on the Johnson *QD* was a lever that allowed the operator to select neutral, forward and reverse. Again, it would at first glance seem to be an elementary addition to outboards, but there were many obstacles to overcome before it could be considered practical. A speed limitation was designed for neutral operations, so that the operators wouldn't over speed the engine without a load on the propeller. Similarly, in reverse, a speed governor was built in so that operators couldn't move astern fast enough to plow water over the transom and into the boat. In normal forward operation, exhaust gases are discharged in back of the propeller, but in reverse this area becomes the front of the propeller, and the exhaust emissions would cause cavitation of the propeller in the bubbly water. Johnson engineers designed an automatic exhaust gas diversion that would re-route exhaust through another outlet in reverse to eliminate the problem. Lastly, an outboard engine in reverse has a tendency to rise up out of the water as the propeller pulls the boat backwards. An automatic lock was designed to engage when the shift lever was in the reverse position. Without these added features, operators would have had to exercise extreme caution when using the engine in anything but normal, forward motion.

From the perspective of the boat operator, the ability to shift from neutral to forward and reverse made an immediate impression. Before the gear shift, the driver had to untie his boat from the dock and aim it at open water before he tried to start the engine, because once it started, he was off and running. If the engine didn't start, then he was stuck drifting in over his head, tantalizingly close to the dock where he would most like to be. If the engine started right away, he was forced to warm the engine underway, and performance with some engines changed appreciably from cold to warm. Leaving the dock without a reverse gear was also a challenge in crowded waterways and marinas, even if the operator was fortunate enough to have an engine that would swivel completely around to give a reversing motion. Unfortunately, each time the engine was spun around from forward to reverse and back again, the propeller would sweep in a circle at the stern, making the boat zigzag back and forth, the bow to slice from side to side, and possibly bang into anything nearby. The pleasures of starting and warming the engine up securely at dock side, and slowly backing out of a slip or away from a finger pier graciously and without pandemonium, made the new gear shift an instant hit on the waterways.

The decision to incorporate a remote fuel tank and gear shift into the Johnson *QD* was not an easy one, and a clash born of viewing the outboard industry from differing perspectives almost doomed the project before it emerged from the drawing boards. Clay Conover, the engineer who was most responsible for implementing the new innovations, had been working at Johnson Motors in Waukegan since 1933. His father, Warren Conover, was one of the original *Four Men From Terre Haute,*[12] Indiana that had formed the Johnson Brothers Engineering Corporation with the Johnson brothers, Louis, Harry and Clarence in 1918. The four started to sell Johnson Outboard Motors in 1922, and by 1928 they would advertise as "The World's Largest Manufacturers of Outboard Boat Motors", and claim that "over half the outboard motors sold are Johnsons."[13] Warren Conover was forced to retire in the fall of 1935, a prelude to the eruption of a storm of personalities and philosophies as Stephen Briggs' Outboard Motors Company, already including Evinrude, ELTO and Lockwood-Ash outboard motors, acquired Johnson Motors Company the following year. Clay Conover, however, remained to forge new engineering and manufacturing strategies for Johnson, and ultimately both Johnson and Evinrude, for 40 years until his retirement in 1973.

Clay had lobbied for ten long years for the remote fuel tank and nearly as long for the gear shift at Johnson, patiently waiting for the right combination of product and political climate to make his move. "I tried to get rid of the gasoline tank on the motor in about 1939," Conover reflected, "and it took me ten years to finally get it. I made the first reverse gear during the war. Clarence Johnson had charge of our experimental machine shop and I said something to Clarence about it. I said 'What do you think about a reverse gear?' He said, 'That would be great.'" With Johnson's help, Clay designed a reverse gear for the Johnson five-horsepower motor, using the tried and true shuttle type clutch mechanism used in washing machine wringers for decades. "We didn't have any fancy lock to hold it down, so we just used a cotton twine to hold the engine down," Conover says, "so if you hit anything the twine would break. But it worked so beautifully that I just couldn't picture anybody being without it."[14] Unfortunately, the large and conservative Johnson organization introduced instead, a five-horsepower engine that offered neutral and forward only, so at least boaters could warm their engines up at the dock.

"I had to fight an up-hill battle with the management," Conover recalled, "because Steve Briggs [then OMC chairman] was in favor of a separate fuel tank and Ralph Evinrude [then OMC president] wasn't. Evinrude was in favor of the reverse gear, but Steve Briggs wasn't, and they both lived in Florida all the time."[15] When the innovative Johnson *QD* was almost ready for production, Ralph Evinrude nearly put an end to the separate fuel tank once and for all. "Ralph said, 'Well, everything is fine, but we can't put it out with a separate fuel tank on it now,'" Conover says. "The engineering was practically completed ... all the research, the styling, everything was done. And he said 'We can't do it.' So old man Rayniak [Joseph G. Rayniak, credited with the design of the first Johnson outboards] was there," Conover explains, "he was executive vice president or whatever it was, and he said, 'Well, maybe we ought to call the old man [Chairman Stephen Briggs] and see what he has to say.'" In Clay Conover's office, Rayniak and Evinrude managed to raise Briggs in Naples, Florida. Conover could only hear Ralph Evinrude's end of the conversation, which left a desperately heavy air of suspense hanging in the room.

> "Steve, we just decided that we can't make the separate fuel tank, it's too risky, chance of failure.
> "Yes, Steve.
> "Yes, Steve.
> "Yes, Steve.
> "O.K., Steve. I understand, Steve. Bye."[16]

Much to Conover's surprise, when Ralph Evinrude put the phone down, he turned and said, "Well, I guess we got the separate fuel tank." "Believe it or not," Conover says, "that's what happened. That's how close it came to being killed." Evinrude was still reluctant, and told Conover to leave the old tank on top, and rig it up so the external tank would pump fuel up to the engine tank, so that if something happened, the operator would have a backup. When Stephen Briggs heard about Evinrude's scheme he said, "Ralph, when are you going to quit putting the crank under the seat?" referring to the obsolete practice of carrying a hand starter crank under the car seat as an emergency backup. "Build them so they work," Briggs said, "you don't build them so you have to depend on a back-up."

But the cautious Evinrude wouldn't give up. To ensure hope of emergency operation, a small thumb actuated pump was located on Johnson's new remote gas tank, which needed to be operated for a few strokes before the engine was started. This pressurized the tank, forcing fuel through the line to the carburetor. Once the engine started, the engine would supply the pressure to keep the fuel flowing. The first *QD*'s weren't infallible, however, and a story is told about one of the early units used on an exploratory expedition in Africa. The engine failed to pressurize the external tank, so a young native boy was engaged to manually stroke the little pump on the tank. For thirty days, the story goes, he kept his thumbs pumping to keep the Johnson moving down the river, dodging hostile tribesmen, hippopotamus and crocodiles. Ralph Evinrude felt vindicated.

The Johnson 10-horsepower *QD* with remote fuel tank and gear shift became so successful, that it would take four hectic years of nearly round the clock production to bring supply in balance with demand. Just as the *Lightning* established Carl Kiekhaefer as a force to be reckoned with in the industry, so Clay Conover's *QD* innovations confirmed the reputations of Johnson and OMC as leaders in outboard motor design and production. So certain was OMC of the QD's success, that over $500,000 was spent on new tooling and dies alone for the new engine.

Clay Conover was also at the beginning of two other epic events in outboard history. When the Johnson brothers built their first batch of prototype outboard engines in 1921, they had fashioned a new kind of starting system to replace the "knuckle buster" system that Ole Evinrude had used since 1909. The "knuckle buster" was a retractable knob on the exposed flywheel of Evinrude's engines, that would be grabbed and spun around in a circle to get the engine started. Often, the engine would catch quite easily, and before the knob would retract and before the operator could clear his hand, the knob would swing around and bash into his knuckles, hence the unfortunate moniker. The Johnson brothers, hoping to improve on this awkward starting maneuver, built their first six engines with a leather strap starter that also had some drawbacks. It used a leather strip of about an inch wide by several feet long, which fit into a flanged drum on top of the flywheel. At one end of the strap was a silver clip that fit into a recess in the flywheel drum. The other end of the strap was brass riveted over a wooden handle. To start the engine, the strap was wrapped around the drum and then the operator pulled the handle to spin the engine. At least once when the Johnsons and Conovers were testing their new engines, the starting strap went over the side into the lake. One of the more ample members of the party volunteered his trouser belt for an emergency strap, but the buckle wouldn't fit in the flywheel groove. They were stranded.

Then a milestone conversation occurred, which Clay Conover remembers vividly. "I remember they said, 'You know, you've got to have something so in case you lose it or break it you can just wrap something around it. It should be something like clothesline,' they said. I was there, and I heard it."[17] From this frustrated group of outboarders, at least one of which was holding his pants up with one hand, came the inspiration that would adorn outboards for generations to come: the rope starter. With a knot in one end and a wooden handle on the other, it could be easily replaced with just about any kind of rope, sash cord, twine, shirt strips, or even fishing line in an emergency. When the first production Johnson outboard motor was introduced in 1922, the new rope starter was standard equipment. Clay Conover, aged eleven in early 1922, driving a lightweight boat powered by the new engine at a speed of fourteen miles per hour, has the distinction of being the first person to ever *plane* a boat using an outboard motor.

Rope starters, still used today on some small outboard motors, slowly gave way to the automatic rewind starter. On the occasion of Ole Evinrude's twenty-fifth anniversary of producing outboards, a new "Simplex" starter was offered as an option on the 1934 Evinrude and ELTO *Lightwin* and *Lightfour Imperial* engines. For an additional $15, the new automatically retracting starter was supposedly designed to make sure the operator never lost his starting cord. Evinrude and ELTO created a minor sensation with the introduction of the simplex starter, proclaiming the new *Imperials* "The Outboard Motor Of The Future," and Ole Evinrude would boast, "We believe no one will glimpse the new Imperial models without feeling a thrill of conviction that outboard motors have taken their greatest step forward."[18] However, there was a difference between how the device was supposed to work and the actual effect on the operator. The brochure said:

"Here is hand starting *made easy*. The finger grip fits your hand comfortably. As you draw it out, the attached pliant cord of airplane cable causes ratchets to engage and spin the completely enclosed flywheel to start the motor. Let go, and the cord slides back into the starter dome -- *rewinds itself.*

"Forgetting or losing the starter rope is now impossible -- *it's an integral part of the Simplex Starter*, securely fastened to your motor. Nor can the starting rope swing free, for the cable is fastened to the rewinding disc."[19]

The brochure introducing the new device shows a man with a white shirt and tie, casually smoking a pipe after using the new starter. But what often occurred is the novice boater would first pull the greasy cable across his favorite shirt, leaving a series of permanent black lines. Once thoroughly frustrated, he might then pull just a little bit too far in panic to start the engine in which case one of two equally horrible things would occur. The handle would come off and the bewildered boater watched in horror as the cable whipped back and forth as the engine, like a child sucking a spaghetti noodle, consumed the cord, never to be seen again. Failing this, the cord might become unattached from the other end, surprising the overextended operator anxious to prove Newton's laws of motion, who continues flying backwards until, if lucky, he lands in the other end of the boat. Either way, it's time to get out the oars because there was no quick and easy way for the average outboarder to fix the fickle mechanism and get underway again.

Again it was Clay Conover to the rescue. Conover, who was chief engineer at Johnson before he turned 25, was working on a new engine design when the question arose over whether to install a recoil starter or leave it as a rope starter. "Old man Briggs [Stephen F. Briggs, Chairman of OMC] was all for the rewind starter," Conover recalled, "but everybody else was opposed to him." A heated discussion ensued at a meeting that included some of the most powerful men in outboard history. Along with Ralph Evinrude and S.F. Briggs was Joseph G. Rayniak, who had helped to design Johnson's first outboard; Johnson's brilliant engineer Finn T. Irgens; Briggs & Stratton alumni Jacob Stern; and OMC's marketing chief Pat Tanner. Though overawed to be included in the high level strategy meeting, Conover ended up saving the day.

"The argument was, 'Well, Evinrude and Johnson both had [rewind] starters and they didn't make any success out of them.' It didn't work, and they said it was too expensive. It cost fifteen dollars and we sold them as accessories. And Briggs says, 'I'm not talking about a fifteen dollar starter, I'm talking about a five dollar starter, and I'm talking about one that works, not one that runs and doesn't work.' And he was pretty pointed about that. He said, 'You've got to have rewind starters on.' He turned to me and he said, 'What do you think, Clay?' And, of course, I felt like I was sitting among giants, and somewhat reluctant to talk. I said, 'Well, it happens I have one on the drawing board in the basement at home right now.' He said, 'Do you think you can make it?' I said, 'Well, it looks like it.' And, so he said, 'How is it made different?' I said, 'It's all stampings.' And, so he said, 'Well, bring it in. Let's see what we can do with it.'"[20]

Clay Conover's solution worked. He discarded the old, greasy cable and replaced it with a waxed cotton sash cord with a bronze inner core for strength. With fewer moving parts and a simpler manufacturing process, the new recoil starter was installed as standard equipment. To test the new starter, Clay had the huskiest man in the Johnson plant pull and yank and crank the prototype until he could hardly lift his arm to eat lunch. Even when the design was proven, the many detractors from the meeting ganged up on Briggs again, telling him, "You're making a mistake here, it's wrong. We've thought about it and we should make it as an accessory." Briggs exploded. "No accessory! You've never made a success out of an accessory. Standard equipment or nothing!"[21]

Other accouterments to outboarding, though seeming like obvious solutions to common problems, failed because of faulty technology. Electric starting was offered as early as 1930 on certain models of ELTO and Evinrude engines, but was unpredictable, heavy, expensive and ultimately unpopular. Remote controls for throttle had been available since the twenties, and many companies offered steering cable systems that would adapt to most popular outboards, using either hand or wheel steering.

Both Stephen F. Briggs and Ralph Evinrude are credited for promoting the use of advanced features and accessories, even though Evinrude originally opposed the remote fuel tank. Briggs wasn't initially in favor of outboard sophistication, and only reluctantly became a born again progressive when it came to anything but the basics. One day Clay Conover challenged Briggs to a showdown demonstration to make his point.

> "I said, 'Mr. Briggs, if I sent you a [specially prepared] cruiser, would you try it out and use it?' He said, 'You know me, Clay, I'll try anything.' So I fixed it up. I put *everything* on this boat that he didn't want. He didn't even want remote controls. He didn't want electric starting. He didn't want a reverse gear in the beginning, but he was converted over. We even took automotive tanks and mounted them in the boat. We put in remote controls. We didn't even make them, we had to buy them from other people.
>
> "I sent it down to him and waited. He came back and said, 'That's better than my Hacker!' He had a special built Hacker Craft with two Chrysler Royal [engines] in it. He said, 'I came back from fishing all day in it and I don't have to grind the valves.' We never heard anything more about remote controls or *anything*. He never said don't build them, he said, 'Just go ahead and start doing it.'"[22]

Like Carl Kiekhaefer and Bob McCulloch, Ralph Sydney Evinrude was born in Milwaukee. The only child of Ole and Bess Evinrude, Ralph was born in 1907, only a few months after his father had tested the first prototype of his new outboard motor. Though other outboard motors had been successfully built and marketed, the simple, practical single-cylinder engine designed by Ole Evinrude quickly captured the marketplace, setting design standards for lightness, power and performance. With the considerable promotional skills of his wife, Bess, the 1.5-horsepower Evinrude detachable rowboat engine had become the industry standard by 1913. That year, Bess took ill and Ole decided to sell the company to his partner Chris Meyer. Among the conditions of the sale were that Ole would stay out of the outboard business for at least five years. Bess recovered her health, and by 1921 Ole had designed a far superior outboard which he called ELTO for Evinrude Light Twin Outboard. Out of loyalty to his old partner, Ole offered the new engine to Chris Meyer at Evinrude Motors. Meyer turned down the engine, and Ole and Bess started production on their own.

Among the most popular features of the ELTO was the unusual lightness. It weighed only 47 pounds, while an Evinrude with similar power weighed 74 pounds. By 1924, after only three short years on the market, the new ELTO was outselling the original Evinrude. When son Ralph was 20 years old in 1927, he started to work with his dad at ELTO, first in the experimental engine department. Ralph's interest in outboards crowded any scholastic ambitions, and after two years at the University of Wisconsin, he left school to work alongside his now famous and affluent parents.

Carl Kiekhaefer, working at Stearns Magnetic in Milwaukee at the time, may have absently read the headlines announcing that the original Evinrude Motors Company was purchased by the cross town small engine manufacturer Briggs and Stratton in 1928. Stephen F. Briggs and Harry Stratton had made quite a name for themselves as pioneers in the manufacture of portable four-cycle gasoline engines. But within a year, the Briggs and Stratton board of directors demurred, convinced that outboard motors had no real future, and recommended that the Evinrude company be sold off. Stephen Briggs protested, and finally decided to invest his own money to acquire Evinrude.

Briggs had always been interested in two-cycle engines, ever since he had built his first one in 1906 while still in college. In order to accomplish the financially and politically ticklish transaction, he formed a syndicate with Ole Evinrude to form Outboard Motors Corporation, merging ELTO, Evinrude and another competitor, the Lockwood-Ash Motor Company. Briggs wanted Lockwood-Ash mostly to acquire the brilliant engine designer, Finn T. Irgens, their chief engineer. Stephen Briggs became chairman of the board and Ole Evinrude became president of the new company. OMC grew quickly and profitably during the next five years, and when Ole died in 1934, 27 year old Ralph Evinrude became president.

Meanwhile, the Great Depression had drastically cut into sales of rival Johnson Motors, located only an hour's drive south of Milwaukee at Waukegan, Illinois, on the outskirts of metropolitan Chicago. A group of industrial financiers, Hayden, Stone & Company, had taken control of Johnson as it began to slide into receivership, victim of the horror the depression brought even to the household names in American industry. Briggs was determined to add the muscle of Johnson to growing OMC. He acquired a letter of introduction to Charles Hayden, the head of the investment house that controlled Johnson. But even before Briggs could see Hayden, it was almost too late.

> "Hayden was ill, living at the St. Regis Hotel, New York, when Briggs called on him and stated his interest in Johnson.

"I've already sold it to Stewart-Warner," said Hayden. "It's sold and it's not sold. I haven't accepted their offer. They want to give me treasury stock which cannot be traded. I've given them until tomorrow morning to get an answer from their board to give me trading stock. If they don't, I'll sell it to you.

"Would you give me $10.35 a share?"

"Briggs said he would. The next day Briggs returned and Hayden said: "It's yours. Stewart-Warner didn't come through."[23]

Stephen Briggs and Ralph Evinrude merged Johnson Motors with OMC in 1936, renaming the company the Outboard, Marine and Manufacturing Company. (Twenty years later, in 1956, the name would be shortened to the current usage, Outboard Marine Corporation.) With the combined strength of Johnson, Evinrude, ELTO and Lockwood-Ash, OMC became the world's largest producer of outboard motors. It also became the monolithic giant that would plague Carl Kiekhaefer in the years to follow, and fuel his paranoid feelings of industrial spying, sabotage and unfair competition. Carl would isolate Stephen Briggs as his competitive nemesis, using every opportunity to sling mud, denigrate and weaken his most hated rival for supremacy in the marine industry.

CHAPTER THIRTEEN

The Race Is On

"The 'factory factor' again got into the picture and it's been our observation that where commercialism enters, sportsmanship flies out the window and the sport degenerates into an exercise between two or more engineering groups with the obvious subsidies of boats, engines and factory accessories available for purchase to no one -- with expendable professional drivers flying in, to become the driver hero, to win out over the remaining patsies who race for sport."[1]

Carl Kiekhaefer

After Carl Kiekhaefer had tried desperately to buy back a freshly released 10-horsepower Mercury *Lightning* from an ambitious new owner preparing to enter the 1947 of Albany to New York marathon race. The owner and engine won their division handily, putting the outboard industry on notice that Mercury was off to the races. Within a year, Kiekhaefer Mercury engines had won a handful of distinguished racing events, and Carl began to see that a new and enormous opportunity was blossoming to publicize engineering achievement.

By the following year, Carl had completely reversed himself, actively promoting Mercury engines as high performance, and working hard to create new venues for stock utility class racing. He successfully lobbied the American Power Boat Association to take official recognition of stock outboard racing, and was pleased to see these new events attracting large crowds and expanded media exposure. During the summer of 1949, Carl reviewed copies of Champion Spark Plug Company's *Spirit of the Champion* magazine, which carried articles on national sporting events. He was disappointed to read that the majority of stories were concerned with events that he felt were far too exclusive. In a hard selling letter to one of the executives at Champion, Carl took the opportunity to establish his racing philosophy, take another shot at Stephen Briggs of OMC, and set the record straight about Mercury's recent victories on the water.

"Your publication carrying articles on National Sporting Events in which your products are used is certainly interesting.

"In glancing through the various copies of the publication, we have noticed that the sporting events attracting large crowds of spectators generally get the most publicity. ... The number of outboard owners engaged in stock utility racing is estimated to be between 5,000 and 10,000. ... Surely you cannot afford to overlook the importance of a group of this size in your over-all picture, because few sports can boast this number of competitors.

"We are proud of the record that our Mercury Outboard Motors have made in this field of stock utility outboard racing. In the first post-war Albany to New York Outboard Marathon, just one lone 10 H.P. Mercury was entered and fought its way down the storm tossed Hudson River to not only win its class, but to be the only engine in that class to finish, a class including 18 H.P. motors. The following year, the percentage of Mercury entrants had increased tremendously and our Mercury engines swept the first 5 places in each of the two classes which they were eligible to enter. ...

"In 1949 the American Power Boat Association took official recognition of the fast-growing sport of stock utility racing and under their jurisdiction the first annual stock utility national championship events were held at Lake Alfred, Florida on September 27. Of the 6 events held, Mercury scored a clean sweep of all places in 5 of these 6 contests for which they are eligible. ...

"As manufacturers of Mercury Outboard Motors and industrial engines, Wizard outboards and Disston Chain Saw engines, we have consistently used and recommended Champion Spark Plugs. ... Since 1939 we have purchased plugs from your company in the amount of $104,000. ... We have often wondered just what might be required before we could crash into that select inner circle, that seems to encircle the Briggs Combine.

"It is true that we do not build spectacular Indianapolis Speedway race engines nor do our products power the exciting Gold Cup boats of the millionaires. We do, however, build a definitely superior outboard motor that brings the thrills of racing to thousands of boys and girls who might otherwise be playing golf or tennis on a Sunday afternoon, and it is in the interest of this class that I make this special plea. ...

"Surely this new class of race drivers are worthy of more encouragement than they have been receiving, and I do feel that it would be good business on your part to give it to them. After all, over 150,000 people watch stock utility racing <u>every</u>

Sunday of the season. There is only one Indianapolis race yearly, and only four regattas for Gold Cup boats."[2]

One of the hard lessons that Carl would learn is that spark plug manufacturers, oil refiners and others that publicize the use of their products in advertising almost exclusively touted the achievements of only their largest clients. In his plea for a place in the sun for the humble stock outboard drivers, Carl overlooked the fact that in many cases the spark plug manufacturers and oil companies were themselves sponsoring expensive and glamorous entries in major events as the Indianapolis 500 or Gold Cup races. Carl would exert almost constant pressure on spark plug and oil producers for decades to come, extolling the limited celebrity of outboarding.

Stock outboard racing steadily rose in popularity, but it became clear that the stakes weren't the honor or ability of the drivers, or even between types of boats. The races were between outboard brands, and with few exceptions the real battle was between Mercury and OMC products. Organizers of the races were careful to isolate different sizes of engines into separate classes and racing heats, but when it came to announcing the results, both OMC and Mercury were thoroughly guilty of disguising the truth to the very best of their abilities.

The 1948 Albany to New York Outboard Marathon was the first opportunity for the two competitors to mutually obfuscate the racing results, each for the glory of their own organization, establishing a pattern and promotional strategy that has endured more or less uninterrupted to this day. On the morning of June 13, 1948, the sixteenth running of the Albany-New York marathon began in a miserable, raw and soaking downpour. The rain and fog was so dreadful that the record number of 180 drivers could scarcely see the 300 feet across the Hudson River at the Albany Yacht Club where the race began. The finish line was a long 134 miles down river at the U.S. Navy float at 72nd Street in New York City, with the big Coast Guard Tug *Tamaroa* anchored offshore to mark the end of the race.

At the 9:00 a.m. gun, six different classes were off and running. Class I was limited to motors of no more than 12.5 cubic inches, which included an even dozen of the Mercury 7.5-horsepower 11 cubic inch *Rocket* outboards, three 7.9-horsepower 12.41 cubic inch Champion *Lite Twin* outboards, four 7.5-horsepower 11 cubic inch Martin *"60"* engines, and a lonely 7.9-horsepower 11 cubic inch Firestone. The Mercury engines, in line with Carl Kiekhaefer's strategy for under rating his engines, were all pulling somewhere around 14-horsepower, while the other engines were actually developing close to their advertised power. The Mercury entries, led by 16 year old Indiana schoolboy Leon Wilton, took the first five places in the class, followed by Champion, Martin and Firestone. It was the smallest engine class in the race, so naturally they took the longest to get to New York City, at around seven hours for first place at just over 19 miles per hour, and over eight hours for fifth place in the group.[3] But the time for young Wilton and his tiny engine was good enough to beat the few token entries of competitive engines in his class.

But there were five other classes racing that day, three of which were restricted to engine sizes that effectively blocked the smaller Mercury entries. In these classes, over a hundred Evinrude and Johnson engines raced against each other for class honors with their higher displacement outboards. Naturally, since only Evinrudes could compete in Class VI, restricted to the 50 cubic inch, 33.4-horsepower *Speedifour*, not too surprisingly an Evinrude won. Of course they won every other place in the class as well. The same thing occurred in the two other high displacement classes, where only OMC entries were permitted. Only in class III, which was populated with nearly fifty of Carl's new *Lightnings* were a handful of token OMC engines represented. The first five places in the division were won by Mercury at a speed of 27.9 mph, with 15 year old Jon Culver holding the tiller arm, the son of Mercury distributor and two-time Jafco Trophy winning outboard racer Merlyn Culver.

Mercury pulled out all the stops to promote the results of this key marathon, making heroes out of the two boys who won their respective classes. In enormous posters produced by Mercury, both the engines and the lads were proclaimed "spectacular", as "Mercury Scores Sweeping Victories in Albany-New York Marathon." Neglecting to mention that his brazen headlines concerned only the small class containing his products, Carl Kiekhaefer announced to his swelling list of dealers that "Lightning Cops First Five Places," and "12 out of 13 Best in Elapsed Time."[4] Word soon spread that Mercury had "swept" the races, and OMC was not pleased.

OMC countered on June 24, 1948 by sending a tersely worded letter to every Evinrude and Elto dealer in America that started out with the words, "To be sure you get the information straight --."[5]

> "Here are the results of the Albany-New York Outboard Marathon Race. ...
>
> "Evinrudes won the first 30 places; the first 7 were Big Fours. ... An Evinrude Speedifour won eighth place. Nine Big Fours followed, then came another Speedifour in 18th place. ...
>
> "And so you'll have complete, irrefutable information, here's the data on the first 71 finishers. Out of a total of 141 (78.4%) who finished within the time limit, at least 47 (49.6%) were known to be Evinrudes.
>
> "We believe we have reason to be proud of the Evinrude showing in this important 'season's opener'".[6]

Both organizations claimed overwhelming victories because both manufacturers, with insignificant exceptions, only raced against their own motors. When opportunities would arise for OMC to enter classes dominated by Mercury entries, they would decline, and Mercury was prohibited by class displacement size from racing against the larger OMC products. Both would continue to claim victories in the same marathon races for many years, though they would seldom race against each other.

The real winners of these grueling marathon contests were consumers. The reliability of outboard motors was being proven, event after event, and both OMC and Mercury were designing products that could endure the punishment of long distance racing, largely at the hands of amateur drivers. The field of nearly 200 entrants in the 1948 Albany-New York race was by a wide margin the largest in the history of outboard racing. Even in the pouring rain and blinding fog that hampered both men and machine the entire length of the 134 mile river course, an amazing 141 completed the race. Only a short few years before World War II this percentage of finishers would have been unimaginable, and a large percentage of outboards wouldn't have even run except in ideal conditions, much less a heavy soaking and a long, wide-open sprint down a river swelling with floating debris, submerged logs, shifting sand bars and muddy shallows. The outboard motor had become a powerful, reliable, and relatively simple pleasure to operate, and Carl Kiekhaefer was quickly learning that racing meant publicity, and publicity meant sales.

From the flood of new buyers anxious to buy outboards following World War II, a steady stream emerged by 1948 and 1949, with greater and greater competition in a market that was showing signs of tapering off. When outboard manufacturers were working around the clock to fill the enormous backlog of orders that built up during the war years, they were comfortable in the knowledge that consumers would purchase virtually any kind of motor delivered to market in quantity. But as the buying frenzy slowly dissipated, the largest manufacturers, which now included fast growing Champion Outboard Motors, realized that engineering innovation, image positioning and headline grabbing promotion were the means to increase market share. Though Carl Kiekhaefer would eventually emerge as the undisputed king of promotion and record setters, the first to capitalize on America's emerging fascination with publicity stunts and record smashing headlines was Earle L. Du Monte, president of Champion Motors of Minneapolis, Minnesota.

As Champion's fortune rose on the swell of post war orders, the image of Champion Outboard Motors took a decided turn following the 1947 season. Earle Du Monte had carefully positioned Champion engines as dependable fishing engines, the "Blue Ribbon Champion," an "unhurried ... certain ...," and reliable companion for a relaxing day on the water. In all of their literature, the word "fast" was conspicuously absent, and slow, smooth trolling was emphasized for even their largest, 7.9-horsepower engine. In 1948, as Carl Kiekhaefer raced to establish Mercury with an enduring reputation for speed and high performance, Du Monte struck the first blow in what would soon become a wild contest for durability and dependability claims.

Du Monte announced in early 1948 that a stock Champion Standard Single cylinder engine had been clamped to a test tank, connected to a continuous feed gasoline line, started, and was still running strong, day and night, six months later. Du Monte was upstaging both OMC and Mercury by demonstrating what the other competitors had been rhetorically claiming about durability. "Could any unsupported claims, or promises," Du Monte would boast, "match this dramatic *proof* of Champion stamina? Instead of saying Champion could do it, or might do it, Champion *has done it*."[7] In a series of dramatic tests called *Motor Magic* tests, Du Monte threw the gauntlet in the face of competition as the public flocked to his dealer showrooms. "Words are cheap," he derisively lashed out at his fellow competitors, "In order to base Champion advertising not on claims but on *facts*, we assigned company engineers to make actual, witnessed tests which would answer these questions:"

> "1. How many hours of life are built into a Champion? How will it stand up under long continued running?
>
> "2. If a Champion should fall into a lake, how much service work would be needed to make it start and run?
>
> "3. What about cold weather starting? Will a Champion start readily on frosty Fall mornings?
>
> "4. Would a heavy rainstorm affect it? Could you depend on a Champion to start and run in a cloudburst?"[8]

"So the tests were made," Du Monte reported, "the strangest, most 'impossible' tests ever conceived."[9] For each of the questions that the average outboard buyer could reasonably ask, Du Monte designed a seemingly irrefutable demonstration, and reported the results. For the durability issue, Du Monte announced "All World Records Smashed by Six-Month Endurance Run." He claimed that the running time of over 4500 hours was equal to 45 years of ordinary usage, based on the average duty cycle of five hours per week for twenty weeks each year. He calculated that the total number of explosions in the cylinder had added up to 650 million, and at an estimated water speed of 7 miles per hour, the motor had "traveled" more than 31,000 miles in the test tank. He calculated that the piston had made one billion, 350 million strokes and that the propeller had turned more than 390 million revolutions. He would swear that, "except for spark plug changes, which take only a matter of minutes, the motor has never stopped," and that "no part of the motor has been replaced."

To answer the question of service work required after submerging an engine in a lake, Du Monte dropped a stock Champion single into a hole carved through 22 inches of ice in Lake Johanna near Minneapolis, on a 15 degree below zero morning in March. Granted, the engine was not running when it was lowered to the shallow bottom on a rope, but Du Monte would claim that, "Dragged up from its arctic bath, drained out and wiped off, this stout-hearted engine was started and running *exactly 107 seconds after emerging from the water!*"[10]

Claiming no service work would need to be performed with a dunked engine, Du Monte slowed the pulse of butter fingered outboarders everywhere when he announced there would be "No tearing down ... no fussing ... no babying the engine ... no hours of waiting for water-logged parts to dry out. The weather-proof, water proof Champion needed only to have the water dumped out of its hood to be started."[11] The following year, a hapless Champion owner dropped his engine into Lake Minnetonka near Minneapolis, and couldn't retrieve it until 21 days later. Champion made good use of the story, for "Wiped free of sand, mud and water it was started in exactly 1 minute and 57 seconds. ..."[12]

Prewar outboard enthusiasts were accustomed to having both their spirits and engines dampened by heavy rains. They listened in numbing shock as their engines sputtered to electrical failure in a downpour, and become hard if not impossible to get started again. Hooded engines and reliable magnetos had more or less eliminated this common concern, but Du Monte staged another test to ease the fear of being stranded in the rain. He took a new Champion from the assembly line, positioned it under a shower head, and let the water run for 72 hours. "With the motor still under the deluge," Du Monte boasted, "the starter rope was pulled and the Champion roared its immediate response. The full force of city water pressure (with water temperature at 54 degrees) pounded down upon the stout-hearted engine for 72 continuous hours, yet it started on the very first pull!"[13]

Lastly, Du Monte contrived to lessen the fear that hunting sportsmen might have of having a cold engine refuse to start at the crack of dawn as ducks and geese flew unmolested overhead. He locked and sealed a Champion in a freezer chest set at 18 degrees below zero. "Three days later," Du Monte reported, "when the motor was lifted out, it was frozen stiff ... white with frost, and so cold it could not be handled with bare hands. Exactly one minute and fifteen seconds after coming out of the freezer, this motor was started and running, with no damage to any part." The next year, Du Monte took four engines, submerged them in beer tubs filled with water, and froze them solid. It presented an intriguing spectacle when Champion engineers ice-picked the giant blocks of ice to slowly and melodramatically reveal the engines in view of a crowd estimated in the thousands. "Then, with icy water dripping from every part, they were set up and three of the four roared into life in less than 45 minutes - one after only 8 minutes, 3 seconds of thawing time." Champion covered up the fact that one of the engines wouldn't run, by advertising in the future that only three engines were used for the test.[14] Du Monte had successfully stolen the headlines for his products, and could claim that he had given his motors the most punishing tests imaginable. The public reacted strongly to this innovative campaign, and Champion sales blossomed.

Unfortunately, Champion management dropped the ball. Though they managed to ship $3.6 million worth of the navy-blue engines in 1948, aluminum shortages, plant strikes and manufacturing problems slowed deliveries to such an extent that $2 million were lost in confirmed sales before the end of the year. By contrast, Ralph Evinrude announced sales of OMC products of over $31.5 million, and Carl Kiekhaefer closed the books on around $11 million in sales for the last of the postwar boom years. For the first time since the Roaring 20s, competition was becoming more of a challenge than production capacity, and the outboard industry was adapting by learning how to flex its newly discovered marketing muscle. Soon to emerge as the industry's most gifted practitioner, without question or peer, would be Carl Kiekhaefer.

CHAPTER FOURTEEN

Staggered Faith

"At this time in particular, with the Government restrictions of material, and no military demands to replace loss of production, complete disorganization in the various government offices, reorganization of sales technique and policies, engineering talent and policies, development engineering and financing, I have been rather busy. ... If you should find someone who would be willing to buy a couple of outboard plants, please get in touch. ..."[1]

Carl Kiekhaefer

Less than five years after the German and Japanese surrender that promised a generation of peace throughout the world, war was declared by South Korea as North Korean forces stormed across the 38th parallel, and raced to overrun the capital city of Seoul. The large scale invasion took the world by surprise on June 25, 1950, and the United States quickly blamed the Soviet Union for precipitating the flagrant incursion. Secretary General Trygve Lie of the United Nations called an emergency session of the Security Council, which passed one of the strongest resolutions of its four and a half year history in sharp protest of the action, demanding the immediate withdrawal of North Korean forces. Within days, United States President Harry S. Truman authorized the use of American armed forces to repel the invasion, in association with other United Nations member peace-keeping forces. Following the fall of the capital city of South Korea, General Douglas MacArthur, the much heralded and controversial corn cob pipe smoking figure of World War II, was summoned to South Korea to assume command of all U.N. forces.

On December 16, 1950, following months of reversing fortunes, and the near total collapse of MacArthur's progress to restore peace in Korea, President Truman declared a state of emergency in the United States, imploring all citizens to unite their efforts to combat "Communist imperialism." Truman also called for "a mighty production effort" for the burgeoning military activity, and declared his intention to safeguard democracy with an "arsenal of freedom." For Carl Kiekhaefer, the meaning of the message was a familiar and frustrating one, and the worst of all possible news: the return of restrictions on the use of aluminum for outboard motors.

In Washington, D.C., the National Production Authority was established, a direct descendant of the War Production Board that had nettled the growth of the Kiekhaefer Corporation during World War II. The NPA board had determined, much as they had during the previous war, that aluminum should be diverted and withheld from outboard manufacturers because the devices were clearly recreational in nature.

But OMC decided to take the NPA to task over their appraisal of outboard use, and dispatched Joseph G. Rayniak to Washington to carry the standard of the marine industry. Born in Czechoslovakia, Rayniak first came to America in 1900 as a nine-year old along with his parents and three sisters. In his youth, Rayniak swept and mopped floors in Detroit to aid his impoverished family. Surrounded by an exploding industrial revolution, Rayniak soon progressed to oiling machinery and performing odd factory jobs. As he grew, he worked at Chrysler, the Pneumatic Tool Company, Burroughs Adding Machine, and Morgan Wright before taking night courses in drafting, design and mechanical engineering. By 1911 he was a first rate apprentice with Studebaker as a die-maker and machinist, and became a locomotive pattern-maker before landing a job as tool room supervisor with Packard Motor Car Company in Detroit. He helped to create the original Liberty aircraft engine, manufactured by the tens of thousands in World War I. As the war raged in Europe, Rayniak worked at the Duesenberg Works, tooling the celebrated 500-horsepower Bugatti engine. Over thirty years later, only blocks from where Rayniak would testify before the NPA, both the original Liberty and Bugatti engines he helped to tool were on display at the Smithsonian Institution in Washington, D.C.

In 1918 Rayniak joined the Johnson Motor Wheel Company in South Bend, Indiana. The Johnson brothers were building gasoline-powered bicycles, but over the following three years Rayniak would help to build their first outboard engine, a two-cylinder, two-horsepower, 2,200 rpm, 35 pound wonder that seemed to nearly fly a few days before Christmas of 1921. He followed the Johnsons to Waukegan, Illinois in 1927, and as the plant went into receivership in 1932, helped to keep the machines turning by mortgaging his home and taking out personal loans. He continued when OMC acquired Johnson in 1935, became a vice president in charge of manufacturing, and by 1949 was named general manager.

As the only outboard industry representative to appear at the NPA hearings, Rayniak prepared his case carefully and conclusively. The NPA had served notice that it intended to curb aluminum use to the outboard industry by an incredible 90 percent, allowing only 10 percent of forecasted outboards to be built during the national emergency. With the aid of a specially prepared pamphlet, *Outboards At Work*, Rayniak explained to the government that over 700,000 outboards were being used for vital commercial purposes, like fishing, logging, conservation, disaster relief, offshore oil operations, water taxi and waterfront construction. His point was that outboard motors were just as crucial to American industry as the many other products left unregulated. Naturally, based on Rayniak's testimony, it seemed to the NPA that Johnson and Evinrude products were the only outboards being used for industry. In a surprising decision, the NPA not only canceled its planned restrictions at OMC, but actually offered to assist the firm in obtaining additional aluminum for future production. But with an almost unbelievable display of partisanship and favoritism, the hearings failed to lift restrictions for the Kiekhaefer Corporation or other manufacturers in the industry. Thanks to J.G. Rayniak, who would be rewarded three years later with the presidency of the corporation, OMC would coast through the Korean War, devoting only about 10 percent of its production capacity to defense production, while Carl Kiekhaefer, for the second time in less than ten years, was nearly pushed out of the outboard business altogether.

Carl became thoroughly disgusted with the conduct of the Korean War, and rarely passed up an opportunity to chastise Washington, and in particular, President Harry S. Truman. In a letter to J. Paxton Hill, the Utility Racing Secretary of the American Power Boat Association, Carl leaves little doubt as to his true feelings.

> "You have no idea as to the amount of travel and working time our entire organization has been put to in the last few months. Six of our executives have been out in the field almost continuously, trying to save the organization and its business from grinding to a complete stop in this era of transition, from

a smooth running peacetime operation to the crazy wartime operation that has only one capricious customer.

"This is the second wartime operation that I am organizing within a ten year period, and we never completely recovered from the first. Believe me, it would be much easier to carry a rifle. I shall not attempt to begin to tell you the troubles that every major manufacturer has today, with the curtailment of materials, and no sizeable war contracts to pick up the difference. The only thing that 'little Harry' [President Truman] hasn't aped Franklin [President Franklin D. Roosevelt] on, is his inability to whip up a good snappy world war having one hundred million men in arms.

"It is not inconceivable that the American public, reluctant to feed cannon fodder through 7,000 miles of pipeline, and reluctant to go along and provoke another world war, might call a halt to all the billions of spending if no world war is immediately forthcoming. Should this happen, all manufacturers may want to scramble back into civilian production on a minutes notice; leaving all of us in a position of trying to guide a 'C' hull [a racing class of boat] with one hand and man a machine gun with the other. By comparison, the last World War was a cinch to operate under, inasmuch as there was enough work to keep all manufacturing plants busy supplying all the armies in the world, with the exception of three [Germany, Italy and Japan]. Presently, we are trying to do that without incentive.

"The gravity of the situation cannot be passed over lightly when one considers just what the future may hold for our two corporations, our 1,000 employees, and our 4,000 dealers. About as ridiculous as the 'Paperhanger' of the last war [Adolph Hitler], we now have a 'Haberdasher' [President Truman] --- a 'Haberdasher' whose picture appeared recently in a newspaper preening himself and boasting of his loud neckties. But then, I believe our janitor could run our business at the present time as well as I can, with the world turning inside out. ...

"Wars always serve as a catching up period for the weaker and more selfish competitors, who, instead of contributing in like measure on the new designs and production of military needs, are content to turn their vast plants over into machine shops to make nuts and bolts for the military while they and their engineering group are planning postwar-wise not only on engineering, but organization-wise.

"I hope that I have not discouraged you in trying to be realistic, but occasionally it appears necessary to curtail our regular work in exchange for a war chore, a nasty chore that we must face as a matter of patriotism and decency, a chore that is not going to escape any of us now that it has gone this far."[2]

For almost a year, Carl had been thinking about selling the business, so the advent of aluminum restrictions only stiffened his resolve to join forces with a larger company to relieve himself of his growing administrative burden, and allow himself more time for research and engineering. At least, that's what he led his inside circle of executives to believe. Actually, Carl was alarmed at the sharp leveling off of sales in the outboard and marine industries, and was tiring of struggling, like David, against the giant of OMC.

In early 1949, Carl began a correspondence and dialogue with Walter F. Rockwell, president of the Timken-Detroit Axle Company, and uncle to W.F. "Al" Rockwell, Jr., president of the substantial Rockwell Manufacturing Company of Pittsburgh. Carl allowed Harry F. Burmester, senior vice president of the Union Bank of Commerce in Cleveland, to whom Carl owed $450,000 in loans, to submit a preliminary analysis of the value of the Kiekhaefer Corporation operations to the elder Rockwell in Detroit. The analysis, in the form of a letter to Rockwell, revealed a trend of rising sales and lowered profits. With an estimated book value of slightly over $3 million, Burmester projected 1949 Kiekhaefer Corporation sales at around $11 million with profits of only $300,000, while noting that the previous year's sales of over $12 million had produced a profit of nearly $1 million.

On June 30, 1949, W.F. Rockwell, Jr. toured Carl's facilities in both Cedarburg and the newly expanded facilities in Fond du Lac. Meeting with both Carl and Fred L. Hall, Kiekhaefer's vice president of sales, Rockwell liked what he saw, and let them know that he was interested in acquiring the operation as a part of Rockwell's growing manufacturing interests. In a follow-up letter to Rockwell, Fred Hall summed up the position of the company, and the reasons for consideration of a merger or sale.

> "The Kiekhaefer Companies have expanded to a point where Mr. E.C. Kiekhaefer can no longer handle all administrative responsibilities and be responsible for financing the operations and continue to do justice to the progress of the art of 2 cycle - high output - lightweight engine advancement. The latter, after all, is his first love, and the development, research and engineering work is his forte - and we feel that with his release from the detail of administrative duties would be insurance that our engines will continue to improve and be demanded for many more industrial, marine and transportation

applications, - thus the willingness of our principal stock holders to consider negotiations with a successful operating company such as yours."[3]

Rockwell replied by saying, "I think there are a lot of mutual advantages in putting the two companies together and I hope we can sit down and talk about this. ..."[4] But when they did, Carl did not like what he heard. The offer made by Rockwell was made verbally, and is unknown, but Carl felt insulted and incensed by the small size of it, and broke off communications with Rockwell for nearly a year. When Rockwell made another offer in 1950, Carl was still not impressed, and in a transcribed telephone conversation with Jim Ashman of Rockwell Manufacturing Company on June 27, 1950, he let them know what he thought of their evaluation of the worth of his business.

Mr. A.: What have you done with our proposition?

Mr. K.: I haven't done anything with it yet. ... I don't think you folks are really interested or you wouldn't offer me the kind of price you do.

Mr. A.: Why?

Mr. K.: Well, you offered three times as much for it this year as you did last year. It's kind of confusing. I just don't know what is going on and there's a lot of unanswered questions. I don't know whether we want to even go into it. It's almost impossible to go into those things. We're getting more for the one plant [Disston had made an offer to purchase the Cedarburg facilities] than you're offering us for two.

Mr. A.: Well, that's good. I wish you a lot of luck. I hope you get a lot more for them. You have some continuing obligations there.

Mr. K.: Well, I'm going to be sowed up, yes, the same way as I would be when I join your organization. You know there wouldn't be any rest; there would be a lot of things that come up and I don't object to the work except there's a limit to what a man can do.

Mr. A.: Sure, I know that.

Mr. K.: It's like selling your right arm. I would like to talk about it some more. You don't usually give up with one contact, but the starting point is so low that I just don't have much hope. ... I have a lot of unanswered questions that I would like to talk over with the Colonel [Col. W.F. Rockwell, Sr.]."[5]

1950 was a frustrating summer for Carl Kiekhaefer, as both the possibility of selling the Cedarburg plant to Disston and joining forces with Rockwell were fast falling apart. When Food Machinery and Chemical Corporation (FMC) acquired the Propulsion Engine Corporation of Kansas City, manufacturers of lawn mowers and two-cycle engines, Carl explored the possibility of a union with the heavily diversified company. During World War II, FMC had reached a peak in sales of over $240 million, due primarily to the manufacture of the Water Buffalo amphibious tank for the Army. Following the war, FMC diversified into agricultural and industrial chemicals, phosphate and shale recovery, food preparation and processing equipment, along with agricultural and industrial equipment. Their Bolens division, acquired in 1946, was a leading manufacturer of garden tractors, building up to 172,000 units a year.

Carl flew to San Jose, California in November 1950, with the intention of selling the Kiekhaefer Corporation to FMC. Meeting with Jim Hait, their vice president in charge of engineering, he toured their plants, and met with key personnel. When he returned to Fond du Lac, he was fearful that he hadn't made a good impression, and wrote to clarify his feelings.

"While I enjoyed our discussions very much relative to the engine business, I am much afraid that we have not begun to tell our story. It is obviously difficult to convey in two hours a sound picture of what took 12 years to build, and we again invite you and your associates to visit our plant at your convenience in order to get the proper perspective to make a decision, be it pro or con. ...

"We are forwarding five of our sales promotion films as produced by our own Photographic Department. I am sorry I do not have a film showing plant operations and history, but we have always considered our plant and laboratory matters of secrecy as far as the public is concerned. ... To get the best results, the film should be shown after dark and on good up to date equipment."[6]

For Carl to tell the chief engineer of a $100 million company like FMC how to project a movie is a little presumptive, but FMC replied within days to say that, "We have discussed your corporation at some length here and are interested in pursuing the matter further with you."[7] But Carl had seen the delays and wasted time that accompanied the Rockwell and Disston negotiations, and wrote a hard selling letter to FMC Vice-President and Controller B.C. Carter.

> "All hastened by war developments, we have a number of decisions to make before the year ends, that will hinge heavily on the extent of your interest in our organization. ...
>
> "Please do not consider this as an accelerated effort to sell; I am trying to give you the facts as they exist and would advise that if you and your associates are interested that we get together as soon as possible so that mutually, we can take best advantage of the times.
>
> "Personally, the possibilities of being associated with your concern and its subsequent opportunities leave me with the feeling of excitement and challenge. It is, however, futile to expect any reaction from your group until you have seen the picture first hand.
>
> "Technically speaking, we are going to have to act fast if we are to take full advantage of the jump we now have on the rest of the industry on military as well as industrial engines."[8]

After reviewing how pushy the letter seemed, and how thinly veiled was his anxiety to make a deal quickly, Carl decided not to send the letter. In the course of his career, Carl would occasionally write emotionally charged letters, often surging with anger, only to reconsider them the following day and scrawl *"DO NOT SEND"* across the face with a bold marker. He would, though, always insist that these letters be filed, so he could be reminded of the circumstances that led to writing the letter in the first place.

In a flurry of confidential exchanges between FMC and Carl, the ground between demands and offers slowly closed, and it looked as if a sale would be made by the spring of 1951. But then, when both sides assumed a response was forthcoming from the other, a frustrating game of wait and see commenced that would ultimately destroy any chance of success. Carl wrote another of his angry letters destined for the *"DO NOT SEND"* pile near the end of February, feeling embarrassed that nobody had bothered to get back to him.

"Owing to the fact that nothing definite has developed from our negotiations during the recent months, we are withdrawing our offer herewith. ...

"We regret that no deal has developed. It was a pleasure working with you ... and hope that we shall all remain friends.[9]

Carl waited another nearly three months before he decided to send virtually the same letter, calling off the deal. FMC admitted their part in the mix-up in a follow-up letter to Carl on May 22, 1951.

"Apparently we have a faculty for getting our signals crossed, as [you] indicate that you were expecting some further word from us while we in turn were awaiting a reply from you. ... Since I have heard nothing from you in the interim, I assumed that you preferred to defer the discussions for the present. ...

"We are still keenly interested in having you and your companies as a part of our organization and, as in the past, will be glad to meet with you anytime to discuss the matter further.

"We believe that your own ability and the excellent products which you produce, coupled with the many advantages which this union would provide would result in a rather spectacular development of the Kiekhaefer Corporations as well as providing a solution to some of your personal and estate problems. ... I will await your reply to this letter to make certain that there is no misunderstanding."[10]

FMC had offered Carl 95,000 shares of stock, with a book value of $4,275,000 for the combined operations of the Kiekhaefer Corporation and Kiekhaefer Aeromarine Motors. No cash was offered, and Carl would have been guaranteed a lucrative though loosely defined position that could best be described as chief research engineer. But Carl vacillated, asking to see a proposed contract that would spell out his "responsibilities, incentives, [and] penalties." He also asked that an option of a cash sale be explored, rather than full payment in FMC stock.

One of the areas most interesting to FMC was the chain-saw market, and Carl disclosed that his new "DA211 saw has been competitively tested against the new McCulloch seven-horsepower saw, and in every instance has been declared by the various logging operators to be vastly superior in horsepower, accessibility, operating ease, and general all around performance." Carl savored this engineering victory over his rival Bob McCulloch, for even though both saws weighed the same, the Kiekhaefer-Disston saw packed an additional two-horsepower.

FMC began the process of preparing a sample contract for Carl, but only as an afterthought in a footnote to a letter did they inform Carl that a cash deal was off. This bothered Carl greatly, for while negotiations were proceeding, Paul L. Davies, president of FMC, delivered a speech before the New York Society of Security Analysts, wherein he estimated the company had an excess profits tax exemption of over $10 million, contracts for over $100 million to build tanks for the military, and was planning on spending over $25 million *in cash* for expansion and acquisition.[11] Further frustrating to Carl was FMC's offer to compensate him for his continuing efforts following the merger at $30,000 per year plus 1 1/2 percent of the profit before taxes. Willis Blank would receive $10,000 plus one-half percent. By the end of the long summer of 1951, no progress had been made, and B.C. Carter of FMC had to write Carl, "With respect to the status of our negotiations, it is my understanding that as soon as you have clarified your personal situation to some extent you will get in touch with me and until such time we will simply hold the discussions in abeyance."[12]

The brief wave of merger madness had finally passed over Carl, and he began to concentrate on the growth of the business once again. In a period of less than 24 months, he had come within a few documents of selling his company to Rockwell Manufacturing, Disston and FMC. The receipt of a contract of over $3 million from Disston for saw engines, combined with a renewed interest by the government in having the Kiekhaefer Corporation build a new, more powerful target drone engine, were the major factors that ultimately convinced Carl to hang on and build for the future.

Another event occurred, which would nearly completely interfere with Carl's plans for the future of the company. On Saturday, October 7, 1950, Carl's father, and still president of the Kiekhaefer Corporation, Arnold Kiekhaefer, was suddenly stricken with an acute heart attack and died. In his memory, both plants in Cedarburg and Fond du Lac were closed for two days. Funeral services were held at St. John's Lutheran church in Mequon, followed by a burial in the parish cemetery. The Common Council of Cedarburg issued a proclamation, saying in part, "... through his initiative and ability he succeeded in establishing the Kiekhaefer Corporation, one of the leading industries in the City of Cedarburg thereby directly and indirectly contributing to the prosperity, growth and expansion of this City and bringing it national recognition."[13]

The truth of the circumstances surrounding the death of Arnold Kiekhaefer would be disclosed in a letter that Carl never mailed, and were quite different from the reports in the area newspapers. Four months after the death, Carl received a short, terse note from an attending physician demanding that a bill for $15 be paid before going to a collection agency. Carl was incensed, and in a blistering reply to the doctor, the facts of the last day of Arnold Kiekhaefer's life were revealed.

> "At times, Doctor, I almost wish that socialized medicine would be put into effect in our country, because from past experience the medical profession well deserves such action. I would not bring this matter to your attention except for the fact that your unethical note has shocked me; particularly after the brutal mishandling my father was subjected to prior to his recent death.
>
> "In this particular case, the doctor was called to attend my father at 8 o'clock in the morning. Without stating there would be a delay, he did not appear until 1 o'clock in the afternoon, whereby an ambulance was immediately summoned and my father was "hauled" to the hospital like a steer going to market. He passed away 5 hours later without even the benefit of a physician in attendance. There was no resident doctor in the hospital and the patient received no attention whatsoever. Not one of the 8 family members knew it was a coronary or that there was anything seriously wrong.
>
> "It is no wonder that the medical profession needs to bind itself together into a powerful association that is legal-proof if not bullet-proof.
>
> "In my particular case, I have contacted your office innumerable times to make appointments, but you refuse to set a definite time. I have left my work at least three times in an effort to see you, (I operate two plants and have created a fifteen million dollar payroll in ten years of business, and am also a very busy man) and arrive at your office to find you are "out of town". On two occasions, I have entered your office to find ten or fifteen people in the waiting room ahead of me.
>
> "Needless to say, all of this is quite understandable, but I certainly do object to the type of note that you have written to me for the sum of $15. It is the type of note one would write to a colored teamster, and will very satisfactorily explain why you have lost a patient."[14]

The funeral for Arnold Kiekhaefer was one of the largest in regional memory. The church was packed with well-wishers, friends and family to standing room only capacity. An overflow crowd spilled out into the crisp autumn morning, spreading over the lawns of St. John's in tribute to one of the community's most respected and beloved friends.

After the funeral, to Carl's total horror and disbelief, he discovered that his father had completely cut him out of his will, leaving everything to his six sisters and mother. Following long passages devoted to the distribution of his worldly belongings and company stock to his mother, Clara, and six sisters, Carl read, incredulous, what his father had left to him. "I make no provision for my son, Elmer C. Kiekhaefer in this my Last Will and Testament for the reason that said son is otherwise amply provided for."[15] Carl was shattered that his father would so cruelly and heartlessly interfere with his plans for the company, and force him to negotiate with his mother and sisters to secure the stock he needed to have a majority interest in the business.

Unbeknownst to Carl, when Clara had first read the will her husband had written, and saw how cruelly Carl was treated, she implored Arnold to revise the will, and soften the blow to her son. "He had said in this will," Carl's sister Isabelle recalled, "that he was not leaving anything to Carl because he had otherwise provided for him and that we girls had received nothing. Mother said, 'That sounds so harsh.' So he threw it in the desk drawer, intending to modify it."[16] But Arnold never did modify the will, and his apparently healthy condition and total control of his faculties up to the moment of his attack served to work against any attempt to dispute the will.

Carl's paranoia swelled with the implications that he might someday again be bent to the will of a collection of stockholders, even if a majority were his own family. He decided to act quickly, and, as his sisters would charge, imprudently. "When Dad died, then of course he went to mother," Isabelle remembered. "Dad died in October, and by Thanksgiving he [wanted to have] an auction on the balance of her stock at a ridiculous price, and we girls said, 'Look, let's wait. ..."[17]

But it was too late, Carl had cornered his mother in her hour of greatest confusion and despair, and persuaded her that she had to sign some documents immediately in order for the company to continue. Clara Kiekhaefer wasn't a businesswoman, and had always left the paperwork and conduct of business life to her capable husband. But now, swirling in a nearly blind fog of sorrow and pain, her husband of forty-five years dying suddenly only weeks before, she assumed the papers were simply the necessary aftermath of tragedy, something Carl had to have signed right away.

In truth, Carl had shamelessly persuaded his own grief-stricken mother into signing an option to purchase the shares which she automatically controlled upon her husband's passing. When Arnold died, he and Clara jointly controlled 44.4 percent of the Kiekhaefer Corporation to Carl's 33.4 percent, and 15 percent of Kiekhaefer Aeromarine Motors to Carl's 40.4 percent. Carl computed what he considered to be the book value of the stocks he needed, and for what would amount to less than $193,000, Carl could purchase a controlling interest in both companies, whose combined sales were over $12,000,000, and whose net worth was then nearly $5 million. Had Carl not callously taken advantage of his mother, she might have signed a more realistic option that could have cost Carl over $2 million, or *over ten times* the amount Carl insisted on.

Even more devastating to Carl's mother and sisters when they discovered Carl's deception, was the fact that the option allowed Carl to purchase the shares anytime within the next three years. So, even though the value of the companies would continue to grow, Carl could be assured of the same ridiculously low price *even three years later.* Meanwhile, the shares could not be sold to any third party, and since Carl had never granted a single dividend on the stock in the past eleven years, Clara and Carl's six sisters wouldn't be able to profit by holding the stock, even down to the last day of the three year option period. "Mother was never a business woman," Isabelle contends, "Carl knew this, he took advantage of her lack of business knowledge."

Once the family realized what happened, they protested Carl's action, and asked the Marshal & Ilsley Bank of Milwaukee, named as co-executors of Arnold's estate, to determine the true value of the stock. But Carl wasn't about to open his books in order to increase the price he would have to pay for the stock. Carl had begun working with Alan Edgarton, a bright and reliable Fond du Lac attorney, when he acquired the Corium Farms property. Now he called on Edgarton to help him untangle the emotionally charged conflict that was growing by the day within his family. The sisters later regretted accepting Carl's legal assistance. "E.C. wanted us to use Edgarton," Carl's sister Almira recalled, "and we went along with it, thinking, alright, he will cooperate. Well, what happened, he [Edgarton] was serving two masters. He was serving Mercury and he was trying to serve our estate. ... We were given $210,000. ... I can remember the banker coming over and showing me the check ... and that's all we got out of Mercury, and it was worth millions of dollars."[18]

To further rub salt into the widening family wound, Carl later hoodwinked his mother into signing an additional document, which would ultimately allow him to keep an additional $30,000 that should have been distributed to the sisters from Arnold's estate. It was to be the final blow to the family by Carl, and one which his six sisters and mother would have difficulty forgiving. "She thought she was loaning him some money," Almira remembered, "and then he later said, 'I had no intentions of paying you back.'"[19] The ruse was later discovered by another sister, Isabelle, who was helping Clara to prepare her income taxes. "[Isabelle] asked, 'Did you loan him some money, or give him some money?' and she said, 'No, I didn't, I loaned him some.' But then it all came out," Almira continued, "and of course she was sorry she didn't understand those things. Mother had to work too hard on the farm to worry about those things ... and worry that much about money affairs."[20]

As if the relationships weren't strained enough between Carl, his sisters and mother, his unshakable paranoia of being whispered about disparagingly behind his back precipitated even more turmoil. Since almost the very first day in Cedarburg, some of Carl's sisters, and ultimately five of Carl's six brother's-in-law, had worked for the company in various capacities, like sister Palma's husband William Moerschel, who became personnel director. When the sisters let it be known they were upset with their brother's highly questionable actions, Carl promptly fired every member of his family, without notice and without reason or discussion. "He let all our boys go," Isabelle remembered, "He said, 'If the girls can't agree with me, I don't want you boys working for me either,' so he canned all the boys."[21]

For Carl, it became all-out war. His sister Isabelle recalled the shock of his heavy-handed action. "That was the beginning of the end there, you know, and I know there were private detectives out here. You know our phone was bugged - I called the telephone company."[22] The sisters were totally disoriented by their brother's brutal tactics, and how he had conveniently forgotten the true origins of the company. "He forgot the days when Dad came to Marion and me," Isabelle said, "We were sitting across the desk from one another, and Dad said, 'Girls, if it doesn't go now, Mom and I don't have a dime'. He had gone down to the Federal Reserve Bank in Chicago and had signed away all his and Mother's assets. ... He and Mother mortgaged themselves to the hilt on this, and you know, Carl always wrote he did it all. But he didn't, you know."[23]

Carl's rash, vengeful and punitive act had thrown virtually every household of his family into chaos. Dismissing his closest relatives without so much as a warning or discussion, as if they were complete strangers, pushed the deepening family crisis to critical mass. Though Carl's sisters practiced the Christian doctrine of "forgive and forget," the truth is that the wounds were so deep, that it was virtually impossible to overlook his actions, even 40 years later.

When the family protested his slight of hand, and demanded more money for the stock than was provided in the option, Carl took it as a threat not to sell the stock at all, and as a move to remove him from authority. To demonstrate to his sisters and mother just how important he was, and how critical his personal decisions were to the success of the companies, he dictated a rather pompous list of inflated duties, designed to overwhelm the sensibilities of his family. He followed up the list with a short note, offering slightly more for the stock.

> "In the future, I cannot continue to take the responsibilities unless control is vested in the writer, because of two estates and eleven heirs being involved in stock ownership at the present time.
>
> "Any death now would complicate matters further with minors, guardianships, additional attorneys, executors, etc., which would make the operations a nuisance, if not unpredictable.
>
> "For the writer to continue operations, therefore, it is necessary that the 105 shares of Kiekhaefer stock now vested with the executors [of Arnold's estate] be sold to the writer. Though appraised at $1,200, I will pay $1,500, an offer that will remain open for ten days only. Without control, there can be no long term contracts on financing, on chain saws, on outboards, on Government contracts, since the writer's signature will be required. ...
>
> "With all the foregoing, we have not as yet mentioned the most important responsibilities of all, namely the responsibility

to 900 employees, the responsibility to the communities, and finally, the responsibility to the stockholders, who have elected me to my office."[24]

Carl's implied threat that his inattention to any of his many duties would weaken the value of the company's stock was not lost on the family. Time and the kindness of his sisters would eventually nearly erode the deep feelings of hostility and suspicion that Carl generated. But between mother and son, the damage was complete, and Carl never forgave her for what he imagined was her duplicity with his father in being cut out of the will. Even five years after Arnold had been laid to rest, Carl and his mother exchanged the most strained and venting letters, beginning with a handwritten note, complete with tear-stained smudges of ink, from Clara to her son.

[Original punctuation intact]

"Dear Elmer,

"Please don't always say that you were disinherited because you are not you got all and more of dad's estate. Dad must have figured he helped you start the business & wanted to give the girls the same as you. They helped us on the farm too, where we made the money that we put in the business. they worked hard from early in the morning till late at night, worked for room & board while going to high school. it was not always so either, maybe it unfair, but that was it way it had to be so that is why I have to be fair now to all of you that is on my conscience all the time. you have received some of my stock which you asked for which is helping you now & has been for some years, which they have not received. They have to be 40 and 45 years [according to terms of Arnold's will] before they receive what Dad left them.

"I think is unfair to me with my stock worth over a million dollars to receive no income from it now. I will have to sell & re-invest it so that have income again.

"I am sorry you don't like it that I went to see [an attorney], but I do go there by myself. but he seems like sensible man, & I explained to him that I wanted to be fair to all my children. He pointed out that when people become elderly they should not sell stock because of the taxes & especially if they do not know what it is worth. I have been to the annual meetings of Thiensville State Bank & Finance Co. & they handed out statements which everybody kept & could study at home. could I please have a statement to see how much my stock is worth. You state that you have no need for my stock but say you can help me dispose of it. I will need a statement to get a value of it first. I have seen papers that investors share if a company grows.

"Now Elmer nobody is happy the way things are now. What has become of my family? & I am saying this with tears in my eyes. why can't we stand together like other families get to-gether have a family reunion once a year. we old sisters & brothers I mean my sisters & brothers we have no hatred for one another. after all we all should be enough of Christians to know better. Any time you want to come & talk things over there is nothing in the way.

Sincerely your

Mother"[25]

Carl's mother had retained a solid block, though a minority interest of stock in the companies, having sold enough to Carl to give her son a clear majority. Now, all she wanted to do was to put the ugliness behind her, reconcile with her only son, find out how much her remaining stock was worth and sell it. But Carl read the letter differently, and assumed that she was closing the door on the embarrassing episode about his father's will. In a heart-twisting rage of defiance, he wrote back to her, pleading his innocence, and even defending his selfish actions. He implored her to remember his own, grueling hardships on the farm, and not to forget who was responsible for the success of the company today.

"Dear Mother:

"Acknowledging yours of August 10th, I agree that you have attempted to right the wrong committed in my being left out of Dad's will, and I am grateful for it.

"Did I not work 15 out of the first twenty years of my life on that farm too?

"Did I not do the work better than any hired man or the girls? Did I not operate tractors, threshing rigs and do all heavy work besides? I can remember the terrible summer "vacations" of World War I when, as a child, I could not rise in the morning with fatigue -- broken out continually with skin rash that prevented sleep. Who repaired the machinery? I think I did my share.

"When the rest of the family celebrated the reunions, I spent years at night school. When the boys played cards and drank beer Sunday afternoons, I designed and invented, and did my college lessons 'till two and three o'clock in the morning -- and yet was at work at Stearns the same time as the others. I worried about earning my future. I did not wait for someone to lay it on a platter. I got to be Chief Engineer and General Manager at Stearns. I could have been contented with that.

"I had ambitions for bigger things. You knew that. Dad knew it. But it was Uncle John Blank who took me seriously.

"The truth is, he came to me about the Cedarburg Manufacturing Company which was going broke. He laid out the organization plans. I bought up - got the creditors to go along with me - on 50 cents on the dollar to cancel one-half of all the debts owed by Cedarburg Manufacturing Company -- on my ambition - on my plans - and my promise. Without this, we could never have acquired the assets of the dying company. Dad was not in on any of this. Dad did not want any part of this venture, but we moved on without his consent. Dad was angry still at my leaving the farm. Dad was the last one to put up any money, and that is the truth. After all the ground work was laid, Dad finally came in -- as a 20% stockholder -- not 100% as the girls always presume. 80% of the stock was put up by others, and certainly that can't be denied.

"No one will ever know the hardships, long hours, sleepless nights, the struggle against strong competitors, the uncertainties of the times, the World War II troubles, help troubles -- saying nothing of my family sacrifices, the miscarriage of Freda's child (due to worry on her part, and neglect on my part). No one will ever know what the responsibilities really were. I was the only one who was always there when emergencies arose, always at the helm of a very under-capitalized, understaffed, inexperienced, young company, pulling it through, crisis after crisis.

"This business did not grow on trees. I gave up all my stock as collateral one time in order to meet the payroll ... another time to pay for materials. When World War II came, we were put out of the outboard business overnight by the Government because we were using aluminum. Overnight, we had to start a new business. Who was it that was then ready to quit? Who prevented the sale in 1940 to Flambeau Outboards? Who was it that prevented Dad's selling out in 1948 to Mehlman and the Hall group? Willis Blank! I didn't even know that Dad had signed an agreement to sell. Please notice how often Dad did things without my knowledge. It didn't make my job any easier. It was no wonder that some of my executives held me up in ridicule when they were conferring with Dad in the background. I was deeply hurt many, many times.

"Whose name goes on millions of dollars worth of notes to finance our business even today?

"Mother, who do you think should take the credit for the growth of the Kiekhaefer Corporation? Who should have the credit for its progress? Who should take the credit for its position today? Who do you think is in the best position to guide it through the always uncertain future?

"Refusal to recognize the necessity of my guiding its course in the future is not being realistic. It still doesn't grow on trees. With rising costs and taxes, I must remain constantly alert and produce new designs and engineer new production savings. With large competitors as Oliver, and Outboard Marine, we cannot make the profit per engine that their financial advantage provides. The net result is --- a continuation of 16-hour days, and no time for my family, which has grown up without a father in the house.

"Don't you understand the extent of my sacrifice, of my effort? Is this not worth some consideration. You are admitting that the efforts did produce -- when you value your stock at $1,000,0000 - when your total investment was less than $50,000 - which I helped to produce. I do not believe you will ever get one-half that amount from anybody, but can you name anyone who, as President, produced such a return on any investment? Is there anyone else that has equaled that record? Admit, Mother, that I deserve some credit!

"Now, then, remember I have also tried to help you, and the girls, even though they have no interest and have not contributed a damn thing to the corporation except trouble.

"I had a solution, I thought, to their money greed. But that opportunity has passed. As I told you something big was in the wind. This opportunity was fantastic. But in view of the heirs' attitude, I let it pass. I have lost interest. I will not again embarrass myself for their sake in seeking further opportunities, of similar nature. Someday, I will tell you exactly what you have missed.

"Unfortunately, no satisfactory sales are made for factories alone. What opportunities arise, call for management, not physical assets. I have turned down a job [Rockwell merger] that was in scope as large as the job as President of General Motors. Why? I have lost faith in people. I don't need the job, nor the money. There is no incentive for working harder. I have carried certain people on my shoulders far enough -- without one word of thanks -- no sign of appreciation -- nothing but abuse and criticism and grief.

"Mother, the biggest favor you could do your girls is to kick their husbands out on the street, and tell them to earn their living for their families. Then they could hold their heads up in pride and self-respect. Subconsciously, they all are ashamed of themselves, with one or two exceptions, and that is what really hurts them inside. If they had made the effort I have, and had applied themselves, they too could be a "big boy" as they sarcastically call me.

"More to settle the family estate situation, than for our own personal ambition, I was ready to accept this new opportunity even though it meant working harder and carrying much greater responsibilities. I was ready to accept this job which had potentialities equaling the Presidency of General Motors. It was a great honor, and I probably would have accepted it had it not been for the power of attorney you vested in Isabelle, appearing at the crucial moment to embarrass me when I had all the principals lined up for a meeting. Because of the size of the venture, I could not speak of it to anyone until the plans were finalized. I had every intent of discussing it with all stockholders and directors just as soon as their proposal was finalized, but this, now, is past history, and not only I, but you, have passed the biggest opportunity which will probably never come in any one man's lifetime, and I am not speaking idle words. I believe it is quite understandable how one does not incline to help people (which I am in a position to do) when one is constantly being kicked in the shins.

"My code of ethics ... my principles ... and my integrity comes from you, my first teacher. I have always felt that you wanted me to do the right thing always, and have always felt that you would never let me down and you must certainly know, in your own heart, that I would never let you down ... come what may.

"Both of us have survived a very unpleasant background that fate put us into. It has left its mark. I, for one, would never want to re-live my childhood in that old homestead, but even you have had a pleasant childhood, which the girls did not have. That, too, has left its mark.

"With the hope that things will change as these girls get older, and that everything will eventually work out to your satisfaction, I remain

With Love,

Your son,

Elmer

Please return this letter to me after you have read it."[26]

The original of the crudely typed letter was dutifully returned by Carl's Mother, and was placed beside the tear-marked letter that he had received from her. Carl was 49 years old as he wrote his reply, still desperately trying to defend his youthful efforts on the farm, and still trying fiercely to persuade his mother of his worth as a son, and as a person.

Carl had effectively insulated himself from love at every juncture; from his six sisters, from his mother, from his wife, and from his children. In a life that was surrounded by people who had a great capacity for affection and love, Carl became an industrial orphan, and an absentee father and a stranger to his wife. Though he recognized the consequences of his rough handling of his mother and sisters, Carl would pretend that the emotional agony didn't matter. All that mattered was the success of his work, and the growth of his enterprise. Insulating himself from both affection and rejection, he had become an industrial Caesar, an intimidating and domineering competitor whose only objective was victory. Victory at all cost.

CHAPTER FIFTEEN

Smart Move

"I have personally always admired a guy that tried to do something big, but this matter of doing something big has many definitions, and it seems that whenever there is anybody doing a good job, not everybody is going to be pleased, and then, it's just a matter of time before the barrage, verbal or vegetable, sets in."[1]

Carl Kiekhaefer

In a letter to Charlie Strang at M.I.T., December 11, 1950

Among the many competitors vying for honors at the 1948 Albany-New York Outboard Marathon was Charles D. Strang, Jr., a young and brilliant research associate at the Massachusetts Institute of Technology. Tall and lanky, with a broad and smooth forehead accentuated by a prematurely receding hair line, Strang's bookish looks disguised a rugged and determined competitor who wasn't afraid to mix guts and gasoline. Strang piloted number U-33, an Evinrude powered Lyman, for about 60 miles during the race before his engine failed near New Baltimore and *U-33* swamped in torrential rains.

When Charlie was 10 years old, his mother, Ann Strang, took him to the 1931 New York Motor Boat Show. It was the beginning of an insatiable infatuation with boats and motors. He had seen a Century *Cyclone* racing boat at the show, and it was nearly impossible for his mother to pull him away from the display. He saw the *Cyclone* as the most important object in the whole world, and pleaded with his mother until she promised he could have one someday. "Someday" isn't the answer to give to a precocious 10-year-old, and Charlie pressured her until she promised he could have one when he turned 14 and graduated from grammar school. His wish finally came true in 1936 at the ripe old age of 15, and by the following summer of 1937, he started racing. He raced like a demon, and with a generous mechanical ability, was able to fine tune his engine and boat consistently into the winner's circle. Before World War II put a stop to organized outboard racing, Strang reigned as the New York State Champion for several years. He attended Brooklyn

Technical High School, taking a combination of technical-vocational and college preparatory courses. As a result of Strang's studies there, 50 years later in 1989, President George Bush presented Strang with the American Vocations Success Award at a White House ceremony. The Award, which was created by the National Council on Vocational Education, "pays tribute to successful individuals whose educational background has included vocational or technical classes." Later, as a mechanical engineering student at the Polytechnic Institute of Brooklyn, he won the last contested Intercollegiate Championships in Class C Hydroplane. The Intercollegiate division never resurfaced following the war, and so, nearly a half-century later, Strang is still the undisputed and reigning champion.

Long before Strang became infatuated with the Century *Cyclone*, his family had been involved with boats, and with the sea. His maternal grandfather operated a spar yard in New York, on the site of what is today the Todd Ship Yard. He had rigged the tall, square-rigger ships of the day, and fabricated masts, booms, and spars. Even a generation before, his great-grandfather had operated a similar enterprise in Germany.

When Strang received his degree in mechanical engineering in Brooklyn in 1943, he joined the experimental engineering staff at Wright Aeronautical Corporation in New Jersey. For nearly four years he was a research project engineer working on early turbojet aircraft engines in the Flight Propulsion Laboratory of NACA, the National Advisory Committee for Aeronautics, which would later become the National Aviation and Space Administration, N.A.S.A.. In the Fall of 1947, Strang joined the Mechanical Engineering faculty at the Massachusetts Institute of Technology, in Cambridge, Massachusetts. It was easier and more fun, Strang explained, to earn an advanced degree by being one of the faculty than it was by being one of the students. For the next four years, from 1947 to 1951, Strang taught both undergraduate and graduate courses in mechanical engineering before earning his own masters degree from his fellow faculty.

While at M.I.T., Strang began writing and contributing articles to boating publications, reporting on outboard racing events and the changing technology of outboard motors. Over the course of the next thirteen years and over 150 issues, Strang would write a monthly column for *Motorboating* magazine, now known as *Motorboating & Sailing*, which was then, as it is today, considered among the very top journals in the marine industry.

Though Strang possessed an intimidating scientific mind, he always had a unique ability to explain the most complicated subjects in a special, down-home and conversational manner that appealed to readers of all levels of sophistication.

In October 1950, 29-year old Charlie Strang was attending the National Hydroplane "Free-For-All" Championships on Lake Alfred, near Winter Haven, Florida. Carl Kiekhaefer had entered a number of new 25-horsepower Mercury *Thunderbolt* engines. One of the Mercury's that had performed brilliantly in one of the heats was driven by Carl's close friend and Mercury dealer, Jack Maypole. Citing a long forgotten technicality, referee Dick "Coop" McFadden disqualified Maypole and unceremoniously moved his number from first to last on the tally board. Though Maypole would later that day establish a new class speed record of around 70 miles per hour, Strang was incensed at what he considered to be an unfair call, stormed up to the judges stand and became engrossed in a loud exchange with referee McFadden, defending Kiekhaefer's driver. During Strang's technical explanation, McFadden signaled with his eyes and hands, and discreetly said "shhhhhhh ... behind you." Carl Kiekhaefer had wandered up behind Strang, and said, "Thanks for sticking up for my driver, I want you to know I appreciate it."

Strang had never met Kiekhaefer before, though as an avid outboard racer and boating journalist, he was aware of the growing reputation, and of the powerful Mercury engines from Cedarburg and Fond du Lac. In fact, only the week before, Strang had written his first letter to Carl, asking for photographs taken at races during the year for an article Strang was preparing on outboard racing highlights of 1950. Later that day at Lake Alfred, Armand Hauser, who in 1950 was Kiekhaefer's sales manager and perpetual booster, introduced himself to Strang, telling him that Carl wanted to speak with him in his suite at the Winter Haven Hotel.

The timing for their meeting was unfortunate, for Carl had just received word that his father had passed away. When Strang reached the suite, Carl was in total disarray, sitting in his jockey shorts on the edge of the bed, furiously trying to pull on his socks and dress as quickly as possible to return to Mequon and his distraught family. Though Carl appeared stressed and in a great hurry, Strang would not have known the reason for the rush if Hauser hadn't tipped him off on the way to the room. Carl wanted to thank Strang again for his defense of Jack Maypole at the judges stand earlier, and to let him know that he would be most happy to assist him with any photographs or information to support his articles. In even the briefest conversations with Charlie Strang, those who meet him are struck with his honesty and with his lightning-quick ease with technical subjects. And that afternoon, even with his great distractions, Carl was quick to recognize the capabilities of the bright young man from M.I.T..

Strang followed up his short meeting with Carl a few weeks later. It was a carefully detailed letter with a proposal to combine a souped-up 75-horsepower Evinrude power head, a Johnson drive shaft, and one of Carl's new *Quicksilver* racing lower units in an attempt to set a new hydroplane "X" class record at over 85 miles per hour. Attached to his letter was a sketch of how his "Mercrude" would be assembled.[2] Carl was once again struck by Strang's technical insight, and impressed by his enthusiasm to break records through engineering innovation.

In the following weeks, the American Power Boat Association held its annual meeting in Chicago, attended by both Strang and Kiekhaefer. When Carl spotted Strang at the gathering, he invited him to dinner along with Dick "Coop" McFadden, the referee that sparked their initial meeting in Florida. Carl ushered them down to Chicago's famed Rush Street, and once seated asked Strang what he thought of his new *Thunderbolt* engine. Strang replied quickly, "Great, but why didn't you build a bigger one," referring to the horsepower edge that OMC still enjoyed over Mercury products in 1950.

"Could we sell it if we did?" Carl asked.

"You bet," Strang snapped back.

"Well, if you're so damn smart, why don't you come out and build it for us?"[3] Carl challenged.

Strang was sure that Carl was only joking, and never gave the comment another thought. A week later, when Strang's article appeared in the December issue of *Motorboating*, Carl wrote to thank him for his positive coverage of the Lake Alfred Nationals, and for applauding the engineering of the *Thunderbolt* that had propelled Jack Maypole to a new class speed record. "We were most grateful for the break you gave the Mercury 25 on the Swift three-point Hydroplane at Lake Alfred," Kiekhaefer wrote. "The number of breaks we got from the boating corespondents in the past ten years could be counted on as many fingers, and it has been somewhat disconcerting."[4] But Strang cautioned Carl in his reply, saying, "You may not be so happy with the article to appear in the January issue, for I took several whacks at engine manufacturers. ..." He went on to praise the Kiekhaefer Corporation, however, writing "I also hope to be able to say something about your reed valves - which seem to be a never-ending source of amazement to most of the drivers. After all," Strang enthused, "these devices of yours are the first new features to hit outboard racing in more than eighteen years."[5] It was a throw away comment in the last paragraph of the letter that must have caught Carl's attention. "I have not taken advantage of your several offers to [assemble the "Mercrude" record attempt outboard] simply because of the current unsettled situation and the resultant possibility that I will be leaving M.I.T. to go into industry or the armed forces. ..."[6]

Actually, Strang was deep in the throes of the most important decisions he would ever make. He was considering an offer to work for DuPont as a research engineer, but he had also been tempted with an enticing package from M.I.T. to remain. The Dean of the mechanical engineering department had offered to have Strang remain to earn a professorship, receive credits toward his doctorate in the courses he had already taught, and also for all the courses he had monitored. Better still, he could even use one of his research projects as a thesis. Essentially, all Strang would have had to do was take about four courses and pass a language exam, and he would have a doctorate and professorship from the most prestigious institution of mechanical engineering in the world.

Strang was currently earning about $4,000 a year as a research associate, and M.I.T. offered to boost his salary to $4,500, while DuPont had extended a robust by comparison offer of $5,500. Strang was thrashing about, weighing the advantages and disadvantages of academic versus industrial life, when Carl Kiekhaefer tracked him down on Christmas Eve of 1950.

Strang was visiting his Aunt at Long Beach, New York, nearly 200 miles from his quarters in Cambridge on the busiest holiday of the year when the phone rang. It was Carl. To this day Strang still has no idea how Carl found him, but it was a Christmas present that he would never forget. Carl offered to give Charlie complete control of a new research department, an atmosphere of unrestrained engineering creativity, and $7,500 a year to start. If that wasn't enough for a young outboard racing bachelor who hadn't left the hallowed halls of M.I.T. for four years, Carl told him that he wanted him to go to Europe first. Following the end of the school year at M.I.T., Strang would spend several months of summer making a complete survey of two-cycle engine manufacturing on the continent, and then return to Wisconsin to take over his new department. Carl suggested that they meet during the New York Motor Boat Show a few weeks later in January. Strang could hardly contain himself.

Carl came down with a bad case of the flu during the first day of the 1951 New York Motor Boat Show, and asked Strang to meet with Kiekhaefer's chief engineer, Reg Rice.[7] Rice described the growing Kiekhaefer manufacturing facilities in glowing terms, and Strang was enthusiastic about the possibilities for the future. But with the worsening of hostilities in Korea, and the subsequent chaos the government was creating among outboard manufacturers, several months went by before Carl wrote in an attempt to close the deal.

> "Under the present unsettled conditions, many corporations find themselves in the delicate situation of an "alley-cat on a backyard fence besieged by hounds on one side and a puddle of water on the other." As far as we are concerned, we do not mind taking on, single-handed, one-half dozen outboard competitors as well as an even dozen of chain-saw competitors, and having a lot of fun. We have also learned to cope (more or less) with all the "Gestapo" departments of the Government; such as the Internal Revenue Department, Wages and Hours Division, N.P.A., U.S. Labor Relations Board, C.I.O., A.F.L., and all the others set upon the destruction of corporations, but when we do not know whether we are going to build civilian or military articles from one month to the next, things are really rough and executive time is at a premium.
>
> "... We are planning an additional engineering building at Cedarburg, primarily for basic research. This program is going to be accelerated in view of excess profit taxes.
>
> "Before planning a new building, I would like to have an opportunity of discussing with you, your future connections with us. If a mutual relationship can be concluded, you could take part in the planning of the proposed new department which would begin modestly but grow rapidly, depending on the results.

"There is a lot to consider when you are planning a move from a position such as you presently have, to go into industry. The large industries such as DuPont and General Motors, at first consideration, may be the most impressive, the most serene, and well-ordered. The small industry which exists by its own capital, its own inventiveness, and shrewd management has far more exciting experiences. Engine manufacture is not easy and competition is keen, but if you really like two cycle engines, there is a good future for the stout-hearted.

"All of the above is, no doubt, obvious to you, but I can speak from experience that the results of individual effort and individual contribution follow the function of X^n where "n" is not less than 2. The limitations on such exponential values are almost non-existent with smaller corporations like ours, whereas in large companies there are factors such as rigid policies, inertia, tradition, politics, and other restrictions; senile and otherwise, which all tend to dilute results of individual accomplishments.

May I hear from you, Charles, as to when it would be possible for you to make a trip to Cedarburg? A weekend would be preferred so that we can have a full day without interruptions to inspect our laboratories, our plants and equipment, and also to meet some of the people you may be working with in the near future. ... I suggest you make it as early as possible."[8]

Strang made the trip to Wisconsin the next month, and during one of his many tours through the Kiekhaefer plants, accepted the challenge and was hired by Carl. They agreed that his official starting date would be June 1, 1951, but that he would spend his first three months on the job in Europe, prowling the two-cycle industry for Carl.

Strang crisscrossed Europe in a mad-dash, non-stop schedule, visiting virtually every notable engine-maker in the process. He hand wrote dozens of letters to Carl during the trip, detailing his findings, from Vespa, Lambretta and Iso motor scooters in Italy, to ball bearing factories in Germany and England. Wherever he roamed, he made note of refinements to combustion chamber scavenging techniques, fuel-injection innovations, metallurgical advances, production capabilities, opportunities for European distribution, and developed a list of the top engineering and design talent working throughout the continent. "Knowing Carl as I did later," Strang laughs recalling the special mission, "He was clever about this thing. He wanted someone to go look at all these European developmental operations who didn't know anything about his!"[9] By August 1951, Strang had finally exhausted his list of manufacturers to investigate, and wrote Carl that he was "Getting itchy to get back to U.S. now with a hot idea or two to try."[10]

Strang worried somewhat about the transition to rural manufacturing, after having been a "city boy" all his life. He told Carl that he wanted to work out of the Cedarburg facility, so he could at least be close to Milwaukee. But when Strang drove out to Wisconsin with his mother in September 1951, the Wisconsin State Fair was in full swing, and they couldn't find a place to stay in Milwaukee. Driving up to Cedarburg proved fruitless, for no rooms were available in the small town either. So they kept on driving and made their way all the way to Fond du Lac, which was the first place they could find accommodations. The next morning when he called Carl, he said "I'm here at the hotel," and Carl promptly replied, "Well, since you're here, you might as well stay here."[11] Gone were his visions of being close to the pulse of a large city. He looked around at the cattle munching the grass next to the two-lane highway and wondered what he had gotten himself into.

He was assigned a desk in the engineering department in Fond du Lac, and after a few days hadn't done much of anything when the manager of the plant stopped him in the hall.

"I notice you're not punching the time clock," he said brusquely.

"I'm not going to punch any time clock. I've never punched a time clock, and I'm not going to start now," Strang retorted.

"By God, you're going to punch a time clock. It's a condition of the job," the plant manager informed him. Strang had a pair of gloves in his hand, and with a flourishing gesture born of the indignity of the situation, began to pull them on, saying "Well, in that case, you can take the job!" and proceeded to march down the stairs toward the exit. Horrified that he could be blamed for ruffling the feathers of Carl's bright new recruit from M.I.T., the manager chased Strang down the stairs, where a tug of war ensued over the remaining glove. "Don't do nothing rash," the manager pleaded, "I'm going to call Mr. Kiekhaefer, just wait for him." After an hour that seemed like a lifetime, Carl came snorting in bellowing, "What's this about not punching a time clock? Everyone punches a time clock!!"

"Do you?" Strang asked hopefully.

"No," Kiekhaefer replied.

"Well, I'm not going to punch one either," Strang announced.

Carl thought about this for a moment, and looking around to see that nobody could overhear them, said "Alright, but don't let anyone know. I don't want anyone to know that you're one over me."[12] Though the incident was over as far as Carl was concerned, it wasn't long before most employees at the plant heard the details of the argument, giving rise to threats of "puttin' on Charlie's Quittin' Mittens" whenever they wanted to get their way in some matter or another.

Strang had been at Fond du Lac for just about three weeks, trying to get his bearings and figure out just what he should be doing as the new director of research. He was ordering measurement instruments and other equipment and thought he was getting the hang of his new life. Then, early on a Sunday morning, the phone jarred Strang out of a peaceful sleep. It was Carl. "Come on down to the plant right away. I want you to meet somebody." When Strang arrived at the plant, Carl introduced him to two men, Tony Bettenhausen and John Fitch. Bettenhausen had recently won the Triple A stock car championships and Fitch had just won the FCCA Sports Car championships. Carl says "Meet John, meet Tony. We're going on the Mexican Road Race." Strang shook their hands and smiled. After a few seconds his eyes glazed over and he turned back to Kiekhaefer and said, "What the hell is the Mexican Road Race?"[13]

CHAPTER SIXTEEN

Operation Mexico

"Winning is not a sometime thing; it's an all-the-time thing. You don't win once in a while, you don't do things right once in a while, you do them right all the time. Winning is a habit. Unfortunately, so is losing."[1]

Vince Lombardi

It seemed like a nearly impossible assignment to everyone except the wide-eyed and passionate Carl Kiekhaefer, who nearly frothed with excitement and enthusiasm. The race, known as the Carrera Panamericana, or Mexican Road Race, was a staggering 1,933 miles long, zigzagging along the most desolate, dusty and hazardous roads of old Mexico. Starting south of the Ithmus of Tehuantepec near the Guatemalan border at Tuxtla Gutierrez, the perilous route snaked along the 9,500 foot spine of the Sierra Madre Oriental mountains, through the mile-high ancient lake bed of Mexico City, and crossed the treacherous and never ending Mexican Plateau through Durango and Chihuahua to the finish line at Ciudad Juarez, across the Rio Grande from historic El Paso.

From the impromptu introduction of Charlie Strang to Tony Bettenhausen and John Fitch on that Sunday morning in October 1951, Carl allowed only three weeks from inspiration to race day, with a starting line that was over 4,000 miles from Fond du Lac. But for Carl, three weeks worked out to 504 hours, and he spread the word that he was prepared to work every one of them in order to arrive at the starting line on time.

Perhaps hoping some fast-lap fortune would rub off, Carl bought a pair of brand-new 1951 Chrysler Saratogas from a rural dealership in Crown Point, Indiana owned by Murrell Belanger, owner of the #99 *Belanger Special*, winner of the 1951 Indianapolis 500. Victory for Belanger's Indy 500 driver Lee Wallard turned to disaster within four days of the race, though, as a broken fuel line turned his Sprint car into a fireball during a "Lee Wallard Day" celebration race. As a result of the horrible conflagration, Wallard endured third-degree burns to over 50%

of his body, and underwent 37 skin grafts during five months in the hospital. Also sponsored by Belanger, Tony Bettenhausen would place ninth in the celebrated Indy classic that year, and went on to win an unprecedented eight victories on the circuit to win his first national AAA championship, earning the most points by a driver in history. In fact, no other driver would even approach Bettenhausen's remarkable record until A.J. Foyt shattered the record books with a spectacular 10-win season in 1964, a full 13 years later.[2]

Bettenhausen drove the first Saratoga straight to Fond du Lac, while John Fitch delivered the other to the plant at Cedarburg. In an article that Charlie Strang wrote for *Speed Age* following the events in Mexico, he describes the engineering melee that ensued in Wisconsin.

> "Things got underway with a bang when Tony appeared at the gates of the Fond du Lac, Wisconsin plant in a new Chrysler Saratoga, fresh from Murrell Belanger's showroom floor.
>
> "The car was completely disassembled by midnight and the engine hauled off to one of the many test cells in the laboratory. Another new Saratoga was undergoing similar treatment at the corporation's Cedarburg plant, forty-five miles to the south, and the engine from that car, too, was sent to the Fond Du Lac laboratory.
>
> "For two weeks between the arrival of the cars and the day they left for Mexico saw a large portion of the Kiekhaefer engineering staff working day and night without cessation.
>
> "Suspension systems were revamped for maximum safety at speeds in excess of one hundred miles per hour. The interiors were fitted with auxiliary instrument panels containing oil temperature and pressure gauges and specially calibrated tachometers. Sturdy roll-over frames were built into each car as a safety measure, as were seat belts.
>
> "It was decided to install the fuel tanks in the trunks in order to increase the proportion of total weight carried by the rear wheels. A standard fifty-four gallon solvent drum was found to fit nicely in this location after a little snipping had been done on the trunk floor. Subsequent fuel consumption tests indicated that a ten-gallon increase in fuel capacity would be necessary so the drums were stretched the required amount by simply hooking a garden hose to the filler cap and letting city water pressure bulge the ends out to a nice radius!
>
> "The engines were tested on an eddy current dynamometer with electronic speed control to facilitate obtaining power

curves ... seemingly endless tests were run on the dynamometer to carburet the engine properly and determine the optimum setting for the spark advance.

"Much work went into the exhaust system as the engine was found to be very sensitive in this respect. A wide variety of exhaust pipe designs were investigated from both the theoretical and experimental viewpoint. The cars were finally equipped with pipes of about three and one-half inches diameter and just long enough to reach the leading edge of the front door, exhausting to the side of the car rather than to the rear.

Interestingly enough, it was found that best output could be obtained with special two-cycle spark plugs developed by the Kiekhaefer engineers for use in Mercury outboard motors!. ...

"As reinstalled in the cars, the engines delivered approximately thirty percent more horsepower to the rear end gears than was available in the cars at the time they were driven off the showroom floor.

"Obviously this added power would result in heavier brake loadings and in an effort to prolong the life of the brakes - and the drivers - a tank of liquefied carbon dioxide was mounted next to the driver's seat in each car. Flexible tubes led from the tank to each brake assembly and by opening a valve on the tank the riding mechanic could cause a snow of dry ice to enter each brake drum with subsequent cooling of the braking surfaces. ..."

"It was night and day, 24 hours," Strang recalled, "We just turned the engineering department upside down ... making race cars out of these passenger cars to run this stupid race. ... In fact, the manager of the Cedarburg plant had a new Chrysler and he drove up one day on some business, and while he was there we went out and stole the exhaust manifolds off the car."[3]

Nearly the entire plant was thrown off balance by the frenzied effort to machine new parts, cast special assemblies and beat the clock. Then, two-weeks later at three o'clock in the morning, just as the exhausted, black-faced and steaming band of mechanics and engineers were reduced to babbling zombies, slurring their words and moving by sheer inertia, Carl says: "Alright, let's take off for Mexico!"

It was a rag-tag entourage that slipped away from Fond du Lac under the cover of pre-dawn darkness, facing a drive of nearly 4,000 miles to the starting line in southern Mexico. Strang and Kiekhaefer shared a standard Chrysler sedan fitted with a spare race engine. Another stock Chrysler sedan carried several members of the Kiekhaefer engineering staff, while the two race cars were driven by their co-driver/mechanics, Ed Metzler and Dick Williams. "The race cars are rumbling away, you know, 4-inch exhaust pipes going bumpity, bump," Strang recalled, "and all kinds of junk piled in the back, tires and everything; we were totally green at this." A large GMC cab-over-truck, crammed to the roof with spare parts, tires and machine tools completed the rolling stock for the trip, while the company's twin engine Cessna "Bamboo Bomber" flew overhead for additional insurance and convenience. "Altogether, fifteen people were involved, including Fitch and Bettenhausen who joined the caravan at El Paso and Mexico City respectively," Strang remembered. "Needless to say, when this array of iron confronted the Mexican customs men at Juarez, confusion unlimited resulted!"[4]

The long journey to the starting line would be punctuated with events that would become grist for Kiekhaefer stories to delight an entire industry for decades to follow. Strang and Kiekhaefer made it only as far as southern Illinois when their race engine developed problems. The Chrysler engines had been having difficulties with the lobes of their camshafts wearing off. "Well, damn if the lobes didn't wear off the camshaft in the car we were driving," Strang remembered.

> "So, Carl sends the race cars on ahead. He and I are sitting in a little garage down there, the mechanics have the engine all apart. It's very hard to get a camshaft because the engines are so new, so Carl sends the airplane to Detroit to get a camshaft to bring back. Meantime, the race cars are getting miles and miles ahead of us. Carl is sitting on the floor, and he says, "Jesus Christ, now all we need is for lightening to hit us!"[5]

Carl had no sooner made his whimsical pronouncement when a gigantic, ear-splitting and flashing bolt of lightning struck the chimney of the garage, causing it to explode in a huge, and hair-raising fireball. Absolute pandemonium erupted inside the garage, as everyone imagined the whole structure was going to be engulfed in flames or simply collapse because of the direct strike. The men scrambled in and out of the garage, and finally attempted to push the sedan outside to safety before it became apparent that there was no continuing danger. "I've told this story a thousand times," Strang promises, "and no one has ever believed it, but by God it happened."

As the entourage passed through Mexico on the way to the starting line, Strang developed a severe case of "Montezuma's Revenge", the nearly debilitating diarrhea commonly contracted by drinking bacteria-laden water or eating less than fresh meat products from roadside stands. "Tony Bettenhausen says, 'I've got a cure for that,'" Strang said. "He opens up a suitcase and pulls out two big bottles of blackberry brandy." Charlie Strang had never had a drink in his 29 years of life until Bettenhausen poured four fingers of the sweet, powerful flavored brandy that day in Mexico. Throughout the long, hot day, Strang stopped at numerous cantinas along the dusty road and ordered a water glass full of blackberry brandy. He doesn't recall if the sticky elixir had any effect whatsoever on his distressing condition, but then he doesn't recall much else after the first two or three treatments. "I don't know what it did for Montezuma's Revenge, but it was a *wonderful* trip."

Along the way, Strang and mechanic Buddy Boyle took turns picking buzzards off of cattle and horses that had been struck by cars during the night, testing the accuracy of a pistol made by a new company called Ruger. "That was my first test," Strang says. It was a strange beginning to a remarkable career that would eventually catapult Strang to the very top of the marine industry.

One of the problems besetting the Kiekhaefer Chryslers gave Strang the perfect opportunity to establish himself as the brilliant magician from M.I.T. who could concoct a solution to any mechanical challenge.

> "The new engines were so powerful that when the car was pushed hard it would burn out the gears in the rear axle, and they burned them out in a matter of ten miles of flat out driving. I had spent considerable time at M.I.T. with surface active agents, very potent lubricating devices, including some that really attacked the surface [of the metal] to make a greasy surface. You would have very, very slow wear. I went out and bought a couple of pints of that stuff and mixed up our own rear axle grease. And after that we never had a failure and Carl thought that was magnificent. That was the nature of our relationship right there."[6]

Ed Metzler smashed into a Mexican cow along the route to Mexico City, crumpling the hood and breaking the radiator of #7, causing another delay for the exhausted crew. When the bleary-eyed group finally reached Mexico City, they had their first shower and bed rest in over a week.

Carl, though, was ready to quit and go home. After arriving in Mexico City following a grinding haul of nearly 4,000 miles, he was appalled at the condition of his cars and crew compared to other entries which had also stopped to rest in the city enroute to the starting line. Particularly upsetting to him was the fact that other participants had neatly trucked their cars under tight covers, and had flown their crews at the last moment to preserve their strength and stamina. Carl had only to see their gleaming, perfectly polished race cars sitting along side of his filthy, mud splattered, crumpled up, guano blasted entries with a thick skin of dead insects glued to the windshield, piloted by the remains of good men turned into zombies by the demanding preparations and trip. He was ready to go home.[7]

But after a good night's sleep, Carl decided to press on. In the morning, he hired a cab driver to lead his parade of vehicles out of Mexico City and onto the road south to Tuxtla. Carl was behind the wheel of one sedan, and lit a giant Cuban cigar as the caravan pulled into heavy traffic. At one of the largest intersections in Mexico City, a huge *glorietta*, or traffic circle, was surging with morning rush hour traffic as the group was quickly swept into the swirling mass of confusion. All the cars made it through on the first orbit except for one. Carl had been sucked into the middle of the stream of cars like a kayak caught in a boiling current of river rapids. With his arms waving madly and cigar sticking straight up by the side of his nose, Carl cursed the drivers around him frantically for revolution after revolution while the helpless but nearly hysterical onlookers from Fond du Lac watched their purple-faced boss going 'round and 'round, breaking free after no less than eight tries. Carl, looking straight ahead, his cigar smoke billowing out the side window, drove to the head of the line and promptly fired the cab driver.

As the group continued on for the starting line, Strang stayed behind with the company pilot Gene Christianson and engineer Reg Rice to follow in the twin Cessna a day later. But as their aircraft neared the southern border of Mexico, Strang recalled the tense situation that developed in the cockpit.

> "Deep in the heart of the Mexican jungle near the Guatemalan border it became apparent that we were slightly lost, air maps for that territory being something less than complete. A mild cough from one of the engines also informed Gene that we were running out of gas.
>
> "Spotting a road Gene set the plane down neatly, gliding in over the head of an astounded ox cart driver. In no time at all dozens of natives appeared out of the countryside bearing long machetes, a Hudson truck drove up, and a bus load of excited Mexicans confronted us!"[8]

The anxious Americans learned that Tuxtla, the starting point for the race, was just over the next mountain. They managed to take off down the road, landing in Tuxtla with mere fumes in the fuel tanks. The following morning the race would begin.

The order of departure from Tuxtla was determined by drawing numbers from a bowl. A hundred cars scrambled into line as the starting time neared. Drivers were given a card with their starting time marked when they roared off at one minute intervals in a cloud of dust for the finish line of the first leg. As they crossed each subsequent finish line, their arrival time would be duly logged on their time card, and then returned to them when their next departure time was recorded. The car and driver with the lowest cumulative driving time on arrival at Ciudad Juarez would be proclaimed the winner.

The road conditions were hazardous even for normal speeds, and roadway safety was a luxury the Mexican road builders had done without. One journalist described the event as "one of the most grueling road races in the world."

> "From near sea level to 9,500 feet and back down. Mile after mile, this highway writhes like a snake, giving neither man nor machine a rest.
>
> "The turns are hairpin, without guard rails and in many places lacking warning signs other than hat size stones painted white and spaced a few feet apart. To slip here is to drop hundreds of feet into a boulder strewn gorge."[9]

During the first leg, from Tuxtla to Oaxaca, champion stock car driver Tony Bettenhausen nearly ended half of Kiekhaefer's chances in the race when he put the #7 Chrysler into a terrific spin at more than 100 miles per hour. The spin was so violent that large flat spots were ground into all four tires, which had to be changed before Bettenhausen could continue.

The other Chrysler, driven by John Fitch, looked like it was going to be the leader following the first leg, having passed all but one car, including the Ferrari's in the line-up, after starting off in 32nd position. But after 325 miles, only four miles from Oaxaca, his oil pressure relief valve stuck open, stopping oil flow to the engine bearings, and putting Fitch out of the race for good.

Bettenhausen, known for his heavy feet on the dirt tracks of America, soon exhausted the carbon dioxide coolant for his brakes in the mountains south of Mexico City, and couldn't stop his car by the time he reached Oaxaca in 66th position. In testimony to years of racing experience, he began circling the public square at high speed, rubbing and grinding his tires against the curb until the car finally slowed to a stop.

In the straightaway race legs north of Mexico City, Bettenhausen surprised the Ferrari teams with a display of brute horsepower, and finished on their tails as they thundered into Durango. Throughout the long night, the streets "resounded to the sharp crackle of the Ferrari engines as the Italians tested for hours."[10] But the next day, Bettenhausen fairly rocketed into Chihuahua, passing everything in his path to set a new record for the leg at a blazing 112.5 miles per hour average across the Mexican flatlands. The following day, during the final leg into Juarez, Bettenhausen again blasted the course record by maintaining an average speed of 114.33 miles per hour, topping the old record by a whopping 14 mph.

Bettenhausen kept Strang and Kiekhaefer amused with his thrilling antics as they flew overhead across the cactus sprinkled plateau. Strang reported Bettenhausen's heroics to *Speed Age* magazine.

> "Always a showman, Tony occasionally dropped one wheel off the edge of the pavement, throwing a high tail of dirt into the air for the enjoyment of the crowds gathered at turns in the road!
>
> "At one point in his wild ride Tony maneuvered the car into a one-lane bridge at better than one hundred per - only to see a dog standing in the middle of the bridge. The problem was solved instantly however, for the dog took one look at the onrushing Chrysler and leaped over the side into the river! Another tight squeak occurred when Tony decided to pass a car

as it went into a spin while on a railroad crossing. The car spun to the right, Tony passed to the left, the glove compartment flew open bombarding Metzler with spark plugs, while Tony roared with delight!"[11]

The Ferrari teams had built up a sizable lead in the mountain legs of the race, and even Bettenhausen's speed in the final legs couldn't make up for the lead enjoyed by the Italians. The Ferrari's came in first and second, while the Kiekhaefer Chrysler came in third according to total accumulated driving time. A loud protest was mounted by the owners of mass-produced American automobiles, insisting that in future runnings of the race two distinct divisions be established, one for standard production models, and the other for special, limited edition sports cars like the exotic Ferraris.

Even though Carl was incensed about finishing behind the Ferraris in his road racing debut, he was so happy to get out of Mexico that he was nearly uncontrollable. It had been Carl's first trip into a foreign country, and he was showing little tolerance for the many inconveniences he was forced to endure. Accustomed to the hearty fare of Wisconsin dinner tables, the spicy foods, tough meats and plates buzzing with flies were driving him positively *loco*. At one restaurant during the race, Carl complained bitterly about the toughness of the steak he had ordered. In order to make his point to the less than bemused Mexican waiter, Carl excused himself from the table only to return with a pair of greasy pliers from the mechanics tool case, and proceeded to rip and tear his steak apart with great aplomb, proclaiming that he liked to tenderize his steak before he eats it. Carl relished the wide-eyed reaction he provoked with his pliers, and would employ this method in the future, even to a tuxedoed waiter in New York City.

Carl hated not being able to communicate with the Spanish speaking people involved in the race, from officials to local mechanics. His frustration peaked in the closing days of the race, and when the crews finally reached the U.S. border at El Paso, Charlie Strang remembers that Carl was almost deliriously happy to be back on American soil.

"I remember the first night we crossed the border into El Paso after being in Mexico for a week. He was so happy we had to immediately go get real American food, so we went to a Chinese restaurant. And then the squirt can of shaving cream had just come out at that time, you know, the pressure can. We stopped in the drug store because he wanted a can of shaving cream. We were walking down the street in El Paso at night, along a long line of parked cars. Carl was so happy, he was kind of jigging around and he'd take the can of shaving cream and make big circles on all the windshields of the cars as we walked down the street. I've seen him do things like that so often ... he was a big kid, and his sense of humor was of the same nature. He enjoyed a good practical joke. Nothing really subtle, but he liked broad humor."[12]

As much as Carl disliked Mexico, he loved the idea of the Mexican Road Race. Considered the ultimate road racing challenge in North America, Carl's tantalizingly close finish in third place behind the Italian Ferrari's had only whetted his appetite to mount a challenge again the following year. Kiekhaefer learned some valuable lessons during the 1951 Panamericana, and when officials agreed to separate the expensive sports cars like Ferrari, Porsche, Lancia and Jaguar from production automobiles, he was determined to win the stock car division.

Carl prepared two cars for the 1952 race, the 1951 Chrysler Saratoga sedan in which Tony Bettenhausen had set two leg records, and a new 1953 Lincoln. Again, Fond du Lac and Cedarburg facilities were pushed to their limits to prepare the entries, and again Carl assembled an impressive caravan to make the long trek to the starting line. Carl selected Bob Korf, an Army Captain from Wright Field to pilot the new Lincoln, and Reginald "Speed" McFee, veteran sprint car driver from Rochester, New York to drive the Chrysler. Among the valuable lessons Carl had learned from the last race was to bring food and water supplies for his crews. Carl had also learned to avoid the long and uncomfortable drive to Mexico, and instead flew to Mexico City to join the race crews. Once again, Carl had prepared meticulously for the challenge, and armed with the hard-fought experience of the last race, he had every expectation of arriving in Juarez in first place.

But Carl hadn't counted on the heavy hand and Detroit muscle of the automobile industry. As early as June 1952, a full six months before the race, a racing group from California was equipped with three of Ford Motor Company's hand built 1953 Lincoln automobiles, cars which had not yet been announced to the public by Ford. The mechanic for the group, Clay Smith of

Long Beach, was considered among the top racing mechanics in the world. As Carl would later learn, Smith and his crews were able to put "thousands of man hours ... into the actual racing cars over and above the original manufacturing time."[13]

Of the 92 total cars that entered the race in November 1952, only 39 would finish. Over half of the entries were either demolished cascading down the steep ravines along the course, or came to a boiling or explosive halt in the blistering heat of Northern Mexico.

At almost every point during the race, the first five cars were the trio of Lincolns from California and the two Kiekhaefer cars. But every time Carl's cars got close, the Lincolns from California were just a little bit faster. Carl began to wonder why.

During one of the closely contested legs of the race, disaster nearly struck the Kiekhaefer crew as the co-driver in Korf's Lincoln was thrown from the car at break-neck speed.

> "The co-pilot, Philip C. Gow, of Fond du Lac, Wisconsin had unstrapped his safety belt to reach for a canteen of water in the rear. While doing this he brushed against and unlatched the door handle. Korf was in a turn to the left at approximately 70 mph and Gow was tossed out, but other than losing some finger nails and suffering multiple bruises, he was unhurt. Korf picked up Gow, and continued."[14]

But Korf and the Kiekhaefer Lincoln had lost precious minutes as Gow limped back to the car. Though not injured severely, Gow was eventually judged unfit to continue, and Korf continued north alone. Further into the race, Korf lost several more minutes when a leaking fuel line caused him to run out of gas.

The California Lincolns, backed by Ford, finished a remarkable 1-2-3 at Juarez, followed by Carl's Lincoln and Chrysler in 4th and 5th place respectively. Astonishingly, the California Lincolns had finished within 4 minutes and 49 seconds of one another. The Kiekhaefer cars finished only ten minutes behind the lead car, and only a few minutes behind third place. Carl smelled a rat, and was enraged when he discovered how the other Lincolns were able to keep just slightly ahead of his own.

Before the race, Carl had questioned race officials about the use of special gaskets that would prevent hot air from entering the carburetor from the engine, so called "heat riser passage" restriction gaskets. The race committee had told him, "No, these are not considered stock." Further, the careful polishing and streamlining of intake manifolds and intake ports was likewise forbidden. But when the three winning Lincolns from California were inspected after the race, all three cars had these "illegal and non-stock" modifications. By the rules, the first three cars should have been disqualified, and the Kiekhaefer cars declared the winners in first and second place. But the power of the Ford Motor Company intervened, and Carl went wild.

Among Carl's deepest concerns in sporting events of any nature, was that fairness of play extend to every participant, regardless of sponsorship. Naturally, Carl was stung by the loss of nearly $15,000 in prize money, but he was more incensed over being robbed of a well deserved and important victory. Venting his frustration, he wrote to Don O'Reilly, publisher and editor of *Speed Age* magazine, who was also the Information Director for NASCAR, the influential National Association for Stock Car Auto Racing, the sanctioning body for stock car racing in America.

> "The reason for my participation in this Road Race stems, no doubt from our inability to race in boating events since we are manufacturers, and under the rules of the American Power Boat Association manufacturers of equipment such as ours are not permitted to participate in the sanctioned A.P.B.A. racing events. Having been a life-time racing enthusiastic and having need for release of this emotion in the form of automotive high jinks and having need for what also may amount to a mental flue cleaning for some of our speed minded engineers, we decided to once more engage in the ... Pan American Road Race. Come what may in spite of the firm resolve made in 1951 that never again would we go to the trouble and the expense.
>
> "We did not expect to run smack into big business down there in Mexico when we took this annual holiday from bucking big business - something we have become quite used to. ...
>
> "But when big business comes way down there to Mexico prepared ... with a militant and organized squad of crack professional racing men, drivers, mechanics, all in uniform of course, with the prize race cars being trucked from Los Angeles to Mexico City on an express haul-away truck, all under the supervision of the Chief of Ford Public Relations, with medical and many other aids in full attendance, all rehearsed and practiced, it certainly had the appearance of Detroit in business. ...
>
> "When the officials, in a quandary as to what to do about the three first place cars that had restricted heat riser passages that had specifically been banned from us by the same officials,

telephoned to the Lincoln factory and asked for clarification on which type of Lincoln was stock, they received that now infamous reply stating that both engines were to be considered stock and that Lincoln would issue a bulletin in the future to justify their reply.

"Confronted with this telegram the harassed race officials had little choice but to declare all of the Lincoln cars to be stock, and as a result apparently none of the cars finishing the race were disqualified for any reason. ...

"Several other drivers who along with us questioned the propriety of these decisions were promptly labeled "poor sports" and "bum losers" by high powered manufacturer's publicity representatives who were anxious to capitalize on the results of the race. Pity the poor individual who ever dare to raise his voice against the smooth working, well entrenched Detroit interests! ... Because a few of us dared to raise our voice against these obvious inequities we have been heaped with opprobrium and vilification on a national scale.

"The very atmosphere of the post race inspection reeked of collusion and heavy handed pressure. The first three Lincolns were constantly referred to as "our cars" by John Millis, termed by the El Paso Times as the "publicity director" for the first three place cars, but actually the Chief of Public Relations of the Lincoln Division of the Ford Motor Company. ... The Mexican Road Race can be the world's greatest sporting event ... right now it sadly needs some real sporting spirit.

"This is an injustice and is insulting to the intelligence of every thinking individual and a disgrace to United States manufacturers, and a low-blow to well-meaning and hard working and sincere race officials. ...

"We are a couple of boys from the country. We are learning fast."[15]

Though seething mad, Carl lost little time in making plans for the 1953 Panamericana. He was particularly proud of the valiant effort made by his drivers, and signed both Korf and McFee to try it again the following year. But the "couple of boys from the country", as Kiekhaefer so piquantly referred to his sophisticated racing efforts, had indeed learned fast.

The audacity of the power play by Ford Motor Company during tear-down inspections following the 1952 race, created a wave of creative thinking in Wisconsin. "We are more determined than ever to have another chance at the 'smart' boys from Detroit and California," Carl said in January 1953.[16] Within weeks, Carl and Charlie Strang were in Detroit, meeting with

Chrysler, and baiting them with the huge publicity coup that Ford had scored with their sweep of the last race. Carl wrote, "Perhaps you are not aware of the extent that Lincoln is publicizing articles ... on their Pan American Road Race successes in the various sports magazines plus the thousands of newspapers that have carried the stories favorable to Lincoln."[17]

In the same letter, Carl threw his new controller, Donald E. Castle, into a panic. In justification of his expensive efforts to win the Panamericana, Carl wrote, "What has this to do with outboard motors? Nothing -- except that the Pan American Road Race event is our annual vacation time. We like the Pan American Road Race like you love to fish; both are expensive hobbies."[18] Castle couldn't imagine a worse thing to put in writing, knowing if the Internal Revenue Service read it, they could disallow all of Kiekhaefer's huge deductions in connection with the races, and classify his expensive efforts as a personal "hobby". "I had to try to recover all those letters," Castle remembered, "because I don't have to tell you what would have happened with Internal Revenue if they'd have gotten hold of that."[19]

Knowing the advantage that the "smart boys from Detroit and California" had enjoyed by having a new model Lincoln available six months ahead of the race, Carl began to pressure Chrysler to keep the playing field level in Mexico for 1953. Carl wanted a 1954 Chrysler "practice car" as early in the year as possible, in order to prepare for the November race. Kiekhaefer could then theoretically share the same competitive advantage with Chrysler as the California team had with Ford. But by the middle of July, Carl reminded Chrysler that they had still not delivered on their promise to provide a "practice car" ahead of production models.

> "We cannot stress too heavily the importance of getting this practice car into 'practice.' ... We respectfully request that everything be done that is possible to expedite delivery of the first race car. We request this first car on the premise that you intend to win this race. No 'half way' measures will win ..."
>
> "We are going into the Carrera Panamericana 'Mexico' with odds enough! How about getting that first practice car by July 31!"[20]

Just as Chrysler agreed to deliver his cars by the end of July, the race committee in Mexico City moved to ban all 1954 cars from the race, explaining, "Models of 1954 cars will not be permitted to run in this year's race regardless of their on-sale date. Allowing '53 cars to enter the 1952 race caused one of the race committee's biggest hassles."[21] It was Carl's own harsh criticism of Ford tactics that forced the rule change, wiping out any advantage he might have gained with the new Chryslers.

Carl then decided to enter four 1953 Chryslers in the race. His drivers included Robert Korf, Reg (Speed) McFee, Frank Mundy and John Fitch, with co-driver/mechanic duties assigned to Donald Ziebell, Robert "Buddy" Boyle, James Blake and George Thompson. Herman Behm accompanied the group as chief mechanic. Almost everything about the 1953 race turned into a major disaster, for both participants and spectators alike.

During the very first leg from Tuxtla to Oaxaca, three drivers were killed and at least six Mexican spectators were crushed to death in the most tragic accident in the history of the race. "A Ford driven by Robert F. Christie of Grant's Pass, Oregon, overturned while speeding around a curve near Tehuantepec," the *New York Times* reported.[22] Spectators lining the roadway when the accident occurred, rushed to aid the driver and his co-pilot, who were not injured. But another Ford, driven by Mickey Thompson of El Monte, California, "rounded a turn and mowed down the spectators. Thompson's car also turned over but he and his co-pilot were only slightly bruised."[23] An Italian co-driver, Giuseppe Escousutti, died when his car ran off the road and overturned, and two Mexican drivers perished during the first leg as well.

Carl was having some problems he could rationalize as an engineer, but others only deliberate sabotage could explain. The race was a complete bust for the Kiekhaefer entries, none of which would finish even the third day of the five day race. As Carl explained later to the *Wizard* buyer at Western Auto Supply, he was out-spent and out-foxed by a well oiled Detroit machine once again.

> "... Thanks for the condolences. We are getting a lot of them and while the kindness is appreciated, we do not have much sorrow since the trouble was not ours but Chrysler's. Not as an alibi but as an engineering explanation, I shall relate to you what our difficulties were.
>
> "We were using a type of disc brake which is not yet in production at Chrysler. On preliminary trials, this brake looked 5 times as good as a standard brake, and subsequently,

all of our cars were equipped with this identical brake. Something went wrong in the material control at Chrysler. Our problem was not temperature but a physical cracking of the discs. When this mechanical failure occurred, the cars were without brakes with the result that transmissions were overworked until either the engine or the transmission disintegrated. Hindsight indicates we should have adhered to the original brakes and the results would have been quite different.

"Of course, we have entered this race primarily as amateurs and in 2 weeks tried to do what the Ford Motor Company, $3,000,000 and 152 servicemen did in 1 year.

"Other strange things happened such as a fuel tank springing a leak from a hole punched in while the car was supposedly under guard in the impound area during the night. Also a seal being broken while the car was supposedly under guard in the impound area with the result that the bearings went out within an hour of racing the following day. This is the type of luck that besets those riding near the top places. I am sure of one thing, next year there will only be Lincolns racing. No one but another multimillion dollar Detroit factory could ever ... compete on an equitable basis. Instead of a sport, this race has become a grim business between factories. ...

"Anything we say, however, will be just considered an alibi. The picture in general is similar to last year, only worse. The Chief of Public Relations of Ford Motor Company made the statement "that this year they were really prepared for Kiekhaefer". You can take that as you wish. At any rate, between the sabotage and Chrysler's brake problem, we've had it."[24]

Carl's first forays into the rarefied territory of auto racing left him with a very bad taste in his mouth. He vowed never to enter the Mexican Road Race again, and he never did. In fact, in the next three decades of his life, he would never again even set foot in Mexico. But the thrill of the competition Carl had felt south of the border left an indelible mark on his personality. The rush he experienced in pursuit of the checkered flag was so well suited to his obsessively competitive spirit, that he began to lay plans to shift his attack to the American track. But much had happened to his growing company and the industry during the three years road-racing in Mexico, which would leave precious little time to pursue his dreams of auto racing glory.

CHAPTER SEVENTEEN

Loyal Assistance

"Loyalty to our team is the first requisite. Loyalty is above all ability and knowledge. Without loyalty, there can be no team work. Without team work, we obviously cannot win. If we cannot win, there is no point in carrying on the hard work [and] the sleepless nights. ... "[1]

Carl Kiekhaefer

Many special people joined the Kiekhaefer Mercury ranks during the exciting decade of the 1950s. They each arrived from diverse backgrounds to make their unique contribution, and some were destined to impact the life of Carl Kiekhaefer in dramatic fashion, both emotionally and professionally. Along with Charlie Strang, who became Carl's most indispensable engineering resource and confidant, fresh talent brought Carl relief from an overload of administrative burdens.

In the late 40s and early 50s, Carl steadfastly refused to shift the growing mountain of administrative chores to subordinates. He insisted on scanning every piece of mail, reading every telegram received by his operations, and personally approving every purchase made, even down to toilet paper and paper clips. Carl had become *King Kiekhaefer*, the absolute monarch of his realm, receiving a torrential flow of papers to sign and approve from his swelling population of subjects. But even as Carl worked seven days a week, 52 weeks of the year, he couldn't keep up with the tens of thousands of details that poured over his desk. His reluctance to delegate administrative responsibility was due in part to his basic mistrust in people, and partly due to his nearly fanatic penchant for perfection.

Carl had run through a succession of secretaries, none of which worked long for the demanding, brusque-talking and intimidating boss. His first real relief arrived in the form of an articulate young man who was hired only three days after being discharged from the service in 1946. Fred "Fritz" Shoenfeldt (shane-felt) was a tall, lanky addition to the assembly line at Cedarburg, whose first job was to assemble air cleaners for Disston-Kiekhaefer chain-saws. But after just three days on the

75-cents an hour job, Fritz figured he "had made enough to last forever."[2] Before entering the service, Fritz had studied to be a court reporter, learning shorthand and typing, and decided to ask for an office job.

Fortunately for Fritz, the secretary for Fred Hall, Carl's vice president of sales, was leaving her position due to pregnancy. Hall had considered the idea of a male secretary, and agreed to give Fritz a try. Fred Hall, Fritz said, "was just about the nicest person I have ever met." In fact, it was just this quality that made Carl quite jealous of Hall. "Fred had the ability to get along with everybody," Fritz recalled. "Everyone liked him. He rented a room down on Main Street in Cedarburg, and he'd walk to work, whistling as he went, and everybody would say 'Hi' to him. Everybody knew him. Carl Kiekhaefer couldn't do that."[3] Fritz worked with Hall for several years before the pressure of Carl's demand for higher sales eventually got to Hall, who was replaced by Armand Hauser.

Hauser had started with Kiekhaefer in 1943, moving into the sales department in 1944. He quickly became assistant sales manager, and by late 1949 was appointed director of sales. (He later became vice president of sales in 1953.) Hauser was also a most affable man, who had the unique ability to sometimes fall asleep at his desk, holding up a report folder to disguise his nap. Carl once passed by the snoozing Hauser, who had propped a newspaper up in front of him, and heard the loud and unmistakable sounds of snoring. "Armand!", Carl screamed as the newspaper flew up into the air, "How can you read the paper with all that racket going on!?" Hauser was also known throughout the corporation as "The Silver Tongued Orator", and "Golden Tongue" for his talent as a speech writer and excellent corporate speaker.

Fritz Shoenfeldt became restless and applied for work at nearby Power Products in Grafton, Wisconsin. But the well established industrial grapevine inside the Kiekhaefer Corporation took only a few hours for word of Shoenfeldt's defection plans to reach Carl's ears. Within a few short years, Fritz had become popular with just about everyone. Since he was adept at shorthand, he was "volunteered" to take the minutes at both the Wednesday evening production meetings in Cedarburg, and also the dreaded Saturday morning meetings at Fond du Lac. A long succession of secretaries had suffered the long and confusing meetings, often keeping them from their families well past midnight on Wednesdays and deep into the afternoons on Saturday. When Fritz became the regular minute stenographer at these meetings, the ladies of both plants who had taken turns at the chore treated him like a hero. On the other hand, Carl and the engineering and production staffs that attended the meetings, appreciated Shoenfeldt for his quick shorthand, and highly organized style. So when Carl heard that Shoenfeldt had applied elsewhere, he buttonholed him and said, "No way Fritz, you're staying here. I need an administrative man, and someone to take this God damn paperwork off my hands. I can't *stand* it!!"[4] Not long after Fritz accepted the position of assistant to the president, he discovered life became very different. "It was a seven day a week job. Day and night. No matter where you were, or what time it was, you could expect a call. And I got them. I mean I got plenty of them. My wife will attest to that."[5]

Fritz quickly gained Carl's confidence, and was soon supervising all expense accounts for the corporations along with engine allocations to distributors. He was cautious to make sure to always obtain Carl's signature on everything before any action was taken. But the sheer volume of details, and the complexity of the operations often put Fritz between a rock and a hard place. One such dilemma occurred during the wartime aluminum restrictions of the early 50s, reducing plant output to only a dozen or so new 25-horsepower Mercury engines a day. "Well Christ," Fritz remembers the problem well, "you've got twelve engines and forty distributors hounding you for engines. So, I would be allocating and I would get one here, one there, two there, three there, and do as best I could. So, one day I come in the office and [someone] says, 'Carl called during lunch hour and he wants you to ship ten of those 25-horsepower jobs to Lee Siebert [a leading Kiekhaefer distributor] in Kansas City.' I said, 'Fine. If that's the case, I'm through!!!'" Fritz had been tearing his hair out debating the politics of giving a single engine to various distributors, and then Carl starts giving them out by the truckload. "I can't sleep nights," Fritz remembered, "I'm trying to keep these God damn orders straight, but if he can give the guys ten at one time, he can take the allocations, I'm going home. I took off and just quit."[6]

Shoenfeldt went home and refused to answer the phone which rang incessantly all afternoon. Finally, about seven in the evening, he answered the phone and it was, of course, Carl. "Who the hell died and made you president?," the booming voice demanded. "Since when do you tell people what I should do?" Fritz, trying to remain calm responded, "I can't perform my duties by giving people the amount of engines you say I should, because there aren't that many."

"What the hell are you talking about?" Carl shouted.

"You're only getting twelve a day," Fritz responded.

"What the hell are you talking about?, Carl repeated, "We're building thirty-five, forty engines a day."

"Then we're working for different companies," Fritz replied, "We've never gotten that many engines in one day."

Carl was flabbergasted. He had been calling the manager of production at the factory and getting a report on the number of engines that went through the line each day. Fritz, on the other hand, was getting his reports from the shipping manager who actually boxed finished engines and put them on trucks for delivery to distributors. In desperation, Fritz said, "Look, I'm in charge of shipping the engines out. I type up the God damn orders, I know what's being shipped out the back door."

Carl, frustrated and irritated at being told his production was less than one-third what he had been told, bellowed, "You get your ass in my office tomorrow morning and we'll get to the bottom of this thing."[7]

At seven-thirty the next morning, Carl and Fritz phoned their respective sources about the previous day's production. Carl called the production manager and was told 38 engines. Fritz called the shipping manager and was told he received only 12. Carl *screamed* at the ceiling, "What the hell is going on around here!!" He then grabbed the phone and dialed the line supervisor, responsible for the actual assembly of engines. The truth of the matter made Carl numb with rage. Yes, 38 engines had reached the end of the production line, but only ten or 12 of those had passed inspection. The engines were pulled off the inspection station, sent back to the line and rebuilt. In some cases, engines were coming through the line *three times* before making their way to the shipping dock. The production manager was counting them as production engines each time they came through. "He called him and said, 'You son-of-a-bitch, you're *fired!*' Just like that." Carl then told Fritz, "From now on, I want you to report to me every night before you go home, the number of engines that go out the back door. I don't want to know how many come off the line, I don't want to know how many are boxed, I only want to know how many are shipped."[8] Every night, for the next nearly 20 years, Fritz called Carl wherever he may be located and gave him the numbers.

Perhaps of all the new additions to Carl's swelling ranks of employees, a young secretary that started in the last weeks of 1949 was destined to have the biggest impact on him, and his family. Rose Smiljanic (smile-jan-ick), was a pretty and quick-witted Milwaukee girl of 21 years when she heard about a job opening while visiting her sister in Fond du Lac. Standing 5-foot four-inches tall in her stocking feet, she had hazel eyes that sparkled when she laughed, and brown hair. Rose wasn't sure what she wanted to do with the rest of her life, and was vacillating about going to college. "I didn't even know what an outboard motor was,"[9] Rose admitted. She accepted a position which turned out to be secretary to Armand Hauser. Hauser, owing to his position as head of sales, spent most of his time away from the office. Consequently, Rose conceded, the job was "totally boring." Hauser's office was in the Corium Farms barn loft, and Rose was relegated to sorting out huge stacks of product literature that had been piling up in his absence.

Rose remembers her first impressions of Carl Kiekhaefer, for whenever the boss was around it created quite a stir in the offices. "Mr. Kiekhaefer's coming!" the cry would go out, and everybody would rush to the windows to see. "I thought, 'how terrible,' and I never would." But she admits that he was quite a striking figure, wheeling up in his brand new Chrysler, stepping out in his flowing Navy blue or black top coat, a Homburg Stetson hat pulled low over his forehead, and a long, brown Cuban cigar pushed over to the side of his mouth. Actually, when Carl came around, everyone couldn't help but sneak glances at the boss who certainly looked the part of a dashing industrial tycoon.

Early in 1950, Rose Smiljanic was approached by the personnel manager, Pete Humleker, and was told that Mr. Kiekhaefer's secretary had become pregnant and would she be interested in taking over the position? Rose had heard throughout the offices that Mr. Kiekhaefer was difficult to work for, "would swear a lot and everything, and maybe I didn't want to be bothered with anybody who was like that."[10] She was still considering the position when Humleker said, "I think you'd probably get along with Mr. Kiekhaefer, and it's more money." She agreed, and started right away. In her wildest dreams she never imagined that she would hold the position for the next 33 years.

Rose was confused by her new job, and had been working in Carl's Fond du Lac office for some time when she realized that she hadn't even met him. Carl was working mostly out of his office in Cedarburg, and rarely worked out of Fond du Lac. "When he'd come up [to Fond du Lac] he had his own private entrance and go into his office and never say *'boo'* to me. All I was doing was opening up the mail," Rose remembers, "take the money to the bank, and it was really a dumb job. I thought, 'How strange, what a bunch of dip sticks.'" She began to miss life in Milwaukee, which, compared to Fond du Lac seemed like New York City to her. "I was going to leave and head back [to Milwaukee]. ... I was just going to go down there and get some action. I didn't like Fond du Lac. They didn't have malls or anything, it was just a very quiet little town."[11] But then the mysterious Mr. Kiekhaefer surprised her by coming up to her one day and saying, "How would you like to transfer down to Cedarburg, because that's where I would like to consolidate the corporate offices." She assured herself that Cedarburg was closer to Milwaukee than Fond du Lac, and so agreed.

It was just before the start of deer hunting season in November 1950, and Rose had been invited with her sister and brother-in-law to northern Wisconsin to join the hunt. She knew that the company required all employees to work five days a week, plus half a day on Saturday, but she rationalized that since she hadn't really started down in Cedarburg that it would be alright. But when Rose arrived on Monday morning, the office staff was already laying odds that Kiekhaefer would blow his stack and send her packing. To everyone's complete surprise, Carl never mentioned her absence. When she arrived, Carl was gone and someone showed her his office. Piled on his desk were dozens of personal details that had been pushed aside by the rush of company business. She quickly organized and squared away his overdue insurance premiums, thank-you notes and schedules, and generally overhauled his filing systems. Carl was grateful, and began to trust her with more and more details of his busy life.

Carl was naturally suspicious of nearly everyone in his organization, many would say to the point of paranoia. He had grown gun-shy about giving his trust to others readily, ever since the attempted sale of the company by stockholders to

Flambeau outboards while he was away in 1940. Fate would deliver to Rose Smiljanic the perfect opportunity to demonstrate her loyalty to Carl, when a wild and potentially disastrous conspiracy began to unfold at the Kiekhaefer Corporation in the spring of 1953.

Rose overheard a secretary for one of the vice presidents discussing a secret plan that involved the stealing of the Disston chain-saw contract, the defection of a number of top engineers and executives, and the secret establishment for Disston of a new engine building facility by a clandestine band of Kiekhaefer's own employees. She was stunned. Carl was on a trip to Florida, supervising operations at the newly opened salt water proving grounds at Siesta Key, near Sarasota, and wasn't scheduled to return for days. She didn't know what to do. If the story wasn't true, she imagined, then she could get into big trouble with the office staff and lose her job for meddling. "If this wasn't true", Rose argued with herself, "I would stir up a bunch of hornets ... they would kill me."[12] On the other hand, if it *were* true, then she knew that her own growing loyalty to Carl would dictate that she tell him immediately, so that he could take appropriate action. "But, I thought, 'I'm going to have to take the chance,'" Rose remembered. She was nervous and shaking as she reached him by phone in Florida, and relayed the information she had learned. The information turned out to be the breaking point to uncover the well-organized conspiracy, and Carl promised Rose that he would never forget what she had done. "He was ever grateful to me for doing that, and he always said, 'Rose, you can have anything you want.'"[13]

Carl described the conspiracy to an amateur private detective he had hired once he had figured out what was going on.

> 'To accomplish their own manufacture, the Disston Company decided to 'play their cards under the table.' Very slyly and subversively, their first step was to employ the confidence of a senior vice president of our corporation; this, according to testimony, happened several years ago.
>
> 'This individual, namely: Guy S. Conrad was in our employ for 12 years. He remained in our employ during the period that the new model development got under way. ... [A] nest of refuge ... is now taking form as the Disston-Wisconsin Corporation. It is to this refuge that our former chief engineer, chief layout

man, chief detailer, chief tool designer, chief tester, etc. all came to roost as they successively 'resigned.' ...

"The net result is a conspiracy on the part of the Disston Company, who were under contract with us not to purchase engines from others; Mr. Guy S. Conrad, our Senior Vice President; and Nelson Pattern Company, who has received as a vender [sic] more than $700,000.00 worth of pattern business from our corporations. It makes quite a story of which I have given you only a brief outline.

"The resulting conspiracy is the subject of a damage suit that may run into $2,000,000.00 With the magnitude of the stakes involved, the Disston Company decided to proceed very slyly and subversively to carry out the 'Trojan-Horse' episode. The fact that we have many loyal employees made them appear somewhat ridiculous, and much evidence has been piled up against Disston. This type of conspiracy suit has been brought to court and upheld many times in favor of the original manufacturer.

"Be all this as it may, our problem is now: Protect the interest of our stockholders, officers and directors, and the welfare of our employees and sales organization; and it is in this effort, we would like to enlist your services at any reasonable fee you may suggest in a manner I would like to discuss with you as early as possible.

"We would be interested in:

1. Knowing as much about Disston's building plans as possible.

2. Installing in their organization surveillances or 'pipelines' of information; stationing employees in strategic spots such as engineering, production and policy departments.

"... If there is any way that you can help us along these lines, rest assured ... we would make it very worthwhile for you. You may not be aware of the fact, but this sort of operation goes on between General Motors and Ford and all of the large manufacturing companies in the United States today; and the stakes in our instance run into figures not less than $20,000,000.00 per annum. It is big business---'bad' big business to some people and good to others, depending upon 'whose ox is being gored.'

Confidentially yours,

E.C. Kiekhaefer

"P.S.: Owing to the nature of this letter, and also to the nature of the business involved, I would appreciate having this letter returned to me the next time we meet. Thanks."[14]

Charlie Strang was perhaps the first to learn of the changing relationship between Carl Kiekhaefer and Rose Smiljanic. Though Carl was depending on Rose more and more to assist him in the day to day blizzard of paperwork and communications, Rose was still, in March 1952, a relatively anonymous administrative secretary. During a trip to Florida that spring, Carl disclosed his hitherto secret relationship with Rose to his most trusted employee.

Carl and Strang were in Sarasota during Easter week to purchase a new parcel of land to expand the salt water proving grounds at Midnight Pass on Siesta Key. Once the transaction was completed, the two engineers headed north in Carl's car, on a long journey across the Florida panhandle, around the coast of the Gulf of Mexico, and on to New Orleans to meet with a distributor. During the long drive, Strang remembers, Carl "just chatted away and he told me all about his relationship with Rose."[15] Rose, then 22, had delighted Carl, 46, Strang recalled, when Carl had brought up the issue of the differences in their ages. According to Carl, Rose had said that "she liked older men," and he was "delighted" and "excited." During the trip to New Orleans Carl would stop "about every hundred miles and call her on the phone and he'd come back and tell me again what a great relationship it was."[16]

As their relationship blossomed, Carl would have a difficult time keeping the wraps on his indiscretion. Slowly, word spread throughout and beyond the Kiekhaefer companies of the affair, and Carl's clumsy efforts to disguise his feelings often made things all the more obvious. "God, it used to get our financial man upset, a guy named Don Castle," Strang explained, "He used to complain: 'The man wants to do this, why the hell doesn't he at least be discreet instead of sending in hotel bills like this.'"[17]

Carl's paranoia spread as a result of his intimate relationship with Rose. He was afraid of being discovered; he was afraid of appearing unfaithful and reckless to his wife and children; but perhaps most of all, he was afraid of losing Rose to another man, especially another Kiekhaefer employee. The thought haunted him, and through the years he checked up on imagined and suspected liaisons between Rose and his executives, sometimes contriving sophisticated plans of cloak and dagger entrapment to satisfy his gnawing suspicion.

Typical of Carl's paranoid escapades, was a trip to the Mead Hotel at Stevens Point, Wisconsin, for a gathering of sports car enthusiasts. With Carl were Charlie Strang, Jim Wynne and Rose. "Carl wound up at the bar talking to some sports car people," Strang remembered, while the rest of the group generally wandered off. Once Carl noticed Rose and the men missing, he stopped his conversation every few minutes to go to the phone. "We had three rooms in a row," Strang laughed, remembering the incident, "it was so funny. First the phone would ring in my room, then it would ring in Rose's room, then it would ring in Carl's room ... calling all three rooms to make sure everyone was in their own rooms!"[18]

MERCURY
THE 1947 MERCURY "LIGHTNING"
The 10-horsepower alternate-firing twin that outperformed competitive models as high as 25 horsepower and set new standards of styling, performance, dependability and durability for the entire outboard motor industry.

The *Lightning*, above, was Kiekhaefer's first completely new design, and was the engine that established Mercury as a high-performance threat in the crowded marine marketplace. The twin-cylinder, 10-hp. engine actually produced over 16-hp.

Below: The Corium Farms barn, purchased by the Kiekhaefer Corp. January 1946.

Rose Smiljanic would become Carl's most intimate associate and friend, and work as his personal secretary and assistant for over thirty-three years.

Ted O. Jones, then the fastest man on the water, dominated the unlimited hydroplane record books for decades. Here, with Carl Kiekhaefer, was head of Carl's target drone engine projects for the Navy at Point Magu, California.

Below: Jones, in dark jacket, supervises the engine start of one the Navy's drone aircraft in preparation for launch.

This page: Carl and his design team developed an impressive number of solutions to the government's need for reliable, high-performance engines to power target drones. Among these were engines from 12- to 90-horsepower. If gunnery training was effective, even his best engines would fly for less than ten minutes.

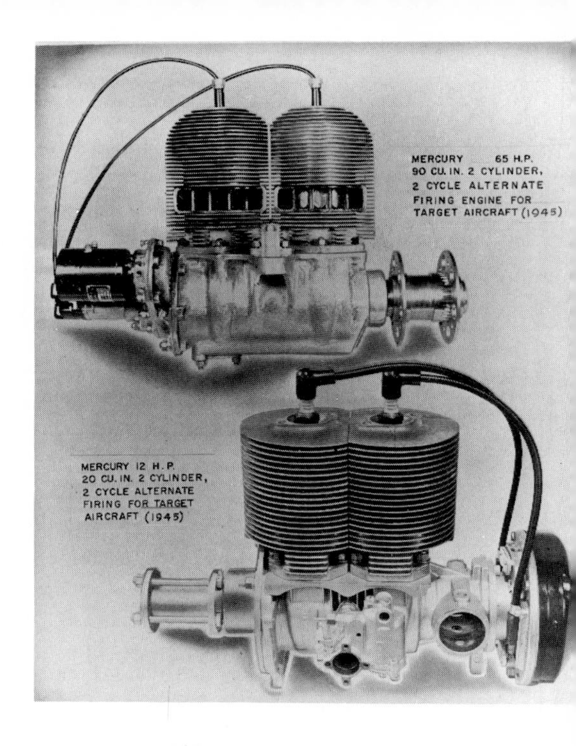

MERCURY 65 H.P.
90 CU. IN. 2 CYLINDER,
2 CYCLE ALTERNATE
FIRING ENGINE FOR
TARGET AIRCRAFT (1945)
MERCURY 12 H.P.
20 CU. IN. 2 CYLINDER,
2 CYCLE ALTERNATE
FIRING FOR TARGET
AIRCRAFT (1945)

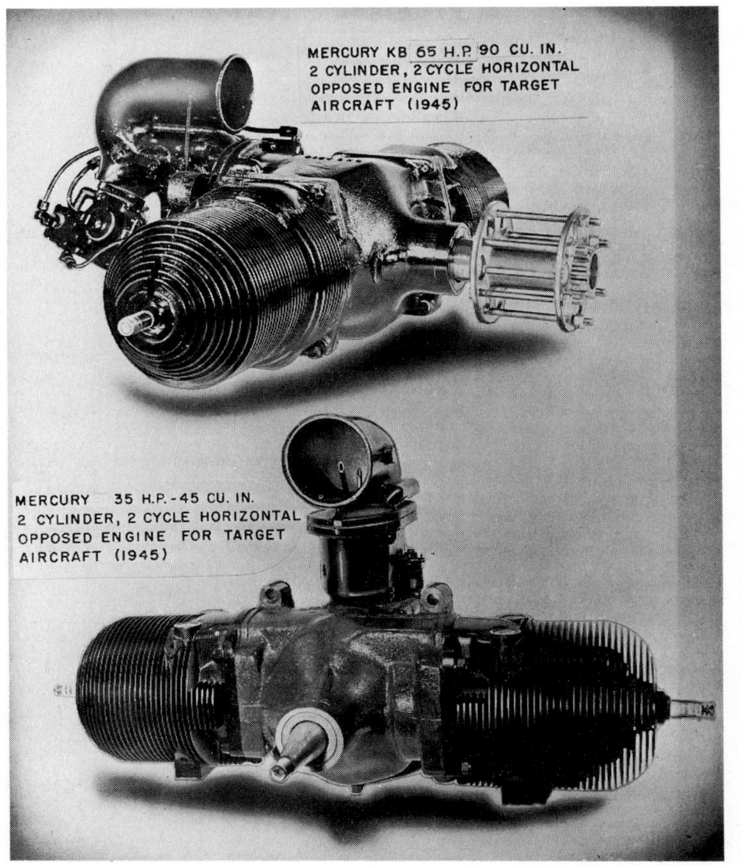

MERCURY KB 65 H.P. 90 CU. IN.
2 CYLINDER, 2 CYCLE HORIZONTAL
OPPOSED ENGINE FOR TARGET
AIRCRAFT (1945)
MERCURY 35 H.P.-45 CU. IN.
2 CYLINDER, 2 CYCLE HORIZONTAL
OPPOSED ENGINE FOR TARGET
AIRCRAFT (1945)

Opposite Page: Above: Ready for the nearly 4,000 mile trip to the starting line of the Mex- ican Road Race, Tony Bettenhausen (#7) and John Fitch (#32) pose front of the Corium Farms Barn assembly plant in Fond du Lac in 1951.

Below: Bettenhausen congratulated by Carl Kiekhaefer after setting a new speed record for the leg from Durango to Chihuahua at 112.5 mph in the 1951 Carrera Panamericana.

7
32

II CARRERA PA
7

Above left: Carl was never reluctant to crawl under the cars with his crews. Here, he helps Captain Bob Korf and his crew during the Carrera Panamericana of 1953.

Above right: Charles D. "Charlie" Strang, Carl's engineering genius and closest friend.

Below: Carl's road racing team departs for the Mexican Road Race in 1953.

They stopped laughing at Carl and his platinum-white, 4,005 pound Chrysler C300 (above) at Daytona Beach in February 1955, when Tim Flock roared over the finish line to win the 160-mile race, Carl's first NASCAR victory.

The following year, Kiekhaefer again dominated the NASCAR circuit, starting with a new record in the Daytona Beach "Flying Mile" at nearly 140 mph in the new Chrysler 300B. His teams would win a phenomenal 80 percent of all races entered, without a single driver injury, and never be charged with an infraction of the rules.

Right: Carl's determination to win stock car races led to his invention of the first paper air filter for automobiles, now standard throughout the world. He made his first ones out of the bottoms of garbage cans, and out of aluminum sauce pans.

Below: When Carl's stock car engines began to disintegrate during races, he collected samples of dirt from each track where he raced.

Carl's oversized Chrysler's swept both the NASCAR and AAA circuits in 1955. Carl poses with Tim Flock (l) and Buck Baker (r), along with the largest collection of stock car racing trophies ever won in a single season.

Below: Among the reasons for such dominance on the tracks, was Carl's insistence on a first class effort every step of the way, like separate trucks for each race car.

Tim Flock, Herb Thomas and Frank "Rebel" Mundy (above) were among Carl's top stock car drivers. Kiekhaefer drivers were among the highest paid in the sport.

Below: Carl's biggest threats were the factory teams from Detroit, like "Fireball" Roberts here driving for the Ford team. But when the checkered flags fell, Carl had beaten them all, and then retired his team from the sport following the 1956 season.

Carl's pit crews could change four tires, fill up with gas and oil, and clean the windshield in 48 seconds. Carl would often direct each pit stop personally.

The onset of winter, an the hazards of testing Mercury outboards in frozen waters (left), led to the search for the year-round proving grounds at Lake X, Florida (below). The 14,000 acre lake, with seven mile shoreline, is situated on a unique property with a net are of over seventeen square miles. Surrounded by natural swamps, access to the lake was nearly impossible except through Kiekhaefer's gates and guards.

Carl loved excavating through the swamps around Lake X, and was considered an expert bulldozer operator. Here, Carl is blazing a new trail that would become an elevated perimeter road around the lake, and would also help to stabilize the level of the lake. It was this Caterpillar D6 that Carl sank in the muck late one night that led to the many popular stories about burying his brand-new Cadillac at Lake X.

The 50,000 mile endurance test at Lake X was the most ambitious record-setting promotion ever conducted by an outboard maker.

Left: One of the endurance boats passes the USAC official's booth.

Below: Endurance boats in the 50,000 mile run were fueled "on the fly." Driving the fuel boat is Jim Wynne, who managed the first 25,000 miles of *Operation Atlas* at Lake X.

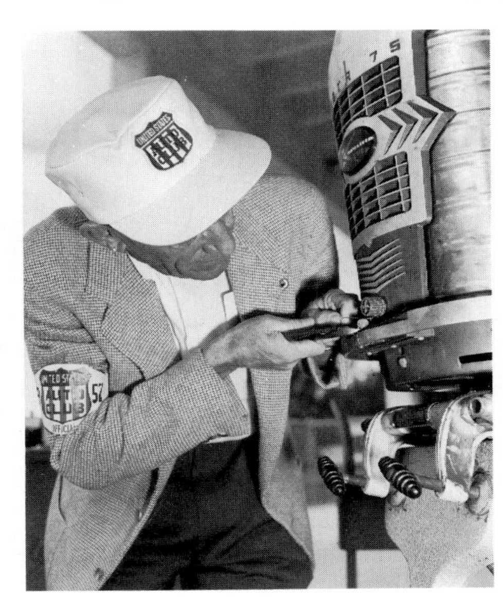

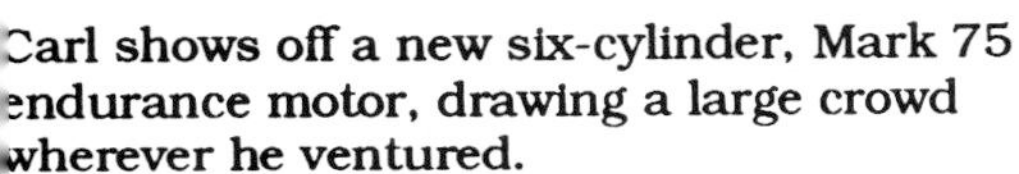

Carl shows off a new six-cylinder, Mark 75 endurance motor, drawing a large crowd wherever he ventured.

Top right: A USAC official breaks the seal on a 50,000 mile endurance engines. Maintenance mechanics at Lake X had long ago figured out how to unseal and re-seal engines after even wholesale components exchanges.

Below: A Mercury outboard shown blasting over a sand bar for a segment of nationally televised "You Asked For It" in 1959. Carl's "Dyna-Shock" absorbers saved many engines and lives.

Carl was ham-fisted at the controls of a boat, knowing only two speeds: off or full throttle. Here, Carl delights some dealers as they fl across the waters of Lake X in 1957, propelled by a pair of his new six-cylinder, 60-horsepower Mark 75 engines.

CHAPTER EIGHTEEN

A Bumper Crop of Talent

"In building up our organization to its present success, ... I have also built up a considerable number of young men. I have had more pleasure in watching these young men and women develop in this organization from ordinary clerks to fine executives and engineers. It seems to put a worthwhile meaning to life to see dreams growing into realities, ideas growing into products, young men and women growing into families, and last but not least, two car garages growing into factories."[1]

Carl Kiekhaefer
November 1955

"The pilot waved toward the small group on shore, a signal that he was ready. Then he tightened the white helmet about his face, adjusted clear-lensed goggles over his eyes, and punched the starter. The engine coughed once, caught and barked into idle. 'Okay, baby,' he muttered. 'Let's show those bastards.'"[2]

Ted Jones had been designing, building and racing boats since he was 17 years old in 1927. In 1950, Jones was preparing to shatter the world's speed record on the water in the excerpt above from *True* magazine, and when his thirty-foot tall rooster tail finally fell back into the water, he had pushed owner Stan Sayres' *Slo-Mo-Shun IV* three-point hydroplane of his own design to an unprecedented speed of over 160 miles per hour. He was the world's fastest man on the water. The following year, after Jones won the American Power Boat Association Gold Cup, and his Unlimited Class Hydroplane design captured the International Harmsworth Trophy, Carl Kiekhaefer showed up at his shop on Lake Washington, near Seattle.

> "I was in the boat house on the lake getting the *Slo-Mo-Shun* ready for a test run. I felt someone was watching me, and yet I was suppose to be alone in the shop and I looked around and there's a fat man with a hat and an unlit cigar in his mouth just standing there watching me. I had a guard on the gate, a Pinkerton man, he'd keep people out 'cause everybody wanted to get in and see the boat and I hadn't time to put up with it.
>
> "'So,' I said, 'How did you get in?' He said, 'Your guard likes money.' He gave him a hundred dollar bill or something, there was nothing that would stop Carl, ever. Nothing."[3]

Jones, 40, strikingly handsome and athletic, was clearly unimpressed with the antics of his portly visitor, and completely ignored him as he continued work on his boat. After a while, without looking up, he asked his uninvited guest who he was and what line of work he was in. When Carl Kiekhaefer proudly introduced himself as the owner of Mercury, Jones took great delight in replying, "Never heard of it," making Carl describe his little outboard engines while Jones finished work on the 1,500-horsepower Allison Aircraft engine aboard the record-setting boat. "I'm trying to think of a way to get even with this guy," Jones remembered, "and I thought the best way to do it is take him for a 200 mile an hour ride and this will get his attention." Jones lowered the *Slo-Mo-Shun IV* into the water and told Kiekhaefer to grab a jacket off the wall and get ready for a ride. "His mouth opened and the cigar fell out of it," Jones says, "he couldn't believe it."

> "So, I stuffed him in the boat. It was a two-place job, and I took him up to about seventy-five, eighty miles an hour and I looked at him. I had gotten his attention all right, but he seemed to *enjoy* it at that speed, so I punched it up to about 175 and that *really* got his attention. So, I moved faster, and he came in and was so thrilled he couldn't talk. He said, 'That was the biggest thrill of my life.' I gave him a card saying that he had attained the speed of a hundred and seventy-five miles an hour on *Slo-Mo-Shun IV* and he used that card for years and years as identification to cash checks and everything else."[4]

After this first meeting with Carl, Ted Jones became curious and learned more about the mysterious Kiekhaefer from his friends around the racing community. "I heard about him that he had the biggest temper of any man in the world," Jones said.

When Carl returned a few days later with an offer for Jones to design record-setting boats in the outboard class, Jones was prepared and turned him down flat. Carl naturally asked, "Why not?" and Jones candidly replied, "Because I heard you had a horrible temper and life's too short to put up with people like that." According to Jones, at this point Carl "turned purple" but contained himself and left, saying "I'll get back to you."[5]

The tenacious Kiekhaefer didn't give up easily, and began to phone and correspond with Jones, slowly wearing him down. "Finally, he sent me $800 and an airline ticket and said, 'Come back and visit my plants.'" Jones relented, and flew to Fond du Lac.

He knocked on the big door with Carl Kiekhaefer's name on it, and heard the gruff voice bark, "Come in!" But when Carl saw Jones at the door, he quickly rose from his seat, rushed past him, closed the door and melodramatically locked them both in the room. Surprised, Jones asked, "What's the idea?" Carl, coolly settling down into his chair, said, "You're not leaving here until you tell me why you *really* don't want to go to work for me." Jones again replied frankly, "I told you, you have too much of a temper." Carl pulled out a pad of paper and quickly hand wrote a contract offer to Jones, including $10,000 a year for record-breaking boat designs, a bonus of $1,000 for each hull when the design was frozen, and an additional bonus of $7,500 for a Gold Cup design that would establish new records in 1952. Carl also wanted Jones available for "special assignments."

A contract worth more than $21,000 in 1951 was impressive, Jones would admit, but there was still the sticky issue of Carl's explosive temper. But when Jones finished reading, Carl gravely stood up, raised his right hand, and with a sober poker-face recited, "I, Carl Kiekhaefer, do solemnly swear to never display my temper to one Ted Jones." Carl's unusual pledge and the plump contract were enough to convince Jones to join the team.

One of Ted Jones' first "special assignments" was to search for a suitable location in Florida for a salt-water proving grounds for outboard motor testing and corrosion experiments. Jones located a feasible site near Pennsacola, and returned with Carl to show it to him. The local mayor, an elderly gentleman, had extended generous terms to lure industry to the area, and even offered to contribute the land for free if Carl would maintain a certain payroll in the area for five years.

Carl rented a car and picked up the mayor to proceed to the site. "I don't know if you know about Carl's driving," Jones wondered, "but if the car would do 110, we did 110. This mayor was about 75, 80 years old and wasn't accustomed to going fast like that. He fainted. I had to shake him to wake him up once we got there. We're going down this dirt road at 100 miles an hour and chickens are flying out all over and finally we got to this land and the old man was too shook to get out and walk around.[6] Carl, though, didn't like the site, and would soon establish a salt water proving ground at Siesta Key, south of Tampa.

As aluminum restrictions imposed by the Korean conflict began to choke off Carl's outboard engine production lines, he once again focused his attention on government contracts. By early 1952, Carl had established a testing program on a new generation of target aircraft engines. The new engines, including a V-4 inverted cylinder, 90-horsepower model, were having technical problems, and Carl was reaching his short limits of patience with both the military and his engineering staff. He pulled Ted Jones away from boat design projects, and shipped him off to the Navy air base at Point Magu near Oxnard in Southern California, where testing had nearly come to a stand-still.

> "I had never seen a target plane in my life. I'm standing in this old hanger with a tiny light bulb in the ceiling, looking at a pile of busted up targets in the corner. Some captain comes in and said, 'You'd better have one of those on the line ready to fly in the morning, and it's not good to defer in the Navy, so get with it.'"[7]

In the dim light, Jones went to work, bolting together the useful pieces of smashed aircraft. He had never even seen the new Kiekhaefer Aeromarine engine before, but he managed to dissect and reassemble one before the scheduled test time in the morning. Jones stood back as the Navy loaded the little target aircraft onto the catapult, and prepared to launch. The engine started quickly, and was throttled up to launch speed. The catapult released, and the little craft went screaming into the air, only to lose power and auger into the ground a hundred yards away. The Navy personnel shook their heads and walked away. Jones knew that Carl's new target aircraft engine program was in serious trouble.

Less than 50 miles to the south, in Los Angeles, another engine testing program was proceeding beautifully. Bob McCulloch, Carl's old nemesis, had secured contracts to produce his own target aircraft engine for the Army Air Corps. McCulloch's engine was performing nearly flawlessly, and was pulling the Army's targets through the sky at almost 200 mph, almost twice the speed of the Kiekhaefer engines. Carl had poured nearly $200,000 into his new engine design, and it seemed like all he could do was bore holes into the California hillsides. Part of the problem, though, was Carl himself.

One of the conditions for releasing Carl's prototype engines for testing on aircraft was that they run fifty hours at high speed without a shutdown. Carl had so intimidated his testing crew by yelling and screaming for results, that they were forced to make the engine pass this crucial test by any means.

> "Some of the people were so afraid of Carl, that if they made a mistake they wouldn't tell him about it, and on this engine, several things occurred. ... The engine would quit, and the mechanics would take the [Navy inspector] out and buy him some coffee or a drink, and repair the engine. They repaired the engine about sixteen times in fifty hours, so the engine was never right when I got it out in the field.
>
> "When Carl found out about that he fired everybody [involved in the testing program] but me, and then hired them back the next day with a slight increase in pay. That's the way he did things. He understood that he frightened the people to a point where they wouldn't admit that they made a mistake, because to make a mistake was to get chewed off something horribly."[8]

After Ted Jones' initial, fatal flight with the new Kiekhaefer engine, he spent the entire night in the hangar preparing a new "bird" for flight test. With five minutes to spare the following morning, he tightened the last bolt on the engine. This time, the target flew 200 yards before crashing into the ground at nearly 100 miles per hour. Jones had concentrated his efforts the evening before on the fuel system, hoping to discover some reason that the engines were stalling after takeoff, but to no avail. Watching the second launch carefully gave him the solution he needed. As the catapult hurled the aircraft towards the sky like an elaborate slingshot, it forced the fuel in the tank to rush against the back of the tank and away from the fuel pump, causing the engine to quit at a critical point in the flight.

By coiling some fuel line as far forward of the engine as he could, he gave the engine about a pint of extra fuel to draw on while the inertia of the launch was trying to starve the engine. Once the aircraft recovered from the jolt of launching, the fuel would stabilize in the tank and feed normally. The next test lasted almost seven minutes, a tremendous improvement over the earlier flights, but then the engine seized and a parachute was radio-deployed from the tail to recover the aircraft and engine.

After every test, Jones was required to phone Carl back in Fond du Lac to report on the results. Carl insisted on knowing the barometric pressure at the time of the flight, the air temperature, along with essentially every technical aspect and result of the test. Week after week and month after month, testing would conclude with seized pistons, as Jones, along with Kiekhaefer engineer Reg Rice, experimented with different cooling and lubricating solutions until they achieved an "out of fuel" flight. A perfect flight would occur when the target was flown for a full thirty minutes until it ran out of fuel, and the parachute was deployed for recovery.

Once the engine began to perform well, the Navy placed orders for production models. Eventually, the Kiekhaefer engine would out-perform the rival engine the Army Air Corps bought from Bob McCulloch. Even though Carl's engine was equally good, McCulloch was paid nearly four times as much for his engine than Kiekhaefer. The Army paid McCulloch $2500 for their engine, while Carl received $675 for his. To make matters worse, some estimates showed that Carl's engine cost over twice as much to build as the McCulloch version.

Carl knew only too well how the game of contracting was played, however, and wasn't adverse to tilting the scales in his favor whenever he had the opportunity. Kiekhaefer had arranged sexual favors, sometimes on a regular basis, for some of the top Navy personnel associated with the drone engine project. Charlie Strang, who was witness to some of these special promotional efforts, found himself completely out of his element on at least one humorous occasion.

> "There was an escort service in Chicago run by a woman. God, we used to get regular monthly billings from her. In fact, the first time I ran into it [was for] a meeting with a group of the Navy people, purchasing people. [We] walked into this hotel room, three or four guys from the Navy, Carl and one of our

vice presidents, myself, with a whole array of young ladies. The meeting was essentially sitting around, everyone drinking merrily. I didn't drink worth a damn at that time, but they had sparkling burgundy [but] because I couldn't stand the taste of it, I was putting sugar in it, which was the worst. God awful. I don't know how I did that.

"I was being very polite. One of my outstanding memories of that was [when] a young red-headed hooker asked me some kind of question and I very politely said, 'No Ma'am.' At that time I was young and handsome, and she *jumped* up and she *screamed*, 'I don't care if he *does* look like Cary Grant, he can't call me a *Ma'am*!' She came *tearing* across the room half-plotzed and everyone disappeared into rooms with their associates. I guess that's how you got the Navy business in those days."[9]

Carl became increasingly frustrated at the blatant displays of military politics and favoritism which took the place of engineering excellence and performance. He anxiously awaited the return of unlimited quantities of aluminum and other strategic materials, so he could withdraw from his government contracts and concentrate on outboard motors.

When Ted Jones had finally accepted Carl's promise not to display his temper to him, Carl added one more condition of his employment: hands off Rose Smiljanic. "He explained to everybody that came to work there that she was off limits to everybody," Jones said, "And he did that to me, too, in a *real loud* voice. He made it very, *very* clear."[10] Perhaps not clear enough, because before the dashing, 42 year old Jones was whisked off to California to work on the drone engine project, he and Rose had fallen head over heels for each other. It was a dangerous predicament for both of them, as Rose would risk an explosion of her relationship with Carl along with her job, while Jones would chance the unpredictable wrath of one of America's most powerful industrialists, not to mention his own job and lucrative contracts.

They corresponded heavily for over a year starting in April 1952, with a flurry of letters passing between Jones in Oxnard, California and Rose Smiljanic's home in Cedarburg, Wisconsin. In Jones' letters he would refer to Carl Kiekhaefer using the code words "Uncle Carl," "K", "Our little friend," "our boy, and "Papa." The last expression must have been in deference to Kiekhaefer's paternalistic attitude about his employees, because Jones was only four years younger than Carl. Jones understood Rose's predicament, and advised her "to do as you see fit or, as your heart tells you with respect to your future," and "whichever way it goes I hope that it turns out to be the happiest for you."[11] At one point, Rose was planning a vacation to California, during which she would rendezvous with Jones. But at the last moment, Carl ordered Jones to go to Kingman, Arizona for additional target drone engine testing, and Jones protested to Rose, writing, "I think he suspects who you were planning to see in California."[12] Another time, Jones hadn't received his expense checks from Carl in Fond du Lac. He considered having Rose apply a little pressure back in the office, but then thought better of it, and wrote, "No, you couldn't do that or someone would get wise."[13] After Rose nearly fainted in the office one day, Jones wrote, "I guess working for 'our boy' is cause enough for anyone to faint," but admitted in the same letter that "You know, under that gruff exterior he is not such a bad guy."[14] Realizing his own danger, Jones accepted the hazards of correspondence with Rose, knowing that Carl might one day discover one of his or her letters. "For you Darling, I'll take a chance on 'our boy's' anger, however I would not want him on your pretty neck on my account. It will be difficult to convince him that love prompted our letters. I'm sure I wouldn't take time to write you 65 letters if I only had 'wolfing' in mind. Do you?"[15]

The threat of discovery reached dizzying heights when Carl asked Jones to teach him how to behave like a gentlemen with Rose. Carl, coming as he did from a rough and unpolished farming environment and years of heavy industry, was afraid that his coarse and clumsy manners would embarrass Rose. "He had an inferiority complex," Jones said, "and this is why he yelled so that you wouldn't detect it. I think this is why he stormed around and yelled and had everybody at bay. ... He was always concerned about his lack of hair and being fat and one thing or other."[16]

> "He took Rose and I to dinner on several occasions and I held her coat for her, and put her coat back on. It was cold country. And I moved the chair out so she could sit, those things Carl never bothered to learn to do. So one day he said, 'I want you to go to dinner with Rose and I tonight, and I want you to tell me ahead of time all these things you did so I could do them.'

"So, we drove up to Johnny's on the lake [Winnebago] and parked. He got out of the car and started for the door and I told him, 'You better open the door for the lady.' And so he went back and opened the door for Rose and he started walking ahead of her and I told him, 'You better walk behind her or along side of her.'

"We got inside and he took his coat off and threw it at the [coat check girl] and started in looking for a table. I told him, 'You take the lady's coat,' and he went back and he took Rose's coat. And then I had him move the chair back so she could sit, and then move it up to her.

"After dinner we started out to the car and he just started out ahead of her and opened the door and got in, and I said, 'Carl, are you going to let the lady in?' and he said, 'Oh, the *Hell* with this, it's too much to do. I'm not going to do this anymore!' So that was the beginning and the end of his attempt to be courteous."[17]

Jones continued work on the target drone engine project at Point Magu, and on March 11, 1952, he finally had the first, 100 percent successful flight. Three months later, after a perfect "out of fuel" flight, Jones was requested by the Navy brass to re-fuel and prepare the target drone for another flight. The Secretary of the Navy in charge of aviation was at the base and was to be treated to a demonstration of the Kiekhaefer engine in person. Take-off and climb-out proceeded without a hitch, but then the radio-control servo units in the aircraft's elevator stuck in an absolute, vertical down position, and "the poor little 'bird' hit the ground at 350 miles per hour." Jones wrote to Rose and asked, "Is 'Papa' unhappy?"[18]

Ever paranoid and vigilant against suspected trespassing, Carl was eventually certain that something was going on between Jones and Rose, and wired Jones to return to Fond du Lac immediately for an urgent meeting. When he arrived, Kiekhaefer angrily confronted him with his suspicions, which Jones fervently denied. Carl blurted out that he had been tipped-off by an informant in the company, a vice president no less, whereupon Jones insisted that the vice president appear to make his accusations in their presence. According to Jones, when the vice president was confronted he couldn't produce any first-hand evidence, so the matter was dropped.[19] Carl was well aware of the subterfuge, though, for he had obtained copies of the many letters Rose had received from Jones. It is unknown exactly when Carl obtained the letters, whether he found them himself, or whether one of his many "spies" had discovered them and delivered them to him. Carl may have in fact been testing Jones, for the "informant" may very well have been Jones' letters themselves. At any rate, the perilous brush with discovery and catastrophe was enough to precipitate an end to the romance.

In the spring of 1953, another strategic recruit arrived in Fond du Lac, in the form of Charles Freeman Alexander, Jr. "Alex", as he was informally known, was discovered working for the Navy at Point Magu by newly recruited M.I.T. graduate engineer Jim Wynne and Charlie Strang. Raised in Kansas City, Alexander graduated from the University of Michigan with bachelor's degrees in mechanical engineering and marine engineering and naval architecture. Within weeks of the surprise attack by the Japanese on Pearl Harbor, Alexander signed up for the Navy's Midshipmen School at Columbia University, becoming a 90 day wonder Navy ensign. Out of 300 engineers in Alexander's class at Columbia, Alex grabbed top honors in the engine course, prompting the Navy to retain him to instruct future officers in the mysteries of internal combustion. After sixteen months, he shipped out on a small aircraft carrier in the Pacific theater, steaming back to the west coast at the end of hostilities. He worked for a year at the Atlas Diesel Engine Company in Oakland, California, before he landed a job as a civilian engineer in the Navy propulsion laboratory at Point Magu in 1947.

At the U.S. Naval Air Missile Test Center at Point Magu, Alex assisted in the construction of a large, five foot three inch wind tunnel for "testing things like target drone engines, sub-sonic ram jets, and things that operated below the speed of sound."[20] When the Kiekhaefer target drone engine arrived at Point Magu for evaluation, Alex was assigned to represent the Navy in the evaluation of the design. When Jim Wynne, whom Carl had appointed assistant director of research, arrived at Point Magu from Fond du Lac in April 1953, Wynne and Alex began to work together, becoming friends. Nothing, though, could prepare Alex for his first impression of Carl Kiekhaefer, who arrived one day with his crack entourage from Fond du Lac.

"The first time I saw Kiekhaefer, is when he marched up into our test area at the wind tunnel, and it looked like a military operation. He walked in front, and single file behind him came the rest of his guys, and everything was 'Yes, Sir,' No, Sir,' it was *really* a well disciplined group. I told my wife that night when I got home, 'I met a guy today that really runs a tight ship. I'd sure never like to work for a guy like that.'

"I was a young fellow, [and] that amount of discipline didn't appeal to me, particularly after three years in the Navy. It struck me as funny that he had such a highly disciplined operation. It made the military around the Navy base there look kind of lackadaisical. It was just his nature and control over things."[21]

After six years at Point Magu, Alex's father, who owned a small business in Fort Worth, Texas, said, 'You're thirty years old now. If you're going to do anything, you'd better get at it!' Alex let Jim Wynne know he was looking for something with more possibilities than the civil service program and government wage scales, and Wynne passed the word along to Strang in Fond du Lac. Strang invited Alex to tour the Kiekhaefer facilities in Wisconsin, and took him for a whirlwind tour in Strang's new blood red Jaguar XK-140 that Carl had given him. "I tore around the roads of Wisconsin that whole day," Alex remembers, and by the end of the day he was hooked on the great future that seemed to be in store for Mercury. Two months later, in the spring of 1953, Carl's new aircraft chief engineer was driving his family and worldly possessions across the great plains toward Wisconsin when he realized, "Boy, this is kind of wild, because I don't even have a letter from Strang telling me what the deal is."[22]

Alex was sucked into the vortex of company operations almost immediately upon arriving. Like being tossed into a blender, Alex joined the others who "worked seven days a week, and when the end of the day comes we have dinner and come back at night. Non-stop work to get the job done." Carl, Alex thought, had so many powerful abilities, that it was hard to narrow down the ones that made his success inescapable.

"One of the main things that really got him ahead is his fierce competitive instinct and his boundless physical energy that just kept him going. He made a lot of mistakes along the way, but that didn't matter, because he corrected them. He once told me in that same connection, 'You know, I like young

> people working for me', he said, 'They make a lot of mistakes, but they've got a lot of energy and they correct the mistakes and you just go on from there.' So, there was a lot of cut and try, and hit and miss, but the activity just seemed to get the job done."[23]

Alexander worked his way up through the Kiekhaefer Mercury organization, and was assigned to special engineering projects suggested by Carl himself. Alex thought of Carl as a unique type of engineer. "His big talent besides being very energetic and competitive," Alex said, "wasn't engineering in the *detail* sense. His talent was sensing what the market would go for and that was also technically possible."[24] Two important inventions, the tilt-up shock absorber and the *Jet Prop* underwater exhaust, were examples where Carl's marketing intuition and practical engineering sense combined to spearhead improvements in outboard motors which became industry standards.

Though Carl is credited with over 200 patents, he often insisted on having his name associated with innovations that he may have had only the slightest brush with, or could claim, "I thought of that *years* ago."

Injuries periodically occurred to boat operators and passengers when the outboard motor would strike a solid object, like a big log, and flip up into the boat with the propeller still spinning. "Put some shock absorbers on the engine so it won't come into the boat when it kicks up," Kiekhaefer bellowed to Alexander one day. "He thought *that* was the invention," Alexander says, "and it damn sure was the assignment to make a marketable product, but the patent office didn't consider than an invention. And he never could understand that. There are many, many examples of that sort of direction that he gave the engineers and then the engineers would figure out some little angle or detail that made the thing function better or function in a unique way that then caused the patent office to issue a patent. And then he'd get mad at the engineer who'd done that."[25] Charlie Strang also realized Carl's engineering limitations, but suggests that Kiekhaefer's unique combination of talents more than made up for a lack of academic training.

> "From the academic sense I would not say Carl had much on the engineering line, but what he did have was a good solid design capability. He knew what, as he used to say, 'Looks good, works good.' He knew what looked good, so he had an

> instinctive sense of 'it's not strong enough.' He would never be one to be able to stress analyze something, or what have you. He would just look at something and say, 'Hey, it's not strong enough.' So, he had an enhanced mechanical capability as opposed to what you normally think of as engineering talent at all.
>
> "The other thing he really had was a real instinct for what the public would like, because it was the same stuff that he liked. I'd tell him that he was no engineer, but he was the world's greatest salesman, and he'd steam over that. But, he was fantastic for seeing what the public wanted."[26]

Another terrifically marketable engineering innovation was the *Jet Prop* exhaust that routed engine exhaust through the propeller hub far below the surface of the water. "Until Mercury perfected the through-hub exhaust, Alexander said, "if you raised the engine a little bit too high on the transom, or if the boat hopped a little bit in turning, it would lift the exhaust free of the water and the thing would howl, *waw-waw-waw* as it would go around the corner." Before the *Jet Prop*, most outboard engines had a small tube or opening just above the propeller that "kind of plowed a hole in the water and laid the exhaust in it."[27] Actually, and unbeknownst to Carl Kiekhaefer at the time, Evinrude had experimented with a form of through-hub exhaust, and secured a patent. "The problem was," Alexander says, "It didn't work. The exhaust would get back into the propeller blades and the propeller would just turn loose and bore a hole in the water and would not push the boat." But again, Carl understood the practical engineering and marketing potential of a successful design, and put Alexander to work once again, telling him, "I want you to work up a design routing the exhaust through the prop." "I found out, Alexander said, "that if you brought the hub straight back, just cylindrical, well behind the blades and cut it off abruptly, you could stop the exhaust flow back into the blades." Alexander's innovation was successful, and is responsible for the look of most outboard propellers with hub exhaust today, sort of a large round cylinder with blades attached to the sides, but he made the mistake of putting his own name on the patent.

> "Kiekhaefer was *furious* when he found out that [the patent] had been issued to me. ... The shock absorbers and the *Jet Prop* are examples of what happened on many, many things where Kiekhaefer sensed the need, the marketing advantage,

thought that it would be technically feasible, and told somebody to do it. Often, [he] would actually take charge and ask someone to get his name on the [patent] application because he considered some of these things just details in carrying out his instructions, and didn't, apparently, didn't recognize that those details were the only patentable part of the whole thing.

"If it was early on [in the application process] and hadn't been submitted, he'd either take your name off and put his on, or if he really felt magnanimous, he'd put his name first and yours second. If the damn thing had already been filed with the patent office, he'd just raise *hell*."[28]

Charlie Strang, who has been responsible for generating hundreds of new innovations in the industry throughout his career, had a more pragmatic approach to Carl's encroachment of ideas. "It didn't bother me at all," Strang said, "It bothered a lot of people. I always had a very practical outlook. If I wasn't making money out of the patent, I didn't care who was [getting the credit]. He would look at something and say, 'Gee, I thought of something like that five years ago. He really ought to have my name on that.' Some of the young engineers who would come up with things would literally turn blue."[29]

Of all the newly arrived talent to the Kiekhaefer organization in the 1950s, Jim Wynne was destined to have the most emotional impact on Carl. Born James Richard Wynne in 1929, Wynne had taken early to the water and began seriously racing boats during his high school years in South Florida. After graduating from the University of Florida in the spring of 1951 with a bachelors degree in mechanical engineering, he entered the graduate school at M.I.T..

Arriving at M.I.T. only three months after Charlie Strang had left to join Kiekhaefer, he was often told that, "You really ought to meet this guy, he used to race boats too, you'd have a lot in common."[30] When the *New York Times* published an article advocating the re-emergence of the inter-collegiate boat racing association that was abandoned during World War II leaving Charlie Strang as its reigning champion, Wynne took the opportunity to write Strang, lending his support to the cause. Strang wrote back to thank him, and asked if Jim would be in a position to help Kiekhaefer Aeromarine with the new target drone engine they were developing for the Navy. Strang needed information on performance of the engine at various altitudes, and knew that Wynne would have access to the Sloan Automotive Laboratory at M.I.T. (named after Alfred P. Sloan, Jr., the brilliant executive and philanthropist who guided General Motors as president and chairman for more than a quarter century, and who graduated from M.I.T. in 1895.). Though Wynne was retained by Strang and Kiekhaefer for the token sum of $125 per month, the true value of the project to Wynne was the ability to use this practical research for his master's thesis.

In the fall of 1952, Jim Wynne made a trip to Wisconsin to meet both Kiekhaefer and Strang for the first time. In February 1953, as Wynne was concluding his academic studies, Strang surprised him by offering a job with Mercury. "I had been aiming for a career in the aircraft industry, primarily in aircraft engines," Wynne recalled, "and I really didn't see where I could get an engine oriented career outside the aircraft field."[31] Wynne considered the opportunity most unique, but was reluctant to locate in the north. "I wanted to get back to Florida, or at least somewhere warm, and the thought of going to Oshkosh, Wisconsin was not very enthusiastic," he said. He was, however, genuinely impressed with both Kiekhaefer and Strang, and in March 1953, reported for work as the assistant director of research assigned to Edgar Rose, the new director of research who had been recruited only a month earlier.

Wynne had barely unpacked his bags in Wisconsin when he was dispatched to Point Magu, California to work on the target drone engine program. "How long should I plan to be out there?" Wynne asked Strang. "Oh, probably a couple of weeks," Strang said. Wynne had just rented an apartment in Oshkosh, and finished unloading all his worldly possessions out of his 1950 Ford when he was summoned to Mercury's twin-engine Beech 18 aircraft for the long flight to California. Aboard the Beech were Strang and Kiekhaefer, along with a couple of other engineers. After a week, Strang and Kiekhaefer returned to Wisconsin, leaving Wynne to supervise the project.

> "To make a long story short, I was there a year and a half. And this was just typical. Carl decided he wanted me out there. The fact that I had an apartment and a car parked in the driveway of the apartment doesn't enter into it. It was about eight or nine months later that I finally got a trip back to Wisconsin on a quick business trip at which time I was able to get rid of my apartment and move my stuff back into my car and park the car behind the research lab. That was typical."[32]

Wynne was bounced around between California, Texas, the White Sands proving grounds and Fort Bliss, tidying up loose ends of the drone engine project and trouble-shooting new outboard engine designs in Washington and Alaska. When he finally returned to Wisconsin in the fall of 1954, he was assigned the task of heading up Mercury's proving grounds, or product and prototype testing departments. In Oshkosh, Carl had expanded his testing operations to include boat transom and clamped down static testing of outboards in the river running through town. At one point, with dozens of powerful engines running day and night along the west side of the river, a rumor was started that Mercury outboards were pushing Oshkosh towards California at a rate of several inches a week, and eager to please engineers helped in estimating when the town would hit the coast. When the river froze in the winter, testing operations continued near Sarasota, Florida, at Mercury's Midnight Pass facility on Siesta Key.

Once Wynne was put in charge of Mercury's proving grounds, his contact with Carl increased dramatically. Carl was impressed with the easy-going and confident young engineer, and enjoyed spending time with him. In fact, when Carl was at Siesta Key, it wasn't unusual for Wynne, along with Strang, to spend six or seven nights a week at dinner with the boss. "I was fascinated with what was going on," Wynne says, "so my hobby was my work. He was fascinating to be around, and I didn't begrudge the time too much. I enjoyed it a lot. But it was tough on the guys who had families, wives and kids. He was equally demanding of those people and I guess that was one of the reasons I got along pretty well with Carl." Proximity to Carl, though, could be unnerving, and Wynne was always grateful that Charlie Strang was near to help keep Carl in check. "Charlie was a great buffer between me and Carl, and he took a lot of the heat off of me. Because Charlie and I were very, very close friends, a lot of the direct punishment that some of the top people took, Charlie absorbed before it got to me."[33]

In the summer of 1955, Jim Wynne underwent a medical examination, and to his total disbelief, a cancerous tumor was detected on a testicle. He immediately called Strang, telling him, "I've got a problem, and I'm leaving for a while." He drove to Dayton, Ohio, where members of his mother's family were physicians. Following more examinations, a decision was made to operate at once. In a few weeks, when Wynne was strong enough to return to work, he was immediately cornered by Kiekhaefer and Strang, who had learned of the advisability of Wynne undergoing supplemental radiation treatments from the Ohio surgeons. Carl had been quietly arranging for Wynne's treatment, searching for the most advanced and experienced facility. Among the leading facilities in America at the time was the Lahey Clinic in Boston, to whom Carl wrote a summation of his deep feelings for Wynne.

> "We here at the Kiekhaefer Corporation are very fond of Mr. Wynne as a friend. We were also very much interested in his future as an engineer. Learning of his condition was an extreme shock to all of us. Not to do everything possible in an effort to restore his health would be a terrific neglect of duty and human relations on the part of the writer.
>
> "In building up our organization to its present success in the past fifteen years, I have also built up a considerable number of young men. I have had more pleasure in watching these young men and women develop in this organization from ordinary clerks to fine executives and engineers. It seems to put a worthwhile meaning to life to see dreams growing into realities, ideas growing into products, young men and women growing into families, and last but not least, two car garages growing into factories. Above all, I cannot see Mr. Wynne being deprived of his chances just when he is emerging on the same with his education completed and just entering the profession of engineering.
>
> "Of course I realize that you, Doctor, along with many in your profession are the greatest humanitarians of all and that you perhaps understand our sentiments better than we to have taken as your life profession the noble work of medicine.
>
> "I trust that I can leave with you the impression that this young man is entitled to the best, the most conscientious and the most thorough attention, sparing no expense, in an endeavor to restore him to health."[34]

Carl ultimately settled on the Mayo Clinic in Rochester, Minnesota, a facility that would become his favored destination for employees or their families that required treatment of nearly any kind. Wynne underwent six weeks of intensive radiation therapy, followed a few months later by a second series of treatments. Carl insisted on assuming all financial responsibility for the expensive treatments, and forever refused to acknowledge his generosity, even to Wynne. It was a pattern of extreme generosity and absolute anonymity that became Carl's trademark with numerous employees and their family members. Wynne recovered completely, and within weeks resumed his duties as the busy head of Mercury's proving grounds.

But Wynne was to have another, perhaps even closer brush with disaster. Often, when Carl would arrive at the Midnight Pass proving grounds on Siesta Key, he was accompanied by Rose Smiljanic. Rose traveled with Carl extensively, and for Carl it was as if he had brought his entire office with him. Wynne was well aware of the relationship between Carl and Rose, but one night he decided to throw caution to the wind. "All of us in the company knew about it," Wynne admitted. "It was discreet and proper, nice, well handled on both their sides. I didn't see much of the other ones [Carl's romantic interests]. I knew that occasionally some existed, but I really never happened to become involved in any of the other ones. Rose was the one that we were all aware of. That's why when I wanted to have a date with Rose, I knew I was courting danger immediately."[35]

Freda Kiekhaefer and the three children had packed their bags and moved to Florida in 1957. Carl bought a house across the road from Cypress Gardens in Winter Haven, where he often worked with the owners of the park to promote Mercury outboards. Freda had decided that since Carl was beginning to spend more time in Florida, that she might as well move there to escape the bitter cold of the Wisconsin winters. An intelligent and perceptive wife, she also was resentful and frustrated at her husband's infidelities, and his almost complete lack of time for her or the children. Carl was visiting the family in Winter Haven on the night that Wynne decided to tempt fate.

"I always kind of admired Rose," Wynne easily acknowledged. "She was always very attractive and I was young and single."[36] Wynne was living on the "base", as Carl was fond of calling the semi-military existence at Midnight Pass, and he and Rose ended up working together late on this particular night. It was nearing seven o'clock as Rose prepared to leave for her hotel, the Plaza, where she and Carl usually stayed. Something subtle had been occurring between the two of them over the past weeks, nothing specific or even alluded to, but the unmistakable first blush of mutual attraction. Perhaps the potential danger of the situation, both would admit, served to promote this unspoken fascination. Only Wynne and Rose were left in the office, when Wynne said, "How about having dinner together?" Rose considered the suggestion for a moment and agreed.

Jim and Rose left the office together in Wynne's flashing Jaguar roadster, assigned to Wynne by Carl as a company car and perquisite. They had no idea where they were going, and decided to figure it out together after Wynne changed his shirt. Wynne parked outside the small cottage attached to the base, while Rose waited in the car. "So, no one saw us leave as far as I know," Wynne remembers, "because there was no one else around at the base." They started driving towards town, and then decided to drive down to Long Boat Key, many miles from Sarasota, to the Buccaneer Inn, a "real charming place kind of tucked away down a dirt road, with a bunch of trees, lots of atmosphere and a nice restaurant."

They saw no familiar faces in the restaurant, found a nice quiet table, and ordered festive drinks. Both, they would admit years later, were busy considering the consequences of this secret rendezvous, and enjoying the rising emotions they were experiencing being alone together. "We were having a delightful time," Wynne recalls, "I was just enjoying the heck out of it."[37] Rose remembers thinking, "This is nice, maybe it could lead to something."[38] Then, the waiter, who had no way of knowing who Jim Wynne was, came directly over to the table and said, "There's a call for you, Mr. Wynne." Wynne froze.

Wynne picked up the telephone receiver, and before he could say, "Hello," the unmistakable gravel and gruff voice of Carl Kiekhaefer snorted, "Jim, let me talk to Rose." There was nothing Wynne could say. Carl had somehow tracked them down, knew they were together, and wanted them both to know *he knew* they were together. Charlie Strang is convinced that Carl had them followed, either by a professional private detective, or by a peeping employee eager to ingratiate the boss.

Wynne returned to the table and repeated what Carl had said on the phone, and Rose's face turned "ashen." When Rose returned only a minute later, she gravely said, "I think we better go."[39] Even though they never had time to even receive their dinners, Wynne paid the check and started driving Rose back towards Sarasota. About halfway back, Carl's jet-black Chrysler shot by them in the opposite direction "about a 105 miles per hour" and just kept going.

After Wynne dropped her off at the hotel, he began to steam. "I was mad. I said, 'Hey! I'm not doing anything for God's sake, I just wanted to have dinner.' Why the hell couldn't I have dinner with somebody. I thought this was all totally ridiculous, so unfair, and so indicative of his personality problem. I got *very* mad about it. I drove back out to the base and I thought to myself on the way, 'I've had it! This is enough! If he starts telling me about this I'm going to tell him to go to hell and I'm leaving. Just wait 'till he brings this up.'"[40] But Carl never mentioned the incident to Wynne. True to a pattern of almost manic jealousy followed by silence and denial, Carl was able to terrorize people with what he didn't say, as easily as by what he did say. Rose, too, was boiling mad over the incident, and realized how tight a grip Carl had over his empire and his subjects. "It wasn't that we were afraid so much of losing our jobs, but the embarrassment of being called on - or like us - under the carpet like a little kid."[41]

Through trials of fire, Wynne had narrowly escaped disaster from both cancer and Carl. But, just when Wynne thought that Carl was about to tear him apart for his imagined infringement on his private domain, Carl handed him one of the strangest and most fascinating challenges in his career.

CHAPTER NINETEEN

Stock Car Fever

"I think Carl Kiekhaefer is truly a remarkable person. ... I believe that he is the most competitive-minded person, with the greatest desire to win, that I have ever known. ... Carl is a dynamic, energetic man, one who 'needs' to have problems to overcome and when problems don't exist, he'll make them."[1]

Bill France
NASCAR President, 1956

The stock car racing veterans, both drivers and spectators alike, laughed until their sides ached when they caught their first glimpse of Carl Kiekhaefer at Daytona Beach, Florida in February 1955. It was during the highly celebrated annual "Speed Weeks" when Carl and his entourage pulled up, and people just couldn't believe their eyes. Country boys and Dixie moonshine runners put their greasy wrenches down to gawk. Scrambling out from under their often filthy, dented and rusted entries, mechanics smeared with oil from hair to shoelaces stared in amazement.

Carl Kiekhaefer, 48 and nearly bald, stepped onto the sandy beach with a freshly pressed long sleeve white shirt, rimless spectacles, a bow tie and a giant hand-rolled cuban cigar clenched between his teeth and shoved to one side of his mouth. He stood with his arms akimbo against his portly frame as his crew, attired in snazzy matching white uniform overalls, slowly opened the doors to the huge white truck emblazoned with the legend: "Kiekhaefer's Mercury Outboard Motors, The Most Powerful Name In Outboards." Then, down the ramp it came. It was a platinum white and gleaming new 1955 Chrysler C300, all 4005 pounds of it. Though it was professionally lettered with the Mercury name and the number 300, to most it looked like a luxury touring car, complete with (unthinkable) an automatic transmission handle poking ostentatiously out of the dash. It looked as out of place as a beached whale, or as one journalist commented, "He might just as well have been an Egyptian pharaoh arriving by barge."[2] And as far as *Mercury* went, the car's lettering might just as well have spelled out "Jim Bob's Wiggly Fish Baits" for all the onlookers cared.

Stock car racing had been born in the South out of necessity during the romantic years of Prohibition from 1919 to 1933. Cars delivering "white lightning" from illegal stills needed the very highest performance to outdistance the "Revenuers" in the heady days of wild, high-speed chases through back woods roads before the dawn of police radio. Looking for all the world like ordinary passenger cars, these early racing stock cars were often meticulously tuned and highly modified to break away from even the fleetest police cars. Impromptu challenges between delivery cars naturally occurred, and eventually, dirt circle and dirt oval racing contests among the hottest cars were held in cornfields throughout the South and Southeast.

The first organized stock car race took place in Langhorne, Pennsylvania in 1939, and the National Association for Stock Car Auto Racing (NASCAR) was founded following World War II in 1947 at Daytona Beach. Daytona had been a favorite amateur racing site for stock cars since the early 1930s, and NASCAR, founded by Bill France, held its first formal, officially sanctioned race there in 1948, seven years before Carl and his bright white Chrysler showed up on the sand. By 1949, when NASCAR had developed a point system to determine a Grand National Championship driver, nine races were being held in five states and the new Oldsmobile 88 was the car of the year with six out of nine victories. The sport began to flourish, with venues doubling over the next several years from 19 races on 14 tracks in eight states in 1950 to 41 races on 34 tracks in 14 states by 1951. By 1954, purses for NASCAR races had swelled to nearly $300,000 for the combined circuit of 37 races. The winning car that year was a Hudson Hornet, owned and serviced by one of the most colorful mechanics to ever crawl under a race car, Smokey Yunick.

Not only did Carl and his spotless entourage look out of place at Daytona Speed Week, Carl didn't even have a driver to qualify his curious car for the 160 mile race that represented the climax of the week's festivities. Late in 1954, Carl had put Tony Bettenhausen behind the wheel of one of the old Mexican Road Race cars, and entered a AAA sanctioned stock car race in Milwaukee. Bettenhausen won the race and a purse of $4,200 before a roaring crowd of over 32,000 fans, and a journalist at the time claimed that from "that moment, Kiekhaefer's interest in stock car racing verged on the fanatical."[3] The complicated rules of rivalry between the AAA circuit and the NASCAR circuit prohibited Bettenhausen from driving for Carl in Daytona. Two of the leading NASCAR drivers, Herb Thomas and Elzie "Buck" Baker (second and third respectively in 1954 NASCAR points), turned Carl down when offered the new car. Too little was known about this new Chrysler muscle car, and drivers were skeptical about both the car's performance and this tough-talking outboard motor man. With the names Kiekhaefer and Mercury prominent on the new Chrysler, Carl would tell any drivers that would listen, "If you enter a race with your name plastered all over the car, you better win."

The new Chrysler was hardly a surprise to Kiekhaefer. He had maintained a close relationship with Chrysler engineers and brass alike over the preceding three years since the last Mexican Road Race. Carl was well aware that the new C300 was a distinct departure from the old Saratogas and New Yorkers that he raced through the Mexican desert. Built for brute strength by Chrysler engineers, it is considered the first factory-built muscle car in America. It was massive in virtually every category, especially the engine. It featured a 331 cubic inch, 300-horsepower Firepower Hemi engine with a pair of enormous four barrel carburetors, special manifolds, special racing cams, solid valve lifters, a heavy duty suspension and dual exhausts. Chrysler couldn't afford the tooling to compete against the fast, two-seater sports car competition from Chevrolet (Corvette) and Ford (Thunderbird). So, the next best thing was to produce an unusually powerful engine and put it into a full-sized luxury sedan. Straight from the showroom, Chrysler could boast that the C300 would whip nearly any car on the street, with a top speed of 130 mph, a zero-to-sixty mph time of only 10 seconds, and a standing start to a quarter mile time of 18 seconds to reach 82 mph. It was a machine of awesome power for the 1955 model year, but naturally, Carl and his retinue of crack mechanics, lab technicians and engineers didn't stop there.

Before leaving Wisconsin, the new car was almost completely disassembled, and a staggering array of allowable modifications were made. In typical Kiekhaefer fashion, the work was undertaken in a relentless drive for perfection, spurred on by Carl himself, seven days a week until it was time to leave for Daytona. "Details make perfection," Carl would grunt between his teeth, chomping down on the omnipresent cigar whose flame had died hours ago, "but perfection is no detail."

Bill France's NASCAR rules allowed generous flexibility where modifications were made to enhance the safety of the car, and allowed equipping the car with accessories and special parts that the auto manufacturers had properly documented and made available through company bulletins and catalogs. "But we never would have won a race," Carl explained in an article he wrote for *Popular Mechanics*, "if our cars had been strictly stock in the sense that the public thinks of stock."[4] Though Carl would claim that his cars were "closer to stock than others on the track because the C300 engine is designed for high performance," the truth is, he was able to accomplish some spectacular fine tuning of the engine at his research facility in Oshkosh. "It is a competition engine when it leaves the [auto] factory. It does not need to be souped-up. We put it in top tune with a few minor adjustments." His engineers must have fallen over on the floor when they read Carl's statement to the press, because even Carl would later admit that just to "fine-tune" three of the Chrysler engines took over 500 man-hours of highly specialized and concentrated labor. "When the 'Old Man' stops to fool with an engine on the dynamometer," one of his crew remarked, "the whole shop comes to watch. Reminds you of a concert pianist."[5] One of the stories that circulated around the Kiekhaefer Mercury plants is, that one night, in a particularly black mood, Carl blew up a Chrysler engine, an outboard motor and an Italian motor scooter engine on dynamometers on the same evening "before he could go home in peace."[6]

Using parts from Chrysler's inventory of heavy duty taxicab and light truck accessories, the Kiekhaefer crews beefed up suspensions, brakes, axles, steering systems, chassis components and exhaust systems. Everything was checked and double checked before the car left Mercury's research garage on Murdock Street in Oshkosh. It was unimaginable for Carl to leave anything to chance.

As Carl methodically searched Daytona for a top-notch driver to qualify his now quite exotic Chrysler for the forthcoming 160 mile race, he experienced a stroke of good fortune. Three years earlier in 1952, a 28-year-old ex-rum-runner from Georgia named Tim Flock had become the NASCAR Grand National Champion, driving a Hudson Hornet to victory in a hotly contested season of races. But at the 1954 Daytona race,

Flock quit in an enormous uproar of controversy with Bill France and NASCAR that would soon change the rules of competition.

> "France knew the life blood of the Grand National circuit surged from the resemblance of the race cars to the customers' own Detroit sedans. Yet the early races sometimes became fiascoes when too many cars fell to pieces in the rough pounding. To match 'stock' appearance with stamina and performance, France had to allow speed kits, beefed-up steering, special hubs, reinforced suspensions. ... What was legal? The question embroiled France in more daily rhubarbs than a baseball umpire in the Cuban Winter League.
>
> "France's problem came to a climax with the 1954 Daytona Beach race. Tim Flock finished first. Thousands left Florida after the race, convinced Flock was the winner. Thousands more heard the news on radio, television and read it on the wire services. Then NASCAR announced that Flock used an illegal carburetor and that [Lee] Petty had won. The reversal brought a storm. 'That doesn't happen at Indianapolis,' said a Los Angeles sports editor. 'We won't print any more NASCAR results.' The enraged Flock quit racing, opened a service station and took the argument to court. Wrote a fan to a newspaper: 'Why go to a race if you have to wait a week to find out who won?'"[7]

Though Flock would never receive credit for the victory in the '54 Daytona race, the owner of the car he was driving cleared Flock's reputation by disclosing that the driver was unaware that the outlaw carburetor had been installed on the engine. But Flock was so disgusted that he loudly protested the unprofessional treatment he had received, and hung up his helmet, he said, for good. "I thought I could hurt NASCAR," Flock would later tell Ray Doern of the Chrysler 300 Club, "but it didn't affect them whatsoever."[8]

Flock had returned to Daytona's Speed Weeks as a spectator, while Kiekhaefer was prowling the beach for a driver. "I was standing on the beach and a Chrysler 300 came by," Flock confirmed to Ray Doern. "I casually looked over to one of the guys I rode down to Daytona with and said, 'Now man, if I had that car there, I'd win this race again.'" Fortunately, standing a few feet away and overhearing the conversation was Tommy Haygood, the Orlando, Florida Mercury outboard dealer. Haygood was aware of Carl's search for a driver, and knew that Flock was among the very best in the business.

"Before Flock knew what hit him, Haygood loaded up the driver into his lettered-up (it said Mercury Outboards) station wagon and delivered him to a rented garage, in Daytona to talk to Kiekhaefer. 'When he told me to get in the 300, I knew he was going to move the seat,' Flock recalled. 'I knew it was going to be my race car right then.'"[9]

Flock later recalled his first impressions of Kiekhaefer. "I knew he was strong and tough right away," he said. "He was smoking a big cigar, and I felt like I was shaking hands with a freight train. He didn't say much, except when he gave orders and he ordered me right into the front seat so they could measure me for it." As Flock soon discovered, he was quickly sucked into a spinning vortex of activity with Carl Kiekhaefer at its center.

"By six o'clock the next morning, Tim was behind the wheel, testing the car. One thing the ex-champ did not like about the 300 -- its automatic transmission operated by a lever jutting from the dash. 'Don't worry about that,' advised Kiekhaefer. 'We'll get a (manual) transmission made for it after the race is over.' At the time, Flock doubted his boss had enough clout with Chrysler to accomplish that.

"Even with the automatic, Flock qualified the car, the following Friday, at a record 130.293 mph. 'They'd never seen a car run that fast,' he said later. "So we sat on the pole with the car and I was the happiest guy in Daytona."'"[10]

The day of the 160 mile race, February 27, 1955, Flock found himself locked in a duel of nerves with Glen "Fireball" Roberts who was driving a '55 Buick Century equipped with a manual transmission. When the huge cloud of dust finally settled over the four and one tenth mile track, Flock was on Roberts' tail as they crossed the finish line ahead of the other 45 cars in the field. Flock was sure that the Buick's straight-stick transmission had won the race over Chrysler's somewhat awkward two-speed automatic. But in a dramatic reprieve and chilling reminder of Flock's fate in the previous year's contest, race officials discovered that Roberts had used illegal shortened push rods in his car, and was disqualified. Tim Flock and the Kiekhaefer Chrysler 300 were declared the winners, and as journalist Jesse Thomas reported, emotions ran high in the Kiekhaefer pits.

"Kiekhaefer broke out the champagne to celebrate. According to Flock, he went 'berserk' over their first NASCAR victory, arranging a big party at Cypress Gardens, where his company supplied outboards for the water ski shows. Thousands of people viewed the winning Chrysler and the ex-rum-runner wound up with a contract he described as 'such a good deal, I couldn't believe it.'"[11]

The victory at Daytona convinced Carl that his strategy of "best car, best driver and best preparation" was an unbeatable combination. Even though the NASCAR Grand National circuit had started three races and four months before the Daytona race, he decided to jump into the rivalry for the crown with both feet. At the outset, he described his rules of engagement: "Eliminate luck, pick the best cars, the best mechanics, most competitive drivers, and then work like hell!" Carl quickly prepared more race cars, hired more drivers and mechanics and decided to take on both stock car circuits at the same time. In the AAA circuit, he outfitted Frank "Rebel" Mundy, the 1953 AAA stock car champion with another new Chrysler C300, number 30, and prepared other cars for Tony Bettenhausen and Norm Nelson for the year long series. Tim Flock started off alone in a bid for the NASCAR Grand National championship.

But something began to go wrong with the Kiekhaefer cars. In race after race, even after establishing new speed records in qualifying, and after easily leading the races from the start, they slowly lost power and speed, often grinding to a sickening halt. Carl had to act fast to find the problem. He described the problem and his ingenious solution in an article he wrote for *Popular Mechanics* entitled "How to Win Stock-Car Races."[12]

"At the start of the '55 season, our Chrysler C300s did not do well. They were fast enough. They usually took the lead and led for much of the race, but then they were apt to slow down or quit. We discovered why: Dirt was ruining our engines.

"After the first Langhorne [Pennsylvania Speedway] race, the air filters on our cars were completely loaded with dirt. The engines were just about to quit when, fortunately for us, the race was stopped because of a heavy rain.

"At Winston-Salem [North Carolina], our two cars conked out almost simultaneously -- so close together that an official accused us of willfully pulling out of the race, an infraction of the rules.

"But we had not pulled out. The cars had just quit. We opened up the engines and there was no oil left in them. Just a cupful of stuff that looked and felt like lapping compound. The engines had no compression. The oil and dirt formed a grinding compound that wore out the rings. All the oil had pumped past the rings; the main bearings were shot and so were the rods. Our 300s were susceptible to dirt. Their two four-barrel carburetors have eight holes to carry the abrasive stuff into the engine. Being a high-performance engine, it breathes faster and deeper, taking in more dirt per lap than other engines. Dirt was killing the 300's chances.

"Back we went to our laboratory to get ourselves off the hook. We found that for racing, the conventional oil-bath air filter is inadequate. It actually collects dirt and dumps it in big gulps into the carburetor. At full throttle we would suck oil and dirt right out of the filters themselves. We'd be better off with no filter.

"But you have to have filters in dirt-track racing. So we drew on our experience in building portable chain saws for the government during the war. We knew that dry paper filters had worked there under dusty conditions. So we tried paper.

"We improvised a filter from the sawed-off bottom of a garbage can. We built one inside an aluminum saucepan. We were on the right track as our hurried tests showed. Finally, with the help of Purolator parts, we came up with the simple, but amazingly efficient dry paper filter that all racers [and conventional automobiles] use today."

Carl collected dirt samples from every race track on the circuit, and had it carefully analyzed to determine the optimum paper configurations to stop the smallest particles from passing into the carburetor, while still allowing the engines to breathe deeply. His remarkable tenacity in developing the new filter system resulted in a barrier that kept out "all dirt particles over two microns in diameter. Dirt under two microns seems to do little or no damage."[13] Virtually every car in the world is today equipped with a paper air filter, the direct result of Carl's burning ambition to win stock car races in 1955.

Carl and his crews also had to adapt quickly to combat the strategic advantages that some of the drivers, with overt support from Ford and Chevrolet, flaunted at every opportunity. As Carl explained, one of the critical obstacles became tire wear.

"With the dirt problem licked, the 300s began to win races. Things looked good. Then at Raleigh, N.C., Herb Thomas beat us. He was driving the same Buick we had been outrunning all season. Something was changed, we knew. He finished the race without a tire stop. We were leading until we had to change tires and we never caught him again.

"The next day at Milwaukee, we were beaten by Marshall Teague in a Chevrolet that went all the way on one set of tires -- this same type of tire, again.

"As we were given to understand, Thomas and Teague were running on special experimental tires given to Ford and General Motors by Firestone for test purposes. Somehow these tires got onto the two cars that beat us. As you can imagine, the tires caused many heated discussions. A pit stop for tires can mean the difference between winning or losing a race. This new tire meant that the races could be won or lost by the tire company.

"We had been doing lots of work with tires. The 300 is hard on rubber, being the heaviest and fastest car on the track. You just can't get oversize tires for it. ... Smaller cars use the same tires we do. ... and being lighter, go farther between changes. Sometimes these smaller cars win on the time they save in tire changes.

"Our first move had been to install wide-base wheels. We have long been aware of the advantages of wide-base rims. Our distributor in New York had gone into the wheel business during the tire shortage of World War II. They proved to me that wide-rim wheels increase tire life sharply on cars, trucks and buses. ...

"The wide-base rim allows the tire to hold more air. And more air means it runs cooler. Wide rims also increase load capacity and stability on turns. ...

"We also developed a tire-curing system. Every tire is broken in for 1000 miles on our station wagons before it goes on a race car. We never trust a new tire for high speed driving.

"We also inflate our tires with a dry gas -- we use nitrogen. The humidity in air can increase pressure to 90 pounds when it becomes steam during the race. So we empty our tires before each race and refill them with dry nitrogen.

"We thought we had the tire problem in hand when along came this fabulous new Firestone racing tire ... that made it possible for Thomas and Teague to beat us twice in a row.

"We and others put so much pressure on race headquarters and the tire company about these special tires that Firestone went into accelerated production on its experimental molds. Just before Darlington [Raceway, South Carolina -- one of the most important races of the year], Firestone officials received a phone call every two minutes and all but went into

> hiding! Finally, we were able to get tires and we continued to win races. ..."[14]

Carl leaned loudly and heavily on Bill France, NASCAR president, to keep the playing field level against the steady encroachment of the Ford and Chevrolet factories. And, as Carl and his stable of cars and drivers won more and more races, Bill France and the NASCAR inspection crews tore down the Kiekhaefer cars more often than any other entry, looking for infractions that the competition was sure Carl was hiding in there somewhere. But Carl played by the book, and remained within both the letter and the spirit of the rules. His ingenious methods, though, often concentrated on what the rule book *didn't* say, as often as what it did say. Smokey Yunick, among Carl's most tenacious rivals and considered by many to have been the most inventive of NASCAR mechanics, summed up this critical difference in an article written for *Circle Track* called "Smokey Tells All: Thirty Years of Fun, Games, and Playing Fast and Loose With the Rules."[15]

> "Trying to figure out NASCAR's rule book threw me at first. Then, after studying the rules from all sides, I realized I'd made a colossal mistake. I'd been reading the rule book to see what it said. And all along, what I should have been doing was finding out what it *didn't* say. ...
>
> "[A] rule called for a certain gasoline tank size. ... But it didn't say how big the gas line could be. I built an anaconda-sized fuel line that held an extra 5 gallons. ... Similarly, a rule said you couldn't improve the weight distribution and handling by moving the engine back in the frame. But the rule didn't say you couldn't move the body forward. I use to move it 3 inches.
>
> "Another rule insisted that the engine had to be centered, left to right, in the chassis. The rule didn't specify where the wheels had to be, though. So I shifted the weight bias to the left side of the car--lengthening up the A-frames on the right side and shortening them on the left. ... Those 'grey areas' of the rules were what I liked best. They didn't say you could do something, but they didn't say you couldn't, either."

Smokey Yunick's reputation for pulling the wool over the eyes of NASCAR inspectors were to make his cars among the most suspicious in NASCAR history. "I went through whole generations of tech inspectors," Yunick admitted, "who knew me by reputation and put me through pure hell."[16] At one point, he was able to pull a fast one on one of Kiekhaefer's own top engineers, Edgar Rose, who is today vice president and chief engineer for Outboard Marine Corporation. Edgar, born in 1926, earned a master's degree in mechanical engineering from M.I.T. in 1948. After completing a two year graduate course in textile engineering at the Lowell Textile Institute in 1950, he joined the M.I.T. Industrial Research Group where he remained until recruited by Charlie Strang and Kiekhaefer in 1953. An exceedingly clever and meticulous engineer of German heritage, Edgar had invented a "camshaft checking machine" that was supposed to verify that camshafts used by competitors on the NASCAR circuit were legal, and unmodified from factory specifications. His "cam checker" would plot out a graph of the profile of the cam, showing variations down to one thousandth of an inch. He would first graph stock camshafts from the manufacturers, and then compare them to the results of suspect camshafts from cars being inspected before and after races. The trouble was, on at least one occasion, at Darlington Raceway in South Carolina, according to Smokey Yunick, a slippery mechanic was able to beat the system.

"... I had been having such a successful season that everybody in NASCAR was suspicious as hell, Buddy [Shuman, chief technical inspector for NASCAR] most of all. Trick camshafts had him almost paranoid. ...

"He'd brought along his own camshaft checking machine. In fact, he even had his own camshaft expert doing the checking. Not only that, the secret expert even had a secret room. He was standing eavesdropping inside it, just out of sight, while Buddy and I exchanged pleasantries and I unbuttoned my little [Hudson] Hornet. Then, when Buddy carried my camshaft back to the secret room, I could hear the secret checker warning him, 'No, no, don't believe him. The sumbitch is screwing you again.' And Buddy soon returned to ask me additional questions. At last, the secret checker himself came out of the secret room to ask me about the camshaft. I immediately recognized him as a guy who worked for Carl Kiekhaefer.

"Now everybody in NASCAR knew exactly what I thought of Kiekhaefer and his militaristic racing team. I even had Kiekhaefer's picture reposing in a place of honor in my garage in Daytona Beach. It was in the men's room, hanging over the john. So I really didn't think it was in 'the spirit of the rules' for Buddy to have made one of Kiekhaefer's guys his camshaft expert, but I didn't say anything. I just expected the worst, anticipating the secret expert to tell me the camshaft had failed the test. Instead, after long deliberation, the secret expert informed me that it had gone through his testing machine and checked out as stock and legal. Now, since I'd ground the damn thing myself, and it was radical as hell, that was quite a surprise to me. But I damn sure didn't argue with him. I just asked him what made him so sure it was legal. 'Oh, it's definitely the right cam,' he said airily. 'And now I've even put my secret mark on it. You won't be able to change it if you wanted to.'"

Herb Thomas, driving Yunick's Hudson Hornet, drove to victory at the Southern 500 the next day, and after the race, Buddy Shuman told Yunick that someone told him he had changed camshafts before the race. "Well, as suspicious as he was of me," Yunick said, "whoever it was hadn't had to argue too hard to convince Buddy of that." Smokey Yunick was going to have to face Edgar Rose, Kiekhaefer's secret cam tester, once again.

"All of us were back inside the inspection station again. Counting spectators, it seemed like there were 2000 people there. Even Buddy's secret camshaft expert -- the Kiekhaefer guy -- put his nose in it. 'Just show us your camshaft again,' he said. 'I just need to see if it still has my secret mark.' ...

"... I popped out the camshaft. ... The secret camshaft tester grabbed for it but, instead of giving it to him, I slammed it as hard as I could into the workbench. It broke into three pieces. He couldn't believe it. 'Well how can I test it *now*,' he complained, looking at the three pieces. 'Wait a minute,' I said. 'You say you're the camshaft expert. If you really are, all you need is one intake and one exhaust lobe to measure. Now, which piece do you want?' By now I was acting out of self-defense. If I gave him the whole camshaft, I worried that he'd take it into his secret room, then come back and say it hadn't passed. He decided he wanted to take the front piece. A few moments later, he came back asking for the middle. 'Why do you want the middle?' I asked. 'Because that's where my secret mark is,' he said.

"I wouldn't give it to him at first. Then, after another hour or so of stonewalling, I did. He reappeared from the secret room. 'There's no question that this is the right camshaft,' he said. 'It's got my secret mark on it, and it checks out exactly.'

My Hornet was re-declared Darlington's winner. I even got a good laugh out of it. ... I even walked over to Curtis Turner [another NASCAR driver] to personally deliver a message of what a no good sumbitch I thought he was too. He swung at me when he heard it, and I swung back. That was sort of the unofficial signal for all the Curtis and Smokey fans in the inspection station to begin pulling boards out of the wall and begin swinging at each other. The result was a riot that, if I remember right, the National Guard had to be brought in to put down. Curtis and I watched it while safely hiding under the workbench. Yes, we patched up our differences, and Curtis later became one of my drivers. But that's another story. I just hope that one of the boards hit that secret camshaft tester who worked for Kiekhaefer."[17]

The Kiekhaefer teams began to win race after race, dominating both the AAA and NASCAR circuits. Tim Flock was joined by his stock car racing brother Fontello "Fonty" Flock by mid-season in 1955, and the "Flying Flocks" were unbeatable. When Buck Baker began to win against the Flocks, Carl lured him away from competition for his own team. He hired Baker by telling him, "I hear you're such an s.o.b. when it comes to driving race cars, I want you to drive for me." As Jonathan Ingram once observed, Baker was one of a kind.

"Baker, a former bus driver from Charlotte, was a tough hombre, indeed. He once dove through a screened window to punch a fan heckling him during a post-race inspection. And he once chased a waiter several blocks for spilling coffee on his wife. 'Buck was a diamond in the rough,' says Kiekhaefer. 'Carl took me out of the kitchen eating hamburger and put me in the dining room eating steak,' replies Baker. 'He put me in a financial bracket I didn't hardly know how to handle.'"[18]

Carl Krueger, Henry Ford, Junior Johnson and most notably Alfred "Speedy" Thompson were also recruited to drive in selected races, usually as a team member with the Flocks. Carl was most particular about who he picked to drive his cars. "Hungry drivers make the best drivers," he preached. "We don't want any prima donnas and we want the highest caliber of driver."[19] Carl's crack teams were so well organized, so well prepared and so disciplined that not even the Ford and Chevrolet teams were able to shake them. Carl was also proud of his pit crews, who were the envy of the tracks. "The boys can complete a scheduled pit stop," Carl boasted, "(gas, oil and four tires) in 48 seconds." His crews were so dedicated, and so methodically driven by Carl, that "a favorite lunch-time diversion at the shop is practicing pit stops."[20] It became commonplace at dirt tracks jammed with screaming fans for Tim Flock to set a new track record qualifying for a NASCAR race, and then lead the actual race from green flag to checkered flag.

But as Tim Flock rose steadily in the point standings in the NASCAR circuit and Frank "Rebel" Mundy began to skyrocket to the lead in the AAA circuit, attention began to focus on Carl Kiekhaefer by the racing press, envious competition, and even racing fans, all to Carl's immense disapproval. Carl, ever paranoid when it came to controversy of any sort, was convinced that just about everybody was out to get him. Following one of the races before he had acquired Buck Baker, Carl protested to NASCAR officials that a combination of drivers and officials were attempting to sabotage his efforts, and were stacking the rule deck against his Chryslers in blatant violation of the rules of good sportsmanship.

> "Instead of running Buck Baker behind the caution flag, who was leading and in first place at that time, [a race official] made Tim Flock follow the pace car, allowing Buck Baker to make another lap. This cost Tim Flock first place, and put Buck Baker in first place. That this was possible was demonstrated by the fact that Tim Flock's car could, at will, pick up two seconds per lap on Buck Baker after Tim Flock's last pit stop. In addition to this, under three different caution flags, the entire field had an opportunity to make up almost a complete lap on Tim on three different occasions.
>
> "From all of the above, it appears obvious that there is severe unfriendliness amongst the drivers (which is understandable) but when this unfriendliness and biased attitude extends to the officials, it can mean only one thing -- that Tim Flock must drive a car of some other make to please the officials. Officials also apparently close their eyes to many of the things happening on the track, deliberate blocking ... refusal to allow to pass ... the drivers going unnoticed and then flip a 'Junker' at the entrance of the pits immediately after Flock signaled his intention to come in for fuel. Fortunately Flock had enough fuel for several more laps until this Studebaker (which qualified at a ridiculously low time) could be pulled away from the entrance. It was also obvious that when the [junker] Studebaker spun in front of the entrance, that the

driver was apparently free of the car at a moment's notice without anyone unbuckling his door. There can be no question about his intent of preventing Tim's pit entrance. If it was not intent, it was certainly a very odd coincidence.

"Of course, such things as razor blades slashing tires can always be expected. This too was found not only at one race, but at several."[21]

Quite a number of instances of suspicious break downs and deliberate sabotage were to hound the Kiekhaefer effort. Later, Buck Baker, then driving for Kiekhaefer, qualified for the front row at Darlington Raceway with a near track record performance. But the next day, as one journalist reported, the race itself was a different matter altogether.

> "[He] discovered all the gaps on his spark plugs had been closed up, or sabotaged. That same year at Elkhart Lake, Wisconsin, all of Kiekhaefer's cars fell prey to jammed fuel lines except [one] ... which ended up stuck in a cornfield. The other cars had been sabotaged by sloppy soldering of fuel lines by a member of the Kiekhaefer crew. 'There was no question that someone had done this intentionally,' said crew member [Andrew 'Chick'] Morris."[22]

Carl blistered with rage when the press called him "a millionaire sportsman", or "outboard tycoon." "Personally," Carl lectured the racing writers, "I like sports rather than the palatial homes, yachts, country estates, expensive clubs, etc., that 'multi-millionaires' wear so comfortably. I prefer to work on the equipment and race it on the track, not in the newspapers."[23] In a vitriolic letter to Ed Otto, NASCAR vice president, Carl let loose with both barrels. Again, Carl never mailed the letter, using it as a catharsis to vent his anger and frustration. The letter was marked with a note indicating Carl had called Otto and delivered his barrage over the phone. It would be the first of many occasions that Carl would threaten to quit stock car racing.

> "I have before me a copy of Page 24 of the June 23rd edition of the 'Pittsburgh Press.' I refer you to the article by Chester L. Smith. I am writing you to point out a few errors. First of all, I am not an outboard tycoon nor am I any more a 'red hot' fan of racing. I have just about had my fill. ...

"I do agree with the paragraph in which you estimated that if I stayed at home and minded my own business the corporation might sell more products, and that is what I am going to do. Frankly, this article sounds like it was originated by yourself. I do not know Chester L. Smith and he is obviously innocent, but this article is very damaging to the writer.

"A copy of this article was mailed to me by a member of an organization we were trying to impress [Rockwell Manufacturing Company]. Instead this article makes me appear like playboy Howard Hughes. The first group of people who will be on my neck are my bankers, next in line will come the Department of Internal Revenue, labor unions and the usual nuisance minority groups of stockholders found in many corporations.

"I have scientific reasons for gaining racing experience; however, I feel very unwelcome and the attitudes expressed by contestants and officials alike have been surprisingly hostile. This would be highly objectionable if I had not already developed a 'pretty good layer of insulation' dealing with our competitors in the outboard industry.

"As mentioned before, our racing activities were scientific. Namely, we felt that the name of our products on the front line winners would be a good advertising gimmick; a gimmick that is in the experimental stage this year and one that is not unusual since it has been used for years by people such as Pure Oil, Mobil Oil, Champion Spark Plug, Firestone Tire and many other accessory and product manufacturers. I fail to understand why the chief executives of Firestone and Champion and all the other manufacturing firms have escaped the personal attention which I have tried to avoid.

"If the attention cannot remain commercial and is getting over into the personal, I will have to withdraw my contracts with the Flock boys and others. ... If I am to be nettled and needled out of NASCAR, it will hurt people like the very nice Flock boys more than it will hurt me. We could probably find other endeavors in the sporting field. I am asking you as a gentleman to keep my name out of the publicity as a personal favor and you can blat the 'hell' out of the name Mercury Outboard Motors all you like. For that favor, I will help the sport of racing with an ingredient which is sorely needed -- sportsmanship."[24]

How Carl could even wonder why attention was being focused on him is incredulous. He was rapidly becoming the absolute center of attraction and controversy wherever he appeared. His generosity with his winning drivers was causing defections from other race car owners, and was setting unheard of precedents in car racing. First, he was putting his drivers on a handsome retainer, up to $1500 a month. Next, he allowed drivers to keep ninety percent of purse money from races, and put the other ten percent into a pool of bonuses for mechanics and pit crews. But it didn't stop there. When drivers won, it was common for him to jam a handful of $100 bills into their jump suits as a bonus for coming in first. He would surprise his drivers with the keys to brand new, off-track pleasure cars and pamper their families with gifts and surprises of all varieties. "He worked his men 24 hours a day," Chrysler's then public relations chief Frank Wylie told Jonathan Ingram at *Stock Car Racing*. "But he paid them so much money they couldn't afford to go anywhere else."[25] He once rented the entire floor of a hotel for his race crews and their families, but then segregated them because he didn't believe in "sex before racing." He thought it might take the razor edge off his driver's aggressive reactions. He actually put his own room between the drivers and their families, and during the evenings he could be seen "patrolling the corridors." In the pits, he forbid his drivers and crews to talk to other drivers and mechanics, fearing that they might divulge some of his racing methods and secrets, or foment associations that might lead to defections. To top it off, the chief executives of Pure, Mobil and Champion were never seen vaulting over the infield pit wall to participate in pit stops like Carl. He was an integral part of many pit stops, checking under the hood, interrogating the driver, pushing the tire change crew and barking orders loud enough to be heard in the stands over the scream of fifty race cars tearing around the track.

Unfortunately for Carl, Detroit was becoming embarrassed by his growing string of victories, and he was quite unprepared for the wrath that was soon focused on him personally. Frustrated at the advantages he perceived were given to Ford and Chevrolet factory teams under NASCAR rules, he lashed out at Bill France.

> "For a number of reasons, I am not anxious to continue in stock car racing. ... I am not complaining about anything in particular, except the presence of out-and-out factory teams backed by a two billion dollar corporation which, from all indications, intends to dominate [AAA] as well as NASCAR. If you have never felt the full pressure of Ford public relations in retaliation of some displeasure you may have caused them, you have something to experience.

"We have entered racing primarily as a hobby and for the pleasure of the sport. If there are prospects of unpleasantness, we shall try some other form of competition. ... Also, the big Detroit interests are soliciting drivers under contract with other people. Who can compete in this maneuver except factory team against factory team?

"... As far as continuing as a public relations and publicity 'gimmick' for Mercury Outboard Motors, we seriously question our again driving heavy cars. As you know, the average race fan considers the Chrysler a villain. The press and sports writers have not been kind toward this particular product. ... At any rate, it looks to us as though racing will settle down to a factory contest between Fords, Chevrolets and Dodges, with the factories maintaining the cars, with drivers running on factory instructions and payroll. Time will tell as to final results."

Though Bill France made no secret of his ambition to lure Detroit further into NASCAR competition, ultimately it was Carl's own aggression that stiffened the resolve of the automakers. As Tim Flock put pressure on the 1955 NASCAR points leader, Lee Petty, Carl pulled off a minor miracle to push his driver over the top. It wasn't particularly newsworthy for a driver to win a race on Saturday night, followed by a race the next afternoon. Unless, of course, the two tracks were 2600 miles apart. With Carl Kiekhaefer's careful planning, Tim Flock was the winner of two Grand National events, racing over 350 miles in competition inside of 20 hours. Flock's first victory was in a 100 mile NASCAR event at the Syracuse, New York Fairgrounds on Saturday night, July 30, 1955. Right from victory circle, Carl rushed Flock on board a chartered TWA airliner to Chicago, and then onto a nonstop, overnight flight to San Francisco. Flock's brother, Fonty, had qualified another Kiekhaefer Chrysler 300 for the Bay Meadows Grand National event while Tim was winning in New York. Tim strapped himself into the other Chrysler 300 and beat a large field in the 250-mile event to set stock car racing history. A clause in the NASCAR rule book allowed points earned in the western division stock car races to apply to NASCAR national points, but it was the first time that a driver had won races on both coasts on the same weekend. Again, NASCAR drivers were enraged that Carl Kiekhaefer's incredible organization and drive was beginning to dominate what was once a somewhat laid-back sport. Smokey Yunick, it is assumed, spoke for most of the racing community when he told veteran stock car racing journalist Jonathan Ingram how the competition viewed the Kiekhaefer strategy.

> "'We never went up against somebody so completely organized and well staffed,' said Yunick, the mechanic behind winning efforts by [Fireball] Roberts and [Herb] Thomas, among others. 'Things we did in four or five weeks he did in four or five days. Most of us had other jobs and we worked our ass off all week and then went to the track and picked up a crew there. He had engine men, chassis men, brake men. We never experienced anything like that, and he would sometimes come with four or five cars. But he was a tough s.o.b. and he worked hard himself. And he was ruthless. He had enough money to pay anybody he wanted, but, he made us all become more professional. Up until then, we were part time racers.'
>
> "'Racing against him,' continued Yunick, 'was like comparing a $50,000 budget to a $500,000 budget. That's how much each of us spent, and how much he spent. But we still gave him all he could handle.'"[26]

Kiekhaefer's nearly complete dominance of the sport in 1955 culminated in an unprecedented string of victories and *two* national titles. His brute strength Chryslers won 22 out of 39 Grand National Stock Car Races entered, and propelled Tim Flock to the Grand National Championship by winning 18 of 39 for an unheard of average of .462. (This record would remain intact for twelve years, until Richard Petty would win more races in a single season in 1967.) His nearest rival was able to accomplish an average win record of .236. Typical of the overpowering presence of Kiekhaefer victories during the season was the 100 mile Grand National event at Spartansburg, South Carolina. His cars swept the event, coming in 1-2-3-4. On other occasions throughout the year, his cars finished 1-2-3. In the AAA stock car competition, Frank "Rebel" Mundy overwhelmed the competition with an equally remarkable record, winning eight of the thirteen races, while Norm Nelson and Tony Bettenhausen each won an additional race to bring the Kiekhaefer score to ten out of a possible thirteen. Carl Kiekhaefer and his unstoppable teams had conquered the whole world of stock car racing in nine short months, coming from absolutely out of nowhere. The outboard motor maker from Wisconsin had stunned the entire racing community, and awakened the sleeping giants of Detroit.

CHAPTER TWENTY

The Fever Breaks

"You're FIRED!! But you can't leave until the car is ready."[1]

Carl Kiekhaefer
Charlotte, NC 1956

Carl's team started off with a bang again in 1956, as Tim Flock piloted a new Chrysler 300B (which Kiekhaefer affectionately called the "Beast") to set a new world's record in the "Flying Mile" event during Daytona Beach Speed Weeks at a blistering 139.373 mph. This latest muscle machine from Chrysler contained an even bigger engine than the C300, at 354 cubic inches and developing a mighty 340-horsepower at 5200 rpm. Another of Carl's new 300B's was barely nudged out of first place in the standing start mile competition by a Dodge 500 and a mere 0.02 mph, screaming to 81.76 mph by the end of the measured mile. Flock went on to win his second successive Daytona Grand National race easily, flying over the finish line nearly a minute before his closest rival. No one was laughing at Carl and his snazzy looking crews on the beach anymore.

As the NASCAR rules began to squeeze Kiekhaefer's hopes of repeat victories with his Chryslers during the 1956 season, Carl covered his bets by adding cars built by Dodge, Ford and Chevrolet to his teams in an attempt to develop the optimum chance for success on any given weekend. At one point during his second season, Kiekhaefer cars had won sixteen races in a row, and it seemed as if no one else even had a prayer. As Kiekhaefer's reputation of track dominance spread, racing journalist Sandy Grady informed *True* magazine readers that race fans began to turn against him, something Carl wasn't prepared for.

"When his Chryslers and Dodges swept to win after win in 1956, hot blooded Southern fans turned verbally savage against the team. Like sports crowds from Moscow to Missouri, the stock car customers treated any overwhelming winner as a villain. Kiekhaefer's huge white vans, his well-known resources and his string of 16 wins made him a juicy target. The fans booed his drivers on sight, screamed with delight at Kiekhaefer's disappointments, and wrote letters to promoters: 'I won't go to no more races until you do something about Kiekhaefer.'

"Kiekhaefer was unable to accept this reaction to his success philosophically. He was deeply hurt. 'What have I done wrong?' he said. 'This is a sport in which I'm not competing with my own customers, as we sometimes did in outboard racing. I'm fighting against General Motors and Ford Motor Co., with all their engineers and money and cars. I'm peanuts compared to the factory teams. Yet the fans almost riot when they see our cars. I guess they want me to quit.'"[2]

Again, the bluff to quit. Some wondered at the time if Carl thought his threatened departure would so seriously wound NASCAR and the promoter's ticket receipts that somehow his sudden unpopularity would vaporize like high octane gas on a hot Southern weekend. But Carl nearly did quit when fans jeered him while campaigning his team in North Carolina, screaming in the pits that "this was our last race."[3] *True* Magazine reported the uproar that surrounded the rumor of his departure from the sport.

"Word spread to the wire services, and the headlines' impact on NASCAR was explosive. 'This is the biggest news of the stock car racing season,' crowed a NASCAR dispatch. Men closer to Kiekhaefer registered skepticism ('the Old Man couldn't be driven out of racing with a stick,' said one). A week later, Kiekhaefer was operating as usual in a major race at Raleigh, N.C. (It was after this race, curiously, that Kiekhaefer, NASCAR and a Detroit team waged their hottest three-way showdown, with an all-night argument over the winning car's legality. The quarrel waxed so emotional that local police had to stand guard.) The winner was declared to be pure as soap flakes, but Kiekhaefer left the race track with five pages of notes on the Detroit car's unlawful parts."[4]

Though Carl Kiekhaefer adamantly denied it at the time, he was often able to garner Chrysler and Dodge cooperation at a higher-- and faster--level than even the factory teams fielded by Chevrolet and Ford could get results through their own long chains of command. Bill Newberg was president of Dodge, division of Chrysler in 1956, and would go on to become the president of Chrysler Corporation within a few years. As an illustration of the spectacular influence Kiekhaefer was able to bring to bear on his racing efforts, Newberg told a tale of virtual overnight and around the clock service to support Carl's racing program. Carl was running a Dodge in a race in the Carolinas, and called to tell Newberg that one of the Pontiacs entered by a competitor had a hot new camshaft that was tearing up the track, and Carl's Dodge couldn't catch it. It was Friday afternoon.

"Carl said, 'I'd like to find out how [this new Pontiac camshaft] would work on the Dodge,'" Newberg remembered. When Carl hung up, Newberg phoned the chief of the Pontiac experimental cam shop to inquire about the new cam. He said, "I don't know that there is anything special," Newberg remembered, so he asked to see their camshaft engineering drawings. Then, Newberg found what he was looking for.

> "I went through their files and I picked out the one that had the latest date on it and I said, 'Well, this has to be it.' It was a pretty wild cam. And, of course, we kept blanks there [un-machined camshafts] for our V-8, so I said, 'Make me up one like that, and I'll stick it in the engine and see how it performs.'"
>
> "I kept a bunch of engines around the Dodge plant there for Carl, and I put one [of the new camshafts] in one of those engines and I put it on the dynamometer, and boy, it did what he wanted. So, we took it off the stand, took it out to the airport, put it on [Carl's] twin Beech, and [Carl] took it down there and put it in the car and won the race on Sunday. Now that's the way you had to work with Carl."[5]

It was Carl's own confidential line to Chrysler and Dodge, and he kept it a well-guarded secret most of the time. "I smuggled him engines and parts and stuff like that," Newberg admits, "and as a matter of fact he didn't have to go through a lot of red tape to get things done with us." Carl and Newberg became close friends because they could each rely on the other's word and discretion. "I know a lot of people criticized Carl," Newberg said, "that he was 'fresh' with them and things like that. But the only way he was 'fresh' with them was 'cause either they didn't 'know' or they didn't deliver. He was impatient with anybody that didn't 'know'. There are only two kinds of guys as far as he was concerned, and that's the way it was with me. Those that knew how and those who didn't know how. And if you knew 'how', that was great-- he'd spend time with you. If you didn't, the hell with you."[6] Smokey Yunick was well aware of Carl's pipeline, though, and complained that "Chevy wanted to be competitive but they couldn't get things done quick enough. We'd ask for something and they'd always say, 'Why do you need that?' We were 14 levels below in their organization structure down here racing and Kiekhaefer could go right to the top."[7]

Indicative of Carl Kiekhaefer's expectations and attitudes about his crew's workload was the advertisement he placed in the top racing magazines early in the 1956 season.

WANTED

Engineer, Mechanical or Electrical, with an intense interest in auto racing and an urge to be associated with America's top stock car racing team. Position will cover the preparation and general engineering of the cars, both at our laboratory and in the field.

Applicants should have a practical outlook on mechanical problems, be unafraid of long hours and extended field trips and be blessed with a rugged constitution. While we don't go in for bugle-blowing at dawn as a general rule, we have been known to work day and night on occasion in order to put a winning team on the starting line. For the most interesting, well-paying and boredom-free position you'll ever be offered, write:

E.C. Kiekhaefer Kiekhaefer Corporation
300 W. Murdock Street, Oshkosh, Wisconsin[8]

On April 8, 1956, Tim Flock won the North Wilkesboro, North Carolina 100 mile Grand National race. Flock was once again the Grand National points leader so far during the season, campaigning Carl's Chryslers. Following the race, Flock decided he had had enough of Kiekhaefer's high-pressure antics and abruptly quit, promptly joining the ranks of Chevrolet factory drivers. Carl was so mad that some people thought he might just explode like an enormous engine out of oil. "I got along with Mr. Kiekhaefer, and I like him," Flock would later say, "but he is too much of a perfectionist. I lost a lot of friends, because Mr. Kiekhaefer tried so hard to win. ... He want's to get you up each morning with a bugle. ... I'm grateful to him, but, gee, I was glad to quit."[9]

Flock was literally falling apart working for the intense Kiekhaefer, and wound up with bleeding ulcers and a driving suit two sizes too big. "I was so skinny I looked like a damn skeleton."[10] Flock surely heard that two weeks later, when Buck Baker and Herb Thomas came in first and second at Langhorne Speedway in Pennsylvania, Carl took his drivers to the local Chrysler showroom and told them to pick out three brand new cars to reward them for their success and their loyalty.

Buck Baker, also driving for Kiekhaefer in the 1956 season, had captured third place in Grand National point standings when Flock called it quits. Nothing incensed Carl Kiekhaefer more than disloyalty or "defection", and he immediately concentrated on moving Baker up the rankings to push Flock out of first place. NASCAR rules allowed a driver to drive anybody's car in computing point standings, so the possibility that Flock could win so many races at Carl's expense, quit and go on to win the Grand National championship for the Chevrolet factory team, drove Carl to pull out all the stops. At one of the races, a Southern promoter offered a $500.00 bonus for any driver that could defeat Kiekhaefer during the upcoming race. Carl withdrew his cars and drivers from the race until "the bounty is lifted from my boys' heads-- it might encourage somebody to knock them through the fence."[11]

Even though Carl was now grooming Buck Baker for the 1956 championship, he demonstrated his displeasure at the Chevrolet team that hired Tim Flock away by hiring champion driver Herb Thomas away from them. Thomas later quit working for Carl, and as journalist Jonathan Ingram explained, was to become the central figure in the most horrifying and controversial incident in Carl's racing career.

> "Points for the national driving championship were awarded to drivers regardless of team association. Thomas won three races for Kiekhaefer and two on his own and was leading Baker in the points derby late in the season as an independent. In an effort to help Baker win, Kiekhaefer arranged to have a race added to the schedule at the Shelby, North Carolina Fairgrounds by guaranteeing the purse for the promoter. At that time, it was not unusual to pick up events on the schedule as the season wore on, but this race became peculiar according to events. Thomas was thought by some to have been knocked off

the track by Kiekhaefer driver [Alfred "Speedy"] Thompson, a [Buck] Baker cohort and a notorious daredevil. This, too, was not uncommon in racing except that Thomas was knocked unconscious and taken to the hospital in a coma. 'They had scraped the track and there was a rim of dirt built up around it,' recalled Thomas' mechanic, Ray Fox. 'Herb had gotten by Speedy [Thompson] on the inside and then his car floated to the outside. Speedy hit him from behind and pushed Herb over a pile of dirt and his car rolled over. It was so bad we found his shoes on the floor. He was unconscious for a long time.'

"Although he recovered well enough to even attempt another race that same season, the accident ended Thomas' championship hopes (he had won the title in 1951 and 1953) and his career and left him with slurred speech and restricted body movement as if he had suffered a stroke. 'It was not unusual for members of one team to go out and run interference for another member of the same team,' said Schoolfield [Hank Schoolfield, Winston-Salem *Journal* and *Sentinel*]. 'You didn't expect to cripple someone. When something like that happens, things get very emotional.' For many members of the old racing fraternity, the Thomas accident was the last bitter straw of the Kiekhaefer era, which had brought so many unexpected changes. 'Kiekhaefer in his anxiety to win nearly killed Herb,' says [Smokey] Yunick. But even Fox acknowledged, 'There was a lot of hanky-panky on the short tracks. No one expected to get hurt. Just spin 'em out and go on. Herb had plenty of skirmishes of his own. Maybe Speedy was just trying to spin him out and didn't know he would get hurt so bad.'"[12]

Detroit took the opportunity to step up its efforts to beat Kiekhaefer, and rumors spread throughout the pits that the new Chevrolet and Ford *Cri de Guerre* was "Stop Kiekhaefer or get out." The remarkable Kiekhaefer string of 16 straight victories was finally snapped on June 22, when one of Kiekhaefer's Chrysler 300B's crossed the finish line in a distant eleventh place. When Jim Paschal, the winner of the race, returned to his pit, Carl was already there, congratulating the Paschal crew on their victory, and telling them that they deserved the victory. Carl had been timing their pit stops during the race, and found that they were beating the Kiekhaefer crew by as much as ten seconds.

But Carl had already won 22 of the first 29 Grand National events, and he wasn't about to let up. Even as the Kiekhaefer crews began to lose race after race -- they didn't win another event for two months -- Carl continued to build the most impressive stock car racing organization in the country. At his zenith, Carl was virtually swimming in hardware and facilities in support of racing, as Sandy Grady described for *True* readers.

> "If Kiekhaefer's methods are exhaustive, his expenditures on his racing stable are also grandiose. At one time he was maintaining 10 race cars, six vans to haul them about the landscape, at least 30 crew members, four drivers, a big parts warehouse and engine shop in Wisconsin and a well-stocked garage in Charlotte. He admits he purchased one Ford last season just to strip its parts for another Ford he hoped to race. His personal plane shuttles racing parts all over the land. When the problem of housing his mechanics came up at Charlotte, Kiekhaefer made a characteristic gesture: he bought two $15,000 homes and stocked them with beds."[13]

Charlotte, North Carolina became Carl's home away from home, as the busy garage he established there assumed the heady responsibilities of Kiekhaefer race headquarters. A young Bobby Allison, who would in later years become a celebrity among race car veterans in his own right, cut his teeth in the Kiekhaefer stables and remembers Carl's legendary temper. "While I was working in Charlotte, at least twenty guys must have quit or gotten fired. His favorite saying was, 'You're fired, but you can't leave until the car is ready.'" Allison joined the Kiekhaefer effort right after graduating from a Miami high school in 1955. He started by "testing motors and boats on Lake Butte des Morts ... [and] nearly froze to death when a boat was swamped and sank under him on a 15 degree day."[14]

Carl began to suspect that the reason he stopped winning was the direct result of sabotage. Since May 30, 1956 at Syracuse, New York, Buck Baker had entered 18 races and only won one. "Speedy" Thompson had entered 15 races, only winning three, and Herb Thomas had entered eight, only winning once. Carl detailed this stunning string of failures at the top of a tersely worded memo to all of his race crews, announcing a cash reward for revealing sabotage against his racing efforts.

> "From the above, you can see that out of 18 races, Baker won only one race. If our competition were paying someone in our group to sabotage Baker's car, they could not have done better.
>
> "Since our championship is at stake, it behooves every man to be on the lookout for any careless assembly or purpose-

ful sabotage. A $500 reward is hereby offered for evidence of deliberate sabotage within or without our own crew.

"In view of our past record in 1955 and the first half of 1956, the present poor showing of our cars is a direct result of our own 'goofing'; the frequency of which is such that it is not accidental. Loyalty to our team is the first requisite. Loyalty is above all ability and knowledge. Without loyalty, there can be no team work. Without team work, we obviously cannot win. If we cannot win, there is no point in carrying on the hard work, the sleepless nights -- and above all -- there is no point in endangering the lives of the best drivers in the business, which we are fortunate to have on our team.

"Just think what a horrible waste of time, energy and money these last two years have been if we cannot measure up to champions. Any information or suggestions helpful to win this championship would be greatly appreciated and loyalty will be rewarded at the end of the season by not only the bonus fund in the racing kitty, but we, of the Kiekhaefer Corporation, will double the amount in this kitty for the crew and mechanics if we win the championship.

"The men who have proved themselves loyal and worthy will be given first consideration with openings elsewhere in the corporation regardless if we continue racing or not."[15]

As Carl suffered this string of defeats during the late summer of 1956, he began to wonder if the publicity value for Mercury connected with his racing program might backfire. He had both his and Mercury's name white-washed over before a race at Chicago's Soldier Field. He fervently denied that it was done to avoid any personal embarrassment of losing a race, but rather the fan and "public reaction to the team was injuring his outboard motor trade."[16] Some of Carl's defeats simply seemed like bad luck, like the 15 cent condenser strap that failed on Buck Baker's Car that cost him a victory in the Memphis 300 mile race.

Carl continued to be concerned that the adverse reactions his efforts were attracting from fans, magazines, Detroit interests and NASCAR itself would eventually overwhelm him. He was so quickly becoming a legend in stock car racing circles, that a comic book appeared that featured a rich, brash talking tycoon who bellowed at his stock car race crews and took dangerous chances to win races, even to the extent of putting his drivers at risk. The appearance of the comic book incensed Kiekhaefer, and he instigated legal action to have it stopped. Then, the movie "Thunder in Carolina" was released, confirming Carl's spreading paranoia about his reputation. It was "a motion picture effort to capitalize on Kiekhaefer's peculiar box office appeal. ... The movie was in large part a cruel caricature of the man, portraying him as a mean and overbearing manipulator of men."[17]

For Carl, virtually every race now entailed bitter arguments with NASCAR officials. He was certain of blatant preference given to Detroit factory teams in order to attract more and more organized money into the sport. Typical of the disputes was a ruling on power tools that infuriated Carl to no end.

> "Kiekhaefer noted that at the 250 mile Grand National race at Martinsville, Va. the ruling that no power tools were to be used during pit stops was waived by NASCAR Commissioner 'Cannonball' Baker. Kiekhaefer's explanation of this decision presented the view that Chevrolet had originally planned to run the race without a tire change, thus making the need for power tools nil. However, when the need for a tire change became evident in practice, Kiekhaefer claimed that NASCAR bowed to pressure from Chevrolet to allow the use of power tools. Neither Chevrolet or NASCAR accepted Kiekhaefer's interpretation of this situation. In an interesting twist of fate, Kiekhaefer cars finished one-two in this event and NASCAR officials rejected a protest that they were shod with illegal tires."[18]

To demonstrate their renewed commitment to winning stock car races, the Detroit automotive fraternity appointed former Indianapolis 500 winners to head up their racing teams against Kiekhaefer. Chevrolet contracted with Mauri Rose, Lou Moore was brought in by Pontiac and Pete DePaolo was commandeered to head up the Ford Team. Detroit plainly had Carl in their gunsights. The new slogan was becoming, "Win on Sunday, sell on Monday." Still, the auto giants couldn't match Kiekhaefer's oversized will to win. Jerome "Red" Vogt, the master mechanic that Carl wrestled away from the Ford team to play a key role in the Kiekhaefer camp, summed up Carl's strength. "Carl was a perfectionist in everything he did. ... He had the best, most sophisticated, and largest setup of the time, of anybody, even the automobile manufacturers. He had it organized exactly like it ought to be. He was ahead of his time. Everything he did had to be exact or else it was no good."[19] Carl flew his race car engines back to the lab in Oshkosh after virtually every race, where they were completely tore down, parts calibrated and replaced if necessary, and then flown back for installation. Richard Petty, only a teen-ager at the time and son of Lee Petty, one of Carl's most tenacious competitors, remembers how most crews viewed Carl's efforts. "He run it like the military. He was the first to come into racing with a first-class operation and treat it like a business. He would haul his cars and five or six engines in big vans while the rest of us were using a dadgum towbar."[20]

Carl's irrepressible cars and drivers won an amazing 22 out of 41 NASCAR Grand National races entered in 1956, with Buck Baker thoroughly trouncing all rivals to capture the championship crown. During the season, Bill France had experimented with a convertible class within NASCAR, in which Carl also competed. After only thirteen starts, Carl withdrew from the controversial circuit, protesting that convertibles offered insufficient protection for his drivers, and weren't designed for the punishment of the track. Even so, Frank "Rebel" Mundy had driven Carl's 1956 Dodge D500 convertibles to six first place finishes, and had Kiekhaefer continued campaigning in that circuit, Mundy and the Kiekhaefer crews would have undoubtedly captured yet another national title in 1956.

Carl was exactly on top of the stock car racing pyramid, having decisively captured three national championships in two short years. Detroit, blushing red from their seeming impotence to beat this determined sportsman from Wisconsin, would collectively budget over $7 million to re-capture the world of stock car racing in 1957. Ford, Mercury, Chevrolet and Oldsmobile were all preparing to stop Kiekhaefer. Carl had spent slightly over $1 million between the 1955 and 1956 seasons, which in itself was more than twenty times his closest independent rival on the circuit. But now, Carl was more than alarmed to learn that the Detroit giants were planning to spend fourteen times even the Kiekhaefer budget to put an end to his total domination of the sport. Rumors spread quickly throughout stock car ranks and publications that Carl, having had his fill of both the high-handedness of NASCAR rules and Detroit muscle, was calling it quits. But like the little boy who cried 'Wolf!!', most merely assumed it was another bluster and bluff by the master of both.

Vicki Wood, who had driven Carl's Chrysler 500B to a record over 136 mph to capture the women's world speed record at Daytona Speed Weeks at the start of the 1956 season, was the first to learn that this time Kiekhaefer meant business. Wood had written to Carl in hopes of repeating her performance at the 1957 Speed Weeks, only to receive the first formal news of Carl's departure from the sport.

> "I dislike very much to have to confirm the rumors, but I think it best you find yourself another ride since we do not plan on entering into another racing season. With the factory teams dominating the racing activities, it looks like we will be out of the picture. ... We understand Ford's plans carry them through 1958."[21]

Carl kept fueling the possibility of competing in 1957 by keeping his drivers under contract. Even on January 14, 1957, only three weeks before the start of Daytona's Speed Weeks, he was unsure of what to do. He called his old friend Bill Newberg, by then vice president of Chrysler Corporation and soon to be president, to debate the merits of continuing or throwing in the towel. In a secretly taped phone call (Carl had virtually every phone call taped and transcribed) with Newberg, Carl made it clear that he was locked in a battle of 'chicken' with NASCAR president Bill France over new rulings that the Detroit interests had forced on the 1957 season.

Kiekhaefer: I just called to check to see whether there is anything new so I can go one way or the other on the racing picture.

Newberg: Well --- I'll tell you. Both the Dodge boys and the Chevrolet boys just wrote down to France [to insist on a 25 percent points penalty on Kiekhaefer entries because of engine size advantages].

Kiekhaefer: I heard about that. It's ridiculous.

Newberg: ... That's the way it stands. The Chevrolet letter and the Dodge letter are saying practically the same thing.

Kiekhaefer: Yes. In the mean time we are running out of time for the Beach. [Daytona Beach Speed Weeks]

Newberg: Well, that's right. Mr. France realizes that ... I thought you were pretty well out of the opinion of going into any racing this year.

Kiekhaefer: Well, the reason I've got to know one way or the other is so I can release these drivers. They are still under contract to me. ... Indirectly I tie into this thing because I'll get the blame for everything. For one thing. Number two, he is discriminatory on anything --- why he changed [the rules] just to handicap me. ... Also, there should be some clearance on what we tell the press. They are coming in here, they are phoning in from all over the United States. They want to know what the hell we are going to do. Someday I am going to have to give them an answer.

Newberg: Yes. What are you going to do? You aren't going to know until France gives the answer what he's going to do.

Kiekhaefer: That's right.[22]

Carl wasted little time in letting the press know that it was up to Bill France and his rule-making that would determine whether he would return to race again. Within minutes after hanging up the phone with Newberg, Carl took a call from Bernhard Kahn, sports editor of the Daytona Beach *Evening News*.

Kahn: Actually, what I am calling about is your plans on Speed Week here, whether you will have cars or not. I've read a piece in the Charlotte paper that you would, one in the Times Union that you wouldn't, and or course, I don't know what's going on. I just wondered what your plans are?

Kiekhaefer: At the moment I haven't any.

Kahn: At the moment you haven't any? In other words, you haven't made any plans for entering yet?

Kiekhaefer: No.

Kahn:	If you were to, would you have time now? I mean, is there enough time left for you to get ready if you made a decision to?
Kiekhaefer:	Time is getting a little short.
Kahn:	Mr. Kiekhaefer, is there any reason why you've curtailed your racing this year?
Kiekhaefer:	Ask Bill France.
Kahn:	I will. ... What driver's ... do you have now under contract? ...
Kiekhaefer:	... Well, I really have nothing for quotation, sir. I'm a pretty busy man. I've got a lot of things to do and I haven't made up my mind whether I am going to run or not.
Kahn:	Is there a possibility that you might run in the Grand National here?
Kiekhaefer:	I think that's up to Bill France.
Kahn:	You mean you have to reach an agreement with Bill or that he has to make an agreement with you. Is that it?
Kiekhaefer:	He's going to have to make it possible.
Kahn:	In what way, sir, is he -- in other words, there is a snag right now?
Kiekhaefer:	Well, I don't think he wants us to run.
Kahn:	Has he done anything to make you think that?
Kiekhaefer:	Oh yes, lots of things. But really, I really have nothing for quotation, sir. I just don't feel like running at the moment.
Kahn:	Well, if you change your mind, I imagine your public relations outfit will inform us, won't they?

Kiekhaefer: Our public relations have never operated on racing. In fact, we didn't want publicity on racing. We raced for the fun of it. We did not intend it ever to be a gimmick of advertising.

Kahn: Well, thank you very much Mr. Kiekhaefer.

Kiekhaefer: After the articles you wrote about us, I'm surprised that you even bothered to call us.

Kahn: Who's that?

Kiekhaefer: You. The article you wrote last year has been rather derogatory. I don't know why you continue to write about us. During the year there have been some derogatory articles.

Kahn: ... I don't think there's anything that we printed that was derogatory.

Kiekhaefer: It seems to be entirely out of definition of good sportsmanship for the press to have written about me as they have. ... You know the public. They just follow you fellas like a bunch of sheep. You write one day that everything's black, why all your fans are black. The next day when you write it in a bright vein, why they feel a little better towards you. They sway back and forth just like they were following a tennis ball.

Kahn: I believe that might be partly true. ...

Kiekhaefer: Well, all right sir, just don't count on us this year. France has his people in there [Detroit] that he wants, and let's see how it'll go. I have no feelings in the matter one way or the other. I have no regrets. I've had my fun. ...[23]

The debate continued in the press as Speed Weeks approached. Newspapers throughout the country contradicted one another as journalists scrambled to second guess the evasive Kiekhaefer. Speculation built to a frenzied crescendo on the eve of competition in Daytona Beach, and Carl, having had enough teasing the press and Bill France, prepared a final pronouncement to settle the issue once and for all time.

"We have noticed recently from several sources, word is going out that we at the Kiekhaefer Corporation intend to make a dramatic last minute entrance at the Daytona Beach Speed Week Races. This is not true!; As I have positively stated a number of times to the gentlemen of the press.

"It is true we have purchased several '57 automobiles including a Chrysler 300C, but not for racing purposes. In fact, no special automobiles of any kind have been prepared for or received by the Kiekhaefer Corp. at any time. Above all things, we hope we have not left the impression with the public that it is anything but a matter of choice that we are not racing. We believe it deplorable that after being out of racing for four months, our name is still being used, possibly as a drawing card for an event in which we have no interest, namely: Speed Week.

"As a matter of fact, our time is completely occupied with the various national boat shows, with production of our Mercury Outboard Motors and with the building of new plants.

"If we should ever again engage in racing, we will be the first to announce it, but at the present time, there are no plans. This means no USAC, no NASCAR, no SCCA, no Indianapolis, no LeMans or any other form of automobile racing.

"Much as we would enjoy to continue to challenge the best that Detroit can offer, and believe me, we have enjoyed it over the past three national stock car championships, we reiterate: we are through. This is final!"[24]

But once again, Kiekhaefer choose to be ambivalent, and stopped the circulation of his news release, and instead issued a terse, industry-wide statement loaded with ambiguity. Even *Motor Guide*, who published an article entitled "Why Kiekhaefer Quit Racing," was dumfounded by the nebulous wording. "Kiekhaefer, himself, at first refused comment. On the surface, it appeared that he would remain completely silent. After much persuasion he finally agreed to make a statement. When released (after much controversial editing), it actually said nothing, as follows:"[25]

"That particular phase of our activity that caused us to race, which incidentally resulted in three successive championships, has passed. With due credit to President Coolidge, we did not choose to run this year. As always, our No. 1 activity is building Mercury outboard motors."[26]

From that moment on, Carl would never speak to the press again about automobile racing without invoking that nebulous statement. Although he was to occasionally, and generally half-heartedly enter a European sports car into a local race, usually something he had bought for his own amusement, Carl was never to race automobiles again. Though a number of articles would be written about Carl's efforts in those two incredible years of conquest, Carl refused to speak to the press to confirm or deny anything associated with his racing days. It was as if it never happened.

Privately, though, Carl was incensed with the treatment he felt he received at the hands of NASCAR and Detroit. In another of his letters he never mailed, he vented his frustrations and his opinions about his withdrawal for the last time.

> "Rest assured, we had no illusions about our popularity, and while we were vilified directly, we feel it was an indirect campaign against Chrysler who did not recognize it as such. In fact, Chrysler behaved almost as though we were a relative that had gotten into a jam and because of the subsequent embarrassment it seemed they sought to disown us, furthermore disclaiming any interest in racing.
>
> "Such is the sport of racing. It is a low-brow business not without its humorous aspects. Now that we are out of the treadmill (we ran in 66 races last year) and have caught up with our rest and business affairs, I can't help but view with amusement how racing has affected the dignity and the decorum of top officialdom in General Motors and Ford. Company airplanes, sometimes two and three of them, loaded with brass come to these races and get into the act like a couple of school boys fighting over a catcher's mitt. Can you imagine a guy like Bill France refereeing a dog fight between the trade billionaires while Chrysler quietly walks off with the bone? I bet there were some embarrassing moments in both organizations, particularly when new budgets had to be set up.
>
> "But be that as it may we are out of racing. We have enjoyed it while it lasted. ... But I gag at the present so-called NASCAR setup where with manufacturers competing, Bill France throws everyone a bone to keep the savage factories from clobbering each other. ... What incentive is there for an individual to race in NASCAR? Surely it isn't much. With the method of selecting a championship now apparently based on overall points rather than by batting average, what point is there in winning a championship. The manufacturer can drop 200 automobiles into the racing circuit and never win anything

better than tenth place and still come home with the gross total number of points, thus becoming a championship car by that method of calculation. ...

"I will not bore you further; but I do want to leave you with this. I believe any sport must be covered by a fixed set of rules, rules that are not constantly revised, ignored, misinterpreted and are not altered by bulletins. Sometimes as many as four bulletins were issued on one race, bulletins that did not become public notice until the day before the race, allowing no time for preparation. ... Commercialism and sports just don't mix. I can't help but feel indignant about this sport that grew by itself to considerable proportion, only to be raped by Detroit. But what makes me more indignant than anything else is that through all these shenanigans the public is the one misled and misinformed by the spectacle. ... Whoever heard of a sport for profit anyway. ...

"P.S. I have marked this confidential since I do not wish to have this get into the hands of the press and rekindle bad publicity. We just cannot afford to have our name further associated with racing."[27]

In those two remarkable years, Carl and his aggressive crews and drivers, mechanics and engineers, had won a phenomenal eighty percent of all the races they had entered. Even more significant, not one violation of the rules was ever attributed to a Kiekhaefer car, pit crew or driver, and none of his drivers received a single scratch from having driven a Kiekhaefer Mercury team car. "Winning a race in a safe and sportsmanlike manner," Carl said, "was the No. 1 objective of the crew and myself."[28]

Time eventually eroded those thunderous days of 1955 and 1956, until in 1980, Carl was surprised to learn that he had been inducted into the National Motorsports Press Association Stock Car Hall of Fame. Most of his driver's had long since been installed, some over fifteen years previously, which only served to confirm Carl's feeling that the fans and drivers of the sport had long forgotten him. Herb Thomas was inducted in 1965, Fonty Flock in 1967, followed by Tim Flock and Junior Johnson in 1973. Interestingly, to the discomfort of many in attendance at the induction banquet, Carl made it abundantly clear that even after 24 years he hadn't ever forgiven Tim Flock, who arrived to honor Carl, for deserting him in the middle of the 1956 season.

The depth of Carl's contributions to the sport were finally recognized, from his pioneering work in chassis setup, wide-rim wheel usage, engine performance, driver safety, driver compensation to the paper air filter that had become a world-wide standard in automobiles. The professionalism he brought to the sport, from crew uniforms to fast pit stops, was finally recognized by his peers who, like Carl, were quickly disappearing from the scene. Carl himself was able to sum up this remarkable phase of his career in three short sentences he wrote shortly after he had decided to quit for all time. "Make no mistake, we have no regrets. We entered and we left when we wanted to. NASCAR stock car racing was one thing when we entered but it was something entirely different when we left."[29]

Prophetically, on June 6, 1957, the Automobile Manufacturer's Association abruptly placed a "boycott on all factory participation in racing."[30] Many considered the withdrawal of the factory teams the result of Kiekhaefer's decision to quit. They no longer had a common enemy. Carl had beaten the giants, after all.

CHAPTER TWENTY-ONE

Lake X And Operation Atlas

"I find dreams of this type to be one of the wonders of the world and the human race. It is a miracle what can be built from a dream with nothing but two hands, and honest heart and ambition. Education is very important too, but even that is unnecessary if the dream is of sufficient strength and the cause is sufficiently noble and interesting to the individual."[1]

Carl Kiekhaefer

Having washed his hands of both the glory and the anxieties of automobile racing, Carl once again focused his undivided attention on the business of Mercury outboard motors. The obsessive diversion had cost his corporations a sizable fortune, and jeopardized their future. The combination of heavy expenditures and reduced product research had been devastating in ways that were known to only a few insiders. Carl spent well in excess of a million dollars in pursuit of his stock car championships at a time when the funds were desperately needed for facilities improvement, promotion and new product development.

But Carl heard the beat of a different drummer. He was pushing the company full speed ahead, feeling that somehow, some way, the money would catch up as he began to scramble again for new products. Fortunately, Charlie Strang had ignored Kiekhaefer's orders to pay 100 percent attention to the auto racing program during 1956. Surreptitiously, Strang conducted a secret development program that was to result in one of the most successful products in Mercury history: the six-cylinder in-line "tower of power", 60-hp outboard engine.

"When he got off on something, boy, nothing else mattered but what he was doing. When we were in the race car business ... it was an around the clock operation. You were virtually forbidden to work on outboard motors during that period. He was on the day shift on the race cars, I worked on the night shift on the race cars. I nutted up a six-cylinder outboard [by taking] some blueprints and cut three cylinders off one four-cylinder engine, and three off another and glued the blueprints together to make a drawing. Then we went out and got a couple of four-cylinder raw castings and sawed them off the same way the blueprints were cut.

"I got one of the race car guys who was a terrific welder, stole enough of his time to weld these things together to make a single six-cylinder block. Then we welded three two-cylinder crankshafts together and bootlegged them through the shop because [Carl] wasn't around. Edgar Rose made an ignition system for it out of automobile parts and this, that and the other thing. We finally got it together and ran it. [Carl] didn't seem to know anything about it, because it was 'verboten' to work on anything but the race cars [and] he didn't want the shop upset.

"We finally put it on a boat and took him down to see it. It was pretty tall. He looked at it, and we pulled the cover off and he started to laugh. I said, 'try it out.' He got in the boat and took one run up and down the river, he came back and said, 'It speaks with authority, let's build it.' That was the decision process. Nine months later we delivered the first one."[2]

The Mercury Mark 75 was the industry's first six-cylinder outboard engine, and at 60-horsepower, the most powerful production outboard motor ever manufactured. Carl best expressed the technical achievements embodied in the design. "An interesting fact is that, although the 60 horsepower six cylinder motor is the largest production outboard in the industry, it produces the most horsepower per pound, the most horsepower per cubic inch of displacement, the most horsepower per dollar of cost and the most miles per gallon."[3] In 1957, the closest thing any other manufacturer had was 40-horsepower, represented by Evinrude, Johnson and Scott-Atwater. Johnson and Evinrude would soon attempt to counter the Kiekhaefer development by announcing new 50-horsepower, V-4 models. The OMC engines were so much more wide and stodgy compared to the tall, sleek-looking Mercury, that throughout the industry they attracted the unfortunate moniker of "fat-fifty," while the largest Scott-Atwater engine, the 40-horsepower *Royal Scott*, was a more conventional-looking two-cylinder design.

Normally, Carl Kiekhaefer had a very accurate sense about what would be attractive to the marketplace. But he committed perhaps his most inaccurate assessment when it came to the method of shifting the new 60-horsepower engine. Large outboard motors of the period, including Carl's own 4-cylinder inline 40-horsepower Mercury Mark 55 engine, had a standard shifting configuration of forward, neutral and reverse. The shifting gears in the lower unit of the 40-horsepower engine weren't strong enough, though, to handle the increased torque and power of the new 60-horsepower engine. Carl, in a fit of serendipity, decided that instead of designing and tooling a new gear case for the Mark 75, they would make the engine *direct reversing*.

Carl was aware that some of the very large marine diesel engines used in ships had what was known as a *silent neutral*. That is, instead of a transmission that would select neutral, the engine was shut down completely. When the helm required reverse, the rotation of the engine itself was reversed, and the engine restarted in the opposite direction. Carl also knew that it was possible to run a two-cycle outboard engine in either right hand or left hand rotation, and that it would be theoretically possible to eliminate a "transmission" altogether if the various modifications associated with a change in running direction could be worked out. From the start, however, his three top engineers, Charlie Strang, Charles Alexander and Edgar Rose, each as diplomatically as possible, hinted that the unusual scheme wouldn't be accepted in the marketplace. Carl, though, was adamant about the clean, engineering simplicity of the concept, and word was passed along that the new engine was to be direct reversing. Carl reasoned that it would be faster and easier to perfect the direct reversing principle in the engine than it would be to design, tool and produce a stronger shifting lower unit. He also wanted it in production as soon as possible, partly because his industrial spies had tipped him off that rival Scott-Atwater would announce a 60-horsepower engine of their own within a year. The engineering staff was stunned by the decision, but dutifully went to work on solutions needed to make it work.

Among the reasons that Carl was against developing a new, stronger shifting lower unit for the new Mark 75 was his basic prejudice about gear shifts in general. It had been OMC that developed the first successful forward-neutral-reverse gear mechanism for outboards, and that was sufficient reason for Carl to avoid them. "That left Carl with scar tissue about gear shifts," Charlie Strang observed. "Just because that damn Outboard Marine introduced this thing, he didn't want it."[4] When Carl was eventually forced to adopt shifting for his larger engines, it still upset him to think that he was relying on an OMC engineering development to satisfy his distributors and the consumer.

Among the problems which had to be overcome for an engine to run in either direction were ignition and reversible starting, operating controls and cooling. The most difficult of these turned out to be reversible starting. After six months of experimentation, Strang, Alexander and Edgar Rose were able to perfect an entirely new electric starting mechanism that would quickly and accurately engage the fly wheel to spin it in the chosen direction. A water pump had to be developed that would pump in either direction, and the team hit a snag when it was discovered that even a small blade of grass blocking a check valve could admit air into the system and starve the engine of cooling water. Strang, drawing on his early days at school in Brooklyn, suggested that what Edgar Rose could use was a good old fashioned "Bronx cheer." A "Bronx cheer" is basically a short collapsed rubber tube that would make the familiar "raspberry" sound when you blew on one end. Strang correctly surmised that the principle would apply to fluids equally well, and wouldn't allow air or water to reverse course in the collapsed tube when the engine rotated in the opposite direction.

When the unit was introduced to the public, problems began to show up. The new motor was supplied with a special control quadrant containing a single lever with a start-button on top. When the operator wanted to start the engine, he had to first choose either forward or reverse, because the engine had no neutral -- except the *silent neutral* of a stopped engine. Starting the engine in either direction, of course, depended on the strength of the battery in the boat. If the battery was weak or discharged, when the operator "shifted" from forward into reverse, as in docking, the engine wouldn't start up in the opposite direction to brake the boat in time. "We called it the *Dock Buster*," Strang remembered. Another problem was related to starting the engine "in gear." Boat operators were quite spoiled by the ability to start an engine at the dock in neutral, let it warm up for a while, and then shift into forward to take off.

But Carl's direct reversing engine had no neutral, so in whatever direction the engine was being started, the big propeller began to turn, slowing the cranking speed and making it harder to start, and once it did, the boat was off and running. Consumer resistance began to mount, even as Carl tried to convince the public that his way was better. "Eliminate the noise of resting at the dock with our exclusive 'silent neutral.'" The trouble was, all too often a slight irregularity in carburetor adjustments or a weak battery also gave the operator "silent forward" and "silent reverse."

Jim Wynne, as chief of Mercury's proving grounds, had to test the units once they came from either the research lab or, eventually, the production lines.

> "I thought it was the most stupid, idiotic, ridiculous idea I'd ever heard of in my life. ... But, I think Carl really thought that this was going to replace the gear shift lower unit. It was lighter and simpler, and I'm sure it was cheaper, but anybody that had any practical experience with operating boats just knew darn well that this wasn't going to be feasible over a period of time in the field. That didn't bother Carl. He went ahead and did it anyway. Forced that thing down everybody's throat."[5]

But the engineers knew that eventually Carl would request a traditional shifting mechanism for the new motor, and so they secretly worked on a design. Charles Alexander dreaded the clandestine engineering, knowing if it was uncovered there would be instant repercussions.

> "We did it secretly because I knew that one day he was going to come around and want me to do it overnight, and I knew I couldn't do it overnight. But when he [finally] saw it, he was *furious*, because we'd been working on it. The dealers gave him enough trouble and he finally had to back down on the thing. [He was furious because] I was not following his directions and he was the boss. It was an affront -- challenging his authority, and actually what I was challenging was his judgment. ... We were all delighted to finally be off of the old direct reversing thing."[6]

It was during this period that even Charlie Strang was getting frustrated in his role of chief "Kiekhaefer handler." Strang was in the often tricky position of buffer between Carl and the top engineering and administrative personnel, all of whom trusted Strang's judgment and had grown accustomed to having Strang shield them from Kiekhaefer's explosive outbursts. Evidently out of respect for Strang's unique abilities, Carl never once fired him, while any number of Kiekhaefer's top echelon would actually count with pride the number of times the boss had "fired" them, only to be back on the job within hours or days. Strang, however, actually *quit* a number of times, and each time he would be totally ignored by Kiekhaefer, an almost complete reversal of roles for Carl during heated arguments.

> "I quit a lot, but he never fired me. He didn't pay any attention to me at all. He just called me and expect me to be there in the morning.
>
> "My most memorable quitting I think was we were having an argument outside of the plant on Murdock Street in Oshkosh. I don't really remember what it was all about but I knew I had a key ring, one of these ball chain key rings and there was somewhere between twenty-five and thirty-five keys on it. I got so damn mad I fired the key ring into the ground and Carl got in his car and drove away. It was dusk, and when I fired the damn key ring on the ground the chain broke and I spent the rest of the night with a flashlight crawling around on the grass looking for my damn keys."[7]

Carl was plagued by his own omnipresent key chain. Even though a standardized master key system was in place for the majority of plant entrances, Carl carried a key ring that was larger than the custodial staff. He was constantly misplacing his ring, or taking off certain keys and losing them. Every time he misplaced a key, however, he insisted that the entire master key system be changed for the plants, forcing a massive re-keying of all the doors and gates, and a re-issuing of new keys to hundreds and hundreds of employees. "Door kicking was his favorite hobby," his assistant Fritz Shoenfeldt recalled. "He had a ring of keys that weighed about ten pounds." When he became frustrated fumbling for the right key, he'd quickly lose his patience, step back a couple of paces, and let fly with all 240 pounds focused on the sole of his size eleven shoe. Doors exploded open amidst flying wood chips throughout his facilities at one time or another. Sometimes it seemed like the maintenance staff followed him around just to fix the broken doors and jambs. "He hurt his foot once on the Cedarburg door," Shoenfeldt remembers, "sprained it pretty bad and limped around for a week."[8]

Gates were another irritation to Kiekhaefer. Since he never worked regular hours, he was constantly arriving and leaving his plants at odd times, and would rarely let the guards know when he was coming. As a result, he smashed open a number of gates, damaging both cars and gates, when the guard was either not in attendance or was too slow arriving to let him in. Finally, gates at the Cedarburg plant were actually spring-loaded by repairmen, so when Kiekhaefer smashed into them, they would fly open without damaging the boss' car. On at least one occasion, though, even Carl found humor in his gate antics. Charlie Strang was with him one day when they laughed so hard their sides ached.

> "One time we had a test operation in Oshkosh on the river, [with] a big high board fence around it with a gate. At that time, Oshkosh was celebrating its centennial, and everybody had to grow a beard or a mustache, or you could be thrown in jail for a day and that sort of thing, which obviously we didn't pay any attention to, but a lot of the guys working with us did.
>
> "One Sunday he wanted to get in the boat house and I was with him and we drove up to the plant. It's locked, so he just blew the horn on the car and held his finger on it so someone came. A guy came along and he grabbed the fence and he pulled himself up to see who it was. And like "Kilroy was here" he pulled himself up and he had a mustache and beard and he was so shocked when he saw Carl sitting there his mouth dropped open. Carl said, 'My God! It's a *snatch*! He was pretty quick with that kind of thing."[9]

As Carl continued to step up outboard motor design, he became increasingly frustrated with efforts to test his new motors in secrecy. He was so paranoid of OMC observation of his testing and production methods, that he once denied a troop of Boy Scouts a tour of his Fond du Lac assembly plant, fearing that among them might be a planted OMC spy.[10] Testing on nearby Lake Winnebago left his new products totally exposed to the competition, and pleasure boat traffic on the lake made it next to impossible to conduct their testing in peace. Rose Smiljanic remembered Kiekhaefer slipping test boats into the waters of the local lake after dark to foil prying eyes.

> "They used to test up on Lake Winnebago, and there'd be boaters around and you'd have to cover everything and sometime we'd go out like eight o'clock at night in a boat and you'd think we were smugglers or something to test a product. We'd have the rear end covered up, and [Carl] would watch so that there was no one in the area. The OMC spies, you know, Evinrude and Johnson. I mean, they were the *enemy* in those days."

At Kiekhaefer's Siesta Key proving grounds south of Sarasota, endurance testing of outboards in the coastal waters was also becoming more and more difficult due to interference from both commercial and pleasure boat traffic. Jim Wynne, in charge of all Mercury proving grounds in the spring of 1957, shared Carl's growing frustration with the lack of privacy and open water to complete even the most rudimentary testing. "We were having more and more problems with interference with other boats ruining our endurance tests," Wynne said. "Carl wanted a place that he could operate twenty-four hours a day, seven days a week and not have anybody around him. So we got in the company plane and scoured the state of Florida looking for a lake."[11] Among the spotters aboard the many Beech-18 flights that crisscrossed the state were Wynne, Rose Smiljanic and Charlie Strang.

Rose Smiljanic remembered how the searches would start with an aerial survey of a certain area, followed by a visit by car. "He was looking for a lake that he could close off and do testing and just chase everybody away," Smiljanic says. Ultimately, promising sites were identified. Malcolm Pope, brother of Dick Pope, proprietors of Cypress Gardens where Carl had for years photographed and demonstrated his engines, knew the owner of one of the sites, a Mr. C.B. "Charlie" and Annie B. Smith of Fort Lauderdale. Smith had recently purchased the land which completely surrounded Lake Conlin, near St. Cloud, Florida, and agreed to take Kiekhaefer and Smiljanic for a close inspection.

There was no real road to get to the lake, only a five and a half mile remnant of bricks laid in the sand that formed a segment of the old Dixie Trail, which had been built before the turn of the century by prison labor. The eight-foot wide trail is considered one of Florida's first thruways, and had been abandoned many years before. Where the bricks ended at the turn to the lake, only a winding dirt road remained to access the remote site. It seemed to Kiekhaefer and Smiljanic that they were driving through the set of an Amazon jungle film, for the tropical swamps and enormous stands of Cypress trees surrounding the lake gave the property a distinct feeling of isolation. Just what Carl was looking for.

When the lake finally came into view at a clearing on the northern shore, the sheer size of the body of water and property was at once impressive. The area of the lake was over 1,400 acres, surrounded by property that made a square (with one corner missing) of five miles by four miles, with a net area of roughly seventeen square miles or 10,462 acres. The shoreline of the lake was about seven miles, with a potentially navigable course for boats of over six miles. The slightly brownish water concerned Kiekhaefer, but he was assured that the coloring was natural for a lake completely surrounded by Cypress trees, the roots of which contribute to the somewhat muddy appearance of the water. The deepest sections of the lake were thirty-feet, and scattered here and there the surface was broken by a towering Cypress tree, seemingly growing right out of the water.

It was completely cut off from civilization, without power, telephone, water or utility services whatsoever. The lake waters and the surrounding lands were teeming with wildlife, including alligators, snakes, bass and catfish, eagles, hawks and herons, wild boars, turkeys, raccoons, armadillos and Florida deer. The lake had no tributaries or inlets, and not a single residence around its rugged perimeter. The lake was even naturally connected to several square miles of swamps, which acted as water accumulators and water table regulators to restrict the effects of yearly wet and dry seasons on the lake water level. The swamps surrounding much of the lake also created a natural barrier to keep out the curious. Carl was ecstatic. The problem was, he couldn't afford it. He had spent so much money on his three championship stock car titles that the company was, according to Kiekhaefer Mercury controller Don Castle, "skating on thin ice."

Carl decided that the opportunity to control this unique and massive property was too good to pass up, regardless of the current state of the Mercury treasury. He successfully negotiated a two year lease with a renewal option for two additional years, with clauses that would give him the first chance at purchasing when the lease expired. Carl wanted the location of the lake to remain a secret, so he referred to it as "Lake X" when speaking with outsiders, and the name stuck. From that day on, Lake Conlin was never mentioned again, and the lake is still known throughout the world of high performance marine products only as Lake X.

As news of Carl's latest gamble spread throughout the organization, Thomas B. King, Kiekhaefer's new director of public relations, was prepared with the perfect inaugural use for the secret facility. Tom King was considered by many to be a real "hot shot" promotions man, who could pull off headline publicity programs with relative ease. He was an "idea" man, who complimented Carl Kiekhaefer's dogged determination with inspiration and finesse. Frank Scalpone, who is currently vice president of the National Marine Manufacturers Association (NMMA), joined King's staff at Kiekhaefer Mercury in the fall of 1957. "He came from Studebaker in 1956," Scalpone remembered, "which had just come out with a couple of bullet nose cars to put them back on the map for a short time." King quickly proved himself to Carl by generating a flurry of positive Mercury publicity around the industry, which brought him close to Kiekhaefer in short order. "Tom had won [Carl's] confidence, which, if [Carl] believed in your loyalty he would stick with you no matter how many mistakes you made," Scalpone recalled. "Tom could get anything he wanted in those days, and Tom came up with the 25,000 mile endurance run."[12]

Tom King was aware of the unfortunate reputation that Mercury products had unjustly received in the industry. Mercury engines were beating everything in the water, but a whispering campaign had been started somewhere -- Carl blamed OMC -- that Mercurys were "fast, but won't last." The consumer who wanted a reliable, trouble-free engine to crank up on the occasional weekend to go fishing was hearing from his local Johnson, Evinrude or Scott-Atwater dealer that he would be better off with a slower, more conservative and dependable product, rather than the fast but allegedly temperamental Mercury. It wasn't true, of course, for Mercury products were enjoying a most enviable service record throughout the industry, but it was one of those slanderous slogans that just fell off the tongue so nicely, that sounded correct -- after all, Mercury's *are fast* -- and as a result, many people began to believe it. King's solution was to propose to Carl that Mercury stage a 25,000 mile endurance run at Lake X, a distance equal to a complete circumnavigation of the world. The resulting publicity would not only help to restrain the "fast but won't last" rumor, but could also showcase Mercury's latest and most powerful engine, the new 60-horsepower, six-cylinder Mark 75 at the same time. Carl was absolutely thrilled by the idea, and immediately assigned responsibility for the enormous undertaking to Charlie Strang and Jim Wynne. Strang would be in charge of planning and strategy for the endurance run, and Wynne would supervise the activity at Lake X and personally supervise the entire undertaking in Florida. It was to be known as *Operation Atlas*.

Among Charlie Strang's greatest strengths was to analyze the most complex assignments from Carl Kiekhaefer and issue a one or two page memo that could make the impossible look rather simple. Strang initially determined that the project would take approximately forty days. In his memo to Carl he calculated that the circumference of the earth at the equator was 24,902 miles, and "it will thus be necessary to maintain a minimum speed of 25.939 miles per hour (let's call it 26 mph)." Strang understood, though, the unpredictability of weather and other contingencies, and so proposed that "it would be wise to shoot for a 30 mph average (34.6 days), so providing a time cushion." He suggested that the new Mark 75 engines be throttled back to a somewhat effortless 4500 rpm (compared to the top speed of 6,000 rpm), and turning a propeller of higher than normal pitch, sort of simulating an overdrive in a car, to "permit operation in the economy range with the possibility of good fuel consumption publicity."[13]

Not wishing to leave anything to chance, Strang also suggested that all major components of the engines selected for the endurance run be "Zygloed and Magnafluxed" before the run, which are tests which would reveal even the smallest cracks or manufacturing defects in crankshafts, pistons, rods or castings. "A local surveyor," Strang suggested, "can measure and certify to such course or courses as we may care to stake out." Twenty-five thousand miles is a brutal journey in a small outboard boat, and so Strang evaluated the endurance of the driving crews that would make the attempt, forecasting that it would be "difficult to get drivers to take shifts of more than four hours at a stretch. With each man taking two such shifts per day it will require 3 drivers per boat on a seven day basis."

Among the problems that Kiekhaefer would need to solve, was how to convince the world that the endurance run was legitimate, and not just a bogus and slight-of-hand sideshow concocted by Mercury public relations. To this end, Jim Wynne searched for an independent group of individuals beyond suspicion that would agree to monitor the lengthy, 24 hour a day endurance run. The challenge was at first greater than he anticipated. Wynne contacted professors of engineering and mathematics at Florida's leading colleges and universities to serve as timers and officials, but the lengthy term of the assignment, in combination with the untamed wilderness of Lake X, made academic participation impossible. Finally, and probably with direction from Carl, Wynne was able to contract with the United States Auto Club (USAC) to monitor the attempt, an organization with national stature and with a generous history of supervising record-breaking endurance and speed events in the automotive industry. For a fee amounting to a few thousand dollars, USAC officials would provide 24 hour supervision, lap counting, and make certain that the engines "would not be modified except for routine maintenance" during the record breaking attempt. Now, all that was left was to prepare Lake X for *Operation Atlas*.

Jim Wynne drove a jeep into a clearing on the north shore of Lake X to begin setting up the record run. With Charlie Strang's logistical assistance, Wynne established a "base" at the water's edge with a house trailer, an equipment trailer and a mobile radiotelephone. When Carl heard that it was illegal to have a mobile phone without having it installed in a car, he promptly secured a junk car with an active Florida registration, stripped it of tires and doors, and sank it up to the floor boards to make a sort of "telephone booth" that would satisfy the letter of the law. Radiotelephone reception at the "booth" was poor at best, and so he ordered a giant, free-standing antennae to be erected next to the derelict car, because the statutes "didn't say how high your antennae could be." Later, Carl would order an even taller antennae for the base as operations expanded. Erecting the huge antennae became a problem when Carl's plan to tilt it up a certain way wouldn't work without twisting and bending the spindly structure. Edgar Rose, quite a polished engineer, proposed a better way, but Carl was adamant that it be done his way. Further attempts to erect the antennae failed, and Carl huffed away in disgust. As soon as he left, Edgar employed his proposed solution to the challenge and it worked perfectly. When Carl returned the following morning and saw the antennae erected, he asked how it had been raised. Edgar admitted that it wasn't Carl's idea that had worked, and

Kiekhaefer exploded, furious that Edgar hadn't followed his orders, and pouted over the incident for several days.

By September 10, 1957, Wynne had moved into the Lake Breeze Motel in nearby St. Cloud as preparations continued to commence *Operation Atlas*. In order to establish a clear course of at least five miles for the endurance run, eleven giant Cypress trees needed to be removed from various locations around the lake. Carl, in a characteristic rush, wanted to blast the trees right out of the water with dynamite. But, as Wayne Meyer, one of Mercury's employees on the scene remembered, *Carl* blew up before the dynamite could even be located. Carl stormed back from St. Cloud one day because nobody would sell him any dynamite to do the job. Meyer, using a much more diplomatic approach, made friends with the supervisor of the local power & light company who *gave* him three-quarters of a case of dynamite for free. When Carl found out that Meyer had succeeded where he had failed, he demanded to know the details and barked, "Well, how much did you have to pay?" When Meyer explained, Kiekhaefer bellowed, "Goddammit ... I can't even buy it and they *give* it to you!"[14]

Meyer and a local scuba diver placed "five or six sticks" of dynamite under the first tree out in the lake, lit the fuse and moved a safe distance away. The huge explosion that followed lifted the giant tree nearly clear of the lake amidst a tremendous shower of water. But then the giant Cypress dropped right back into place as if nothing had happened. "So we went back and got a long rope and tied it to the tree and got on this big barge we had there, and while we blasted, we was pulling. Then we just dragged them to the side of the lake."[15] The course thus cleared was officially surveyed and measured, coming out to exactly 5.5366 miles for a complete circuit.

The endurance attempt was to be *continuous*, meaning that the exchange of drivers, the addition of fuel and the accomplishment of minor repairs would all have to be made on the run, 24 hours a day, in daylight and dark, rain or shine, as the boats were flying around the course at over 30 mph. Strang, Wynne and his crew of drivers and mechanics, developed a host of unique solutions to overcome every obstacle. Strang decided that two identical boats would be specially prepared to go the distance, with two additional boats standing by in case of hull failure or accident. Fifteen foot Raveau family runabouts designed by Marcel Raveau were selected for the attempt, and adapted for the unique assignment. Windshields and windshield wipers were added, along with a 30 gallon fuel tank to reduce the frequency and danger of refueling operations. Automobile style headlights were fitted to either side of the runabouts for nighttime operations, complimenting lighted buoys that were placed throughout the course. Two special refueling and crew shuttle boats were built, featuring a 50 gallon fuel drum elevated five feet above the deck on a cradle of angle iron. As proven in practice runs, the fuel boat could synchronize with the speed of the endurance boat to be serviced, and a three-inch flexible hose would be passed over to the driver who would insert it into the fuel tank. Gravity pressure from the elevated supply tank would quickly fill the smaller tank, although the often erratic motions of the boats on the water accounted for a certain degree of spillage. A fresh driver would then cross over into the endurance boat and grab the wheel while the previous driver crossed back into the supply boat. According to the endurance run log books kept by officials and Kiekhaefer employees, only once did a tired driver lose his footing and go overboard during the crew transfers.

A special observation stand was constructed for the USAC officials led by Chief Steward Charles W. McDonald, under the direction of Duane Carter, director of competition for USAC. Carter had written to Wynne three weeks prior to the scheduled start of the run to lay down the conditions for authentication of the attempt. Among the requirements was that "all the component engine parts which may be required as replacements during the run must be carried in the boat during the run." The most important condition of USAC certification, however, concerned inspection of the engines following the run. "After completion of the run, the engine will be disassembled and compared with another disassembled engine selected from stock. All the parts must pass and be certified by the USAC steward and his technical representatives, to be stock as advertised and available to the general public."[16] This is the area where the ingenuity of Kiekhaefer's crews and engineers were to be tested to the limit.

Two minutes before seven on the morning of September 11, 1957, USAC Steward McDonald punched his stop watch as the first of the two boats roared away from the starting dock, followed by the second boat within minutes. Thirty-four days, 11 hours, 47 minutes and 5.4 seconds later, the lead boat crossed the finish line, having completed 4,516 laps of the Lake X course and 25,003.286 miles. The second boat finished only moments later. But what happened between the start and the finish was only partly observed by USAC officials.

By all accounts the engines performed amazingly well under the circumstances, but quite a number of unsupervised and wholesale replacements of parts became necessary as the laps added up. The first problems to surface were excessive build-up of carbon deposits in the exhaust ports of the engines, even though "white" gasoline was being used to reduce lead deposits on spark plugs. The engines would slowly lose power, so that eventually the drivers were forced to use wide-open throttles just to maintain 4500 rpm over the course. Periodically, the heads were surreptitiously removed while the USAC officials were distracted and assumed that *routine* maintenance, like spark plug or distributor points replacement was occurring. The heads would then be completely decarboned. But while the heads were off, Carl's mechanics also took the opportunity to replace complete distributors, complete sets of pistons, rod bearings, main bearings, crankshafts, carburetors, gear case components and at one point even exchanged the entire power head. Carl, unwittingly, would actively assist in the diversion, for much of the clandestine replacements were done when Carl would arrive and invite USAC inspectors for lunch in St. Cloud. Once, when Jim Wynne reported to Carl that he didn't think the engines would make it under the guidelines for the program, Carl icily replied, "Yes they will, Jim. You just make damn sure that they do!"[17]

USAC had placed a crimped seal and wire around part of the engine that was supposed to prohibit overhauls, but mechanics quickly learned how to drill and saw around it when necessary, or even move the seal from engine to engine. Naturally, when USAC tore down the engines at the finish, and the parts were compared with a stock engine taken randomly from Kiekhaefer's line in Fond du Lac, the components matched perfectly, for only stock parts were substituted during the run. Carl was never fully aware of the lengths his loyal crews went to guarantee his success in *Operation Atlas*.

Nighttime operations were the hardest on the crews and the equipment. The lights mounted on the bows of the boats were very unreliable, as the constant jarring would break either the filaments or the glass lenses. Quite often, a second driver or mechanic had to lay across the speeding deck in the middle of the night to replace lights as the boats sped around the course. Eerily, the pink eyes of Lake X's many alligators would reflect the lights as the boats made their laps, and drivers were able to estimate the size of the 'gators by the distance between the eyes. One night, one of the drivers inadvertently ran over a rather large alligator, kicking the engine up against the shock absorbers. The entry in Jim Wynne's official log for the evening was, "ran over alligator which surfaced just in front of boat! Engine kicked up. No damage to boat or engine. Damage to alligator unknown."[18]

Carl reaped a public relations bonanza when the results of the endurance record were announced. Strang prepared some calculations for Carl to throw around, attesting to the brutal punishment endured by his engines. Each engine, Strang figured, had made 225 million revolutions, and the total number of sparks or explosions per engine was a staggering 1 billion, 350 million. Carl confirmed to the press that "there were no mechanical or electrical failures and the batteries were still fully charged at the end of the run." Kiekhaefer and Tom King, supported by savvy new public relations arrival Frank Scalpone, were quick to capitalize on this "evidence" that Mercury was both fast and would last. Carl was so pleased by the success of his new Mark 75 engines, that he instructed USAC to keep them in official quarantine, because he wanted to run them *another* 25,000 miles to shut down once and for all the vicious, allegedly OMC inspired rumors that Mercury outboards were temperamental, delicate and fickle.

Jim Wynne, for one, had experienced enough of the mosquitoes, snakes, spiders and near jungle environment of Lake X to last a lifetime, and asked Strang to have someone else manage the second half of Carl's proposed 50,000 mile, *twice around the world* endurance run. He returned to overall supervision of Kiekhaefer's proving grounds at Siesta Key and Oshkosh, and was delighted to encounter civilization once more.

The second 25,000 miles at Lake X was pretty much a repeat of the first effort, though 17 foot models of the same boat were used for more stability, as a more windy and harsh season for boating was fast approaching. One night one of the endurance drivers fell asleep at the wheel and before he could recover, crashed into the densely wooded shoreline and was killed. Though Carl and the Lake X crews were deeply saddened by the event, the engine was put on a backup boat and completed the distance. This tragedy, along with the potential for future injury in normal testing activity, was among the reasons that Carl contributed heavily to the construction and renovation to the nearby St. Cloud Hospital. Altogether, Carl would donate over $300,000 in funds and services to the hospital, to insure that whenever Kiekhaefer Mercury test crews were injured, they would receive the very best of emergency medical attention and treatment.[19]

Carl Kiekhaefer and his able group of engineers, drivers, mechanics and public relations professionals had pulled off the largest, most organized and exotic of all endurance feats in marine industry history. It was a truly remarkable undertaking and represents a record which is yet unbroken, and given the extreme difficulty in reproducing the feat today, may likely stand for all time. Yet, even in the face of the banner headline publicity, the reams of test results and the sworn eyewitness testimony of USAC officials, the whispering campaign of "fast, but won't last" wouldn't go away. Of all of Carl's business and engineering frustrations throughout a long and celebrated career, he would be most disappointed by this unfair, unjustified and perpetually nagging accusation.

CHAPTER TWENTY-TWO

The Enemy

"I think it is about time we laid the cards on the table from both sides. ... We've got enough ... evidence that both sides are guiltier than hell. And I think it is about time we got together and straightened this thing out. ... Your men are guilty. Our men are guilty."[1]

Joseph G. Rayniak, president, OMC
To Carl Kiekhaefer in 1957

From the day that Carl Kiekhaefer boxed and shipped his first Thor outboard motor in 1939, he considered Outboard Marine Corporation a threat to his survival and a uniquely personal enemy with which to do battle. Over the years, Carl would become embroiled in skirmish after skirmish, both within and out of court, in an inexhaustible campaign against the outboard giant. Certainly, OMC would learn to consider Kiekhaefer an able and enthusiastic competitor, but without developing the paranoid, nearly diabolical hatred for their closest competitor as Carl would.

OMC, it was obvious to Carl, enjoyed the many fruits of a three decades head start in the outboard business. They had established secure reputations for their brand names of Evinrude, Johnson and Gale products, and their private label engines were distributed widely under a variety of well-known and trusted names. Their products were well designed, well built and reliable. In many undeveloped areas of the world, the Johnson brand name was so well established that it became a synonym for outboard motors, much as Xerox would one day enjoy this distinction among copying machines. Even if an African native was guiding his Caille, Elto or Mercury down some rain swollen river, it was called a Johnson. The advantage and power of these brand names was nearly magic. OMC, however, was not adverse to taking advantage of this privileged position, and it would literally drive Carl to distraction.

By 1950, Carl had set into motion a campaign against the OMC "myth" that Ole Evinrude "invented" the outboard motor. He flashed with anger whenever he would read the oft-repeated story of how Ole invented the outboard after rowing desperately to get his wife-to-be Bess Cary some ice-cream on a hot afternoon in 1909. In 1949, Carl contacted Cameron Beach Waterman, who, in 1905, successfully built and marketed an outboard known as the *Waterman Porto.* After connecting a small air-cooled motorcycle engine to a pair of bicycle sprockets and a propeller, Waterman tested the contraption.

> "In the winter of 1905, with big chunks of ice floating in the Detroit River, we drove a boat across the river with our motor. Then we knew it would work. ... In 1906 we made twenty-five of these motors and sold twenty-four. In 1907 we sold 3,000 and about the same number in 1908. Then in 1909, when the Evinrude motor hit the market, our sales doubled, because that convinced people we had a practicable machine and not a silly gadget."[2]

Cameron Waterman coined the word "outboard" in conjunction with his new device, and his advertisements are the first place the word appears. Clearly, Waterman was successfully marketing his mass produced outboards years before Evinrude built his first engine in 1909. Evinrude advertising, though, still promulgated the mythical "ice-cream incident" story, and proudly proclaimed Evinrude as the inventor of the outboard. This, understandably, was driving Carl Kiekhaefer crazy.

Carl convinced Waterman, then 73 years old, to be Mercury's guest at the 1950 National Motor Boat Show in New York's Grand Central Palace, and to be honored as the true, undisputed inventor of the outboard motor. At special Kiekhaefer sponsored events around town, Waterman told how he was on the rowing team at Yale ("We never lost to Harvard") and how the back-breaking competition started him thinking about a portable gasoline propulsion system for small boats.

Daily headlines greeted boat show attendees and exhibitors, proclaiming Waterman as the inventor of the outboard motor. Carl presented Waterman with a bronze plaque of recognition for his achievement, and Waterman was deeply touched. "I feel something like Rip Van Winkle, being dug up after all these years and plunged back into all this. While I did start the business, I certainly didn't realize what it was going to become."[3]

Carl's tactic had an immediate and dramatic effect. "The truth is," the New York *Herald Tribune* would report, "that the

Kiekhaefer Corporation, manufacturer of the Mercury motor, showing a deplorable lack of reverence for tradition, has been nudging Bess Carey's ice-cream cone out of the place it has held in history alongside Mrs. O'Leary's cow, Deliaha's scissors and King Arthur's shiv, Excalibur."[4]

In 1955, Carl declared the "Golden Anniversary" of the outboard motor, based on Waterman's 1905 device, and again trotted out the aging inventor to the New York show. Waterman, 78, received additional accolades from both Kiekhaefer and the American Power Boat Association, along with a brand new Mercury outboard. When the whirlwind round of appearances was concluded, Waterman wrote Carl to say, "I certainly am grateful for all you have done for me and hope the Evinrude myth is now finally put to rest."[5] These were to be his last appearances, as Waterman died on April 19, 1955.

On June 26, 1955, The State Historical Society of Wisconsin dedicated a marker and plaque to Ole Evinrude at Lake Ripley. The plaque identified Evinrude as the "inventor" of the outboard motor, and once again Carl was beside himself. The brief chronology of Evinrude's life was cast in bronze, and concluded with the statement, "... he invented the outboard motor and founded a new American industry."[6] Kiekhaefer badgered the society to correct the erroneous fact, and was even given support by W.J. Webb, vice president and division manager of Evinrude Motors. Webb, a popular and highly regarded historian of the marine industry, represented the Evinrude family at the dedication. Before the plaque was revealed to the public, Webb told officials "that his company would like to see the last sentence of the marker-text changed, and offered to pay for the cost of change. ... Ole Evinrude not only did not invent the outboard motor, he did not even produce the first outboard built in America."[7] The plaque was changed to read "... invented *an* outboard motor ...," rather than *the* outboard motor. Carl was pacified, at least for the moment.

The controversy was still smoldering as late as 1957, when Evinrude again used the slogan, "First in Outboards." Carl was infuriated, and dashed off a strong letter to his attorneys to initiate action to have it stopped. He also made reference to the Johnson legend that their motor was invented so the family could get to the best walnuts upstream.

> "As you may recall, Cameron Waterman produced thousands of outboards some three or four years before Ole Evinrude invented the 'Ice Cream Special' or before Johnson invented the 'Walnut Special'; all of which is rather ridiculous since in those days, they had launches, top buggies, bicycles, motor cycles and automobiles, domestic and foreign. If the ... Johnson and Evinrude boys were tired of ... bending over the oars with their sweaty, brawny muscles, it would have been much more romantic to use the 'bicycle built for two' or the 'merry Oldsmobile' for hauling the ice cream and the walnuts to go with them, it seems."[8]

Among the many perceived injustices that frustrated Carl Kiekhaefer in his battle for market share against OMC, was the Johnson and Evinrude policy of exclusive dealerships. OMC representatives and distributors would regularly tell their dealers that they were to handle OMC products exclusively, and were not allowed to handle any other outboard motor lines if they wanted to keep their valuable Johnson or Evinrude franchises. Since OMC manufactured the oldest brands in America, Evinrude since 1909 and Johnson since 1922, they had been developing dealerships in the best locations in the most desirable cities for over three decades before Carl Kiekhaefer started business. They had more or less acquired the cream of the crop over the years, and by the early 1950s had developed the largest, most visible, and most favorably located dealerships in America.

These dealerships handled many different brands of boats, many different brands of fishing tackle, many different brands of water skis -- in fact -- many different brands of everything else except outboards. They handled only Johnson or Evinrude outboard motors. These elite OMC dealers were constantly besieged by competitive outboard manufacturers, particularly the aggressive representatives from Champion, Martin, Scott-Atwater and Mercury, only to be told that they couldn't risk losing their OMC franchise by displaying an additional line of outboard motors. Many in the industry felt that OMC was enjoying an unfair advantage over the other manufacturers, none of which insisted on exclusive dealerships. As it happened, so did the United States Government.

On May 15, 1951, the Federal Trade Commission filed an action to conduct hearings into the "exclusive dealer" policy of OMC, to determine whether or not competitors had been unfairly restricted from trade, and if so, to what extent they had been injured. The FTC charged that OMC had 7,421 dealers in the United States, and [during 1951] sold over $20.5 million in outboards, or "more than 50 percent of the dollar value of all outboard motors sold in the United States by the entire industry." They further alleged that OMC created a "clog" in the industry, and that it was the duty of the FTC to remove this unfair restraint on the trade of the industry.[9] OMC stockholders were alerted to the hearings in the 1951 annual report:

"The Federal Trade Commission has filed a complaint against your company, which is based on Section 3 of the Clayton Act and is intended to prevent the company from continuing its policy of selling Johnson and Evinrude outboard motors only to those dealers who do not handle competing lines. Your company denies that its actions are in violation of law and is in the process of defending its position."[10]

In an attempt to prove that competitors had flourished while OMC had maintained exclusive dealerships, attorneys for OMC processed subpoenas for virtually every manufacturer of outboard motors to produce records of their production, sales, advertising, dealerships and distributors. The Kiekhaefer Corporation, along with the rest of the industry, was required to turn over their most confidential records to their most dreaded competitor. The industry was collectively aghast, and actually started to quietly discuss amongst themselves the possibility of resisting the order of the court, and refuse to deliver their records. Guy Conrad, Carl's in-house legal consultant at the time, sent a memorandum to Carl, briefing him on the growing tide of discontent.

"... Mr. Henry Smith of Scott-Atwater called, and wanted to know whether we were going to submit the material called for by the subpoena. He was told that a decision had not been made.

"Mr. Smith then explained that hearings were scheduled in Minneapolis for Martin, Champion and Scott-Atwater, starting today, February 11, 1952, and that he was of the belief that the companies in the industry other than Outboard Marine Mfg. should take a common approach to the subpoenas. He meant by that, that either all of us should submit the requested information or that none of us should submit the requested information. From his conversation, he is also, I believe, of the opinion that if such information is submitted by the companies

> in the industry outside of Marine Manufacturing Company, that then Outboard Manufacturing Company should likewise be required to submit identical information. ...
>
> "It is obvious that Outboard Marine Mfg. Co., thru the accumulation of the data called for by the subpoenas hopes to prove that companies other than Outboard Marine have prospered in the industry despite Outboard Marine's attempt to exclusively control their dealer set-up. Their thinking probably also is that in the event other companies in the industry refuse to submit the requested information, that their position before the Federal Trade Commission would be stronger than as if they had not subpoenaed any of the other companies."[11]

When Earl Du Monte, president of Champion Motors Company, and the first manufacturer to testify, was asked during the hearing to produce the records which had been subpoenaed, his attorney, Edward J. Callahan, Jr., jumped up and addressed the FTC examiner.

> "May I interrupt a moment? I don't know quite the proper procedure, so if I am out of order will you please correct me?
>
> "Mr. Du Monte has appeared here in response to a subpoena ... requiring him to bring certain records and documents of the company into court, and these documents contain information which is confidential company business and documents and information which is not generally available to his competitors.
>
> "Before ... the documents are admitted, I would like to be sure they are germane to the issue and that this is not just a fishing expedition on the part of the defendants [OMC].
>
> "I also wish to state that I realize there are a number of other companies which are under this subpoena, and I would like to ask the attorneys for [OMC], if this is going to be just a general expose, to put the worst word on it, of the practices of the company, or [what] ... are they trying to get at by these documents?"[12]

Following a lengthy off the record discussion, the FTC decided that the competitors would have to show summaries in tabular form, without having to show the entire records. Carl Kiekhaefer, though, remained adamant against handing over his confidential records and "trade secrets," like his dealer and distributor lists, for the whole industry to see. He investigated the possibility of legally refusing, and found he could be fined up to $5,000 and imprisoned for up to two years. Thirteen companies were ordered to surrender their records, and by May 19, 1952, only Carl Kiekhaefer had so far defied the order. His attorney wrote in desperation, pleading with him to comply with the court order. "The situation is becoming rather embarrassing to me, as the Trial Examiner has been pressing us for action. Your Company is the only one which has not yet filed the pertinent data, and I should very much appreciate your arranging to have this done at once."[13] Carl finally relented, only under the threat of imprisonment, to reveal his records. He delivered the records on the condition that the information was to be considered a "trade secret", and that none of his competitors be given access to his lists of dealers and distributors, and that the court give him adequate assurance of protection against the prying eyes of his adversaries.

In late 1952, following the revealing testimony of thirteen active competitors in the outboard industry, the FTC examiner ruled against OMC, and ordered them to remove the exclusive restriction from their dealer and distributor network. But OMC quickly appealed the decision, and the commission reversed its decision with orders to take additional evidence in the matter. More hearings were held and more testimony taken three years later in 1955, and an FTC examiner again ruled against them. OMC made a final appeal, but on June 27, 1956, the FTC "ordered the Company to cease and desist from selling Johnson and Evinrude motors only to those dealers who do not handle competing lines."[14] But by the time the order was issued, OMC outboard motor products were being sold by over 17,000 retail dealers throughout the world, with about 10,000 in the United States, 3,000 in Canada and 4,000 in other countries. It represented a numerical dealer advantage of more than 3-1 over Mercury, the second largest volume producer in the industry. OMC was, without question, the most powerful name in the marine industry.

In 1957, Carl hired a private detective in an attempt to trace back to their source the many derogatory rumors circulating about Mercury and the Kiekhaefer Corporation. He prepared a draft of a complaint to the FTC.

"The outboard motor industry, at the present time, is peculiarly susceptible to the impact of misleading and monopolistic practices. Not only is it a market already dominated to an alarming degree by a single company, but it is at present an

> enormously expanding market offering an exceptional and most dangerous opportunity to any company which might succeed in effectively eliminating competition and achieving a substantial monopoly."[15]

Once he had approved the language of the complaint, he sent his detective out on the road to gather evidence, calling on Johnson and Evinrude dealers to elicit responses about Mercury outboards and the company.

On a sweep of three states and over two dozen cities, Carl's undercover detective, J.B. Klein, received a wide array of disparaging remarks from OMC dealers, that Mercurys were overpriced, needed special tools, were over-produced, lacked parts and service, were "fast but won't last," burned out spark plugs, couldn't carry a load, were gas-guzzlers, had no resale value, had over-rated horsepower, were only built for racing and would soon discontinue pleasure motors, had inadequate cooling, would burn out within 100 hours, the Kiekhaefer Corporation was in financial difficulty, their plants are shut down, were losing dealers, Carl Kiekhaefer had resigned, among other damaging statements. This well established and consistent rumor campaign was slowing Mercury sales, even though overall Mercury production was expanding.

Frustrated by his efforts to get the FTC to act against the alleged OMC rumor mill, Carl authorized attorney Tom Fifield to lobby certain Washington politicians, including powerful Senator William Proxmire, hoping to pressure the FTC into action. This effort also failed, as even Fifield would admit that maybe Carl should back off. "It worries me ... [if] we start trying to put any pressure on that, we may end up in a position where somebody is going to say 'Is Kiekhaefer Corporation so hurt that it is going out of business?' In other words ... they tend to measure the validity of a complainer by how hard he is being hurt here and now. Well, the fact is we can't show that, what we show is just in reverse."[16]

The harmful rumors continued to enrage Carl, and he bellowed about the list of "Rayniak's Ten Commandments" that he assumed OMC chairman Joseph G. Rayniak had compiled to embarrass him. Though Carl called these the "... Ten Commandments", each time he recited the list, there were always eleven or twelve items of OMC propaganda that he assumed were carved in stone at the outboard giant's headquarters in Waukegan, Illinois.

1. All Mercurys are racing engines, not fishing engines.
2. Mercurys are fast; won't last.
3. Kiekhaefer's had another coronary.
4. Kiekhaefer's broke again.
5. Mercurys need special mechanics to tune them and keep them running.
6. Mercurys require special tools for disassembly and reassembly.
7. Mercurys are built in a barn and if you don't believe it, remove a spark plug and smell it.
8. The Coca-Cola truck driver being fired.
9. Mercurys require special engine oil to keep them running.
10. You can't use leaded fuels in Mercurys or they will burn up.
11. The "Fat Fifties" can out-pull a Mercury.
12. Kiekhaefer buried a car at Lake X.[17]

Carl would continue to deny the "Coca-Cola driver" story, as it became known, for most of his life. The story, widely circulated, and changed from version to version, actually had a basis in fact, but Carl would never admit it because it made him sound quite eccentric, and an "odd ball." The most popular version of the story is still being repeated today, and recently appeared in both *Boy's Life*, and *Reader's Digest*. Adding a few minor changes, it becomes the Kiekhaefer "Coca-Cola driver" story.

> "Carl Kiekhaefer had decided to make a surprise tour of one of his factories one day. Walking through the plant, he noticed a young man lazily sitting on a crate of soft drinks. 'Just how much are you being paid a week?' the boss angrily asked him.
>
> 'A hundred bucks,' answered the lounging guy.

> The boss pulled out his wallet and peeled off five $20 bills. 'Here's a week's pay,' he shouted. 'Now get out and don't come back!'
>
> Wordlessly, the young man stuffed the money into his pocket and took off. The plant manager, standing nearby, stared in amazement. 'Tell me,' Kiekhaefer barked, 'how long has that guy worked for us?'
>
> 'He didn't work here,' replied the manager, 'He was just delivering Coca-Cola for the soft-drink machines.'"[18]

Actually, Kiekhaefer fired more than one person who wasn't working for him, as confirmed by one of Carl's hapless victims. Ted Karls was driving a truck for Anchorage Transfer and Storage in the early 1950s, and made a scheduled delivery of packing cardboard to the Kiekhaefer Mercury facility in Fond du Lac. After he had backed his tractor-trailer rig to the loading dock, Mercury personnel told him they would unload the shipment, and he was to wait for his signed delivery receipt. "Because I didn't have to do nothing, I was just standing there by the rail, with my foot up there, smoking a cigarette," Karls said. He had been waiting about a half hour when Carl Kiekhaefer came through the loading dock, moving like a freight train. Even though Karls didn't know the busy industrialist personally, he recognized him from his periodic deliveries and said, "Good morning, Carl. How are you?"

Carl snorted, "Good -- what are you doing?"

"Nothing," Karls responded.

"When are you going to do something?" Carl demanded.

"When that trailer's empty," Karls said, motioning to the truck.

"You ain't 'gonna stand here that long -- you follow me," Carl said brusquely. Karls followed Kiekhaefer through the plant, to an office where a young lady was seated behind a desk.

"How much do you make a week," Carl demanded.

"I make $140 a week," Karls said.

"Okay, write him a check," Carl told the girl, "YOUR FIRED!" he told Karls, and then continued on his journey through the plant.

Karls took the check from Kiekhaefer's obedient employee, and returned to his truck which had been unloaded in his absence. He told the shipping dock workers that Carl had fired him and gave him a check and they told Karls, "You're the second one." Karls left the plant and stopped at a nearby grocery store to cash the check. The cashier, who knew that Karls drove for Anchorage Transfer asked, "When did you start working for Mercury?" Karls explained that he had just been fired by Carl Kiekhaefer himself, and the cashier responded that he had cashed another check only two weeks before for a driver with Railroad Express trucking in Oshkosh, as Carl had "fired" him also. The cashier ribbed Karls a little, because the other driver had received $160, $20 more than Karls. Word spread amongst drivers at both Anchorage Transfer and Railroad Express about Carl's hair trigger and no-questions-asked dismissals with pay. "The guys all laughed," Karls remembered, "and every time somebody would go out there, they'd just stand around and wait for Carl to come and fire them. But nobody else ever got fired."[19]

Carl would deny the stories that circulated over this embarrassing incident for the rest of his life. "Why am I still being disparaged and ridiculed?" he protested over twenty years later, "with these oddball stories which you know have no foundation. One could feel perhaps that it is a bit of defamation of character -- to paint of you an image that would appear to be ridiculous, if not half-witted or an odd-ball ... like firing a loafer in my plant who turned out to be a Coca Cola salesman, or a Coca Cola truck driver perhaps, as though I didn't know my own employees by their first names."[20]

Another of the nagging stories that wouldn't leave Kiekhaefer in peace was the "burying the Cadillac at Lake X story," an original Carl Kiekhaefer classic. This story, too, has many derivations from being repeated around the world by so many current and ex-Mercury employees, and yet also has a basis in fact. The most popular version of the story occurs at Lake X during the improvements that Kiekhaefer undertakes to make the base more livable, adding living accommodations and engineering garages. Carl had arrived with a brand-new Cadillac to supervise construction and testing at the lake, and while driving around the primitive dirt perimeter road encircling the lake and surrounding swamps, became mired in the soft earth, and was unable to extricate himself. His stormy temper, the popular version of the incident suggests, led him in a blinding rage back to the base, whereupon he mounted a giant Caterpillar DA bulldozer and returned to the stubborn Cadillac. In his haste to push the Cadillac from the muck, he damaged the new car with the blade of the bulldozer. With peaking anger, the story continues, he scraped out a hole along the side of the road, and pushed the focus of his frustration to the bottom, and filled it over.

Charlie Strang is certain that he was with Carl at Lake X when the nucleus of this Kiekhaefer folklore began to grow with a life of its own. It seems that one of Carl's new Chryslers was involved, not a Cadillac. Carl wasn't particularly fond of Cadillacs. John Hull, who started with Kiekhaefer Mercury in 1957, recalled how once Carl was getting out of a brand new Cadillac with less than a hundred miles on it, and he knocked his hat off in the process. "Get rid of the son-of-a-bitch," Kiekhaefer screamed as he picked up his hat from the frozen slush on the ground. "Nobody thought about a man wearing a hat when they engineered that car!"[21]

"Do you want the truth?" Strang began with a mischievous grin. Kiekhaefer and Strang were sharing the same small room at Lake X, in a complex of five rooms that became known as the "motel". "He and I shared the end room in the darn thing," Strang said, "two big beds and that was our room." Expansion of the facilities was underway during the rainy season, and "there was a lot of muck there, it's very soft earth." Strang and the other employees on the base had gone to bed, but "Carl was roaming around and he decided to see how they were doing on the excavating." He drove out to the area under improvement, got out of his car to take a closer look, and "while he's out wandering around, the Chrysler is sinking in the mud."
Carl went to get the giant Caterpillar DA bulldozer, thinking of making short work of this minor embarrassment. Soon, Strang is wakened by a breathless Kiekhaefer, whispering to him in the "motel" room.

> "'Shhhhhh ...,' Carl says. 'I told those guys not to get that Cat stuck in the mud,' he says, 'and I just stuck it in the mud. Come help me get it out.'
>
> "So the two of us, it's now about 3:00 in the morning, the two of us are sneaking around in the dark in the jungle, and there's this *gigantic* DA Cat up to the driver's seat in the mud. So, he goes and gets another car and ties a rope -- one end to the Cat and the other to the back of the car. Well, needless to say, this is like pulling a tree with a motorcycle. Nothing happens. I said, 'Carl, we need more help.'
>
> "'Yeah, yeah, I'll get them,' Carl says.

"He goes around and wakes everybody up. 'Shhhh, Shhhh.' The whole crowd is creeping around, going 'Shhhhh,' trying in the dark to get this giant thing out. That was so funny. Everyone sneaking around in the middle of the night, all going, 'Shhhhh,' and *everyone* in the place is on the job. Needless to say, we didn't get her out. The next day, I got a couple of more Cats and finally dragged everything out. The car, too."[22]

Carl never did "bury" a car at Lake X, but the story survives yet in any number of versions, and Carl himself best described the basic faults in the story. "There is no question about the fact that I could bury my Cadillac, my Mercedes and maybe some of my higher priced construction equipment if I wanted to," Carl explained, "but a person would have to be out of his mind, don't you think, to bury something you could fix and I haven't had any equipment that I couldn't fix. This is in line, of course, with some of the other things they have said about me."

There were far too many eyewitnesses to the eccentric antics of Carl Kiekhaefer for a rumor mill to feed upon. Charlie Strang remembers one day at the engineering lab on Murdock Street in Oshkosh, when Carl went to get chocolate milk out of the vending machine in the building. It happened to be out of Carl's selection, so he "ripped it all loose and he had it carried out and put on the lawn," and then phoned the vending company to have it taken away.[23] It was precisely this brand of spontaneous, anger induced and impulsive behavior that wouldn't let the "bury the car at Lake X story" die. He loved to drive bulldozers, he loved to drive big cars, he had a quick temper, and things did get stuck at Lake X. The story just fit Kiekhaefer so well, and made too much sense to ever go away. Distressed over the quantity and consistency of the rumors in the field, Carl issued a memo to all Mercury dealers to rally his forces against what he considered to be OMC's campaign to embarrass him.

"During recent months there seems to have been a concerted effort on the part of some of our competitors, including Outboard Marine and Scott-Atwater to circulate damaging rumors regarding Kiekhaefer Corporation, its products and, in some cases, its executives.

"Some of you have experienced similar assaults in the past and are well able to discount them as being the tactics of competition which cannot compete with the quality of our product. However, some of these rumors which have falsely described our situation, our products and our program can be seriously damaging to our business and we do not propose to stand by and let this business slander continue without taking appropriate legal steps to put a stop to it."[24]

Carl finally decided to pick up the phone and talk directly to OMC President Joseph G. Rayniak about the malicious rumors, in hopes of working out a truce. It was the second time that Carl and Rayniak had discussed industrial warfare.

"Now we've got a fist-full of evidence," Kiekhaefer warned Rayniak, "and I am just calling you once more as a friend Joe, to see if you can't stop some of this unfair advertising or we will reply in kind -- we will canvass your dealer organizations with facts."[25]

"I think it is about time we laid the cards on the table from both sides," Rayniak replied. "I have heard a lot of adverse publicity the other way. I haven't said anything because I thought, well, all these things will work out themselves automatically. ... But evidently, there's something gone haywire and I think what we'll have to do now, is to open this thing up and put on the table ... so we can discuss these things in a friendly way and because both sides are guilty. Your men are guilty. Our men are guilty."[26]

"Well, we've got a lot of automotive people with us," Carl retorted, "and they said we have seen a lot of things go on in the industry, but we have never seen it on a personality basis where you take each other's executives apart."

"I don't think we should crucify either one or the other," Rayniak suggested, "unless we bring all the evidence together and put it on a table and say, 'Now, here's what we've been doing to one another. Now then, we've got enough witnesses.' We've got enough paraphernalia here, evidence that the both sides are guiltier than hell. And I think it is about time we got together and straightened this thing out."[27]

"It is crap," Kiekhaefer responded, "and we ought to get rid of it."

"That's right," Rayniak agreed. "It is crap, yes, and it doesn't belong, but I've seen evidence and I have seen papers, literature, where your organization, if I may say and I don't want to say it unless I have to, that they are just as guilty as our people. ..."[28]

In fact, Carl was just as actively knocking OMC in his advertising and especially dealer bulletins. But what made Carl hopping mad was that OMC had such a stronger voice, such longer tentacles, and such greater power in the marketplace, that his own protests were largely overshadowed.

Rayniak and Kiekhaefer never did meet to compare anti-competitor literature, because neither wished to face the enormous bulk of damning evidence the other would bring to the table. Instead, a week later, OMC executive vice president and general manager, William C. Scott, wrote to Carl with a long list of denials to Carl's accusations. In his blunt and irritated reply, Scott said, "I can find no excuse, within the framework of a supposedly friendly competitive relationship, for your intemperate and unpleasant telephone call to our President. ... Also, I find it difficult to discover any common ground of consideration of ethical practices in the light of derogatory, and in some cases, misleading, statements recently made by your company concerning our products. In short, I can see no useful purpose to be served by a further personal meeting or by prolonged correspondence."[29]

Carl was furious. He blasted out at Scott in one of his letters he never mailed. "I was sincere," Carl protested. "We may have been competitive but we have never attacked personally members of your organization. ... Believe me, Bill Scott, I would not want to take the responsibility of closing the door on any further discussion of matters that certainly have been unpleasant and unethical. It is conceivable that cooperation might at some time in the future be quite expedient and useful. ... It seems to me that I would recommend to you to proceed with the greatest care and caution. If anyone is vulnerable, I believe it is you and I do not believe you realize just how vulnerable you are and the position that you are placing your company in."[30]

Carl was pleased to break at least one of "Rayniak's Ten Commandments," when a public showdown was staged between a 66 cubic inch Mercury Mark 78 and a 70.7-cubic inch Evinrude 50 in a test of power. The match resulted from a challenge made at the Chicago Boat Show in 1958, when Andy Fisher, an Evinrude dealer, told Tom Ward, a Mercury dealer, that the Evinrude "fat fifty" could out pull the "fast but no power" Mercury. Two boats were lashed together, bow to bow on Cincinnati's Licking River, and on a mutual signal, opened their throttles. Hundreds of spectators, alerted to the contest by local newspapers, watched as the Mercury slowly started pushing the

Evinrude upstream. The Evinrude started moving backwards "reaching a speed of five miles per hour until the water coming in over the transom of the retreating boat caused Fisher to wave the test to a halt."[31]

Among the policies that Carl Kiekhaefer stubbornly enforced was his refusal to ever release statistics or financial information to any governmental department or private boating association or publication. In guide books dating from 1939 onward, even listings of the models and horsepower ratings of Kiekhaefer Mercury engines are conspicuously absent. Carl, in a letter drafted by his sales staff, once spelled out his reasons for refusing to cooperate with, in this case, a Department of Commerce report on outboard motor sales in America.

> "The Kiekhaefer Corporation can be seriously hurt competitively with this knowledge in the hands of the Outboard Marine Corporation. As you know, this company with McCulloch Corporation, (owned by the son-in-law of the Outboard & Marine Chairman of the Board), claim to sell 80% of all outboards. The Kiekhaefer Corporation has made consistent progress in combating the domination of this group, but because of their size, is to a certain extent at the mercy of their merchandising techniques.
>
> "In spite of this situation, Kiekhaefer Corporation has developed an increasing consumer acceptance for its product and a continuously stronger industry position. There is always the possibility, however, that factual information relative to the strength of Mercury may stimulate our competitors to take drastic action in the form of price reductions or merchandising techniques to stunt the growth or eliminate Kiekhaefer Corporation from the outboard picture."[32]

Carl went to great lengths to see that his "no data" policy was enforced, and was continually paranoid that OMC would somehow steal information from himself personally or one of his employees. John Hull had just started working for Kiekhaefer Mercury, and was attending his first sales meeting when Carl blew up over confidential information. "We were making notes about our new products," Hull explained, "and Carl was always suspicious that Evinrude and Johnson were spying on him and they would send people inside to try and get information. I left my notebook out on a shelf in the hallway where I hung up my coat at a meeting down at the Retlaw Hotel in Fond du Lac. He said 'Those notes are privileged information and competitors would like to get it.'"[33]

Another time, Carl couldn't find his briefcase on a business trip to Chicago. He screamed, "Somebody stole my brief case out of my room," John Hull remembered. "We're looking all over for it, and we call the house detective and everything. And he says, 'There are important documents in there, the competitors -- THE ENEMY's got it!' We found out he left it in Florida. He called down to Florida, there was his brief case sitting on his desk, empty."[34] On another occasion, Carl blew up at Hull for putting out Mercury's new brochure an hour before a national boat show was going to open to the public. "He came in and just raised all kinds of hell because we had the literature out. Well, my God, the world was going to see it in another hour. I don't know what our competitors could do to change the design of their product in the hour before the show opened."[35]

Carl was often visibly upset when he would read competitive advertisements in newspapers and trade journals. "What has happened to the free press and the freedom of speech in journalism generally?" Kiekhaefer once wrote.

> "I wonder if it is affecting others like it does me. I almost hate to pick up a magazine. Firstly, because of the high proportion of advertising to the 'meat,' and secondly, because of the false advertising, which I have learned to detect in the past 22 years in the outboard business. These ads are written completely oblivious to facts and take liberty with the laws of science.
>
> "I sometimes wonder if we should not call this the 'Age of the Big Lie.' The after effects of these ads are frightening. It makes one wonder if this isn't the time to dispose of one's business and move to a nice quiet country like Mexico or Cuba."[36]

Worse still, Carl was out-gunned on the covers of boating publications by the ubiquitous photos of OMC products. He recoiled at what he considered the collusion between OMC and Bob McCulloch's Scott-Atwater company. Bob McCulloch married the daughter of OMC chairman and significant stockholder S.F. Briggs. McCulloch, a perpetual thorn in Carl's side, purchased Scott-Atwater outboards in 1955. Carl complained vociferously to the FTC of the potential for illegal and monopolistic practice between the two organizations. "It is recognized that the mere fact of this relationship between chief executive officers is not by any means conclusive in its effect. ... At the same time, the potentialities of such a relationship are serious and should be considered ... the possibility of their combining, for their mutual benefit, in measures to undermine the competitive position of [Mercury] is disturbing."[37] He was often frustrated, as in the excerpt to the following letter to his attorneys, to the point of giving up and selling out.

> "... We see nothing but a further encirclement by the great Mr. Briggs [S.F. Briggs, chairman of OMC] with the help of McCulloch and friends at Briggs and Stratton, having a monopoly not only on the small industrial engines but on the outboard motor as well as the chain saw business in the United States and Canada.
>
> "Coupled with the fact that the advertising, publicity and public relations programs by OMC are so stupendous, we hardly know where to turn. Pick up any trade magazine of the marine industry, pick up any sporting or fishing magazine, pick up any newspaper and you will see nothing but Johnson, Evinrude, Buccaneer or some splinter of the OMC Company represented in ads. Either Evinrude or Johnson has the front or back covers of all the sporting or marine publications. Paging through the magazines in coverage by articles, photographs or on boats, their ratio of five to one on photographs alone is evident, with Mercury representing one.
>
> "We fear strangulation with further help from OBC (Outboard Boating Club of America), which is nothing but a public relations body for OMC, although it is disguised as a happy family boating club, a happy family water ski club and a happy family trailer club. Also somewhere in this confusion is a so called happy family of outboard manufacturers. How ridiculous can this thing become with the manufacturers association in particular where Johnson, Evinrude, Montgomery Ward, Goodyear, Gamble Stores, Buccaneer, Atlas Royal (which is Standard Oil) each carry a vote, though each represents the same company.
>
> "Is the Federal Trade Commission really aware of the type of encirclement we are facing; not just in outboards, chain saws and lawn mowers, but on small industrial engines in general?
>
> "I believe some real study should be given as to how we can cope with the situation or what course to follow. Perhaps the best course is to sell out while we are ahead."[38]

In retaliation, Kiekhaefer developed a counterintelligence campaign to keep tabs on OMC. Just around the corner from Sea Horse Drive in Waukegan, Illinois, a few short blocks from OMC's secret engineering labs, is a restaurant known as Mathon's. It was, and still is, a favorite hang-out of OMC executives and engineers. Charlie Strang, who today is chairman of the board of OMC, remembers that Carl would send spies down to Mathon's to sit at the bar and keep their eyes and ears open for whatever information they could pick up. The espionage program worked better than anybody ever dreamed, as once Strang and Kiekhaefer were able to get a hold of an experimental casting before OMC's research group even had it. Knowing this from his own first hand experience, Strang has since warned OMC engineering to assume Mercury is still listening to conversations at Mathon's.

Perhaps the best example of the lengths to which Carl would demonstrate his hatred of OMC and their products, was his annual sales meeting, and the sacrificial bonfires that would sometimes follow. Carl would construct an enormous, and often dangerous bonfire in the forest behind his home near Fond du Lac, and hang a Johnson, Evinrude or Scott-Atwater engine by its "ankles" and lower it into the flames. Carl would sometimes hook up a few wagons behind a tractor to "bus" his dealers to the bonfire, and Carl would proudly drive the rig to impress his audience. "He always had to let everybody know he could run anything," Wilson Snyder remembered. Snyder, Carl's all around helper and farm hand, was in charge of preparations for these pagan industrial rituals, and described the magnitude of the inferno that Carl demanded to get his message across. First, Snyder would stack twenty or thirty bales of hay in a clearing of the woods, and then pile nearly a cord of green oak over the dry hay. Carl always wanted to soak the entire pile in gasoline, but Snyder talked him out of it. "Christ, you'd blow that shit all over everything and kill somebody," Snyder told Kiekhaefer. Instead, Snyder would soak the whole works in less explosive fuel oil. A giant tripod of steel girders was put in place over the wood, and a stainless steel cable was guided through a pulley at the top.

> "And then we'd light that fire and you could see it sitting west of town here, they'd see that big blaze. And then he had that tepee was built out of metal, you know, and that would be just *red hot*, and then with the cable they'd pull that fucker up,

a bunch of guys. ... I was always afraid somebody was going to get hurt."[39]

Carl would always ignite this funeral pyre for the enemy engine personally, and tell the attendees, "Look what theirs is made out of! You can't burn ours!"[40] With plenty of liquid refreshments to get his dealers and sales troops in the mood, the melting of the "enemy's" engine was greeted with nearly neolithic delight by his entranced audience. The heat from the huge conflagration would soon start to melt the new outboard, and globs of molten aluminum would spatter and drop off into the roaring fire. It had a hypnotic effect on those in attendance, and many would admit that they were somehow overcome with the passion of the ceremony. Carl loved the ritual burning, and continued to electrify his sales force with this menacing demonstration of his passionate hatred of OMC for several years. But unlike the dying embers of his sacrificial fires, Carl's fierce competitive spirit would continue to blaze unabated.

CHAPTER TWENTY-THREE

Merger Mania

"As a business man who grew up on a farm, who has created his own industry of 2,500 employees and has created some 75 million dollars in tax money to say nothing of payrolls, all from creative ambition, I am concerned with the future. ... I know from my own experience, the combined tax loads of the State of Wisconsin and the Federal Government make it impossible for us to carry on the necessary research and expansion to continue long in business. It is one thing to meet normal competition, but it is quite another to be taxed to death. ..."[1]

Carl Kiekhaefer
To U.S. Senator Alexander Wiley

Even before Carl's financial splurges into auto racing, two of the banks that Carl had been doing business with requested in 1952 that an outside controller be brought inside the corporations to help monitor Kiekhaefer financial operations. "Well, it was more than a mess," said Donald E. Castle, the controller chosen for the assignment and who would remain for the next five, turbulent years. "I came into a company that was in really bad financial straits -- just *real* bad."[2] Carl was working with both the Marshall & Ilsley Bank of Milwaukee and the Union Commerce Bank of Cleveland when cash flow problems began to hamper operations. "Carl had *absolutely* no respect for money," Castle complained.

"You just couldn't seem to get it across to him. He had this real penchant for [auto] racing, and he'd spend the money right and left and we didn't have the money to spend. I couldn't meet payrolls, and you know, he was spending a million dollars for racing -- Christ, for automobiles! So, it wasn't all the most enjoyable period of my working life."[3]

Castle had been vice president and controller for Northwestern Steel and Wire company, pioneers in the field of electric furnaces in the steel industry. He was accustomed to large expenditures for the hardware of manufacturing, and was well versed in the relationship between raw materials and finished goods in heavy industry. But he wasn't prepared for the circus of emotional and political decisions that stood in the way of financial planning in Carl's burgeoning empire. Carl tried unsuccessfully to intimidate Castle, who had the unqualified support of the banks, even telling his company pilots to avoid airports where Castle would be waiting for a meeting. "I'd [phone him] and say, 'Carl, I've got to talk to you.' He'd say, 'Well, I'll be flying up to Oshkosh, I'll stop the airplane and you can talk to me on our way up.'" Castle would dutifully wait at the Fond du Lac Airport, hoping for his chance to speak with Carl, even if only for the ten minute flight from Fond du Lac to Oshkosh. But then he would watch the plane fly right over the airport, droning on to Oshkosh without stopping. "It was a hide-and-go-seek game with Carl."

Fortunately, Carl had a wonderful relationship with Al Puelicher, president of the Marshal and Ilsley Bank. "If it hadn't been for Al Puelicher and the Marshall and Ilsley Bank, there would be no Mercury company today. ...," Castle maintains. "I'd call Puelicher many times, and say, 'Al, I've got a payroll of 50 some thousand dollars coming up Friday, and I don't have any money as you well know.' Because they had our bank account. He'd say, 'Well, I'll put some money in for you.' So that's how close."[4] Slowly, Mercury and the Kiekhaefer empire had begun to unravel.

To give the bank more confidence, Castle developed a clever strategy for Kiekhaefer to pledge receivables, in combination with production engines for collateral. "I rented warehouse space in Milwaukee, and we shipped engines in there and put them under the warehouse receipts, which gave the bank plenty of collateral. Receivables were pledged, the finished inventory was pledged, and that came out of my experience in the steel industry where we warehoused a lot of things."[5] Though this banking practice of receivables financing and inventory as security would in later years become common banking practice in industry, in the early 1950s it was a rather bold and maverick concept.

Over the years, Carl fielded dozens of casual inquiries from ırger organizations that were searching for solid companies to :quire in programs of diversification. As a matter of course, ırl politely responded to these overtures by telling all comers ıt he was quite happy to be in charge of his own destiny, and

Carl is unwittingly dubbed *Chief Swift on the Water* by a disguised Johnson dealer in Canada (above). He would have fired everybody involved if he had discovered the ruse.

Left: Carl could bring his dealers to fever pitch during the ceremonial burning of *the enemy*'s engine. Here, Carl attacks an Evinrude with a tomahawk for his mesmerized tribe.

Left: Carl has been compared to a concert pianist when engrossed in tuning an engine. He also delighted in performing destruction tests to evaluate potential flaws.

Below: Engineering meetings with Carl were anything but dull. Here, Carl holds forth on some new design, while Charlie Strang, to his left, awaits the decision.

Above left: June 7, 1958, Hubert Entrop sets a new world's speed record for outboard motors at Lake Washington at 107.821. Carl, irritated at Entrop, wanted to announce the record without naming the driver.

Above right and below: Carl signals the to send Burt Ross on his way to a new world's speed record of 115.547 mph at Lake Washington near Seattle, May 4, 1960.

Left: Jack Hanigan, Brunswick president, was the target of Carl's most intense anger and profuse criticism. Hanigan, Carl was certain, was taking him for a ride in more ways than one.

Below: Following the sale of the Kiekhaefer Corporation to Brunswick on September 30, 1961, Brunswick Chairman Benjamin "Ted" Bensinger nervously leans on one of Carl's engines for a publicity photograph.

During the ground-breaking ceremony of the new *mile-long plant* in 1964, Carl, standing next to Jack Hanigan, first turned the soil with a silver spade (above).

He then surprised the large audience by mounting a giant Euclid earthmover (right) to cut the first swath across the site of the world's largest outboard plant (below).

Left: The 1930 "Johnson Tilting Stern Drive," that was expected to "obsolete the past," here shown on a Ludington runabout. Though this effort solved some of the problems associated with inboards, it created others.

Below: A portion of Charlie Strang's M.I.T. inventions ledger for 1948, clearly shows the very first depiction of the modern stern drive. Of particular interest are the use of the hookes coupling with tilt pin and swivel pin (universal joint) which was the basi for the later patent awarded to Jim Wynne. Also present is the tilt-trim adjustment, which i also a critical feature of contemporary stern drive design by all manufactur ers.

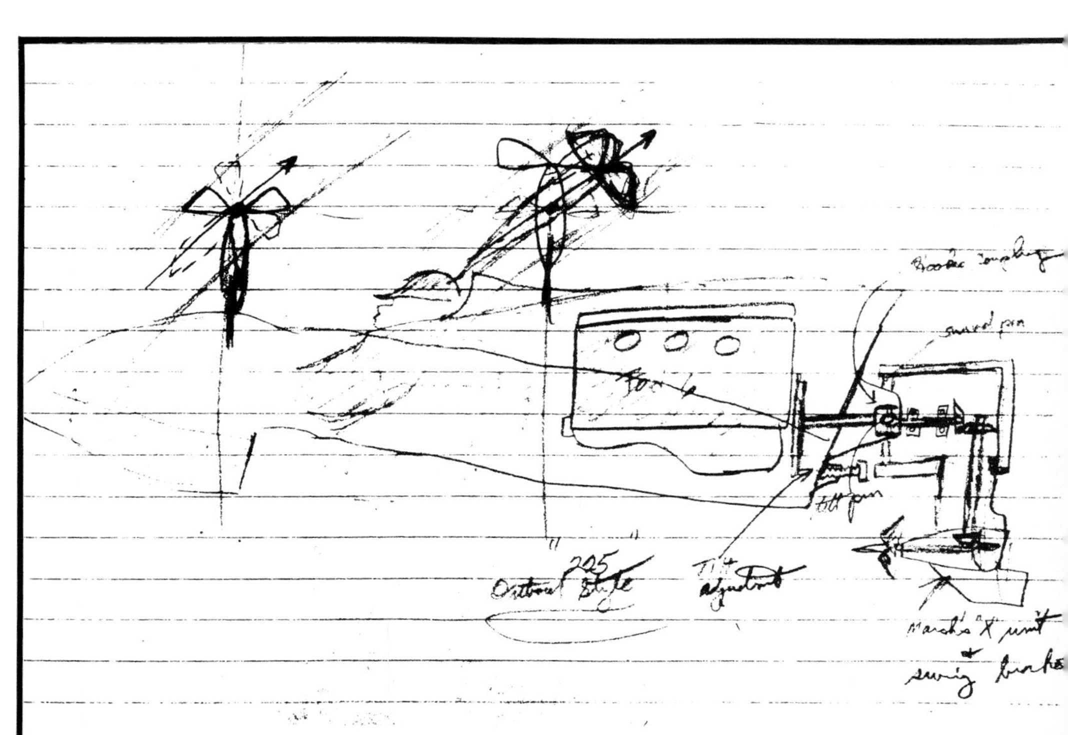

Jim Wynne and Carl are all smiles, before their stormy and emotional split during the *Christmas Mutiny* of 1957.

Below: Jim leans on the first prototype of the Volvo Penta *Aquamatic* stern drive at the New York Boat Show in 1959. The new drive was destined to change the course of the marine industry. Wynne raced the prototype without permission from Sweden, setting four new world's records in the nine-hour Orange Bowl Regatta marathon, aboard this 16-foot Ray Hunt-designed deep-V (below right) in June, 1959.

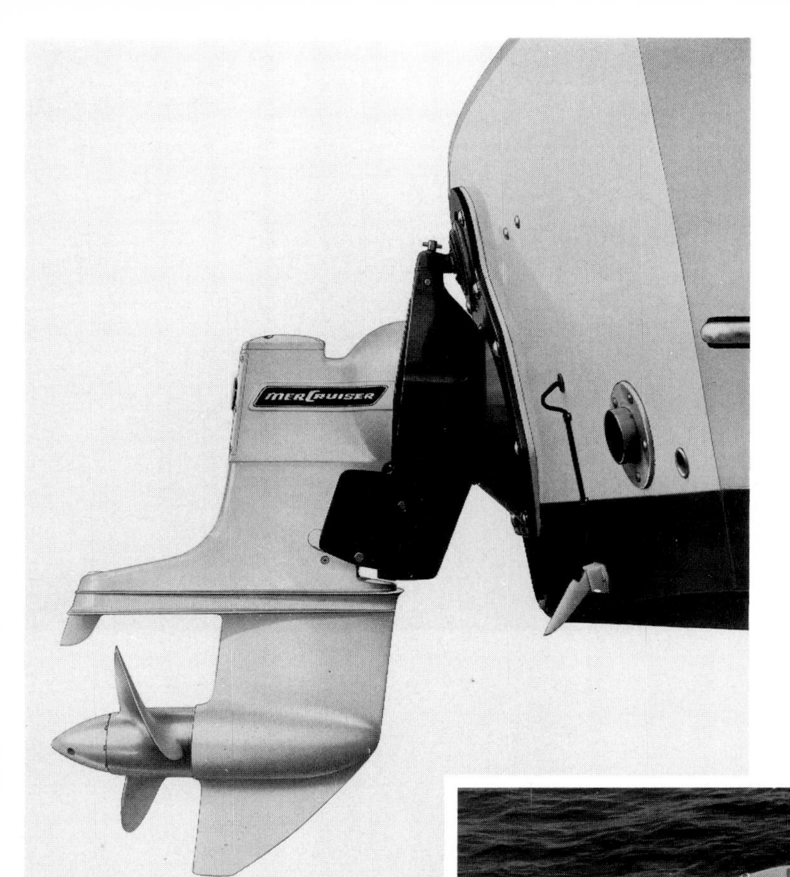

The *MerCruiser* stern drive made its first appearance in 1961, two years after Volvo introduced Jim Wynne's Aquamatic. Charlie Strang felt free to use his original invention concepts, without fear of reprisal by Volvo.

The clean lines, superior horsepower and performance of the Mercury version, soon captured the majority of the stern drive market, easily surpassing both the Volvo and OMC's initially under-powered effort. Stern drive sales by Mercury, OMC and Volvo are responsible today for a critical portion of the worldwide $18 billion marine industry.

wouldn't consider forfeiting his freedom and leadership just to be a part of a larger company. But as severe cash shortages began to disable his operations in the mid-50s, Carl began to flirt with the idea of merging Mercury with a larger, financially stronger partner.

Besides the obvious increased opportunities that a sizable investment in manufacturing facilities and product development would bring to Mercury, Carl had other reasons to consider a merger. In his ramrod, bulldog fashion, Carl had fashioned Kiekhaefer Mercury into the second largest outboard manufacturer in the world, and his management style had evolved into a unique form of benevolent dictatorship. As industrial Caesar, Carl had neglected to develop an organization that could survive without him. He had acquired a cadre of bright and talented engineers and manufacturing management, especially Charlie Strang, but Kiekhaefer was still the absolute nucleus around which all activity at Mercury would revolve. Carl began to fear that if something were to happen to him, Mercury might not survive.

Carl's anxiety over the succession of leadership at Mercury was magnified extensively one afternoon in the fall of 1956 when he became convinced that he was going to die. Mercury was introducing the newly announced six-cylinder, 60-hp Mark 75 outboard motor for the 1957 model year at the annual Mercury dealers and distributors meeting. The announcement banquet was being held at a large restaurant between Fond du Lac and Oshkosh, and Carl was to be the featured speaker during the evening ceremonies. Carl and Charlie Strang were together in the engineering department following the meal, when Kiekhaefer came to Strang and said, "Oh my God! I think I'm having a coronary!" Pete Humleker, Carl's executive in charge of industrial and personnel relations was in the lab at the time, and Strang told him, "Get Carl in the car - get him to his house and I'll have a doctor waiting with everything he needs by the time you get to Fond du Lac."[6] True to his word, by the time Humleker and Kiekhaefer arrived at Carl's house, a physician was waiting, complete with EKG machines, oxygen and a cardiac emergency kit.

Strang and the public relations staff, meanwhile, were caught in the middle of last minute preparations for the very important new line announcement to the distributors and dealers, and were "off chewing our nails wondering what's happening to Carl." Tension mounted as the starting time for the meeting approached. Kiekhaefer was the star of such gatherings, and "everyone was saying, 'Where the hell is Carl?'" Finally, someone came to Strang with a message that Kiekhaefer was on the phone.

> "I go to the phone. 'Carl, how are you doing? A very, very weak voice says, 'Charlie, I don't know what's going to happen,' he said, 'but you know you always expected to be president of this company -- looks like you're going to get your chance. But what I'm going through is what the stress does to you,' he says. 'Don't let the guys know I'm sick, and carry on, boy!'
>
> "I go out and everyone is still wondering where Carl is, and I do the whole damn presentation and tears are running down my cheeks while I'm speaking and everything else. Who's in my office the next morning, but Carl! He says, 'It was the cucumbers I ate at lunch. It was gas.' He was a heartbreaking man."[7]

Even though Carl's "coronary" was a false alarm, it triggered both his own sense of mortality and widespread rumors about his health. From New York to Los Angeles, word circulated that Kiekhaefer had a "coronary", and with the aid of Johnson, Evinrude and Scott-Atwater dealers earnestly repeating the rumors, the public began to wonder about the fate of Mercury. The harmful rumors enraged Carl, and he bellowed loudly and often that OMC Chairman Joseph G. Rayniak was behind this latest campaign to embarrass him and weaken the image of Mercury.

Another, equally strong inducement to considering a Mercury merger, was the exposure of Carl's personal estate to federal and state taxes. As the owner of 90 percent of the stock of both the Kiekhaefer Corporation and Kiekhaefer Aeromarine Motors, any outright sale of his holdings would subject his estate to severe tax liabilities, with some loss estimates approaching 80 percent of his net worth. Tax laws have since changed, but the penalties in the mid-50s for owning such a large percentage of a growing business could be devastating not only for an outright sale, but also in case of death, the surviving family estate would also be taxed heavily.

Carl was against taking Mercury public, due to the constant agitation he had experienced with stockholders, both initial investors and family alike, before acquiring all but 10 percent of his company's stock. He began to seriously consider a merger that would allow him to align his companies with a much larger and more powerful organization, and reduce his tax exposure by taking a combination of cash and stock from his merger partner. If a smaller portion of a transaction involved cash, his tax liability would be appreciably smaller, while taxes on stock received would be deferred until sold. Lastly, Carl knew that for Kiekhaefer Mercury to grow fast enough to compete equally with OMC, he needed to invest significantly in new plant facilities, new die casting equipment, and undertake greater research into new technologies and product development.

By 1954, Carl's empire was beginning to lose money, just when he needed it most to expand his operations. When the contract with Disston for production of chain saw engines was terminated in early 1953, over $460,000 in annual profits disappeared overnight. Even though sales of Wizard motors to Western Auto Supply increased from 1953 to 1954 by nearly $1 million, only a nominal profit was produced because of production problems encountered in converting Mercury models to Wizard models. On top of these losses, sales of Mercury outboards fell by well over 11,000 units in 1954 alone, and when calculations were completed for the bottom line, Carl had lost nearly $200,000 on sales of almost $16 million for the year. He even cut $10,000 from his $60,000 salary. Though he guarded knowledge of this financial disaster with nearly maniacal secrecy, rumors of "Mercury is going broke" began to circulate throughout the industry, and the flames of this latest rumor were fanned by obliging OMC and Scott-Atwater dealers from coast to coast.

Carl's first serious attempt to merge occurred in 1955 when he came within seconds of closing a hard-fought agreement with Daimler-Benz of North America. Carl was enamored of the engineering excellence so evident in Mercedes automobiles, and was fond of comparing his own engineering style to the German manufacturer. One day, according to Charlie Strang, a very impressive emissary from Daimler-Benz in Germany came to speak with Kiekhaefer. He was a Russian born German who had once been a General in the Russian Army. "They came in and expressed an interest in buying Kiekhaefer," Strang remembered, "because they wanted to do some manufacturing in the U.S., and they thought this would be an ideal company to tie up with." The two organizations had so many things in common, that a rapport was established almost immediately. Both companies were very high-performance minded, both were operated by chief executives of German ancestry, and both were proud of their respective records for engineering and manufacturing excellence.

Daimler-Benz was interested in having Carl build Mercedes diesel engines, and to manufacture a newly conceived Mercedes automobile designed to compete with the four wheel drive Jeep, called a *Unimog*. Discussions centered around a two-stage involvement between the two companies, the first step to be a joint venture in manufacturing, with an option for a full merger within 24 months. Financial statements were exchanged by the end of March 1955, and Daimler-Benz prepared a formal proposition for Kiekhaefer's consideration. Carl countered with a letter to the board of Daimler-Benz in Germany that he would sell up to two-thirds of the Kiekhaefer Mercury companies for $4 million in cash to be paid out over three and a half years. Any further payments Carl might expect from the anticipated second-stage merger he would accept in Daimler-Benz stock.[8]

As negotiations continued throughout 1955, the parties came closer and closer to complete agreement, and it was assumed on both sides that a deal was imminent. German engineers and administrators flew to Wisconsin to tour Carl's plants, and Carl took the opportunity to make his second trip out of the United States, and his first to Europe to visit Daimler-Benz in Germany. "He was like a little boy," Rose Smiljanic remembered. "They got along well and they really treated him royally ... it was like a little boy going off on a vacation."[9] By the end of 1955, virtually every wrinkle in their prospective agreement had been ironed out, except, as Charlie Strang remembered, one large one. "They wanted to build their version of the Jeep over here. ... Carl wanted to build a sports car. But they never did get anywhere with their little Jeep, and of course their sports cars went wild. So he had the right plan, they had the wrong plan."[10]

Finally, nearly a year later in January 1956, verbal agreement of all major issues was made, and a closing meeting was scheduled to take place at the Carlyle Hotel on East 76th Street in New York. On board an American Airlines Flight from Chicago, Carl was still skeptical about proceeding with the joint venture and merger. He took an airline post card, and made two lists on the back, one marked "Advantages" and the other marked "No". When he finished, the advantages, at least in number, seemed to outweigh the disadvantages by a comfortable margin.

Advantages	NO
Security	End of Kiekhaefer Corporation
Prestige	End of independence
Cash	End of personal liberation
Expansion Opportunity	End of freedom of expression
Travel Experience	
Opportunity of a lifetime	
More experimental	
Consolidation of plants[11]	

It is interesting to note that every reason on the "No" side of his list had to do with his independence and freedom. One of the most enticing elements of a proposed Daimler-Benz/Kiekhaefer Corporation was that Carl would be president, earning $100,000 a year with absolute freedom to execute his responsibilities as he saw fit. If he wanted to give up all administrative duties and assume a role as head of research and development, he was granted this freedom under terms of the agreement, retaining his full salary and benefits in the bargain.

Perhaps nagging Carl on the flight to New York was the lack of enthusiasm that his law firm, Fairchild, Foley & Sammond, had expressed about the venture. When Carl asked senior partner Frederic Sammond for his opinion, Sammond was carefully noncommittal.

> "While our discussions ... consider major question as to whether you should retain a minority or majority interest and whether a cash purchase is desirable and so forth, we do not reach the question of whether or not you ought to make a deal at all. This is a question that no one but you can answer because the answer is one that you will have to live with one way or the other.
>
> "Undoubtedly Kiekhaefer Corporation could raise additional working capital if necessary for the growth of the outboard business without any combination with Daimler-Benz. Entering the larger field relative to its products and associating

yourself with them are the steps on which none of us can do more than submit our analyses and suggestions, leaving the ultimate decision to you."

By the time Carl entered the closing at the Carlyle, he was in almost total limbo. A large contingent of people from both sides of the Atlantic had assembled at the hotel to close the deal. Carl had brought along a large group of financial, legal and engineering consultants, all of whom assumed a deal was already made, the only thing left was to sign the papers. Donald Castle recalled Carl's blow-up moments before he was to sign the final documents.

"We spent over a week in New York at the Carlyle Hotel. We had attorneys from all over ... as a matter of fact we had attorneys from England, Mercedes being world-wide. That was a sad day.

"We had a battery of attorneys from [Fairchild, Foley & Sammond]. Fred Sammond the senior partner and I were down there together, and we had everyone from Mercury. It was Charlie Strang, [Armand] Hauser, [Willis] Blank, everyone, and Mercedes had their top people there. And Carl's heart and soul was not in the merger. He just didn't like the idea. He didn't understand it, I think.

"It wasn't a buy-out, it was more of a merger situation. And we were all going around the table signing the documents. Carl was never really sold on the deal. Always a question in his mind whether this was a good deal, and [the president of Daimler-Benz North America] made an off-hand remark. 'Our Board of Directors, Carl, have said, 'We're getting into the outboard business,' meaning that these documents we were signing were going to put them in there. And Carl straightened up and said, 'You don't frighten me that you're going to get in the outboard business, the HELL with this! Come on!' And with that, the meeting was over and that was it. And in spite of everything they could do to get a hold of Carl, that was it. That wound up the deal. We all gathered our pencils and paper and flew back to Fond du Lac."[12]

Fourteen months later, in early 1957, Carl Giese of Daimler-Benz of North America sent Kiekhaefer a telegram announcing that agreements linking themselves with Studebaker-Packard and Curtiss-Wright had forged an alliance involving "107,000 people, with sales in 1956 totaling $1,266,000,000." The telegram disclosed agreements "providing for a fully integrated program of engineering, production, sales and service of automotive vehicles; automotive, marine and industrial gasoline and diesel engines; and diesel and gasoline fuel injection systems."[13] Though Carl may have entertained regrets over his actions, he was busy investigating fresh merger possibilities.

For the second time, Carl considered a merger with Food Machinery and Chemical Corporation (FMC). Years earlier, as Mercury was still growing by leaps and bounds, Carl had entertained the possibility of an alliance with the giant industrial firm. But in May 1957, Benjamin Carter, the executive vice president of the machinery divisions, wrote Willis Blank to tell him that Mercury had grown too big to be of interest to FMC any more. "... We have given a great deal of thought to the advisability of resuming negotiations with you, and have reluctantly come to the conclusion that your present size is such that a merger of the two companies would materially change the character of our business, which we do not wish to do at this time. In retrospect, it is too bad, from our point of view, that we did not get together during prior negotiations."[14]

Many people within Carl's inner circle of executives were convinced that Carl was being unduly influenced by Willis Blank in his merger considerations. Though Blank controlled only 10 percent of the stock, he often exerted pressure on Carl wholly disproportionate to his holdings. "Willis was literally in a turmoil," Charlie Strang recalled, "because he never knew what his stock was worth. There was no public value of the stock, and he really wanted out in the worst way. He never cared much for the business, and what he really wanted was for Carl to go public with the company."[15] Carl, though, was adamant about his disdain for dealing with stockholders. Eventually, in the spring of 1960, Blank was able to convince Carl that going public with the company stock was the best solution for all concerned. But Carl said, "Well, I'll be *damned* if I'm going to screw around with stockholders -- the *HELL* with that!" Saying this, he turned and looked at Willis Blank, as if to say, "You do it." Willis quickly said, "Well, *I'm* not going to screw around with that!" Strang remembers that Carl then turned to look at him, saying, "O.K., Charlie's going to screw around with them." Carl had decided that if he had to go public, he would make Strang president of the new public company, and let *him* deal with stockholders. Carl could then remain a somewhat aloof and isolated chairman of the board. They contracted with Merrill Lynch to take the company public and make the initial offering. "I had to go buy a charcoal grey suit," Strang recalled, "I had never had a suit before. I was taking lessons from Merrill Lynch on what had to be done and so on and so forth. Right in the middle of this whole process the market went [sour], so Merrill Lynch advised not to go public at that time, and the whole thing came to a halt."[16]

During the years 1957 through 1961, Carl and his chief executives were almost perpetually distracted by wave after wave of prospective merger partners. Willis Blank reported to Carl that over two dozen sets of financial reports had been released to various corporations and brokers during the period.[17] Among the organizations that Carl was courting was giant Borg-Warner Corporation of Chicago. In October 1957, Chairman R.C. Ingersoll politely warned Carl about potential estate problems.

> "I do feel that you have a serious estate problem, because if anything happened to you there would be a large state and federal estate tax and inasmuch as you have been the guiding light and genius of this company, the market value of the company would shrink very materially, probably as much as 50%. ... I do feel that your company should become associated with some larger company with a ready market for its securities and should be headed and operated by the type of people with whom you would like to be associated."[18]

Carl prepared an illuminating reply that he never mailed. In the letter he outlined his reasons for merger, and his feelings about the relationship between potential parent and subsidiary.

> "Like the automobile industry, the outboard industry has had its attrition which boiled the field down over the past 30 years to two basic companies, namely the Outboard Marine-McCulloch combination and the Kiekhaefer Corporation. Some 20 companies have come and gone in and out of the outboard industry ... only Mercury has been able to stand off the Outboard Marine group. ... [W]e are not particularly worried. I believe history speaks for itself.
>
> "It is true perhaps that if money were the only consideration, it would probably be wise to be looking for an umbrella and to liquidate and transfer as many personal assets to my children and to my family while I am alive. However, I feel I still have 20 active and useful years to contribute to my

business and to be the driven rather than the driving gear somehow or other does not appeal to us.

"We have been asking ourselves, 'What did Steve Briggs have that we don't have?' 'What did Food Machinery, Textron, and yes, Borg-Warner have that we don't have?' Perhaps we have just finished the first rather than the last chapter of the Kiekhaefer operation.

"Mr. Ingersoll, let me sum it up this way. ... We are cognizant of the times and the danger signs on the horizon. We also realize what it means to be the 'bride' rather than the 'groom' in any urge to merge and having reached our 18th year, we believe we have just reached the age of consent. We are not ready to accept your proposition at the offered price."[19]

Carl was careful to conceal his need for a merger. Within two weeks of the collapse of negotiations with Borg-Warner, Carl told a cash rich and anxious to merge National Automotive Fibres of Detroit (NAFI) that, "Well, we haven't thought about anything like this. We are in a very good position industry-wise and financially, and I see no reason for entertaining any idea of selling or even merging."[20] NAFI had grown wealthy by manufacturing automotive interiors for both Chrysler and Ford, and was controlled by Paul Shields, the one-time chairman of Curtiss-Wright Corporation. Again, Carl was not satisfied with the match. NAFI had annual sales of less than $50 million, and Carl feared that his companies would quickly outgrow any usefulness a merger would provide. Determined to gain a stronghold in the marine industry, NAFI would soon purchase Chris-Craft Corporation from the founding family of Christopher Columbus Smith for $40 million in cash and stock.

In early December 1957, Motorola Corporation took their turn at bat, saying "The management of Motorola has been having continued discussions about the possibility of taking on a line of outboard motors ... and using your company as the nucleus for this division. We understand that this would solve [your] estate and inheritance tax problems."[21] Negotiations never really got off the ground. A week later, Kiekhaefer and Strang began meetings with various divisions of Chrysler Corporation, with the intent of developing a working agreement for the development and manufacture of horizontally opposed eight-cylinder automobile engines, and to carefully feel out the possibility of a merger. Though no immediate interest in either possibility surfaced, a little over a year later, in January 1959,

Chrysler engineers from Detroit were dazzled by a tour of Kiekhaefer's assembly lines. In a taped telephone conversation with Willis Blank, a Chrysler production executive said of the tour, "It was one of the most interesting days I have ever spent in my life. ... Boy, that was one of the cleanest plants I think I have ever been in. ... I wish we could get ours that way. ... You must have some real good talent in there, because you just don't do that by waving a wand. There is a lot of sweat, blood and tears in that operation."[22] Chrysler had been making dramatic headlines throughout the year, due to a tremendous, 380 percent increase in profits, sales of over $2 billion and record production which would nearly double their 1956 output. Intermittently serious discussions with Chrysler were to continue for the next three years.

Carl divided responsibility for dealing with prospective mergers between Charlie Strang and Willis Blank. The job of courting Chrysler fell to Strang. Strang assembled a giant picture and data-filled book detailing the history and achievements of the Kiekhaefer Corporation for presentation to Chrysler, and when Carl saw it he was overwhelmed.

> "Carl looked it over and said, 'This is *great*! This is *beautiful*!' He starts scribbling and hands me a personal check for $5,000. He says, 'This is for the extra effort.' He said, 'Cash it -- Don't tell anyone where you got it, so you don't have to pay any tax on it.'
>
> "I mulled this over for about 24 hours and finally went back and said, 'Carl, you know this is part of my job, I simply can't take this -- here's your check back.' He thought I was crazy."[23]

Mercury had, as Carl advised Borg-Warner Chairman Ingersoll, "reached the age of consent." Carl was torn between attempting to continue the struggle on his own, or acquire the financial support he needed for dramatic expansion. He was confident that with the right partnership he could threaten and eventually overtake OMC for supremacy in the marine industry. Strapped for cash, his prospects for expansion without a merger were almost nonexistent. He was also certain that his illness or death could mean the end of Mercury. He became much like a nervous bride, slowly marching toward the altar, but with his eyes and concentration keenly focused on the exits along the way.

CHAPTER TWENTY-FOUR

The Bowling Bride

"Despite persistent rumors, the Kiekhaefer Corporation is not for sale, nor are there any negotiations under way for sale or merger ... we repeat, we have no interest other than staying strictly on course in our own field, which has been the design and production of better outboard motors. ..."[1]

Carl Kiekhaefer
February, 1961

In the closing years of the decade, a three way race for making a deal with Carl Kiekhaefer began to shape up between Chrysler, American Machine and Foundry Company (AMF), and Brunswick Corporation. Of the competition, most interesting was the quiet battle for control of Kiekhaefer Mercury that ensued between bowling industry rivals AMF and Brunswick.

AMF had become among the most popular stocks on Wall Street due to a combination of engineering ingenuity and executive talent. In 1943, Morehead Patterson took over control of AMF from his father, Rufus Lenoir Patterson, who had founded the company in 1900 and invented the first automated tobacco handling machines. These complex machines could weigh and bag tobacco, as well as roll cigarettes and cigars. By 1940, AMF machines were rolling over a million cigars a year. The company also developed baking and stitching machines, but was in danger of stagnation by the time the younger Patterson stepped in. Following World War II, Morehead Patterson "decided that the company had to grow or die."[2] In his search for new products, he came upon a crudely built prototype of an automatic pinsetter for ten-pin bowling.

Bowling is an ancient sport, with roots as far back as 5200 B.C., evidenced by the discovery of stone pins and balls in the burial tomb of an Egyptian child. From the fourteenth to the seventeenth centuries, various forms of bowling flourished as a pastime in central Europe and Great Britain. One of the reasons for the term "club", as bowling pins are often called, is the fact that in the third and fourth centuries, German citizens often carried a club for protection, known as a *Kegel*. As part of a religious ceremony, the *Kegel* was stood on end in a church cloister, while the practitioner tried to knock it over with a rolling stone. A successful "strike" meant that the person was cleansed of sin. Even today, some bowlers are known as *Keglers*. Through the centuries, many variations of the sport were developed, and one version, called nine-pin bowling, was introduced into the United States by Dutch settlers in the 1600s. Unfortunately, it became so popular and associated with gambling that it was outlawed in many areas. According to bowling tradition, a tenth pin was added to the sport to cleverly circumvent all the laws that been passed against nine-pin bowling. By 1895, most of the rules and sizes of balls and pins familiar to modern bowling had been standardized, and an organized national championship was sponsored in 1901. But one of the biggest drawbacks to public, and particularly family acceptance of the sport, was the seamy and vulgar atmosphere created by bowling establishments, centered around the ruffian antics of the pin boys charged with setting up the pins and returning the balls. "Before World War II," claimed the *New York Times*, "many bowling places had reputations as hangouts for rugged characters who kept the air above the lanes blue with tobacco smoke and rough language; too many pin boys sassed back the customers or heckled the ladies."[3]

An automatic pinsetter is a remarkably complex device, and many inventors had failed to conquer the combination of accuracy, speed, safe pin handling and ball return. An inventor named Fred Schmidt, working in a henhouse in Pearl River, New York, had patented a device using strong vacuum gripping of pins. Robert E. Kennedy, a salesman for Brunswick, learned of the new device and formed a partnership with the inventor. Brunswick was already heavily involved in the manufacture and sales of pins and balls for the sport, and should have recognized the potential of the new machine. But when Kennedy approached Bob Bensinger, Brunswick's president, he was surprised to be turned down flat. Bensinger later recalled his justification in a book on Brunswick history prepared by Chicago journalist Rick Kogan. "Who needed an automatic pinsetter? Not the bowling proprietors -- they didn't have the money. Even if they did, why should they buy? There were plenty of pin boys around."[4] Kennedy promptly took his remarkable device to rival AMF, where Morehead Patterson quickly made a deal.

It would take AMF six years to make the cantankerous contraption work reliably, but when they introduced the latest version in 1952, the AMF *Pinspotter*, Brunswick was stunned. "When we heard that, it was like a death pall hit the company," a Brunswick executive remembered. "What the hell are we going to do? Fold up our tents and walk away?"[5] Brunswick began to work along the lines of previous experiments into pinsetting, and developed their own device, but fell nearly four critical years behind in getting a competitive device to market. By 1961, AMF would install 68,000 *Pinspotters* under lease, for an average annual gross of $68 million. The international potential for the new machine was also realized, and AMF leased an additional 3,000 machines in 17 foreign countries.

By the time AMF was ready to approach Carl Kiekhaefer with an offer to merge, AMF had diversified into a giant, stable and forward-looking company with a great deal to offer. They had bought their way into in other sporting goods lines, such as W.J. Voit Rubber Corporation, manufacturers of scuba gear and tread rubber, and Ben Hogan Company, a leading maker of golfing equipment and accessories, as well as Wen-Mac Corporation, builders of engine-powered toy airplanes. Substantial profits were also being made by AMF technical divisions, as the company designed and built launching silos for military Titan and Atlas intercontinental ballistic missiles, and rail-car launching systems for solid-fuel Minuteman ICBM's. AMF was even competing head to head with General Dynamics in a race to become the world's largest suppliers of nuclear reactors for research, having installed 21 reactors throughout the free world.[6] By 1960, AMF chairman Patterson and president Carter L. Burgess would report burgeoning sales and lease payments of $362 million, and profits of over $24 million.[7]

Various brokers and agents had since 1957 attempted to get Kiekhaefer and AMF in the same room to explore a merger, but serious discussions didn't get underway until 1960, when one broker told Carl what had happened to another AMF acquisition. AMF had acquired the J.B. Beaird Company of Shreveport for $6 million in AMF shares. "[The merger] has turned out extraordinarily well for the Beaird family, as today [with stock splits and rising stock prices] the vendors have a marketable value of something like $15 million for their company. This is the sort of thing that can happen."[8] As it turned out, AMF had been carefully watching Kiekhaefer's progress for years, waiting for the right moment to make an offer.

On March 16, 1961, Charlie Strang met with AMF Chairman Morehead Patterson, and the following day met with President Carter L. Burgess. When Strang learned that AMF had "been keeping tabs on our company for a number of years," he asked why they hadn't made contact directly. One of the AMF executives replied with "a Shakespearean quote ... to the effect that what a person thinks and desires most becomes an unapproachable object."[9] Knowing that Brunswick was also courting Kiekhaefer Mercury, AMF plied Strang with comparisons between the two companies, such as "AMF has a sounder financial program," and "they claim to have quite a jump on Brunswick in European markets." Carl, feeling growing confidence in the value of his enterprise since three well-known corporations were anxious to merge, was steadily raising the stakes. One of the notes that Strang made during his meetings was, "They would like to make a deal, but wince at the price they feel they will have to pay."[10] By May 1961, AMF presented Carl with a most tempting offer which would amount to over $30 million in cash and stock over a four year period, and leave Carl in place as president of Mercury. Under his photograph in the 1960 annual report he sent to Carl, AMF President Carter L. Burgess wrote, "With best wishes for the possibility of a strong future for us both."

Meanwhile, Brunswick Corporation had for several years been struggling for Carl's attention. In 1958, Carl had arranged to entertain Brunswick-Balke-Collender Company President Benjamin Edward "Ted" Bensinger at Tommy Bartlett's spectacular water ski show at the Wisconsin Dells. Bensinger was impressed by the power and beauty of the Mercury outboards used exclusively by Bartlett's highly regarded shows, and commented, "The water ballet and show was terrific."[11] He also spoke with Willis Blank about the potential of getting their two companies together, and both agreed that "it might not be amiss for us to 'touch base from time to time.'"[12]

John Moses Brunswick, a 26 year old Swiss immigrant, began building carriages and fine furniture in Cincinnati, Ohio in 1845. In this same year, young Brunswick saw his first billiard table, an elegantly crafted example imported from England. Though billiard tables arrived in America before the revolution, they were mostly considered instruments of bawdy establishments, and soon gained a reputation for attracting drunkenness and gambling. When U.S. President John Quincy Adams had a table installed in the White House, he was soundly criticized for having "gambling furniture."[13] Like bowling, the early history of billiards was wrought with many variations in both rules and equipment, but before Brunswick built his first table in 1845, slate beds and rubber cushions had become the norm. In the early part of the nineteenth century, a French infantry captain named Minguad, "whiled away his sentence playing billiards in prison."[14] He discovered that by gluing a patch of leather to the end of his tapered cue stick, he was able to manipulate the path of the ball considerably, by hitting it somewhat off center. He became so enamored of perfecting what would eventually be referred to as applying "English" to the cue ball, that he asked to be kept in jail a month longer than his sentence.[15]

When news spread of the high quality billiard tables built by John Brunswick, his business mushroomed. By 1858, Brunswick had expanded to 75 employees, including two half-brothers, and was building $200,000 worth of tables a year. By 1873, Brunswick merged with Julius Balke's Great Western Billiard Manufactory of Cincinnati, and by 1879 with Phelan & Collender of New York, his two largest billiard table competitors. The company, which moved to Chicago after the great fire, would be known as the Brunswick-Balke-Collender Company until the name was shortened in 1960 to Brunswick Corporation.

In 1912, Brunswick anticipated the onset of Prohibition sentiment, and discontinued a $4 million sideline manufacturing elegant bar and tavern fixtures. In the 1920s, Brunswick diversified into automobile tires, phonograph cabinets, phonograph machines, and eventually, phonograph records featuring the voices of such musical idols as Al Jolsen, Duke Ellington and Cab Calloway. By the eve of the Great Depression, Brunswick sales were nearly $30 million. Well timed divestitures of music division assets helped to nurse Brunswick through the worst years to follow. Following World War II, the great bowling pinsetter war with rival AMF would dominate Brunswick thinking for decades to follow.

When AMF unveiled their latest and most reliable *Pinspotter* in 1952, a flurry of pinsetter development began at Brunswick. Benjamin E. "Ted" Bensinger, the third Bensinger president of

Brunswick since his grandfather became president in 1890, borrowed $16 million from a private investor to fund a massive race to develop a competitive device, and in 1956, the first Brunswick pinsetter was installed. Unfortunately, AMF had already installed over 9,000 rival *Pinspotters* throughout the country. Brunswick's new machine proved so popular, though, that $30 million in orders flooded the Chicago headquarters. By the end of 1956, Brunswick had installed 2,000 machines, 7,000 by 1957, and when Ted Bensinger began feeling out the Kiekhaefer Corporation in 1958, over 11,000 Brunswick devices were in use. The company's growth was nothing short of spectacular. Sales of $33 million and profits of $700,000 in 1954 swelled to sales of $422 million and earnings of $45 million by 1961.

Near the end of 1960, rumors spread that Brunswick was also considering a merger with Kiekhaefer's perpetual rival, Bob McCulloch and his Scott-Atwater company. When Carl heard that Bensinger was holding talks with McCulloch, he refused further discussions. Brunswick, though, was on a shopping spree, and between 1958 and 1961, purchased 18 companies, ranging from MacGregor Sports Products to Owens Yachts and Larson Boats. Bensinger, avoiding confrontation with the unpredictable Kiekhaefer, would phone Willis Blank in Fond du Lac when he was certain Carl was out of town. Blank would record his conversations with Bensinger, and play them for Carl when he returned. Bensinger attempted to convince Blank that Brunswick was to be a force in the boating industry, and that he was determined to acquire an outboard manufacturer.

> "One way or another, we have come to a corporate decision, and that is that we're going to have to be in that business. ... [W]e're going to be a dominant factor, or try to be, in that marine operation. ... This is going to be, as you can imagine, over the next few years, a battle of survival. You folks are one of the two or three well established in the field on the outboard end.
>
> "... [W]e haven't had any serious negotiations with anybody, including yourselves. But that doesn't mean we aren't going to be having plenty of discussions with others, and I hope, as well as yourselves. ... What's best for Brunswick? Should we try and merge with Kiekhaefer or Outboard Marine or West Bend or McCulloch? Should we import some motors? Should we do some things on our own? We don't know. I'm candid to admit it. ... The only thing we have a firm conviction on is one way or another we're going to be in the game."[16]

Proposals and counter proposals began to flow from all three Kiekhaefer suitors in the spring of 1961. AMF, Chrysler and Brunswick jockeyed for emotional position with the volatile founder of Mercury, and a considerable portion of Carl's time was spent evaluating and ultimately rejecting various offers and suggestions. Early in the year, Carl removed Chrysler from consideration, feeling that he would be swallowed by the giant automaker, and become an insignificant and powerless division in the bargain. More to the point, Chrysler had strategically blundered by stipulating that Carl would have to step down as president of the company, and take orders from a Chrysler appointed executive. The battle for ownership of Kiekhaefer Mercury narrowed to AMF and Brunswick.

In November 1960, Ted Bensinger offered Kiekhaefer $25 million for Kiekhaefer Mercury, $18.75 million in cash and $6.25 million in Brunswick stock over a six year period. By January 1961, Bensinger changed the offer to $25 million in immediate cash. The following month, Bensinger raised the offer to $30 million, $25 million in cash and $5 million in stock. But on March 7, 1961, Brunswick financial analysts discovered that Mercury's entire net worth was only $15 million, and that 1960 profits after taxes were a paltry $42,000. The offer dropped $5 million overnight to $30 million. Carl angrily broke off negotiations, and began to look toward AMF. Robert C. Anderegg, Carl's new controller, remembered the differences in approach between AMF and Brunswick. "Brunswick didn't go near into the analysis that AMF did. I think Ted Bensinger was just determined he was going to get Mercury one way or another, and paid a lot less attention to the details. He just plain wanted to buy the company."[17]

Carl had become alarmed by the swirl of rumors circulating the country that he was planning to merge. It seemed that every time he spoke to the press, all they wanted to know was who he was negotiating with, and when a deal would become final. Carl correctly assumed that Mercury sales would suffer as a result of this widespread speculation, and issued a terse press release, attempting to convince the public that his company was not for sale, even as he was in the midst of almost daily negotiation.

> "Despite persistent rumors, the Kiekhaefer Corporation is not for sale, nor are there any negotiations under way for sale or merger. ...

"We feel we can serve the industry best by continuing in our present form and concentrating on our own field of endeavor.

"It is true that in recent years we have been approached by a number of reputable firms who have suggested merger or sale; however, we repeat, we have no interest other than staying strictly on course in our own field, which has been the design and production of better outboard motors. ..."[18]

Brunswick lost little time in attempting to soothe Carl's frayed nerves. Brunswick vice president F.E. "Gene" Troy wrote Carl a note of apology, trying to explain why Brunswick could often seem cold-blooded in its approach to making a deal. "I feel that through our wide experience in the field of mergers and acquisition we have become somewhat callous to the personal feelings of others, and while this is not intentional, the impression sometimes created has worked to our disadvantage. If we seemed over-enthusiastic in our negotiations with you, it was an honest enthusiasm for a great product and greater opportunities."[19] Carl ignored the overture, and continued discussions with AMF.

AMF offered Carl approximately $30 million in cash and mostly stock, with a payment schedule that stretched over nearly four years. Disturbing to Carl was the lack of flexibility he perceived in an AMF merger. The language of the proposals, and the inferences AMF would make in negotiations, led Carl to believe that they were insincere in promising him the freedom to run the company in his own fashion following a merger. He began to consider the benefits of establishing a separate engineering consulting company to raise capital, rather than yield to a Mercury/Brunswick merger.

In a scrawled out list comparing his options, Carl revealed the factors influencing his decision. Among the advantages he listed for a Brunswick merger was the "personal enjoyment of the excitement inherent in this business," and a "field in which I am experienced and like." But in the disadvantages column, Carl scrawled, "No likelihood of advancement, either financially or other," "constant barrage of emotional crises," "somewhat prison-like life with no known free time - ever," and finally, "wide variations in ratio of responsibility to authority from one time to another."[20] Among the advantages Carl listed for starting an engineering-consulting business were, "More direct control of own future - less reliance on other people," and an "opportunity to relocate in a more salubrious climate (last winter was rough)." Finally, exhausting all the feasible alternatives available to him, he grudgingly admitted that his best opportunity was with Brunswick.

By June 1961, Carl had received four different proposals from Brunswick headquarters in Chicago. Among the various plans was one that would provide Carl and Willis Blank with $34 million in Brunswick stock, based on the average price of the stock ten days before and ten days after a formal closing. Among the various terms defined in the prospective agreement was Carl's ability to personally purchase the Siesta Key salt water proving grounds property at Midnight Pass near Sarasota, the house at Winter Haven, a house in Fond du Lac, and most remarkably, the enormous properties of Lake X. In the case of Siesta Key and Lake X, he would be able to lease these facilities back to Brunswick on favorable terms, while maintaining a most lucrative investment opportunity.

Carl agonized over every detail of the proposed agreement, and almost daily vacillated between acceptance and rejection. His moods would swing widely, one moment beaming with pride over a condition of sale that favored him, the next moment scowling and agitated over a paragraph that limited his freedom. Even when the hundreds of nuances involved in the transaction were ironed out, Carl was still unsettled about giving up control of the empire he had carefully built for over 22 years. He demanded of Ted Bensinger, and received, a separate and confidential letter to guarantee his autonomy as president of Mercury, and for the job security of his top executives following the closing. "It is contemplated," Bensinger wrote, "that the present Kiekhaefer operation will be continued as a separate subsidiary of Brunswick, and, without commitment as to term by either party you will be made its president and act as its chief executive." Carl would also be elected a vice president of Brunswick Corporation, and be given two seats on the board of directors. It was never Willis Blank's intention to remain with the organization following a merger, and it would be up to Carl's discretion to fill the second seat on the board with an individual of his choosing.

The preliminary *Agreement and Plan of Reorganization* between Brunswick Corporation and Kiekhaefer Corporation was typeset and delivered to Carl near the end of July 1961. Closing of the merger was actually a two-step process. His signature on the agreement would authorize both corporations to prepare all of the necessary closing documents, and for Carl to prepare deeds of title to Kiekhaefer properties and facilities for delivery at the closing. Charlie Strang remembered when the moment of truth arrived for signing the preliminary agreement. Charlie was sitting with Carl in his office on July 29, 1961, when he was preparing to sign the papers. "I said, 'Carl, it takes you ten minutes to get rid of what took you 22 years to build.' He looked at me a minute and says, 'Yeah,' and went ahead and signed it."[21]

Within 24 hours after signing the preliminary agreements, Carl received a warning from Arthur A. Burck, a prominent Wall Street consultant who had tried in vain to get Carl to consider a merger with Studebaker-Packard earlier in the year. The letter cautioned against a merger with Brunswick, suggesting that the price of the stock was artificially high.

> "Was lunching today with a group of investment bankers, and was jolted by a rumor that you were selling for $33 million in Brunswick shares.
>
> "This jolted me because in the opinion of those present -- as well as my opinion -- Brunswick stock is way overpriced. In other words, the important thing is not the immediate market value you get, but where the stock will be a year from now -- this is what counts."[22]

It was the only formal warning that Carl received against a merger with Brunswick, and so he discounted it, assuming that the consultant was simply sore that his proposed merger with Studebaker-Packard hadn't materialized. After all, Brunswick was experiencing record sales, and the tremendous popularity of bowling seemed to insure high profits for many years to come. Another, more subtle warning came from his long-time advisor and attorney Alan Edgarton in Fond du Lac, who said, "Carl, if I know you at all, I don't see how you'd ever work with somebody unless you were the boss. I'm not trying to hurt your feelings, but you're that kind of fellow."[23]

There were sixty days between the signing of the initial agreement and the scheduled closing meeting at Brunswick's Chicago headquarters. "I'm sure from the first of August on, Carl had begun already to have second thoughts," Mercury's controller Bob Anderegg remembered. "He talked to his attorneys and actually tried to find ways that he could wiggle out of the deal. Between the first of August and the end of September, he tried every which way to wiggle out of it." According to Anderegg, Carl at one point told his attorneys, "The HELL with it, *breach* the contract!" but cooler heads prevailed and he calmed down.[24] As the closing day approached, Carl became more anxious and unsettled, and was nearly impossible to be around.

On the morning of September 30, 1961, Carl cried in his bedroom, comforted by Rose Smiljanic. "I won't sign," he wept, "I'm not going to go through with it." But then Al Puelicher, his influential friend and president of the Marshall & Ilsley Bank called and said, "You've got to go through with it."[25] Carl washed his face and prepared to leave. Two cars left Fond du Lac for the long drive to Chicago. Carl and Rose Smiljanic drove one car, while Bob Anderegg and Alan Edgarton followed in another. Assembled in the Brunswick board room were Ted Bensinger, a phalanx of Brunswick attorneys, and rows and rows of neat stacks of papers to be signed by both parties. Following a few nervous minutes of greeting, Brunswick handed Carl the first set of papers to sign. Nobody really knew what Carl was going to do, and the tension was nearly suffocating to everyone in the room. "It was a question of 'go' or no go,'" Edgarton remembered. "If you knew Mr. Kiekhaefer, you'd know it was a possibility [that he could walk out]. I knew it because I'd been into some situations ... many of them where he just had enough and walked out. But I think that in back of it all was the [question], 'What will I do? Where will I get the funds to keep this whole organization going, and do what I want to do worldwide?'" Carl was unhappy, Edgarton remembered, but it was too late. "Well, if you knew him ... you could read it as well as you could a newspaper. Because he was one person when he was happy and normal, and he was an entirely different person when he was in a mood of that kind. He was just *mad*, and he wanted to put it off, but they weren't going to put it off."[26]

Carl slowly began to sign. Anderegg and Edgarton sat on either side of Kiekhaefer, and as each document was handed to him, they would briefly explain what it was for and where he had to sign. "But after a while," Anderegg recalls, "there were just so many of them he just finally got to signing the papers and throwing them on the floor. He was just exasperated with the whole procedure."[27] And suddenly, it was over. "Then, the truth really hit him," Anderegg observed. "He had sold his company. I'm sure he had tears in his eyes. I never saw a man so dejected as he was on the way home from that meeting."[28] Anderegg and Edgarton again shared a car for the return trip to Fond du Lac, and both remember being affected by Kiekhaefer's obvious agony. "We both thought the same thing," Anderegg explained. "It was kind of a torture to see what Carl was going through and what he must be thinking. It was a quiet ride back for everybody, I'm sure."[29]

Carl was in shock. The company didn't belong to him anymore, and there wasn't a thing he could do about it. On the way home, he cried.

Within days of signing the documents giving Brunswick control of his empire, he received a memo from his new "boss", Ted Bensinger. It was a boiler plate announcement to all Brunswick divisional presidents concerning an upcoming Brunswick board of directors meeting. Carl then realized where he stood in the scheme of things. The terse memo gave Carl only fifteen minutes to explain to the board a lengthy list of subjects, including his original and revised budgets for 1961, "What you actually expect to have your subsidiary attain for 1961," "The reasons - economic, personnel, or otherwise - that were responsible for your attaining the final objectives set; or the reason why the objectives will not be attained. And, <u>most important</u>, what remedial steps you have taken during 1961 to get your house in order to assure a 1962 satisfactory result." With the time left[!], "be prepared to discuss your Management Team or organization (a) as it was; (b) as it now is; (c) and as you plan it to be for 1962."[30]

Carl would soon experience an emotional and even life-threatening trauma as he realized the true consequences of his merger with Brunswick. Not since he left the family homestead over forty years ago had Carl suffered such a crisis of identity and self-worth. Though he proudly displayed the stock certificate for 600,000 shares he had received from Brunswick, he had difficulty savoring his good fortune. Among the terms of the agreement signed on July 31, was that Brunswick would pay 600,000 shares to Carl for 100 percent of Kiekhaefer Corporation, after which Carl would distribute 10 percent of the shares to Willis Blank. The agreement also stated that if the price of these shares of Brunswick stock amounted to less than $34 million when the closing took place in Chicago, that Brunswick would give Kiekhaefer additional shares to make up the difference. But, the agreement went on to specify, if the price of Brunswick stock were to rise during the interim period, Carl would still receive the whole 600,000 shares. In fact, on the day of closing, Brunswick stock had risen to nearly $63, a gain of $6 from the signing of the preliminary agreements. On the day of closing, Carl was handed Brunswick stock certificates worth $37,725,000. He and his family would keep $34,357,000 while Blank would receive around $3.3 million. The recent rise in stock value was an extra personal bonus of over $3.5 million for Carl alone.[31] Unfortunately, the disagreeable task of informing the world of his decision destroyed any sense of financial victory he may have felt.

He directed Tom King, who had become Mercury vice president of marketing, to immediately wire all distributors of the merger, before word would leak to the press. Of greatest concern to Carl was that distributors might feel they would be forced to represent Brunswick's boatbuilding interests, or that Brunswick outlets would become Mercury dealerships. Carl outlined his "official" reasons for merging, including:

"1. You automatically become a partner in a well-diversified, 116-year-old corporation internationally-known for its products for the leisure-time market and with gross sales, which will, with the addition of Mercury, be approximately $450 million in 1961.

"2. The addition of 90,000 Brunswick stockholders and employees as Mercury boosters.

"3. The opportunity for a combined Brunswick-Kiekhaefer program of engineering research."[32]

If Carl's announcements seemed unconvincing, it's because Carl himself wasn't sure about the future with Brunswick. Within days of the merger, a picture of Ted Bensinger standing with Carl next to a Mercury outboard was printed in hundreds of newspapers across the country. Carl is glowering at Bensinger, who displays a decidedly nervous smile as he tentatively rests his arm across *Carl's* engine. The picture captured the essence of the newly emerging and uncertain relationship between the two men.

Worst of all, almost immediately following the merger, Brunswick stock began to fall. After Carl had received a handsome dividend check in the months following the merger, the dividends abruptly stopped. Carl was prohibited by Securities and Exchange Commission regulations from selling more than 1 percent of his holdings each six months, because he was now technically *an employee* of Brunswick Corporation, and subject to insider trading rules. Ted Bensinger offered to assist Carl in selling additional shares, if he wished, through a secondary offering which Bensinger could legally arrange. But Carl was suspicious of Bensinger's motives, and assumed that Brunswick wanted him to reduce his holdings to reduce his potential influence as a large stockholder and as a Brunswick director.

In fact, Carl held the single largest block of Brunswick stock, a position of strength that he felt was more important than selling off shares in a declining market. What Carl didn't realize, is that Brunswick shares would plummet from $63 on the closing day of the merger to only $6 in less than a year. The bowling boom that had catapulted Brunswick stock ever skyward had busted, and pinsetter customers stopped making payments. Carl's value for 22 years of effort, and for selling his company which was producing over $35 million in products each year would be reduced to $3.6 million, an unbelievable loss of over $30 million. Relative to his holdings on the day of closing, Carl had been wiped out. Willis Blank, Carl's loyal counselor who eagerly pushed for the sale, quietly sold his stock without even informing Carl, managing to liquidate his shares before the stock's dramatic decline.

Carl was devastated by the combination of financial loss and the emotional crisis of not owning his own company. Though Carl always enjoyed a few drinks "with the boys," he was never considered an excessive drinker. But his habits changed shortly after the Brunswick stock began to plummet, and Carl began to drink heavily. Only a few people realized how ravaged Kiekhaefer had been by his misfortune, and how desperate he had become in such a short time following the merger.

His paranoia flared with unprecedented vigilance, and he soon suspected nearly everyone around him of duplicity with Brunswick. He began to drink excessively, pouring down the driest of martinis in an attempt to drown his sorrows. "Well, it was getting out of control," Rose Smiljanic remembered. "He was very competitive and he knew he had made a mistake that he couldn't change, and he had to live with it." Carl drank so heavily for a few months that he began to have "blackouts", and wouldn't remember what had happened.

> "After the merger he was getting so that he and the guys used to always meet at these different places after work. I guess he needed their companionship, possibly to kind of hold him up. And it got to be that always about five-thirty or six o'clock they'd head down to one of those pubs. At one point he blacked out twice when he was driving and he had an accident, and he had to have it covered up. ... I learned about alcoholic amnesia. I'm not saying he was an alcoholic or anything, but I said, 'I'm quitting the job. I can't stand drunken people or things like that.'"[33]

At the worst stage of his depression, he threatened to kill himself. "He just couldn't take it. He'd be sobbing ... and have his head in his hands in the office." Rose took his threat seriously enough that she contacted Dr. Meyer, the head of the University of Wisconsin Medical School. She was genuinely worried that Carl might carry out his threat, and was further alarmed by Dr. Myer's comment that "People who say they are going to [commit suicide], sometimes do. Stay with him and watch him."[34]

During one of Carl's blackouts, he had picked up a woman in one of the bars in Sarasota. "Everybody knew him," Rose learned later of the incident. "She had been drinking, and apparently there were these women who were after a good time, or money, and I didn't know about it. But one of the people told me at work, and when I found out about it I really refused him, and he just felt so bad. And then I discovered it was this alcoholic amnesia."

Even though Brunswick was in the throes of a precipitous avalanche of debt and despair, the Kiekhaefer Corporation was having a banner year, spurred on by the quick acceptance of important new products, and by increased sales of standard lines. Within a few months of Carl's binge of alcohol and thoughts of self-destruction, he started to recover. At Carl's home near Fond du Lac, on a naturally occurring "ledge" overlooking Lake Winnebago, he had an unusually well stocked liquor cabinet. One day, perhaps following Rose's threat to leave, he "goes into his cupboard and takes all these bottles and dumps every one of them down the sink," Rose said. "I would say like sixteen bottles of different things." The next morning, when Wilson Snyder, Carl's number one farm hand and all-around helper, came by, he saw the barrel full of empty bottles and exclaimed, "Jesus Christ, there must of been a *hell* of a party!"[35] In a way it was Carl's final toast to his depression. He was ready to face the world again, and was preparing to take on OMC for dominance in the marine industry.

CHAPTER TWENTY-FIVE

The Christmas Mutiny And Lessons Of History

"There comes to every mechanical contrivance used by mankind that epochal improvement representative of the highest embodiment of human achievement, that obsoletes the past and opens a vista of tomorrows extending as far as human conception can visualize.

"Such a revolutionary achievement, a product of human aspiration and skill immeasurable in terms of utility alone, now comes to motor boating. All that can be said about the ... Stern Drive -- all the facts of superior performance -- are but the simple truth of an engineering accomplishment which will change all present conceptions of basic motor boat construction and performance.

"So long as 'mankind goes down to the sea' in boats -- particularly in the medium size, faster motor craft -- so long will the new Johnson Tilting Stern Drive principle provide that reliability, safety, high efficiency and amazing maneuverability which heretofore has been the characteristic of Outboard motor driven boats only."[1]

Johnson Motor Company, 1930

Jim Wynne was still smarting from the embarrassment and anger of being caught by Carl Kiekhaefer having an intimate dinner with Rose Smiljanic. Even though the expected outburst and tirade from Kiekhaefer never came, he had already considered leaving the company. He enjoyed his job as head of Kiekhaefer's various proving grounds, but he was increasingly upset with what he considered to be Carl's juvenile emotional antics. Now based at the Siesta Key salt water proving grounds, Wynne had been placed in charge of the "base" over the head of Joe Anderson. Jim was an engineer, and Joe was a mechanic. Anderson had been doing a wonderful job as plant foreman at the remote testing facility, and it made Wynne feel awkward and uncomfortable to push Anderson out of his job, just because he had more education.

Salt water testing of outboard engines was the mission of the Siesta Key proving grounds, and after his stint at Lake X supervising the first 25,000 miles of the 50,000 mile endurance runs, it seemed like paradise to Wynne. A small fleet of yellow test boats would run most of the day, building time on engines in the warm waters of the Gulf of Mexico just south of Sarasota, Florida. Other operations included static testing of components hung by strings into the tidal salt water, to observe the effects of exposure to both sea and sun. Mercury learned much from these operations, from protective paint finishes, the proper balance of sacrificial zinc anodes, to component metallurgy to slow the advance of electrolysis from contact with the sea.

On December 16, 1957, about a dozen men were working at the base, mostly young endurance drivers, mechanics and engineers. The next morning, many of them were scheduled to fly back to Fond du Lac and Oshkosh to be reunited with their families for a few days of holiday at home. Wynne decided to have all the men together for a "little going away party," a simple dinner at one of the small local restaurants.

"It was a tiny little place," Wynne remembered. "It had only six or eight formica tables, bright lights, and nothing fancy at all." The group settled in to the restaurant about seven o'clock and were enjoying a round of beer after placing orders for dinner. "All of a sudden," Wynne says, "the door opens and Carl walks in." Wynne and the other crew were surprised by Carl's entrance, and didn't know what to say at first. Kiekhaefer must have picked up on the awkwardness of the group and immediately went to another table across the room to sit by himself. "So he's sitting right there, and we're sitting right here, fifteen, twenty feet away. And he's just kind of sitting there, very glum looking."[2]

Wynne walked over to Carl's table and said, "Hi Carl. Didn't know you were back in town. How about coming over and joining us -- we're going to have dinner because the guys are leaving tomorrow to go back to Oshkosh. Why don't you come and join us for dinner?"

"Well no, Jim," Kiekhaefer said, dejected. "You're having *your party*, and I'm not included, and I don't want to interfere with *your party*."

"Carl, this is not a *party*," Wynne tried to explain. "We were just going to have dinner together because these guys are leaving tomorrow."

"No, no. You go and enjoy yourself, Jim. I wasn't invited, you just go and have a good time."

Another version of the story would eventually filter back to John Hull, who reported that Donald A. Henrich, a talented propeller designer who was one of the unfortunate crew, claimed that Kiekhaefer had actually thrown food up against the restaurant wall, and demanded to pay for the check, screaming, "I want a bill for this meal here, my people don't freeload off anybody."[3]

All agree that the festive mood in the restaurant completely disintegrated. "It wasn't a very pleasant dinner at that point," Wynne says. Through the whole affair, Carl just sat at his table by himself, looking much like a wounded dog that had been kicked. "I thought, 'Holy hell, this is just totally ridiculous!'" Carl finally shuffled out the door, and the table conversation erupted with expressions of disgust over Carl's insensitivity and lack of tact.

"This is the last straw," Wynne blurted out. Another at the table said, "This is the worst performance we've ever seen."

> "I've made up my mind," Wynne said, "I'm leaving."
> "We're all going to quit," another chimed in.
> "We just don't want to work for this guy anymore."
> "No! We're all going to quit. We'll all quit together."
> "We can't work for this son-of-a-bitch anymore."
> "We're going to show *him*, by God!"

Wynne was alarmed by the overwhelming rush of sentiment from the men at the table, and started to feel guilty for having said he was leaving. "I didn't want to influence you guys," he said, but sensed it was too late. Each agreed that Carl's action had been childish and totally out of place with the circumstances of the evening. Nobody knew he was going to be in the area. Everyone thought he would be spending time with his wife, Freda, up in Winter Haven, and nobody expected him to show up at the base.

One of the group was a Kiekhaefer spy, and when the crew returned to the base, he quickly alerted Carl to the insurgency in progress. "Everyone's going to walk out," the snitch revealed. Carl was alarmed enough over the mass insurrection that he called his offices in Fond du Lac to ask for help. "Mutiny," he called it, and phone calls for peace and calm began to issue from cooler heads in Wisconsin.

Meanwhile, Carl, perhaps in revenge over Jim Wynne's clandestine dinner with Rose Smiljanic, carefully placed the blame for the entire incident on Wynne. He began to call each member of the Siesta Key crew into his house on the base, and one by one tried to plant his version of the incident in their minds. "Well, I'm not blaming you for all this," he explained to the surprised crew. "This is Jim's fault, because he arranged this party and he didn't invite me. I wasn't feeling well. I wasn't myself, and I'm sorry I broke up your party, but you have to understand it was all Wynne's fault."[4] He asked each of the men to reconsider their decision to resign, and told them he "wouldn't hold them responsible" for what Wynne did to create the incident. Wynne, though, was never called to speak to Kiekhaefer that evening.

Another version of the confrontation that ensued later at the base was provided by John Hull, who remembers one of the crew telling him that "they threatened to knock [Kiekhaefer] on his ear, I mean, according to Don Henrich, he was just behind his desk and they were going to slug it out with him."[5] According to Hull's recollection of Don Henrich's youthful boast, Kiekhaefer then said, "What do I have to do to make things right with you guys?" to which "one of the guys said, 'Well, you've got to apologize to the restaurant owner, and he agreed to do that. ... I don't think he ever did, according to the story that I got."[6]

Though a couple of the crew did resign over the incident, by far the greatest loss to Carl was Jim Wynne, who swore he was leaving and would never return. It was a deeply emotional conflict, one that would haunt both men for many decades to follow. Not too long before the "Christmas mutiny," Wynne had crashed and totaled the sleek Jaguar XK-120 coupe that he had bought in Seattle while still working on the drone aircraft engine program.

"I was coming back out to the base out on the winding road from town. Another group of the guys had taken one fork in the road, and I'd taken the other, and we're 'gonna see who gets to the base first.

"I was doing about seventy or eighty down this little narrow road. I hit a patch of fog and the road turned and I didn't. I get it off in the soft sand shoulder and it rolled upside down in a ditch in three feet of water. Two of us were in it, the other guy was Don Henrich, my room mate.

"Somehow we got turned as the car landed upside down in this ditch. We had our heads up by the floorboards of the car. I don't know how in the world two people could turn around in that car but we did. The windows and all the glass came out, of course, and we were in water up to our necks and we got our sense about us and crawled out the window on each side. Neither one of us got a scratch on us. Totaled the car, just totaled it completely, glass in our hair, but never got a scratch on us. That was the end of that car."[7]

Carl knew how much the Jaguar had meant to Wynne, so he went to Charlie Strang for a solution. Carl would go to great lengths to supply the right cars for his favored executives, and he had a swap in mind that would surprise both Strang and Wynne. Strang had been at Kiekhaefer Mercury only two or three weeks, driving a new Buick, when Carl approached him and said, "You can't be driving around in *that* thing -- it's an *old lady's* car. What kind of a car would you like?" Strang wasn't aware of Carl's whimsical penchant about automobile gifts and thought he was just joking. "I'd like a blood-red Jaguar sports car," Strang quipped back.

"He came in about two days later and he says, 'Here's a couple of airplane tickets. You and your mother fly to New York and pick up that blood-red Jaguar. So we flew to New York and went to Max Hopkins' place on Park Avenue, the only one handling them at that time, and there is a brand new 1952 Jaguar XK-120, blood-red, waiting for me.

"So we got in the thing and drove it home. I get to the Fond du Lac office, and he and Armand Hauser are waiting for me. Carl has a big hat box and he said, 'Well, you've got a British car, now you need the hat to go with it.' He opens it and he's got a deer stalker cap [like Sherlock Holmes'].

When Jim Wynne demolished his Jaguar, Carl offered Strang a brand new Mercedes 300SL Gull Wing, probably the most exotic and sought after sports car in automobile history. Strang was delighted, and Carl presented Wynne with Strang's blood-red Jaguar to use as a company car. In the week before Christmas of 1957, about the only thing Wynne was unhappy to leave behind at Siesta Key was the Jaguar. "I quit," Wynne shouted as he finished packing. He had called a friend to come and pick him up, and reluctantly left the keys to the Jag behind as he pulled out. By one o'clock in the morning, Wynne was in

Sarasota. "I was off the base." He returned the following day, just to drop off a formal notice of resignation. Wynne stayed with a friend in Sarasota for a few days, and Charlie Strang called, saying, "Carl would like to talk to you. He thinks that was a pretty silly thing that happened the other night. He wasn't feeling well and he'd like to talk to you. He'd like you to reconsider."[8]

"No, I really don't want to talk about it," Wynne said. "I want to leave." He left Sarasota for his parent's home in Miami, and soon after arriving, Strang was on the line again. "Carl would *really* like to talk to you about this thing," Charlie pressed for Kiekhaefer.

"I don't want to talk about this anymore," Wynne blurted out in exasperation. "I've had it."

"Well, Carl would really like you to have that company car," Strang insisted. "He wants to give you that as a little going away gift. We'll send you the keys." Shortly thereafter, Wynne received the keys and the title for the car, and returned to the base to pick it up. The outright gift of the Jaguar would be the first of many attempts to lure Wynne back into the Kiekhaefer fold.

Within a few weeks, Strang called again to offer Wynne a boat to drive in the upcoming Mississippi Marathon, as part of the Kiekhaefer Mercury team effort. The fast catamaran hulls entered were set up with a pair of Mercury's new six-cylinder, 70-hp outboards, and would fight their way against the steady current of North America's longest river, from New Orleans to St. Louis. Wynne, bored from weeks of lounging in the Florida sunshine, agreed to drive one of the boats. Even though he and his co-driver only placed third in the grueling event sometimes known as the "nightmarathon" because of the treacherous shoals, sand bars and floating debris hazards of the river, it was the perfect opportunity for Kiekhaefer forces to lobby Wynne to return to Mercury. Wynne steadfastly refused, saying he was enjoying his leisure time, and there was nothing left to talk about.

The truth is, Wynne was clandestinely working with a secret partner on a new product that would prove to revolutionize the entire marine industry, and it was the real reason why he couldn't return to Mercury under any circumstances. Carl Kiekhaefer could have offered Jim Wynne *the moon*, and it wouldn't have made a bit of difference.

Until the late 1920s, there were really only two ways without wind or oars to propel a boat through the water: inboard engine or outboard engine. Both had drawbacks. The four-cycle inboard engine, usually converted from an automobile, was big and heavy, and had to be mounted at an angle so that the propeller shaft would place the propeller far enough into the water to do some good. This meant that the propeller was usually pushing away at an angle of some ten to fifteen degrees down in the water, rather than pushing directly in line with the direction of the boat. This imperfect angle of attack reduced the efficiency of the propeller considerably. V-drive transmissions helped inboard installations to move the bulk of the engine back towards the transom, but the angle of the propeller in the water remained about the same.

The biggest advantage of inboard engines over outboard engines was horsepower. In 1957, Carl Kiekhaefer made the most powerful outboard engines in the world, at 60-horsepower. Inboard engines, available from about 65-horsepower to many thousands of horsepower, could be installed in boats ranging anywhere from 14 foot runabouts to ocean going freighters, all relying on the same basic installation. Inboard engines, though, also had to have a rudder to steer, and a strut to hold the propeller shaft in place under the water. Most also needed a skeg, or narrow plate of steel that would strike the bottom before the propeller would be damaged. Each of these steering, strength and protection elements represent a performance penalty in the form of parasitic drag under water. For large, slow boats, the effect of this penalty was negligible in terms of overall speed, but for small boats, all this iron and steel dragging through the water could really slow a boat down, no matter how large the engine or how efficient the propeller. Lastly, the propeller shaft of an inboard engine had to protrude through the bottom of the boat through a big hole, which requires vigilance to keep it secure and not leak a lot of water into the boat.

The outboard motor, lacking only large horsepower, did away with many of the troublesome features of the inboard engine. The underwater components were all streamlined, the weight and bulk of the engine was far aft, out of the way, and steering the outboard-equipped boat was accomplished by turning the propeller itself, rather than by dragging a big rudder through the water, so the boat didn't slow down in a turn. A skeg to protect the propeller was already built into the outboard, and if it hit the bottom or an object in the water, the whole works would tilt up to protect the propeller and shaft. The only drawback with outboards was the limitation of power.

By 1927, the Johnson Motor Company, builders by then of the world's most popular outboard engines, developed a special line of boats to go along with their products. These Johnson Flyers or "matched units," as they were called, were disguised to look like an inboard boat, emulating the popular lines of Chris-Craft, Gar-Wood or Hackercraft. The outboard was mounted a foot or so forward of the stern, and then covered by a somewhat rotund deck hood. Even though it was an outboard powered boat, it served to answer the objections of some inboard enthusiasts looking for smooth lines from bow to stern, without the bumpy and often dirty outboard engine hanging off the back end. Though Johnson's matched unit program was a dismal failure in terms of unit sales, it was the stimulus for thinking of new ways to propel their boats using the best features of both inboard and outboard motors.

In 1930, Johnson engineers unveiled their very practical solution, called the "Johnson Tilting Stern Drive." A four-cycle gasoline engine was mounted nearly up against the inside of the transom, connected through a shaft to what was essentially a lower unit of an outboard motor protruding from the back of the boat. The four-cycle automobile or marine engine could be up to 60-horsepower, whereas the most powerful outboard motor available then was the Johnson 34-horsepower Sea Horse. The new *stern drive* seemed to do everything an outboard could do -- and more. Warren Ripple, president of Johnson Motor Company, would conclude that the new device was going to revolutionize marine transportation, and the reason for its acceptance would be the ability for the propeller to tilt up like an outboard when an obstruction was encountered.

> "Get the significance of this -- the outboard motor boat has mastered the hazards that disable motor craft with stationary propellers and has thus introduced speedy travel in waters hitherto impassable or perilous to motor boats. ...
>
> "The tilting propeller is an epochal contribution to transportation. Already among the most important, it is destined to

so to increase in greatness that its future application to large craft as well as to small, is beyond conception."[9]

Johnson literature waxed poetic about the "revolutionary," "amazing," and "epochal" new stern drive that would "obsolete the past." The public, however, remained skeptical, and sales were almost nonexistent.

By 1931, another "stern drive" unit was marketed by the Morse Chain Company, the Marine Division of Borg-Warner Corporation, under the name "Silent Chain Drive." The Columbian Bronze Corporation, a leading supplier of propellers to boat and engine builders, also produced a version of the device, as did the American Outboard Drive Corporation with a new design by Joseph Van Blerck. The K.E. Ahlberg Company of Culver City, California, introduced a new "Inboardoutboard Motor" that used a tractor, or pulling propeller, rather than a traditional pushing propeller. A similar, though larger and slower version of this tractor drive was also built and marketed by Gifford-Wood company of Hudson, New York. The engineers and corporate executives of these various stern drive producers seemed ecstatic about the obvious benefits of these new propulsion systems, but again, the public wasn't interested.

For one thing, almost nothing was easy about getting one. The interested buyer had to supervise the marriage of a boat, an engine and a drive system, all supplied by different people, each of which adamantly refused to take responsibility for the outcome of the combination. Boats weren't built with stern drive units in mind, and so owners had to modify the interiors of boats to accommodate the new principle. Interior seating configurations had to be moved, engine beds and hatches had to be modified and relocated, transoms had to be strengthened, the old inboard propeller shaft holes had to be plugged and sealed, and tricky alignment holes had to be drilled through the transom -- usually right through the name on the stern of the boat -- and the stern drive mounted. None of these steps were easy, and all of them had to be figured into the cost of the new drive.

But even once each of these time consuming and costly installation steps were completed, the owner ended up with less speed than the original inboard installation, and less flexibility than an outboard boat, at an ultimate cost approaching twice either one. When engine problems developed, the engine builder blamed the stern drive. When stern drive problems were encountered, installation errors or the engine builder were blamed. When boat performance was questioned, the boatbuilder blamed both the engine and the stern drive. The public quickly caught on to this closed circle of design and performance finger-pointing, and kept away from the new product.

To eliminate some of these problems, the Hydro Division of Ludington Aircraft, Inc. of Philadelphia began to sell complete boat, motor and stern drive packages under the Ludington name in 1930. Two models of a 17 foot *Stern Drive Sportship* runabout were marketed, built of airplane spruce, white oak, Mexican mahogany, copper nails and brass screws. Ludington first selected the Columbian inboard-outboard drive for their boats. It was a special, heavy duty, all bronze unit that featured a complete 180 degree swiveling propeller that eliminated the need for a reverse gear on the engine transmission. It was slow and lacked the all-important tilt-up feature that made the stern drive concept attractive in the first place. Again, sales were dismal, and Ludington would soon offer Johnson Tilting Stern Drives on their boats as well.

When the Great Depression stalled all manner of boat and engine sales in the early 1930s, the stern drive products offered by Johnson and the others were among the very first casualties. Johnson's miraculous new device "that obsoletes the past and opens a vista of tomorrows extending as far as human conception can visualize," was out of production in only two years. By the time America entered World War II, the concept had been abandoned by all manufacturers. The brilliant new idea had become a complete failure.

In 1958, while enjoying his independence from Carl Kiekhaefer and the Kiekhaefer Corporation, Jim Wynne, then 28, the story would be told and re-told, invented the stern drive. Between 1958 and 1991, an estimated over $20 billion have been spent for stern drives and parts in the worldwide marine industry. If one takes into consideration the value of the boats which have been designed for the stern drive market, the figure is considerably higher. If the value of trailers, accessories, fuel and other expenses which naturally accompany ownership of this type of boat are figured in, the impact on the world economy of the introduction of the modern stern drive is conceivably approaching $75 billion. It is the most financially significant product introduced in the marine industry since the internal combustion engine. The popular public story, and the carefully guarded secret story concerning the development of the modern stern drive, are altogether different. First, the public story.

"In trying to build a stern drive I first looked at all the inboard-outboards that had been built since way back in the 1920s," Jim Wynne would repeat the basic elements of the public story many times.

> "They all had something that didn't work. For example, some would steer the boat like an outboard motor but they didn't tilt. Some tilted but didn't steer properly. It seemed to me that if a sterndrive was going to be successful, it had to do everything that an outboard motor did because that was the competition.
>
> "The idea was to get something that steered and tilted just like an outboard motor but with the engine inside the boat covered up like an inboard with a full transom. Use of a four cycle automobile engine was also more efficient at that time. Essentially what I wanted was the bottom half of an outboard motor mounted outside the boat so that it would swing back and forth for steering and also tilt up and down. The drive shaft had to go through the transom of the boat which meant that it had to bend.
>
> "I had seen the front wheel drive mechanism of a car, which does the same job, and this was the tip-off that a certain type of universal joint that I used originally -- a constant velocity joint -- was the answer.
>
> "The steering and tilting axes ran through the center of the universal joint and this allowed it to move in two directions at the same time."[10]

In another interview, Wynne explained just what, in 1958, was the breakthrough concept that would make the modern stern drive more successful than the Johnson drive of 1930.

> "The basis of the patent was the use of a double universal joint in the horizontal shaft behind the transom, so that the drive could both swing back and forth for steering, and also tilt up about the center of the universal joint. This allowed all the torque being transmitted by the shafts and the gears to be completely encased inside the housings. It was a practical way of making an outdrive.

> "Other outdrive units had attempted to steer the lower end while holding the upper portion stationary. This required the steering mechanism to restrain all the torque that was in the vertical shaft going down to the underwater unit. There were a lot of units around [back in the 30s] that had that problem."[11]

In simple terms, what Jim Wynne was telling interviewers was that the difference between the modern and the old stern drive was a pair of universal joints, like when you make the O.K. sign with your hand, and then join the sign with an O.K. sign from the other hand. Sort of a ring connected to another ring like a link in a chain. When one turns, so does the other one, even though the motion of one can be started from almost any angle to the other. Every car has one between the transmission and the drive shaft if you look underneath. In his interviews with virtually every boating publication in the free world, Wynne responds to the question, "How did you come to build the first one?"

> "I left Mercury at the end of 1957. As we know, I had a little run in with Carl at Christmas. I started working on this inboard-outboard idea in February or March of 1958.
>
> "I started developing a prototype, just a crude working model using old outboard parts [mostly Mercury], in my parents garage in the spring of 1958 and applied for some patents.
>
> "I purchased an engine that I had seen at a Volvo car dealership and that Volvo Penta was marinizing [modifying for use in boats] -- a little 1.6 liter, 80-horsepower unit that seemed to me just the right size. I ordered it without the reverse gear because I wanted to bolt it directly onto the stern drive."[12]

The public story continues that after Wynne ordered the Volvo engine without a reverse gear, the Volvo representative in the area became curious about the intended application. They met, and after Wynne disclosed his new prototype, the enthusiastic Volvo representative arranged a meeting with Volvo headquarters in Sweden, from which an exclusive deal was struck for use of the new drive. Volvo introduced the new stern drive at the 1959 New York Boat Show, and the course of marine industry history was forever changed. Wynne was hailed as among the most significant inventors in the marine industry, and was mentioned in the same breath as Ole Evinrude -- or even Carl Kiekhaefer. In 1989, Wynne was inducted into the National Marine Manufacturer's Association Hall of Fame, an honor which had only been bestowed on eight other individuals, including Ole Evinrude, Carl Kiekhaefer, Christopher Columbus Smith (Chris-Craft) and Gar Wood.

The real story will surprise the entire marine industry, Wall Street, and even the closest friends of those involved.

CHAPTER TWENTY-SIX

The Great Stern Drive Conspiracy

"That is a *horse shit* idea! Now, if you took the engine and you stood it on end, and put it on top of the gear box, you'd have an outboard motor. *Then* you got the right touch!"[1]

Carl Kiekhaefer on the stern drive, 1951

When Charlie Strang was lecturing and working towards his masters degree in mechanical engineering at M.I.T., he was also an avid outboard racing enthusiast. He grew interested in mounting an assault on the American outboard speed record, then held by Clint Ferguson of Boston since 1938 with a run of 78.121 mph. Strang knew that the only thing holding back an American record was the current horsepower limitations of outboard engines. In 1948, as Strang began to address the problem, the largest American production outboard engine was the 33.4-horsepower Evinrude *Speedifour*. It was impossible for Strang, as a student at M.I.T. to attempt an independent design and prototype of a larger power head, so he began to look at the more powerful engines used in auto racing. "There was an engine in England that was very popular," Strang says, "called the Coventry Climax. It was a fire pump engine during World War II, and after the war they converted it for racing. It was aluminum, very light, and it was only about a one liter engine. They used it in a lot of race cars in Europe."

Strang's idea was to mate this light, horizontal engine to the lower end of an outboard, and then mount the entire thing *outside* of the record attempt boat, hoping that officials would still consider it an "outboard". When he discovered that it wouldn't be legal, he abandoned his idea of the outboard record, but continued to refine his combination of an automobile engine with an outboard lower unit, but this time with the engine inside the boat, and the drive unit outside. In his engineering "invention ledger," Strang would draw different versions of his design, and then date and sign the pages, which was the suggested M.I.T. procedure. One of his entries during 1948 was for his final version of a modern stern drive, complete with the "universal joint" linkage which would ultimately be patented, not by Charlie Strang, but by Jim Wynne. In his explicit drawing, Strang even identified the universal torque transmitter by its scientific name of "Hookes Coupling," and further labeled and identified the tilt and swivel pin features within the coupling, exactly as they would appear later in the modern stern drive, and in the patent application submitted exactly ten years later by Jim Wynne.

When he had finished tinkering with his idea in the pages of the ledger, Strang didn't know what to call his new drive. He was an avid fan of comic books in his younger days, and remembered a voluptuous siren named Apacinata Von Climax from one of his favorites. Since he had considered joining his new drive to the Coventry Climax engine, he named the new stern drive after this imaginary seductress, and the *AVC* drive -- the modern stern drive -- was born.

Strang was convinced that the new stern drive had an enormous financial and practical potential in the marine industry. He was in exactly the correct position to analyze the possibilities for the new concept; as an engineering scholar, an avid outboard racing enthusiast, and as a highly regarded journalist within the marine industry. From Strang's unique and well-rounded perspective, the AVC drive would fill a critical need in boating propulsion left wide-open by the extreme difference in horsepower available between the largest outboard and the smallest typical inboard engine.

In the summer of 1951, after only two weeks in the Kiekhaefer Corporation's employ, Charlie Strang disclosed his idea to Carl Kiekhaefer. These were the heady days in transition between Carl's two-cylinder *Lightning* style models, and the beginnings of his new, and significantly more powerful, four-cylinder *Thunderbolt* years. Carl was flush with success over the introduction of his new 25-horsepower masterpiece, and had his marketing gun sights firmly on OMC and their even larger outboard models. So, when Strang proudly explained the enormous potential he imagined for his new AVC drive, Carl wasn't the least bit interested. "I was passionate about the idea," Strang remembered. "And the first thing I did was show Carl this concept, and that's when, you know, he said I was *nuts*. I showed the drawings and everything to Carl, and he said, 'Oh, that's *ridiculous*,' and he wanted no part of it." But, he had some tongue-in-cheek advice for Strang. "That is a *horse shit* idea," Kiekhaefer said, "Now, if you took the engine and you stood it on end, and put it on top of the gear box, you'd have an outboard motor. *Then* you got the right touch!"[2]

Charlie was swept into the maelstrom of Kiekhaefer activity, which would soon include the Mexican Road Race and a blizzard of engineering challenges, both on the track and on the water. He kept thinking about his AVC drive, though, and his first-hand experience with the powerful Chrysler, Ford and Chevrolet automobile engines used in stock cars, left a lasting impression on him.

One evening in 1955, aboard Kiekhaefer's slow and onerous Beech-18, Charlie Strang and Jim Wynne were flying down to Texas to testify in a product liability case concerning a Mercury outboard motor. It was a long and boring flight in the drafty twin engine aircraft, and they listened intently to the Sugar Ray Robinson vs Carl "Bobo" Olson middleweight boxing fight on the radio. After Robinson recaptured the crown from Olson, the conversation drifted into engineering, and the potential speed of outboard motors. It was aboard this flight that the greatest conspiracy in the history of the marine industry began.

"It was a long, slow flight down to Texas," Strang recalled, "and I started sketching this thing up to Jim, and showed it to him, and he got very excited about it. He got excited about it and he talked to Charlie Alexander about it." Charles "Alex" Alexander had been promoted to engineering vice president of Kiekhaefer Mercury, reporting directly to then executive vice president Charlie Strang. Wynne also reported directly to Strang. A secret series of conversations ensued between Strang, Alexander and Wynne about the possibilities for the AVC drive. Strang remembered when the decision was made to develop the new product themselves, secretly and without the knowledge of Carl Kiekhaefer. "Jim and Charlie and I said, 'What the hell,' maybe we could start a company and make this thing since Carl didn't want it."

Eighteen months before Carl would catch Rose Smiljanic and Jim Wynne sharing a romantic dinner together in Sarasota, Wynne and Carl had become close friends. Wynne, as a measure of the great confidence Carl placed in him, had been appointed chief engineer of Kiekhaefer Mercury proving grounds. Charlie Strang had become Carl's closest friend and most trusted confidant, and was second in command as executive vice president. Charles Alexander occupied the most sensitive engineering position in the company, that of vice president of engineering. These three men, arguably Carl's most critical and most highly trusted executives, were actively conspiring to build what would become the most significant product in marine manufacturing behind Carl's back.

In the spring of 1958, Jim Wynne, gone from Kiekhaefer Mercury for less than 90 days, "invented" the stern drive. Charlie Strang and Charles Alexander remained at Kiekhaefer Mercury, and continued to support his development of a prototype AVC drive, through surreptitious meetings and phone calls. A company had been formed, Hydro Mechanical Development, headed in principle by Wynne, for he was the only one of the three on the outside. Lacking sufficient funds to mount the expensive tooling and manufacturing of the drive, they decided to seek outside assistance. In a brazen move, the three met in Indianapolis in early 1958, in the office of John Buehler, president of Indiana Gear Works, manufacturers of various types of bezel and other gears, as well as the U.S. licensee of the Hamilton Jet Drive of New Zealand. The three had settled on approaching Buehler because of his existing gear production abilities, a critical element in the manufacture of the stern drive, as well as his well known foray into boat jet propulsion, a possible indication of a willingness to invest in novel ideas.

Buehler was quite novel himself, as Strang remembered the visit. "We walked in and this huge fat man wearing a Boy Scout uniform, shorts and all, says, 'We're a Scouting family.' He gave us the scout signal [salute] and everything else. Then he took us into his office, and I'll never forget his office because it was a gorgeous office, enormous, and he had mounted animals all around the walls -- but it was the ass-end of the animal! And there were arrows sticking out of them. He was a real weirdo, funny as hell."[3] Strang, Alexander and Wynne tried their best to persuade Buehler, sitting in his Boy Scout shorts and hat under the rear end of a moose with arrows sticking out of it, that the new stern drive would revolutionize the boating industry. But try as they may, Buehler was stubbornly convinced that his Hamilton Jet Drives were going to take over the industry, so "why waste time on anything else?" An executive from Warner

Gear Division of Borg-Warner once went to lunch with Buehler and asked him if he would like a drink. Buehler quickly replied, "Well, let's see what time it is. I never have a drink before five o'clock." Saying that he showed his watch to his host, and it was one of those watches that has the number five on every hour of the day, so no matter what time it is, it's after five. Unfortunately for Strang, Alexander and Wynne, it wasn't Buehler's time for the new stern drive.

Back in Florida, Wynne completed the "cobbled together" prototype of Strang's AVC drive in his parent's garage. One of the lessons Wynne had learned from Carl Kiekhaefer was the value of strict secrecy. He covered over the windows of the garage, and spoke with no one about his project until the prototype was completed and tested. He borrowed a 20 foot fiberglass boat from Woody Woodson, the founder of Thunderbird Boat Company in Miami, and installed the new drive system. Wynne carefully covered up his new unit and trailered it to Pelican Harbor on the intracoastal waterway in Miami before dawn. After a few trial runs and some adjustments, he was satisfied that the concept would work.

Shortly thereafter, the New Jersey based general sales manager for Volvo-Penta, John Jarnmark, made a routine sales call to Wynne, as a follow-up to the sale of an 80-horsepower engine delivered without the reverse gear. Wynne told Jarnmark that he was working on a special marine application of the engine that could conceivably produce a great many more orders for Volvo. Jarnmark was interested in finding out more, but Wynne was at first reluctant to fully disclose the idea. But once Wynne had filed the patent applications, based on Strang's drawings and universal joint concept, he invited Jarnmark to Miami to see for himself. Jarnmark was most impressed with the concept and installation of the Volvo-Penta engine, and sent photos, diagrams and descriptions of the device to the home office in Sweden.

In the meantime, a friend of Jim Wynne, Ole Botved, the manufacturer of Botved-Coronet outboard boats and cruisers in Copenhagen, Denmark, invited Wynne to join him as one of three co-pilots in an attempt to cross the Atlantic Ocean in an outboard boat. Wynne, always ready to tackle an unusual challenge, agreed. The voyage, from Copenhagen to New York, was designed as a publicity venture to generate business for Botved's boat line, and to establish a new world's record for outboard powered boats. Since Wynne had already helped to establish Carl Kiekhaefer's first 25,000 mile endurance run at Lake X, Botved wanted Wynne's expertise as an outboard mechanic as well as a driver. A small freighter planned to steam alongside the trio, providing fuel for the crossing, and stand by to pull them aboard during severe storms or in case of mechanical failure.

Jim flew to Copenhagen, and during the period devoted to preparations for the transatlantic attempt, he traveled to Sweden and met with Harald Wiklund, the president of Volvo-Penta. Wiklund is a remarkable individual, who would be president of Volvo-Penta for 28 years, from 1949 until 1977. When he took on the job of president in 1949, the company was generating sales of a few million Swedish Kroner per year, and when he left, Volvo-Penta, except for outboard motors, was the world's largest supplier of marine engines, with sales of over a billion Kroner annually. When Jim Wynne walked into his office in 1958, Wiklund had been president for nine years, and had already put Volvo-Penta on a course of growth that was making the Swedish marine engine the envy of the world. Wiklund was quite impressed with the idea brought to him by Wynne, and it quickly looked as if a deal might be struck. Wynne was still operating under the assumption that Charlie Strang would leave Mercury -- perhaps any day-- and join him in the stern drive venture. Strang was leaving all of his options open at the time, and kept delaying his ultimate decision which was making Wynne more and more nervous. Wynne knew that Strang's direct involvement would be critical to the success of any manufacturing venture, and he was concerned that he alone didn't have the expertise to create a company to build the drives without Strang. But now that Volvo was showing a sincere interest, he phoned Strang to find out his intentions once and for all. Would Strang really leave Carl Kiekhaefer? If not, what was he supposed to tell Volvo?

Strang was in Fond du Lac when the call came from Sweden. "I remember I was eating dinner at home one night," Strang recalled, "and Jim called. He said, 'I'm in Sweden,' he said, 'There are some people over here called Volvo. I was out with them last night and I told them about the AVC,' he said, 'and they're *real excited* about it. ... They want to do something about it.'" Strang's keen mind was spinning. Carl had just recently proposed that Strang head up a public Kiekhaefer

Corporation as president. The war with OMC was entering perhaps the most crucial phase in Kiekhaefer Mercury history, as Carl declared, "The Mercury organization, this year, is like a boxer in a ring who has just scored and is moving in for the knock-out. There is no stopping our organization at this time. Never before have we seen dealer morale and plant morale so high, but we must move with aggressiveness and intelligence even though our competition may appear to be groggy at the moment. We want to be on the lookout for sleepers. It may be a lot tougher in the next round."[4]

Strang was busier than he had ever been, and his prospects for advancing into the presidency of Mercury were tantalizingly close. On the other hand, he knew the enormous potential for his new stern drive. Wynne was waiting on the other end of the line. If he said he would leave Kiekhaefer, Jim would merely negotiate to buy Volvo-Penta engines at a volume discount. If he told Wynne to go ahead alone, he would negotiate a license agreement with Volvo under the pending patent, and Volvo-Penta would manufacture the drive and combine it with their engines as a package. Strang took a deep breath and told Wynne, "Jim, do what you want with it."[5]

Jim Wynne met again with Volvo President Harald Wiklund, and proceeded to negotiate the license agreement. "We spent two days together, and I guess I sold him on the idea," Wynne said. He was very astute, very receptive, and we signed a letter of intent for them to build this thing under the ideas of the patent that I had filed for."[6] Wiklund remembered that Wynne was actually not very well prepared. "You know, the only thing that Jim had to give me was an idea. He had no drawings or nothing. We had to make all the drawings. I bought the *idea* and the patent rights from him, but he had *no idea* about the construction of the whole thing."[7] Wiklund guaranteed Wynne $7 U.S. for every unit that Volvo would produce over the lifetime of the patent. After sales initially faltered, Wiklund renegotiated with Wynne and a figure of $3.50 per unit was agreed upon, plus Wynne would receive a generous 12.5 percent of all future license income Volvo might receive from other builders, which would eventually and ironically include Kiekhaefer Mercury. "In the beginning," Wiklund says, "he got *too much* money for it, you see, because the only thing he had was the idea, and we had to do the whole job. ... We sent him many millions from Sweden."

A tacit and somber understanding was reached between Strang and Wynne, that Wynne would claim credit for the invention of the drive, to protect Strang's position with Carl Kiekhaefer. If Carl were ever to find out that his most trusted friend had secretly given perhaps the most significant invention in the marine industry to a competitor, their relationship would, of course, be shattered. More than this, however, as Volvo quickly began to make plans to manufacture and promote the new stern drive, Strang and Wynne both understood that Strang could very well be held liable for the disclosure, as he had continued to develop the idea while on the Kiekhaefer payroll, even involving Wynne and Kiekhaefer head of engineering Charles Alexander. It would foment a scandal of enormous proportions, and so the bond of secrecy was made, and a conspiracy of friendship was entered that would endure for over thirty years.

Flush with success from the signing of the Volvo licensing agreement, Wynne returned to Copenhagen to prepare for the outboard motor transatlantic crossing attempt. The idea for the crossing had started as a joke between Ole Botved and Robert O. Cox, the world's largest dealer for Botved boats, located at the Lauderdale Marina in Fort Lauderdale, Florida. Cox, who would in later years gain national recognition as the popular and long-time Mayor of the city, had teased Botved about delivering the first of his new outboard cruiser designs under its own power. Botved surprised Cox by agreeing, and the challenge was on. Cox flew to Copenhagen to supervise preparations on the boat, which would eventually end up at his marina in Florida.

The 22 foot Coronet Explorer outboard boat was rigged with a pair of Johnson V-4, 50-horsepower engines, the *Fat-Fifty* as Kiekhaefer was fond of calling them. Accompanying the 29-year-old Wynne on the attempt were boatbuilder Ole Botved, 32, and Sven E. Orjangaard, 37, the first mate on the 7,500 ton Swedish freighter *Clary Thorden* serving as tender for the crossing. On July 14, 1958, the trio of drivers and the freighter departed Copenhagen. Ahead were 4,177 miles of open Atlantic.

The crew sustained itself with dark bread and beer, "and an occasional can of spaghetti if the water was calm."[8] The 22 foot *Coronet Explorer*, built of plywood with a fiberglass coating, made good progress before heavy seas began to concern all aboard. Two men would man the helm while the third slept, and at six in the morning and six at night, the *Clary Thorden* would lower a 300 foot rubber hose over her towering rails to refuel the *Coronet Explorer*. According to Wynne, at only one time was the crossing in great danger, when the boats were 1300 miles at sea, and waves whipped by 70-mile an hour winds "made it difficult to refuel."[9] Average weather conditions for the crossing were 25 knot winds and 40 foot ground swells. The boat finally tied up at Woods Hole, Massachusetts after an unforgettable ten days on the high seas. When the *Coronet Explorer* motored into New York harbor, the three adventurers received a hero's welcome, complete with a parade of fire boats pumping lofty fountains of water into the harbor sky, and the raucous horn blasts and screaming sirens of small craft assembled to greet her. Carl Kiekhaefer was enraged.

Not only had Wynne abandoned the Kiekhaefer organization, an act always punishable by Carl's perpetual cold shoulder and the honor of being categorized with *the enemy*, but the brash young deserter had the temerity to directly aid the enemy by driving their engines to a supposed new world record. Carl acted quickly to discredit Wynne, launching an independent investigation of the trip. Johnson issued a press release to herald the crossing as proof of the toughness and reliability of their 50-horsepower engines. Carl, though, spoke with captain Gote Gustrin, master of the *Clary Thorden*, the mother ship that escorted the outboard cruiser across the Atlantic. He was both shocked and thrilled to learn that the *Coronet Explorer*, with her two Johnson outboards, was hauled aboard no less than five times during the voyage, and actually rode along on the deck of the freighter for an average of one hour out of every five for the entire trip. "Since there will be some talk about this," Carl wrote in a press release to Mercury distributors and dealers, "we want you to know the true facts because 'the first successful Atlantic crossing <u>by outboard' has not yet occurred!</u>."[10]

With the help of the captain's log, Carl calculated "that of the 10 days, 16 hours, and 18 minutes they say it took for the trip, the boat and its outboard motors rode piggy-back on the freighter for about 20 percent of the time and from 600 to 1,000 miles of the 4,100 mile distance. "Certainly," Carl concluded, "this is not what Mercury would call success -- riding piggy-back 20% of the time and then trying to take credit for the whole distance as if they had gone all the way by water." When the *Coronet Explorer* would fall behind the *Clary Thorden* due to high seas or heavy winds, the freighter would lower a cargo derrick capable of lifting over 5,000 pounds, and pluck the smaller boat right out of the water and drop it on a deck cradle.

Botved had approached Mercury, Evinrude and Johnson for donation of the engines and partial sponsorship of the crossing. All three had evidently turned Botved down, but, according to Carl, "Johnson, even though previously they said they were against it, suddenly up and ran with it, no doubt feeling cute about out-foxing other manufacturers." Naturally, Carl was against any sponsorship of the trip because Wynne was on board, but more than this, he cited image and safety concerns.

> "We turned them down because we do not consider this to be the right kind of boating to promote. Whether successful or not, such trips encourage other people to make open water trips across large bodies of water such as Lake Michigan, Lake Huron, the Gulf of Mexico, etc. Too many lives have been and will be lost in these attempts. Outboard motors and boats have come a long way, but they are not ready for this sort of thing until boats of sufficient length, beam and heft are built.
>
> "We know you will agree we did the right thing. We would turn it down again if it were offered to us. We consider the Johnson trip to be very bad for all outboarding and an extremely unfortunate publicity stunt."[11]

Even spending a generous amount of time aboard the freighter during the crossing, the three-man crew was exhausted by the time the trip was over. When the *Coronet Explorer* finally tied up to a dock, curious spectators asked Wynne what he wanted first after the trip. "I'll take a solid bed that doesn't move up and down," he answered without hesitation. And when asked whether he would be attempting a return trip to Copenhagen, Wynne replied, "Not for a million dollars."[12] Within weeks of the trip, Wynne appeared on the nationally televised "I've Got a Secret" program, and stumped the panel when they tried to guess his secret. Little did they realize, Jim Wynne and Charlie Strang both had an explosive secret that would remain hidden for thirty long years.

Volvo-Penta began a crash program to complete engineering drawings and prepare production tooling for the AVC drive, which they called the *Aquamatic*. Though a capable engineer himself, Wynne was unable to answer the many technical questions that Volvo President Harald Wiklund and their chief marine engineer, Neil Hanson, were asking. As a consequence, Wynne had to disclose the true origin of the invention to Wiklund and the engineer, and swore them to secrecy. A series of clandestine meetings took place during the late summer of 1958, between Wynne, Wiklund, Strang and Hanson. Strang would correct errors in the engineering specifications, and make direct changes to the drawings spread out on motel beds located off the beaten path. Charlie Strang remembered the secret encounters.

> "I had a half a dozen surreptitious meetings with the chief engineer of Volvo, who'd come over with the drawings and we'd go over the layouts together. [For example], I was at the boat races in Lakeland, Florida, and Jim and the chief engineer of Volvo tracked me down, and we were sitting on the floor of the motel room going over the drawings, and so on."[13]

It was a most dangerous business for Strang, who was risking his entire career by conducting these secret engineering briefings with Volvo and Wynne, both considered *the enemy* by the volatile Kiekhaefer. Strang knew well that Carl Kiekhaefer had spies everywhere within the company, and all it would take would be one report back to Carl and the game would be up. Harald Wiklund felt he spent more time in aircraft that summer than behind his desk in Gothenburg, Sweden. "I was Scandinavian Airlines' biggest customer. They say Columbus discovered the United States. I discovered it too, in my own way."[14] With Strang's brilliant and total guidance, the Volvo-Penta *Aquamatic* drawings were finalized, and Swedish engineers rushed to complete tooling for the new stern drive in the fall and early winter of 1958.

The *Aquamatic* was unveiled with great fanfare at the 1959 New York Motor Boat Show in early January. It had taken Volvo-Penta engineers less than six months to produce the tooling for the prototype unit, and it drew some of the largest crowds of the show. Jim Wynne was a guest in the booth to help answer questions about "his new invention," and hundreds of photos were taken with Wynne, looking most distinguished in his carefully trimmed beard, leaning on the *Aquamatic.* Ingemar Johansson, the Swedish heavyweight boxing champion of the world, also made an appearance to pose with Wynne and the new drive. Carl Kiekhaefer was stunned to see the crowds formed around Wynne and the new stern drive at the Volvo booth, and was miffed at continually fielding questions from his own dealers and distributors about the merits of the new drive. It took him almost a year, but Carl finally got his hands on an *Aquamatic*, had it installed on a 18-foot Dunphy Boat, and *tested it himself* at Lake X.

His report, sent to Charlie Strang before the end of January 1960 in Fond du Lac, was perhaps the most biased, unfair and clouded evaluation that an engineer could ever have made. Carl was steaming over the introduction of the Volvo drive, not only because he considered the concept without merit, but also because Jim Wynne was involved. He was still stinging from Wynne's loud resignation, and support of Johnson outboards in the Atlantic crossing. Carl hated the Volvo drive before he even got in the boat. He never once recalled talking to Charlie Strang about the AVC drive years earlier, and he never would.

Carl, who was known among the test crews as perhaps the most ham-fisted, indelicate and clumsy of boat drivers, got the boat going around the Lake X test course at a little over 31 miles per hour. His comments, all contrary to just about everyone else's concerning the new unit, would come back to haunt him. Kiekhaefer's test report on the new stern drive is a classic example of a totally prejudicial evaluation.

> "Volvo engine: Extremely noisy, even though compartmented with sound-absorbing material as liner. Noise is combination of intake and mechanical. Gear whine noticeable at part throttle although not at high speed.
>
> "Steering extremely dangerous. Spun out boat at first hard left turn. Except for center position, steering force so violent as to twist wheel out of hand. ...
>
> "Installation costs must run considerably higher than an outboard since one large hole must be cut into the transom to take the engine mount. ... A water pick-up, in addition, must be installed on the underside of the boat. ...
>
> "The Volvo outboard-inboard drive, aside from its cost and weight disadvantages, has all the other disadvantages of an inboard installation and while a certain segment of the public might go for it, I do not believe it is a threat to outboard motors at this time. ... Gone too is the stimulant of annual model

changes. Styling plays no part. The product does not advertise itself, being hidden, and has all the romance of a 371 diesel power plant!

"I believe a destruction test at wide-open throttle at normal engine rpm is in order."[15]

For Carl, it was only part of a nearly two-year-long tirade against the new stern drive. Wynne was helping Volvo write and produce their *Aquamatic* advertising campaign and brochures, and the strategy was really upsetting to Carl. "Don't by an outboard," Volvo's exhibit signs shouted, "before you have seen and tested the revolutionary Volvo-Penta from Sweden ... that combines inboard efficiency and safety with outboard flexibility and speed."[16] This approach was irritating enough to Kiekhaefer, but then Wynne invaded two of the territories long claimed by Carl: speed and endurance.

Wynne had learned very well from Carl the promotional advantages of racing and endurance. He urged Volvo to ship him one of their ten, hand-built and precious prototypes, following the boat show introduction in January 1959. Guessing correctly that the very conservative Volvo organization might forbid him from entering the new unit in a race, Wynne secretly prepared another of Woody Woodson's 18 foot Thunderbird Boats to accommodate the new *Aquamatic*. In April, Wynne entered the boat in the Miami-Nassau race, and took aboard as his co-pilot the editor of *Popular Boating*, Bill McKeown. The magazine, which today is called *Boating*, was and is the most widely read journal of the sport, and McKeown's presence aboard was a stroke of promotional fortune for Wynne. Harald Wiklund, Volvo president, later told Wynne that, "Jim, you're *stupid*,"[17] and "If I had *any idea* that you were going to put that thing in a race, I would have personally come over there and taken it away from you."[18] Wynne and McKeown won their class in the race, and came in fourth overall, a remarkable accomplishment for the unknown drive system. In the process, Wynne garnered a landslide of positive publicity for the *Aquamatic*, and set his sights on future races.

In June, Wynne established four new world records in the nine-hour Orange Bowl Regatta marathon in Miami, handily beating Mercury entries from Carl Kiekhaefer, along with the most powerful engines available from Evinrude, Johnson and Scott-Atwater. Wynne had installed the stern drive in a little 16 foot deep-vee fiberglass hull designed by Ray Hunt, and built in

Massachusetts by Marscot Plastics. As if Wynne wasn't satisfied with setting four new world speed records, averaging better than 31 miles per hour for the whole nine hours of the race, he also upset nearly every other contestant in the marathon by making only a single pit stop for fuel, an Orange Bowl Regatta first. The combination of Charlie Strang's design and Jim Wynne's promotional abilities had united to thrust Volvo, the small, sleepy Scandinavian automobile and marine engine manufacturer, into the headlines of the marine industry press. Sales of the Volvo-Penta *Aquamatic*, sluggish at first, began to rise, and Carl began to realize that some Volvo stern drive sales were made at the expense of his new 70-horsepower Mark 78A motors, the world's most powerful outboard. Both Wynne and Volvo-Penta were becoming well know from these adventures, and Harald Wiklund feels that Volvo-Penta made Wynne a star. "You see, Jim Wynne was *nothing* before we started with the whole idea of the *Aquamatic*. Then, we built him up. We paid him a lot, of course, because he gave us the idea, but also, we built up his name. The *Aquamatic* has meant a lot for Jim Wynne."[19]

Many of Carl's dealers and distributors began asking when Mercury would come out with a stern drive, and Charlie Strang felt obligated to approach Carl on the concept once more. Strang, as the actual inventor and principle engineer of the Volvo stern drive, was intimately familiar with the potential for the device, but now he had to be most careful when discussing the idea with Carl. Finally, in early 1960, over a year after the *Aquamatic* had been introduced, Carl authorized Strang to at least begin thinking about the possibility of Mercury stern drive. Grudgingly, Carl began to admit that some kind of a stern drive should be in the Mercury line-up. Carl, though, was absolutely convinced the idea would still fail in the marketplace. In a strange sort of reverse psychology, Carl was allowing Strang to begin his own design work, while looking forward to telling him later, "I told you so."

As late as June 1960, Carl tried to negotiate an agreement with Harald Wiklund for Mercury to be the exclusive sales agents for the *Aquamatic* in the United States, selling the Swedish drive at Mercury dealerships. Not even Volvo themselves would be able to make a single sale in the U.S. under the outline of Carl's agreement. Carl wanted Volvo to actively enforce any infringements on the basic patent, so that OMC, or others, wouldn't be able to build a stern drive either, and thus potentially cut into Kiekhaefer Mercury outboard sales in the future. The deal began to unravel when Carl balked at the guarantee of purchasing a minimum of 10,000 units per year for three years.[20] He was sure that the marketplace wouldn't absorb this many of the new drives, especially because at that time Volvo was considering attaching no engine larger than 100-horsepower to the *Aquamatic* package. Had Carl been successful with this agreement with Volvo, he reasoned, he could have avoided the major design and tooling costs of a competitive drive, and at the same time stalled the entry of other stern drives into the market, thus protecting his own large-horsepower outboard sales.

But sales of the *Aquamatic* were in trouble. Harald Wiklund had tooled Volvo-Penta up for an initial run of 10,000 units. At the end of 1959, he had sold less than 3,000, and was embarrassed to be sitting on an enormous inventory of 7,000 stern drives. Though highly touted in the press, and boatbuilders were candidly enthusiastic about the promise of the stern drive, they were reluctant to buy. "All of the boatbuilders said, 'This is something we need,'" Wiklund remembered. "It was easy for them to sell a boat without *any* engine, and the consumer could buy an outboard engine. So I had a *big* problem in the beginning."[21] Wiklund turned to his old friend Captain Botved, the father of Ole Botved who had crossed the Atlantic with Jim Wynne in the *Coronet Explorer*. He made a deal with Botved to make *Aquamatics* available, interest free, for a full year if he would install them in his line of *Coronet* boats. "They started installing the *Aquamatic* in their boats, and stopped selling the boats without engines. It was a success, and after two years, we began to have a very good success with the inboard-outboard. It seemed like the Americans were afraid to go with it at first."[22]

At the Mercury sales meeting in the fall of 1960, Carl admitted the possibility of a future Mercury stern drive, but again, he characterized the stern drive concept as an almost sure loser. "The outboard-inboard may well come into the picture next year, though the Volvo undoubtedly will not be a factor. Our plan will be to let someone else go first, someone else to try the rotten stairway, someone else to walk into the haunted house, someone else to walk first out on the thin ice."[23] It is clear from his remarks that Carl had absolutely no confidence in the prospects for public acceptance of the modern stern drive. Yet, as pessimistic as Carl was, when he spoke to his distributors a few months later on January 13, 1961, he threw down the gauntlet to consumers. If they wanted it, he would build it.

> "Surely it's no great invention to put another kink in the drive line of an outboard motor and make a zigzag drive. ... I don't think much is going to happen to the outboard industry over night after it has been developed to this point over a period of some 50 odd years. I think a little wash-out period will clean out the industry. I don't believe at this time it is necessary to completely change the method of propulsion. ...
>
> "However, when the customer really starts wanting and ordering these things, that's the time to look into it. But up to now it seems to me that most of the interest has been generated by just someone else [Volvo] who wanted to get into this over-advertised, over-promoted marine industry. In view of that, do you still want an outboard-inboard drive? If you folks tell me that you want one, we're ready to build one.
>
> "Perhaps we ought to bury the hatchet on the outboard-inboard drive and market it. Maybe we can save ourselves a lot of headaches. And if the public wants to buy it without any engineering reasons like they buy golf bags and like women buy hats, perhaps we shouldn't knock the idea. If the public wants to buy it, we should build it. If this is the age where people do such things, perhaps we're wrong in condemning them. We've debated it many times. Should we blast this thing out into the open? With a few well-placed ads I'm sure we could cure the whole idea of the outboard-inboard drive. But is that a good way to do business? Is the public tired of outboard motors?"[24]

Charlie Strang continued development of a Mercury stern drive. Fortunately, Carl didn't think it was unusual that Strang could move so quickly with the drawings. Strang wasn't the least concerned about infringing on the Wynne patent or on the Volvo license, because Wynne and Volvo were clearly aware that the drive was Strang's idea in the first place. Carl, of course, was unaware of the subterfuge, and when Strang's mechanical design for the outdrive unit started to look a lot like the Wynne-Volvo design, Kiekhaefer could only assume that Strang knew what he was doing, and that somehow, Kiekhaefer Corporation would be protected. What Carl would never know, however, was just how completely protected they really were. "He was concerned about patenting it," Strang recalled. "Well, I wasn't the least bit concerned about patenting it, 'cause I knew damn well that Volvo would never open their trap if I was involved in this thing, because they knew where it came from."[25]

Through the Kiekhaefer network of spies and informants, Carl learned that OMC was working on their own version of a stern drive. Though most details about the OMC drive were lacking, he discovered they would announce it to their dealers at the Chicago Boat Show on March 25, 1961. Carl then moved up the dates for the announcement of the Mercury stern drive to a press luncheon March 23, followed by a large display in the Mercury booth on the opening day of the show, March 24, beating OMC to the punch by two days. A week before the press announcement, the name *MerCruiser* was selected for the Mercury stern drive.[26]

When OMC announced their new V-4, two-cycle, 80-horsepower stern drive, dubbed the *OMC-480*, Carl was jubilant. His *MerCruiser* was designed to attach to four-cycle automobile style engines ranging from 125- to 200-horsepower. Even his new 80-horsepower outboard could easily out perform the new OMC stern drive. OMC added a number of deluxe features to their stern drive, such as electric shifting and an automatic oil mixing system, in order to make up for the small power, which in turn raised the price to $900, a price comparable to the 80-horsepower Mercury outboard.

Carl, still uncertain of the future for the stern drive, elected to take a most conservative approach by offering the *MerCruiser* only to engine builders, who would then package engine and drive together for sale to boatbuilders. Kiekhaefer Mercury would then warrant the drive to the engine builder, who would then be responsible for the consumer warranty. So certain was Carl that the stern drive was a fad that would soon go away, that he didn't even want his own warranty department involved. The only places where boatbuilders or consumers could buy the first *MerCruiser* units scheduled for delivery in the late fall of 1961, would be from eight different engine builders, such as Chrysler, Daimler-Benz, Crusader Marine, Dearborn Marine and Gray Marine. Mercury dealers, of which Carl had nearly 3,500 devoted merchandisers, could care less about the new drive, because they wouldn't be able to offer the *MerCruiser* themselves. It seemed to Charlie Strang and Tom King that Kiekhaefer was going out of his way to make the *MerCruiser* marketing effort fail.

Strang remembers when he was finally able to convince Carl Kiekhaefer to take the new drive seriously.

> "My mother and I were walking down Main Street in Oshkosh one night. There was a new little Chevrolet in the window of Gibson Chevrolet, and I wandered in and looked at it. It was a little four-cylinder engine, the first Chevy II.
>
> "I dragged Carl in there the next day, and I said, 'Hey, we ought to package this with the stern drive. ...'
>
> "'Why don't you get a hold of somebody at General Motors,' Carl asked. So I called the guy who was then in charge of Chevy, Ed Cole ..., he had not yet moved up to the presidency. We knew Ed from the auto racing days. A fellow ... called back that afternoon and Carl and I flew over [to Detroit] and brought home the first engine on the plane. That's how the *MerCruiser* package came about."
>
> "Carl still didn't like the thing, even after we had it going pretty good. He'd bitch and he'd rant and rave about it. Finally, one day I said, 'Carl, if you don't like this thing, give me a price on it. I'll find someone with the money and I'll buy it from you.' Never heard another complaint from him. He figured if I was ready to do that, it was all right."[27]

Little did Carl know, Strang and Charles Alexander had already tried to raise capital to develop the drive with Jim Wynne. Kiekhaefer was also unaware that his top executive had now designed the two most popular stern drives in the business: both Volvo and Mercury. Carl, though, was going to take advantage of the strategic errors committed by both Volvo and OMC. Both had married their stern drive device to low-horsepower engines, limiting their marketing appeal considerably. Once a deal was consummated with General Motors, Carl was able to offer his "*MerCruiser* Stern Drive Power Package" with dependable 110- and 140-horsepower marinized Chevrolet engines, designed and built by the best minds in Detroit.

To consolidate his hold on the high-power segment of the industry, Kiekhaefer had Strang designed a second drive, somewhat stronger and with advanced features, to handle 225- and even 310-horsepower engines. The two drives became known as the *MerCruiser* I, and the *MerCruiser* II. Over two and a half years late in entering the market with a stern drive, Mercury, by the end of 1961, had captured the bulk of the market by offering two models and a wide, powerful range of engines. Orders from boatbuilders began to mushroom, and within the first year on the market, *MerCruiser* orders began to pour in from virtually every established boatbuilder in the country.

Though Carl had been consistently opposed to the stern drive from the moment Charlie Strang first disclosed his idea in 1951, Carl was now swimming in the glory of having the hottest new product in the industry. By 1962, as OMC would later document, only three years after Volvo-Penta's introduction of the *Aquamatic*, no less than "sixteen manufacturers were producing stern drives ..."[28] Once Carl opened up the sale of the new drive directly to the over 2,500 boatbuilders in the United States, the *MerCruiser* would outsell all other stern drives combined, and eventually capture an incredible 80 percent of the world-wide market. For all of his work in the outboard motor trenches against *the enemy*, it would be the stern drive, the product Carl never wanted, that would eventually push Mercury ahead of OMC as the world's largest manufacturer of marine propulsion. OMC would maintain their lead in strictly outboard production, but because of the *MerCruiser* stern drive, Mercury products would power more boats than any other brand in the world. Carl's dream of overtaking OMC was finally coming true.

The conspiracy of friendship between Jim Wynne and Charlie Strang had actually worked to Carl's advantage. This secret relationship made possible Carl's own entry into the stern drive market, after letting someone else "go first, someone else try the rotten stairway, someone else walk into the haunted house, someone else to walk first out on the thin ice."

Wynne's patents wouldn't actually issue for ten long years, because remarkably, another individual, unknown to Strang, Wynne or Alexander, had filed an application only two weeks after Wynne. More confusing to the patent office, the other inventor, C.E. MacDonald of Seattle, had actually constructed a similar prototype *before* Wynne had tested his in Miami. After 10 years of engineering and legal debate, the patents, based on Charlie Strang's universal coupling, were issued to Jim Wynne on April 9, 1968.[29] Wynne would be rewarded handsomely by Volvo under terms of the licensing agreement reached in the summer of 1958. But, during the entire seventeen year life of the patent, which didn't expire until 1985, Charlie Strang would never accept or receive a cent for his invention. "Charlie once said to me," Harald Wiklund recalled, "You gave Jim a lot of millions, but I didn't get anything."[30] Strang would eventually become chairman of the board and chief executive officer of OMC during Carl Kiekhaefer's lifetime, and swore an oath of silence to never disclose the incredible story behind the invention of the modern stern drive while Carl was still alive.

CHAPTER TWENTY-SEVEN

Crisis of Control

"No one more than I can attest to the wisdom of 'walking alone' and 'not running in packs.'"[1]

Carl Kiekhaefer, 1962

From all appearances, it should have been known as the golden age of the Kiekhaefer Corporation. In one, glorious twelve-month period, Carl Kiekhaefer introduced the *MerCruiser* stern drive, sold the company to Brunswick, and then capped it off by introducing the first production 100-horsepower outboard motor in the world. Everything seemed to be going Carl's way. OMC would announce that they would discontinue the manufacture of private brand outboards, shutting down the production lines for Gale, Buccaneer and Sea King for all time. Carl had long predicted that OMC would abandon private label outboards, feeling that Johnson and Evinrude dealers didn't appreciate competing against these virtually identical OMC products sold through catalogs and chain stores without regard to territory or price restrictions.

Yet, within weeks after Carl closed the sale of the Kiekhaefer Corporation to Brunswick, a crisis of control began that would kindle Carl's paranoia to a roaring emotional blaze. Brunswick Corporation was a large, unwieldy, diversified organization with an entrenched bureaucracy managing such unrelated divisions as bowling, health and science, sporting goods, school equipment and defense products. As a management principle, Brunswick subscribed to the conservative, methodological and slow-moving canons of the American Management Association. In fact, Ted Bensinger was sufficiently enamored of AMA techniques, that L.A. Appley, the president of the New York based organization, was elected to Brunswick's board of directors. Carl's organization, at the other extreme of management practice, was a lightning fast, lean and aggressive organization, highly disciplined in the rigors and tactics of industrial and competitive guerrilla warfare, skills that had effectively broken a thirty year lead and virtual monopoly enjoyed by OMC in the marine industry.

Carl made intuitive decisions based on his decades of experience in the volatile marine marketplace, and each decision took the form of an unquestionable pronouncement and mandate from the king to his subjects. Brunswick, by contrast, submitted recommendations to committees for evaluation and analysis, and engaged in long-term planning and policy conferences. To Carl, they seemed to operate in a kind of heartless, direction-less and anonymous vacuum filled only with endless meetings, administrative panels and discussion groups.

Carl was, of course, well aware of the potential for conflict when he finally agreed with Ted Bensinger to become a part of Brunswick. Next to the actual price of the transaction, the most important bargaining issue between them was the status that the Kiekhaefer Mercury organization would enjoy following the merger. Carl demanded, and received, a side letter from Bensinger promising that as a wholly owned subsidiary of Brunswick Corporation, the Kiekhaefer Corporation would continue to operate as it did before, independently and without interference from Brunswick management. Carl intended to have Brunswick live up to their end of the bargain completely.

One of the most apt descriptions of Carl Kiekhaefer's management technique is contained in a reluctantly drafted letter of resignation from R.A. McCarroll, who quit while the ink on the Brunswick agreement was still wet. McCarroll had joined the Kiekhaefer Corporation in early 1961, having diligently worked his way up through the production ranks at Chrysler to become head of the Detroit giant's engine division, then becoming Carl's general manager, with authority over nine Kiekhaefer plants. But after only seven months of reporting directly to Carl, he was flabbergasted. Nothing at the carefully controlled, layered bureaucracy at Chrysler could have prepared him for the rough and tumble, iron-fisted control of Carl Kiekhaefer.

> "Personally and socially you are one of the finest and most considerate friends I have ever had; but business-wise, you become a tyrant with little or no patience or understanding of people's feelings or problems. I get the impression that you wish an employee of yours to heart-and-soul belong to you, twenty-four hours per day, seven days per week and you do not hesitate to let him know you are unhappy if they aren't available to you at any time. This is adverse to our free

American way of life and modern democratic management philosophy. Free American people do not like to feel owned -- they do not mind, if they are at all conscientious, working all hours when absolutely necessary if they could feel it is voluntary not mandatory. It is necessary in today's modern, highly competitive business world for managers to have free time to relax or they become stagnant. It is top management's job to plan their work and schedules to accomplish this necessary factor for the mental and physical sake of their employees.

"Another very serious problem is the feeling of lack of trust and confidence that you put across to your people, either knowingly or unknowingly. It is quite common for you to ...

1. Give two or three people the same assignment.
2. Give assignments to executives out of their areas of responsibility or specialty.
3. Give assignments to subordinates without the superior's knowledge (sometimes all the way down to hourly employees).
4. Reprimand an executive or supervisor in front of others or in front of their subordinates.
5. Criticize other executives or management people to others without proper information or knowledge of the facts.
6. Criticize or countermand an executive's decision, right or wrong, to show you are the boss (usually in front of other people).
7. Deliberately call meetings, necessary or unnecessary, on Saturdays, Sundays, holidays and weekends to upset any possible plans some executives may have -- primarily for 'roll call.'
8. Lose your temper in front of all levels of employees -- executives, managers, salaried and hourly.
9. Lack recognition of levels of management responsibility and equality in regard to assignments, salary or benefits.

"All the above mentioned items are certainly uncalled for, but maybe you have your reasons -- I do not know what they could be. I do know that any experienced, professional, matured man in today's day and age will not tolerate these conditions unless 1) he is paid more than he could get somewhere else, or 2) he is tolerating it due to close personal feelings for you. In any event, this will only lead to demoralization, unhappiness, lack of confidence and self-respect, and distrust of the management team. ...

"I do not see any sense in elaborating any longer on the subject. I have tried only to hit the highlights. ... [T]he road ahead for this corporation, with its size and complications, is going to be rough and rugged, particularly with the operating requirements of a publicly-held corporation. You will find their way of operating completely different than the way you are accustomed to operating. We do not have either the management philosophy, organization structure or, in some cases, the personnel to handle the requirements they will need, in my opinion.

"In closing, I would like to say I am certain, if you, for your own happiness, will just have more confidence in people and practice the 'golden rule' (treat others as you would like to be treated) and take the time to listen and help them with their problems and decisions without losing patience and flying off the handle, everyone would be much happier. Because, after all, you're not the big roaring lion you would like people to believe. Way down deep you have a big heart and warm feelings for people and they already know your bark isn't as bad as your bite and you are only fooling yourself if you think they don't."[2]

With the introduction of the MerCruiser stern drive and the 100-horsepower outboard, Charlie Strang was convinced that Mercury was finally in a position to challenge OMC for supremacy in the industry. Only one week after Carl had signed the Brunswick merger papers, Strang was jubilant in describing how OMC, displaying a growing engineering weakness, had copied many Mercury design features.

"We find that they are still trying to climb aboard the Mercury bandwagon without admitting it. Once again they have copied or imitated some of our older features in one form or another. Now -- in the Russian manner -- they are claiming these features as 'firsts' and are devoting most of their sales pitch to these so-called 'firsts.'...

"If this report of new competitive outboard features seems brief, it is solely because the competitors don't really have much that's new to talk about. Contrast this with the Mercury line for '62. The whole line -- not just one or two models -- offers single lever controls, fixed jet carburetion and the exclusive Jet Prop exhaust. New horsepower in four of the seven basic models. Major changes in the famous fours. A brand new, ultra-compact, 10 that will completely capture the ten-horse field. And, despite competitors' propaganda saying it is impossible to build outboards greater than 75 or 80 horsepower, we have broken the barrier with the Merc 1000 - a 100 horsepower beauty that is lighter and more compact than their 75's!

"1962 will be our year!"[3]

It was uncharacteristic for Strang to wax so ecstatically about Mercury's success against *the enemy*, but OMC's engineering hibernation coincided perfectly with Mercury's own string of successful advances and he could hardly contain himself. Strang's own star continued to rise, and in October 1961, he was elected president of the American Power Boat Association (APBA). Capping off a 25 year association with the APBA, Strang had for years been the APBA's chief measurer and inspector, and had been chairman of the stock outboard technical committee and a member of the outboard racing commission. Jimmy Jost, Strang's close friend and fellow Mercury employee, was elected senior vice president of the APBA. Again, Kiekhaefer Mercury had scored a decisive victory against OMC, for the visible prestige that would accompany the Strang and Jost elections was further proof of Mercury's continued dominance in racing.

In the basement of the engineering research building in Oshkosh, a tradition began in 1961 that would literally change the face of Mercury outboard motors for all time. The stark-white Mercury outboards had been the standard signature for Kiekhaefer Mercury products, since Carl had abandoned his program of offering a wide range of outboard colors to consumers during the mid- to late-50s. This program reached its zenith in 1958, when Mercurys were offered in such exotic decorator colors as Marlin Blue, Gulf Blue, Sunset Orange, Tan, Sarasota Blue, Sand, Mercury Green and Silver combinations. Carl had advertised Mercury as "The Most Colorful Name in Outboards." Carl also claimed that Mercury had "taken outboards out of the dark ages and put them into the bright, beautiful age cf color -- and you know color is here, and *will* stay."[4] OMC had also gone through a similar, nearly psychedelic phase of colorful paint schemes, and were by the early 60s standardized on white for Johnson outboards, and white and blue with red trim for the Evinrude line.

The introduction of the six-cylinder in-line Mercury engine had spawned a number of industry jokes about its tall, nearly soaring appearance compared to the stodgy and rotund products of the competition. "Tower of Power" was the most complimentary of these characterizations, while the "U.N. building with a girdle" was the description of the new engine most favored by OMC. The competition delighted in chiding prospective six-cylinder Mercury buyers with the possibility that the new engine was too tall to tilt up properly on some models of boats, and could even tip over if you tried to turn too sharply. Of course, none of these rumors were true, but the sting of competitive criticism weighed on the minds of Charlie Strang and Carl Kiekhaefer.

As the original 60-horsepower engine grew to 70-, 80- and finally to 100-horsepower by the summer of 1961, the engine seemed to grow taller and taller, until even to Mercury designers the new model began to look somewhat out of proportion. Strang and a small crew of engineers and stylists were trying desperately to figure out a way to make the new 100-horsepower outboard prototype look smaller and more compact. Their meeting lasted well into the night, and as the midnight hour approached, Ann Strang, Charlie's mother, "wandered into the building to find where her long lost son" had gone. Charlie was explaining to her what they were trying to accomplish with various cowling configurations, when Ann galvanized the group with a simple observation. "Well, a large woman always wears a *black* dress. Why don't you paint it *black*?" "Just for the heck of it," Charlie said, "we painted one black, and the engine looked like it shrunk about twenty percent!"[5] To prove his point, he prepared a demonstration for Carl. He had one painted completely black, except for chrome highlights around the cowling and shock absorbers, and another was prepared all white with the same trim accents. He covered both engines with cloth and invited Carl to the basement without telling him the purpose. "I said, 'I want you to see something,' Strang offered mysteriously. "We ripped off the white one, and then we ripped off the black one. He said, 'My GOD! Let's paint *them all* black!' That's how they became black, and it has never changed since."[6] It actually took over two years to eliminate the last of the white models in the Mercury line-up, but with the introduction of the 1964 engines, Mercury's 25th anniversary, every Mercury was black.

Introduced in the fall of 1961 as a 1962 model, the world's first 100-horsepower production engine was originally named the Phantom, due to its very mysterious and somewhat sinister appearance and power. But after double-page ads were placed in the nation's leading magazines announcing the great leap forward in the race for outboard horsepower, an insurmountable legal conflict with the name ensued, and thereafter, the trailblazing engine was simply known as the *Merc 1000.* It was Mercury's first black engine; it would be the last engine Carl would build before he sold his company to Brunswick Corporation, and conversely, the first engine built under Brunswick ownership. As the first 100-horsepower engine available to the public, it was also a significant milestone in the evolution of the outboard motor. For these many reasons, it was selected as the engine depicted on the dust jacket of this book.

Two years prior to the introduction of the *Phantom,* when the *Merc 800,* 80-horsepower engine had become the world's most powerful, Carl confidently announced that the sheer size of the motor would make a manual starting rope obsolete.

> "I'm sure you will agree with us that there must come a time when the engine size is such that not even a strong man can crank it over, much less a woman or youngster. It is many years since automobiles, trucks and tractors were cranked and you realize, of course, that inboards have never had cranking means.
>
> "Hand-cranked engines have always been a phobia to me. In my younger days on the farm, I remember going to the funeral of a man who lost his life hand-cranking a farm engine. I saw my father painfully injured hand-cranking a farm tractor. ... I saw my grandfather with a broken arm after having tried to crank a Model T. I saw an enlisted man's hand shattered trying to crank a 35 h.p. target plane engine. I still remember almost losing my pilot's license for hand-cranking an aircraft that had no one in the cockpit.
>
> "Many an operator of a 22 h.p. outboard has slipped, lost his balance, and has fallen out of a boat, hand-cranking an engine that might not even have been balky. One of the biggest dangers of a two-cylinder 30 h.p. engine was its hand-cranking, because its automobile-sized pistons had the kick of a mule. I personally have experienced a backfire ripping the rope out of my hand, leaving me with the feeling that my arm had come out of its socket. Put this possibility into the hands of inexperienced and young operators and you have a real hazard. ...
>
> "If an engine won't start with a good battery, certainly it won't start with hand-cranking. ... Let's remember this. The manual starter and the auxiliary sheave for a starter rope are

as ridiculous on a high powered outboard motor as a hand-crank on a high powered inboard. ... Throw away the crank!"[7]

As logical as his arguments against hand-cranking large engines were, Carl's distributors and dealers were equally adamant that consumers wanted the back-up insurance of manual starting in case of dead batteries. So, when the *Phantom* was rolled out in the fall of 1961, poking out of the top of the massive black cowling was a diminutive pull handle, a temptation avoided by all but the strongest professional wrestlers and football linebackers. Not until the fall of 1964 was the decision finally made to eliminate the starting handle on all six-cylinder engines, starting with the 1965 line.

The first outward signs of the coming clash of management styles between Carl Kiekhaefer and Brunswick surfaced less than two months following the merger. On January 5, 1962, Carl sent a terse memo to all department heads, mail clerks and switchboard operators at all plants.

> "This memo is being issued as a reminder that all phone calls and correspondence from the Brunswick Corporation must be cleared through the Office of the President. This means incoming and outgoing calls and mail. After calls and correspondence have been screened they will be referred to the proper parties for handling.
> "This policy must be adhered to without exception."[8]

Carl was effectively shutting Brunswick out of Mercury. No communications with the parent organization were allowed, except through Carl's office. After paying over $34 million for Kiekhaefer Mercury, Brunswick wouldn't even be allowed to speak to any of the over 4,500 employees except for Carl. Brunswick executives, though naturally stunned by the announcement, were intimidated by Carl's overpowering presence, and carefully acquiesced to his unusual request. A few weeks later, Ted Bensinger informed his own administrative staff in a cautiously worded memo to "advise Carl who your deputies are and give him a brief description of their functions. This will help to expedite and improve communications between our two groups."[9] Soon, Carl would even forbid access to all plants by any Brunswick personnel, unless first cleared by him personally.

An avalanche of memos, phone calls and requests for reports poured into Carl's office starting within days of the merger. Carl anticipated the routine demands for progress reports on Mercury sales and production, but he wasn't prepared for the diversity of time-consuming and seemingly worthless requests that began to rise in a great pile on his desk. Suddenly, it seemed to Carl, everyone at Brunswick had a better idea for an outboard, or how to distribute, or how to save money. Carl began to lash out at intrusions into his empire, with growing levels of frustration and anger. "While repeated requests have been made that all communications between the two companies be through [myself]. ...," Carl wrote, "attempts at direct contact with Kiekhaefer operating personnel continue. This creates conflicting demands upon executives' time, raises questions as to project priorities, and in the end slows down completion of whichever project should have priority. ..."[10]

Soon, thoughts about how Carl's manufacturing facilities could benefit other Brunswick companies began to appear on Kiekhaefer's overflowing desk. One such suggestion, that Carl die cast aluminum or magnesium bowling pins for the Bowling Division required Carl to prepare a long and technical reply. Among the many reasons why Carl didn't want to get involved, was that Brunswick policy put a very low ceiling of nine percent on any profits that one part of the organization could make on another. Dozens of studies were prepared by Carl for dozens of Brunswick generated ideas for manufacturing cooperation, and each time Carl would politely but firmly refuse. "Is there anything we can do to stop this useless and time-consuming correspondence?" Carl would ask. "I'm sure if I sent a copy of all my replies to the attention of Mr. Bensinger he would never find enough time to read them."[11]

Most irritating to Carl was Brunswick's attitude that since they were so much bigger than Mercury, that they must have greater buying power and influence with Mercury's many vendors. Boxes and boxes of correspondence and purchasing studies resulted in nothing but wasted time and effort. After hundreds of hours of materials auditing and report compilations, Brunswick saved Carl $28.97 on steel strapping from one of his vendors. In another instance, one particularly adamant Brunswick purchasing agent demanded to "help" Carl save money with his corrugated box purchases, and requested that one of every kind of box used by Mercury be sent to him. Most Brunswick personnel had absolutely no idea how large and sophisticated were Mercury's manufacturing operations, involving tens of thousands of parts and a very complex and highly structured assembly operation. To prove his point, Carl filled an entire semi-trailer with box samples and gave the driver instructions to deliver them to the pesky Brunswick staffer at his tiny office cubicle in downtown Chicago. Still, the purchasing agent pressed on, demonstrating after weeks of paperwork involving at one point over six people, that Brunswick buying power could have saved Kiekhaefer Mercury a grand total of $138.00 on office staple orders of over 4 million staples.[12]

Carl began to wonder if the Brunswick purchasing agent, Chester F. Teeple, might really be working for *the enemy*. Carl hired a team of private detectives to investigate his background, checking on any possible ties with OMC, and to see if he wasn't really a communist intent on disrupting industrial America through a hideous plan of bureaucratic waste.

Tom King, saddled with coordinating research for many of the Brunswick inquiries, finally threw up his arms in disgust, asking Carl "... the question must certainly be raised as to how Brunswick wishes our key executives to use their time. It can be devoted to such areas as selling effort, market research, modernizing facilities, research and development, or it can be devoted to defending our policies, steel strapping studies, paper box studies, freight bill reviews, studies on inbound and outbound freight costs, etc."[13] Carl summed up his feelings by writing, "If any studies are to be made, the entire Brunswick organization should be analyzed, laundered and ironed. Unless this is accomplished in the reasonably near future, the prospects with Brunswick look dim. While Brunswick may not fail, it will have a long hard climb back."

Brunswick was indeed in serious trouble. In the 24 months following the merger, Brunswick sales plummeted from $426 million and profits of over $44 million to $315 million and a net loss of over $10 million. The value of Brunswick stock, the only compensation received by Carl for the sale of his company, was falling at a steady and sickening rate. Shock began to set in.

Brunswick, in a headlong dash for pinsetter sales against rival AMF, had been eagerly helping to finance bowling centers across America. For six years, Brunswick amassed a mountain of debt, asking as little as ten-percent down from chain after chain of bowling establishments, and giving them eight leisurely years to pay. In 1962, the bowling boom finally crashed, pinsetter sales fell dramatically, and hundreds of bowling centers informed Brunswick that they could not meet their payments. Brunswick had borrowed wantonly against the expected payments by bowling establishments, and by the end of 1962 was overwhelmed with nearly $360 million in debts. Carl watched in horror as the value of Brunswick stock fell from a high of $75 in 1961 to $13 and still falling within a year, on its way to a low of $6. Carl's dreams of uniting with a strong, rich partner to build the manufacturing facilities he needed to overtake OMC were shattered.

Carl began to look for ways to reverse or somehow nullify the sale of the Kiekhaefer Corporation to Brunswick, alleging that Ted Bensinger was aware of the coming financial crisis and would be unable to live up to their promises. Carl's attorneys began to work furiously on this strategy by the beginning of 1963, attempting to prove that Brunswick had artificially maintained the price of their stock until the merger was complete, and then allowed it to plummet. Following a thorough investigation, however, they concluded that "... it could be argued that during negotiations with Brunswick, Mr. Kiekhaefer could have apprised himself [of the declining fortunes of Brunswick] by investigation of these records,"[14] really placing the blame on Kiekhaefer's own lack of homework before the sale.

Before long, Carl began to hint at total rebellion against Brunswick, or failing that, resignation. He drafted a scathing letter to Brunswick President Ted Bensinger in early 1963, exploding with animosity and feelings of betrayal.

> "In my judgment, I do not need to be crammed with a lot of management ideas and theories to make this division perform better. I do not agree entirely with your theory on 'cross-fertilization' between divisional managers where the divisions are completely different types of businesses with no common grounds of product, marketing, manufacturing, engineering or research and development. ...
>
> "I am sure that this will not be your opinion, but in my opinion, I feel there has been a breach of good faith. ... I am convinced that the future of Brunswick looks pretty glum for some time. ... I feel that a thorough and complete discussion must take place between yourself and the writer in the very

near future if there is anything to be salvaged of our relationship."[15]

The following day, Carl further revealed his agony in a letter never mailed to Howard F. Baer, chairman of the Brunswick Aloe Division and fellow director. "I also realize the tremendous ground one treads on when one doesn't agree with 'Nero.' I am of a disposition that if I cannot be proud, I cannot work at all. I am taking the only way out open to me, so I may again work with dignity, pride and the joy of accomplishments." Baer was among the first of the Brunswick directors that Carl began to brief on his dissatisfaction and suspicions of having been duped.

> "We listened to a lot of glowing dreams of expansion so we could really 'clobber' Outboard Marine, not knowing, of course, Brunswick's precarious cash position. ... Not knowing the tremendous borrowing with its interest drain. ...
>
> "As a couple of country boys, without too much guile, cunning or shrewdness, we were somewhat enchanted by the smooth courtship music. Being somewhat intoxicated with the success of Brunswick and our own, we failed to detect a few false notes. ...
>
> "By now there must be a question in your mind, why did we do it? We did it because we believed that Brunswick was an organization of high integrity and that they were people of their word. That they would not take unfair advantage and that they were sincere when they said, 'Carl, we don't want your brick and mortar, we want you - your engineering and manufacturing, marketing and sales experience.' ... We have kept all our promises. We did better than we said we would do. We owe Brunswick nothing. ...
>
> "As far as my personal angle is concerned, in the past 35 years I have become rather used to adjectives and personality connotations. The fact that I have lost some 26 million dollars on paper in less than a year is, of course, a bit disturbing to say the least, but nothing fatal I can assure you. The fact that the Kiekhaefer Corporation with a net worth of some 18 million is now worth only 12 million on paper, which, of course, is also disturbing. It makes it extremely difficult to re-establish confidence and as a result, I have been forced to attend countless dealer meetings to bolster sales and sagging interest in Brunswick.
>
> "If there are any flustrations [sic] on my part and this I do not believe I am exclusive, is to put on appearances before thousands of people and with ringing confidence, wear my sales

> hat when, at the same moment, you feel helpless -- so completely helpless -- in trying to turn the tide. A feeling of helplessness stems from a lack of communication, from a lack of incentive, realistic incentive, to come forth and do battle with all the forces that are now detrimental to the welfare of Brunswick."[16]

By the summer of 1963, as Brunswick continued a dramatic slide to their worst loss in 120 years of business, Ted Bensinger began to be inundated with hints from bankers, associates and family members, that he needed outside help to get through the deepening crisis. One of the names that kept cropping up was that of Jack L. Hanigan, the executive vice president and general manager of Dow Corning Corporation. Hanigan, the short, husky and tough, 52-year-old administrator, had spent the last 26 years at Corning Glass Works and Dow Corning, working his way up through the organization in various assignments, including one as general manager of their Electrical Products Division. "My initial reaction was not to join Brunswick," Hanigan would admit. "A mutual friend set up the meeting between myself and Ted. I liked him a lot, and his company's problems presented some interesting challenges."[17] Just as Ted Bensinger had worked his persuasive magic on Carl Kiekhaefer, so too was Jack Hanigan soon won over to the Brunswick cause. On November 5, 1963, Hanigan was elected president of Brunswick, while Ted Bensinger moved up to chairman and chief executive officer following the retirement of his brother and Chairman R.F. "Bob" Bensinger. It was a traumatic moment for the Bensingers, for except for a short time before the turn of the century, Brunswick had been guided by 120 years of exclusive Brunswick family leadership.

The tone of any potential relationship between Jack Hanigan and Carl Kiekhaefer was unfortunately set in concrete within moments of their first meeting near the end of 1963. "So," Hanigan said brusquely as he shoved his hand in Kiekhaefer's direction, "You're the son-of-a-bitch that's been making all the trouble here." Carl was absolutely stunned. While the Mercury division was setting meteoric sales records and breaking all profitability predictions, here was the new president of Brunswick -- a company falling on its corporate face -- calling him a son-of-a-bitch. According to onlookers, it was all Carl could do to keep himself from hauling off and punching Hanigan.

It was only the first of many clashes between the two, whose similar personalities were to provide a constant shower of fireworks between the two organizations. Hanigan was just as stubborn as Kiekhaefer, and would often go to great lengths to impress on Carl just how important, or how deserving he was of his position. Carl had been negotiating with the Rover Company in England concerning a possible license for manufacturing and distribution of small turbine engines for use with *MerCruiser* stern drives. When Carl was able to persuade Hanigan that a personal visit was necessary to iron out details of an agreement, to his horror he was told that Hanigan would accompany him on the trip. The trip turned into an extended tour of European centers of industry, and Hanigan drove Kiekhaefer crazy with constant sight-seeing and a penchant for long, complicated and rich meals of every description. Carl scribbled a quick note to Rose Smiljanic aboard Lufthansa Airlines en route to Berlin from Frankfurt, Germany, that after only a week, he was ready to strangle Hanigan, sell his stock and quit.

Dear Rose,

"I would have paid $5,000 to avoid this trip. Hanigan would rather travel than eat. I have been dragged all over hell, thru all kinds of resorts, spas, eating places he's been before. I'm so tired of hearing of other places he's been -- people he knows, I'm about ready to blow my top! And funny, he thinks he's doing me a great big favor!

"I must quit -- I cannot stand much more of it! He calls his wife everyday at Co. expense. I sure wish this week were over. Diet gone to hell, weight up 15 lbs. - back on sugar - it's just G.D. awful. I just can't & won't travel like this anymore.

"I'm out of laundry, Vitalis, etc. I had 3 kinds of money today, Swiss, Italian & German & English.

"The trip is every bit as bad as we expected. I'm so dam tired of 5 kinds of champagne, wine, cheese and these god dam dining excursions & late meals after which you can't sleep. The sleeping pills no longer have any effect.

"Pray that I will survive. Have Anderegg check advisability of selling another 200,000 shares before it drops below 8. I'd pay $500 for a good nites sleep!

Sincerely and with love,
E.C. Kiekhaefer[18]

The longer he interfaced with Hanigan, the deeper his hatred became. The rivalry for macho dominance between the two powerful executives became so intense that staffers in both organizations would literally run down the hall whenever the two men approached one another, scrambling to get out of the line of fire. Carl had written some deeply critical letters about various people during his long career, but when he sat down to organize his thoughts about "Black Jack" Hanigan, pure poison flowed from his pen.

A QUICK PROFILE OF JACK HANIGAN

1. In his glory, traveling.
   a. Has been everywhere.
   b. Has eaten everything.
   c. Drank everything.
   d. Met everyone.
2. Tourist
3. Gourmet
4. Linguist
5. World Shopper
6. Echo of Ted Bensinger
7. Has Built-in Inferiority Complex
   a. Lives in the past. Mostly Corning Glass. Butts into every conversation. Can't remain quiet in any technical discussion.
   b. Must dominate every conversation, bluffing if necessary.
   c. An Irish Bully.
   d. International Playboy who never gets to bed before midnight.
   e. Dinner is always a project which gets started about 9 p.m. with martinis on the rocks with hors d'oeuvres, 2 or 3 kinds of wines, 3 or 4 kinds of cheese, finishing with a cordial.
8. Is critical of others in front of subordinates.
   a. Criticizes division managers.
   b. Criticizes directors.
   c. Criticizes staff subordinates
   d. Criticizes corporate officers (all behind their backs)
9. Not behind one's back, but to one's face, criticizes:
   a. Me.
   b. Members of my family.
10. Is extremely nosy about my family and my family affairs.

11. In periods of remorse after, no doubt, feeling he over-played his hand, makes much small talk, repeating old jokes and trying to make himself liked; and depending upon the amount of alcohol seeking to be a good 'feloo.'
12. After an evening on the town, wakes up grumpy as a bear and expects all of his employees to be front and center like buck privates.
13. Understandably, of course, since he is a walking drug-store, he takes pills as:
    a. Tranquilizers
    b. for gout
    c. for heart palpitations
    d. diaperticulosis [sic]
    e. sleep
    f. pep
    g. cathardic [sic]
    h. diarrhea
14. After 10 days of traveling with this man, he was supposed to be on a diet, one is apt to become a complete nervous wreck, if not nervously exhausted.
15. While he is accusing one of being 'foggy' in logic, he has the habit of asking questions, yet never waits for an answer before asking the next. If anyone is foggy, it is he.
16. In conclusion, I am convinced:
    a. He has never had many friends.
    b. He has a terrific inferiority complex.
    c. He is trying desperately to make new friends.
    d. He is yearning to make his job at Brunswick last and in moments of weakness, admits as much. I feel genuinely sorry for this man, a man who has no talents, no skills, no technical or scientific knowledge, a man who is very unsure of himself, a man who is bullying others around, a man who has to stand on the shoulders of others, a man who makes fun of money and those who have it, a man who ridicules anyone like myself who loses sleep nights worrying about the future of Brunswick, a man who I have learned to dislike thoroughly as a phony, as a bully and a man who has never done a stitch of creative work in his life.
17. I do not know how long I can stand this person but most important, God help Brunswick. It is but a matter of time before I must leave, unless he leaves first. ..."[19]

Given this incredibly caustic opinion of Hanigan, it must have been a source of unimaginable pain for Carl to beg Hanigan for money to modernize Mercury's manufacturing facilities, and to continually lobby him for the expansion capital that he was originally promised by Ted Bensinger. Carl assembled long, powerfully substantiated letters to Hanigan, pleading with him to release funds for new die casting facilities, more room for manufacturing and inventory, and for a host of essential machines crucial to new production techniques and model changes. To Carl, it seemed like Hanigan delighted in delaying every conceivable request, or worse, endlessly vacillating between approval and rejection.

Carl was incensed at Hanigan's refusal to release funds, when Mercury was having record year after record year, pouring profits into the parent company. Hanigan would occasionally receive a letter of some minor complaint from an owner of a Mercury outboard, and would seemingly delight in sending it along to Carl with some cryptic note scrawled across the bottom, like, "Better look into these problems, Carl." Carl would then reply by lashing out at Hanigan, sending bitter and flame-fanning letters in return that would, naturally, make matters even worse.

> "We would appreciate and would probably react accordingly if a pat on the back would come this way occasionally rather than the well-aimed kick in the pants that makes us feel as though we have been guilty of great oversight and that we have been lax and careless in inspection and manufacture and utterly stupid in engineering."[20]

It eventually got to the point that even when good ideas were put forth by Brunswick, Carl would routinely dismiss them rather than get any more deeply involved with their bureaucracy. Charlie Strang remembers when Ted Bensinger, then chairman of Brunswick, was desperately trying to get Carl to approve a pension plan for Mercury employees. Carl had never established a pension plan of his own, and Bensinger felt that it could only benefit Mercury employees. Strang recalled the ruse he had to employ to get Carl's attention on the matter.

> "Carl didn't believe in pensions, so you worked there but there was no pension whatsoever. And Brunswick had a very good one. [Bensinger] was a kind of a philanthropic type of guy, and thought this was a good thing for the company and the employees. Carl would say, 'Get out of here with that damn circular thing!'

"I remember Ted called me one day and said, 'Charlie, you got to get that pension plan in order.' I thought it was a great idea too. Carl was in lake X doing something with the bulldozers and I had a stroke of genius. I said, 'Ted, have someone calculate what Carl would get at retirement.' He said, 'I'll be back in twenty minutes.' He said, 'At retirement, Carl would get $55,000 a year,' which was great money in those days, early 60s. I said, 'O.K., I'll get back to you.'

"I called Carl. They finally dragged him in from wherever he was. I said, 'You know, that pension plan is kind of interesting. Do you know how much you'd get on retirement?'

'No, how much?' I said, '$55,000 a year.' He said, 'Forever?' I said, 'Yeah.' He says, 'Why don't you get a hold of Ted and tell him to go ahead.'"[21]

Strang also remembered Bensinger's attempts to involve Carl in the American Management Association seminars, a mandatory pilgrimage for Brunswick divisional managers. "Bensinger and his boys would go around saying, 'We are pros, we are managers,'" Strang says. "You didn't have to know what you were doing as long as you understood management." One of the meetings was scheduled at the headquarters of the AMA in New York. Bensinger first asked, then pleaded with Carl to go. Finally, when Bensinger *demanded* that Carl attend, Kiekhaefer volunteered to send Strang instead.

"I got flown off to New York. This is in the middle of no where, the middle of winter, and the snow is eight feet high. They had a bunch of little cottages and a big round building in the middle with a huge round table. It must of been thirty-five feet across. We sat around this round table and the guy who was head of the American Management Association had a home near by. Well, he didn't come to the meetings, but he had a loud-speaker up in the ceiling and a microphone [at his house] and he would talk to us!

"Carl wanted me to call him every night and tell him what was going on, which I'd do, and he'd love what I was telling him because I felt the same way he did about this thing. We were flying home from that meeting, and Brunswick had a psychiatrist on the staff. He said to me, 'How can I get with Carl? I think in two years I can make something out of that man.' I'm looking at this jerk psychiatrist sitting there, and *he's* going to make something out of *Carl Kiekhaefer!*"[22]

Actually, Brunswick was quietly beginning to consider ways of getting rid of Carl, and because of Carl's unique personality, they imagined that they could somehow discredit him through a negative psychological profile. The first approach by the psychiatrist was made in the middle of the New York Motor Boat Show, when Carl couldn't have possibly been busier, or had more things on his mind. Naturally, Carl brushed off the "headshrinker," and no doubt left a wonderful impression on the man. Carl wrote notes to himself about this persistent and nosy Brunswick doctor, saying that he would "make his own appointments, without explanation by anyone in Staff as to nature or purpose of interviews. Unknown also at whose direction the examination, causing anxiety, distrust and anger. It knocks out spontaneous and normal creative thinking for long periods of time." This psychiatrist admitted once that "Kiekhaefer is an enigma to Bensinger. Can't understand Kiekhaefer's good performance."[23]

Carl's spirits had nearly bottomed out when he scribbled a long handwritten note to himself, obviously racked with despair, summing up his deep feelings of resentment and remorse.

> "All of these factors, coming from so many directions, are of such magnitude, and have such a titanic effect on the personal life, liberty and welfare of an individual that they not only drain enthusiasm and incentive and dilute the concentrated effort required to operate a successful business, but have literally destroyed lesser men. ...
>
> "We were promised complete autonomy of operation -- which we had before the merger. In fact, to sum it all up, we are doing nothing more now than we could have done alone, without Brunswick. We would have had no difficulty borrowing short-term money as we always had from our own banks; we could have gone through with our preliminary discussions for long-term financing for any major capital expenditures <u>and</u> we still would have owned the company.
>
> "I can not think of <u>one</u> single advantage which the Brunswick merger has produced."[24]

CHAPTER TWENTY-EIGHT

Days of Darkness

"We have been running like an 8-cylinder competition engine here at Fond du Lac, and you should hear the screams if one of the cylinders misses occasionally. ... The dividing line between utter despair and the highest exultation is a razor-sharp line, and I am getting cut up a little crossing it. It is but a matter of time before I can no longer take the pressure."[1]

Carl Kiekhaefer

Carl had assembled his key executives for the weekly staff meeting. It was the often dreaded ritual that could expose anyone in attendance to the deep cutting and flashing criticism from the increasingly discontent leader at the head of the table. His ongoing battles with Jack Hanigan and the Brunswick organization continued to rage, and the scars were apparent. He was often preoccupied with preparing lengthy reports to Hanigan, justifying the equipment he needed, explaining the expansion that was required, and informing the board of directors how further delays were hurting the company and aiding *the enemy*. His frustration was peaking. During the meeting, the telephone rang.

He was at the head of a long, oval shaped conference table, surrounded by a dozen of his top administrators and engineers. Carl didn't like to get phone calls during these hectic and wide-ranging meetings, swelling as they were with dozens of complex engineering, manufacturing and marketing issues. The phone was on the wall, a few feet behind Carl, and so he pushed himself away from the big table, riding along on the small swivel wheels under his chair. As he reached back for the receiver, something in his calculations went wrong, for the chair flipped over backwards, spilling the 225 pound engineer flat on his back, his legs flying straight up in the air.

To John Hull, representing the sales department, who was sitting at the table next to Don Graves, the export manager, it all seemed to go in slow motion. "Here's a millionaire with *holes* in his shoes -- *both shoes* -- and on the seat of his pants it's sewn across where it had split up."[2] Carl struggled to get to his feet as those at the table went deathly silent, expecting the worst. He grabbed the chair and turned it over. To his utter shock it was *made by Brunswick*! Brunswick had a school furniture division that was struggling to stay in business, and someone had evidently decided to buy some of their chairs for Carl's conference room. "God Damn lady-finger engineering son-of-a-bitches," he shrieked at the top of his lungs, throwing the chair down in disgust. Hull and Graves quickly ducked under the table, pretending to tie their shoes, "to keep from laughing." But then Graves started to chuckle in almost explosive bursts, and then they just couldn't hold it back any longer. Carl heard them giggling like school kids under the table and lashed out at them. "You guys think it's so God damn funny, I could have broken my back." It must have seemed to Carl that Brunswick had designed the chair for the express purpose of embarrassing him in front of his key executives.

As Carl continued to worry about his fate and the fate of his company under Jack Hanigan and the Brunswick Corporation, he decided to secretly prepare for the worst. Should Brunswick's financial condition continue to deteriorate, Carl reasoned that it was at least possible that they might slide all the way into bankruptcy, taking the Kiekhaefer Corporation with it. At the same time, of course, his Brunswick stock would become virtually worthless, leaving him without a company and without sufficient start-up capital to begin a new venture. As a hedge against this possible doomsday scenario, Kiekhaefer and Charlie Strang set up a secret room in the basement of one of the buildings, and moved in a leased photocopy/microfiche machine. They "swore a couple people to secrecy," Strang confirmed, "and they did nothing for months but go through every print the company ever had and make photocopies and presented Carl with a [huge] stack of film."[3] Carl could then rest easier, knowing that a complete back-up of all Kiekhaefer Corporation engineering drawings was safely tucked away in his basement at home on the ledge above Fond du Lac.

The more depressed Carl became, the more he called on Charlie Strang and Tom King for solace and for a sympathetic ear. His agony over having sold his company kept growing and gnawing at him until it was clear to Strang that Carl was reaching the point of no return. By the middle of May 1964, Kiekhaefer's unhappiness began to snowball, until Strang finally realized that any prospects of reconciling Carl with Brunswick were futile.

> "Carl was so unhappy. He would just sob and groan and moan all day long, and every night he'd ask Tom King, who was our marketing vice president, and myself to come down to his house on the ledge. And he'd just sit and bitch all night, and this went on night after night. ... Hanigan was running the firm, and Carl and Hanigan were two of a kind, you know, and they were batting heads.
>
> "Finally, one night, I said, 'Carl, if it's this bad, why not get the hell out? Why don't we go start something else?' He says, 'Why not? *Why not?*'
>
> "But -- I think to be smart -- we sat down right now and typed out our resignations. The three of us sit down on Rose's portable typewriter and hunt and peck out resignations. The language was pretty simple, because we didn't know how to type that well. Carl takes them all. And Carl says, 'Now, here's what we're going to do. You two go out and get this thing started, and as you get it under way, I'll follow.'
>
> "Fine. We stood, talked about a number of businesses to go into, and he'd bankroll it, and we'd get it going and he'd join us when it was rolling because he had such a substantial stake [in Brunswick], and he had to guard his stock. The next morning the phone rings. It's Carl.
>
> "'I've been thinking about this,' he says. 'I really can't go through with this. I just got too much of a stake in Brunswick stock. I can't walk away from it, forget it all. Forgot it for about a week, and then he calls in the morning. He was like a kid who had just been hit on the head. He says, 'I think I did something bad.' I said, 'Like what?' He said, 'I had a fight with Hanigan on the phone this morning, and I told him that you and King were leaving because you couldn't stand his policies.' I said, 'Oh my God!'"[4]

Strang and King had been trapped by Carl's ongoing feud with Jack Hanigan. Carl knew how important both King and Strang were to the company, and decided to use their "resignations" as a device to bluff and injure Hanigan, and at the same time hinted that he might even join them. To Brunswick, the loss of any of the top three executives at profit-making Mercury would be embarrassing, but the loss of two, or heaven forbid, all three at once, would be catastrophic. Carl had never really prepared a second level of executives to step in and take over the reigns in case of emergency vacancies in the front office. Tom

King was the architect of the Kiekhaefer Corporation's successful marketing efforts; Strang was the undisputed engineering genius behind a long string of technical triumphs; and Carl provided the powerful personality and legendary drive to inspire the whole organization to achieve great things. Carl had indeed struck a nerve.

Carl was prepared to embarrass Hanigan not only to the Brunswick board of directors, but also to stockholders at large, exposing Hanigan as a tyrannical "bully" whose insensitivity was driving away the corporation's top talent. Strang and King were caught in the middle, and decided that the only way out would be to meet with Hanigan, verify the story that they both wanted to leave, but try to negotiate a consulting contract to give them a valuable head-start on the outside. On Monday, May 24, 1964, Strang and King sat down in Jack Hanigan's office in Chicago, but were completely unprepared for the surprise that awaited them.

After the brief exchange of pleasantries, Hanigan got right to the point.

"Why are you *really* leaving?" Hanigan demanded. Strang anticipated this question and replied confidently, "We are leaving to start our own company and take advantage of the capital gains tax."

"The *HELL* you are!" Hanigan erupted. "You're leaving because you can't stand Carl! Well, I'll tell you what's going to happen. I'm firing Carl Kiekhaefer!"

Both Strang and King were absolutely stunned. "Well, Tom sits there with his jaw hanging," Strang said. "I'm sitting there with my jaw hanging, listening to this, and [Hanigan] says, 'Here's how we're going to do it. This man is crazy and he could try to *kill* you, so I want you to leave tomorrow for three weeks in Europe. Tuesday we're having a board meeting and I've already talked to every member of the board and gotten their approval, and while Carl is at the board meeting, I'm going to fire him. And while he's being fired we're changing all the guards at the plants and we're moving all of his belongings out of his office into the parking lot. He'll never get into the plants again!'"[5]

What had started as an adolescent bluff was turning into a disaster for Carl, and Strang was quick to ask if their joint resignation was the cause of this drastic proposal. "Then he started bitching," Strang recalled. "'Nope, it's been in the works for a long time,'" Hanigan declared. He told them that he should have done this a long time ago, but it had taken until now to get the approval of a majority of the board of directors.

Hanigan then launched into a full-blown tirade against Carl that left little doubt about his true feelings.

> "Carl can't work for *anybody*! I don't see how *anybody* could work for Carl. Carl doesn't understand that he's *sold* the business. It isn't his to do with as he pleases. He's not a team man. I can't get the facts from him. I can't believe what he tells me because I've caught him in too many lies where I've had the facts from other sources. I don't get any cooperation on speeches or other corporate programs. All I get is excuses, phony excuses and lies. The other division managers see this and then they question my authority. Is he even *sane*?"[6]

Tom King and Charlie Strang were sworn to secrecy by Hanigan before they left his office. They were also told to think about what he had said before pushing their resignations or proposals for a consulting contract. They left the Chicago offices of Brunswick Corporation in a daze, their minds flooded with the many implications of Hanigan's treacherous plan against Carl. Two days later on Wednesday, May 27, 1964, Hanigan came to Oshkosh to give a speech to a group of employees in the basement meeting room of the engineering research building. The room had become famous throughout the marine industry, due to a momentary outburst of Carl Kiekhaefer's paranoia. When a major addition was made to the engineering building, a full basement was also excavated. Edgar Rose suggested that it be finished off so that it could be used for medium-sized sales meetings and other group activities. Edgar, as usual, completed his assignment with great efficiency, and transformed the vacant basement into a well-appointed and impressive meeting room. Strang recalls what happened next.

> "Carl hadn't seen it yet. I came into work one morning and the janitor said, 'Mr. Strang, Mr. Kiekhaefer was here last night. He went downstairs and he looked into that room and he *screamed*, '*Jesus Christ!* That God damn Strang is building an *empire*!'
>
> "So, I went out and had a brass sign made with etching -- '*Empire Room*' -- and had it mounted over the door to the basement. And Carl came in later that afternoon, and when his eye hit that thing he stopped. His cigar went straight up,

and then he got a big grin on his face and never said a word about it."[7]

When Hanigan finished his speech in the *Empire Room*, he took Charlie Strang and Tom King to lunch. He was eager to brief them on the new timetable for his move against Carl. He said it was going to be the following Tuesday, June 3, less than a week away, during a regularly scheduled meeting of the board of directors. He announced his plan with a confident finality, and then said, "Someone is going to have to take over the Kiekhaefer Corporation after Carl is gone." He claimed that he had asked Carl who his successor would be some day, and that Carl had replied, "No one is worth a damn." Hanigan then asked Strang if he had ever thought about being president, to which Strang replied, "Carl had often promised me the post before the merger, and had more or less trained me for it, but I hadn't thought about it since the merger."[8] Strang frankly told him that he "wasn't sure it was such a hot job now." Hanigan, though, told him that he had thought about it, and had decided that Strang was the logical choice for the job. When Strang told him that he didn't feel like he wanted the job under these circumstances, Hanigan brusquely told him, "Get it through your head. Carl is finished, and will be gone next Tuesday! The only question is whether you will leave or become president!" He cautioned Strang about becoming emotionally involved with his decision, and told him that if he didn't take the job, that he would put in Charles E. Erb, a Brunswick executive that had recently been recruited to be corporate vice president of marketing. Hanigan then told Strang and King that time was short, to sleep on it and come to Chicago the next day with "any questions we may have."

The following morning, Thursday, May 28, 1964 Strang and King flew to Chicago aboard the Kiekhaefer Corporation twin-engine Cessna. Again, both Strang and King were prepared to push for a consulting contract with Brunswick, but once more Hanigan brushed off their request, and began to loudly address his own agenda. This time, however, Hanigan was intent on turning both Strang and King against Carl. It was, according to Strang, "a real hate campaign." Hanigan told the two that when he asked Carl who would eventually replace him at Mercury, he replied that "neither King nor Strang were competent to be dog catchers," and that Carl claimed "he personally engineered and marketed everything because of our incompetence." Following his opening attacks, Hanigan then outlined his full strategy for disposing of Carl Kiekhaefer.

His plan was based on crucial timing. While Carl was being informed at the board meeting of his summary dismissal, Hanigan would order that all the locks of all the plants be changed, even those in Florida, and that regular Kiekhaefer Corporation guards would be replaced by Pinkerton guards so that Carl couldn't talk them into opening the gates or doors. Hanigan expected a violent reaction from Carl at the board meeting, and planned on having special guards standing by to make sure Carl didn't attack him physically. Strang quickly pointed out that what Hanigan proposed wasn't legal, and even "aside from the moral side, Carl was a director and had access to all plants as such." Undeterred, Hanigan summoned Edward A. Stephan into the room, a member of the board, Brunswick's general counsel, and partner in the law firm that had represented Brunswick since the nineteenth century.[9] Hanigan told him, "Ed, I fired these guys from Kiekhaefer Corporation and hired them for Brunswick. Strang will be the new president of Kiekhaefer Corporation." When Hanigan asked Stephan about the legality of a "lock-out," the attorney told him that he "could not legally lock-out a director." To this unanticipated obstacle, Hanigan replied, "I'll lock him out anyway. I've got enough on him to have him removed as a director!"[10]

Strang had heard enough. He told Hanigan and Stephan that they were making a mistake, and that considering the low price of Brunswick stock, there must be a large number of unhappy stockholders, and "it wouldn't take much to get a stockholder's suit going. Kiekhaefer's a good hater," Strang continued, "and would probably try something like that. You know he's thought about it in the past." Stephan, unimpressed, said, "We're prepared for any such action on his part, and anyone else's part."[11] King agreed with Strang, declaring his disapproval of these drastic actions against Carl. To Strang, Hanigan's thinking had gotten out of hand. Why not offer Carl a fair settlement of some kind, like an generous consulting contract for example, rather than carry out an emotionally charged, dramatic scenario of eviction, deception and devastating embarrassment? Hanigan, though, had made up his mind. Kiekhaefer was out. It would be a clean break, and the discussion was over.

"My God! Strang exclaimed, "I don't think I want anything to do with this." Hanigan surprised him once again by saying, "Well, you won't have to. I've arranged to have you leave immediately for a European trip, and it will be all over and done when you return. How does the job sound?"[12] Strang, eager to get out of the room, said, "I'll let you know tomorrow." Hanigan then said he was sending him a letter anyway, informing him that he was president of the Kiekhaefer Corporation, and would include travel instructions for his European trip. Hanigan would never send the letter.

"So, we go home," Strang explained, "and we go to a little bar somewhere and have about three drinks and King is saying, 'Do it! *Do it!*' He said, 'You are not your brother's keeper.' When I got home, Hanigan is calling me about every two hours and, having learned one trick from Carl, I'm recording all the telephone calls. He's telling me, 'I know how you feel about Carl, but it's not going to do any good. You'll lose out and he'll lose out this way if you don't come with me, and I'm going to give the job to somebody else. He's going to be fired Tuesday, *period!*'"[13]

Once Strang and King informed Carl that they were required to attend another meeting with Hanigan in Chicago, Carl quietly slipped away and no one knew where to find him. "Can't even talk to him," Strang said. "He disappeared. He didn't want to be within hearing range of anything I guess."

In Carl's absence, the pressure was starting to build on Strang to accept the presidency, and to put aside his deep personal feelings and thirteen years of close friendship with Carl. Charlie's mother Ann, encouraged him to take the job, as did his close friend Jimmy Jost. According to Strang, it was a rough night.

"What kind of a dirty deal are we being suckered into?" Strang asked Tom King.

"I don't know, but I don't like it one bit," King replied. They began to carefully analyze the situation Hanigan had forced them into, and a few basic facts were agreed upon. "If Hanigan is telling the truth, and Carl is definitely out it would be foolish for Strang to throw away the presidency after working for it for thirteen years. But, if by any chance, Hanigan took his action on the assumption that King and Strang would be here to take over the Kiekhaefer Corporation, then the acceptance by Strang will ensure that Carl is fired. A refusal by Strang might mean that Hanigan will back down on firing Carl. So, it isn't worth the risk that Strang's acceptance may ensure Carl's firing, and Strang must refuse."

The following evening, May 29th, 1964, Strang called Jack Hanigan at his home outside of Chicago. Strang clearly understood the importance of the call he was about to make. His entire career depended upon the decision he would reach. He had evaluated the insistent pressure applied by his family and friends, and had pondered the set of morals and business ideals that had served him well throughout his life. He was 43 years old, and was in the very prime of his career. The answer he would give Hanigan that early summer evening would irrevocably affect the entire balance of his business life. More than this, the celebrated career of his closest confidant, Carl Kiekhaefer, lay heavy in the balance. In a paradoxical twist of fate, Strang's decision to accept or reject the job he wanted more than anything else in life, might very well mean the end of a prestigious and heralded career for his best friend.

The transcribed telephone conversation which follows has never before been publicly revealed. It is faithfully reproduced from the original dictabelt recording belts that were stored for over twenty-five years in Strang's private archives. Special handling and new technology was required to extract the audio imprints contained on them, but a most successful transfer was obtained. Because of the importance of Strang's decision, and because of the profound articulation of moral principles which his announcement to Jack Hanigan represents, this single conversation is perhaps the most important of his career.

Hanigan: Hello.

Strang: Hello, Jack.

Hanigan: How are you?

Strang: Fine. I have something to discuss, and knowing of the tight schedule of your program, I thought I'd better ring you tonight, though I hate to disturb you at home.

Hanigan: That's O.K., I just got here.

Strang: How's that for timing? Well, what it boils down to is I want to make a little speech, more or less, and I hope you'll hear me out.

Hanigan: Alright.

Strang: Since we discussed your plans yesterday, I've really been doing some soul searching here most of the night. As you know, I'm extremely enthusiastic about the economy and expansion programs we talked about yesterday ...

Hanigan: Yeah. ...

Strang: ... and carrying them through to completion. And I'm really honored that you offered me the opportunity you have to be president of Kiekhaefer. The thing that's bothering me though is that even though you told me that your decision to get rid of Carl was made before he brought you my resignation, I can't escape the feeling that my acceptance can only expedite his departure. And as much as I long wanted this thing for thirteen years, I don't think I can accept it under these very unfortunate circumstances. If Carl was leaving voluntarily -- or even if I had no advance knowledge of your plans for his dismissal, it would be a totally different matter. Operating the corporation profitably is one thing, but this matter involves a personal friendship of some thirteen or fourteen years. I really like Carl. I can't help it. And I don't want to have any part in hurting him. And what's coming next Tuesday is going to be brutal for him, and for me, of all people, to take his job at that time would make it that much worse for the fellow, and I just can't do it to him. Now obviously ...

Hanigan: Well, what do you intend to do now?

Strang: As I was going to say, obviously, this little speech is not the way to get ahead in industry. But I do want to say that if there is any way, you know, in which I can help you, the corporation -- and Carl -- I'll be glad to do it. I hope you can understand.

Hanigan: I understand fully.

Strang: 'Cause, there's a lot of my blood in this company. I've bled for it for a long while, but there is a certain long-standing loyalty to Carl which comes through -

- I can understand your problems completely, and I can sympathize with it in every way, and yet I personally can't be involved in hurting him as I would be in this case.

Hanigan: What I'd like to propose is that you stay in the job you're on and Tom stay on the job he's on, and I'll put Erb in there as general manager.

Strang: I feel if I'm doing what I'm doing, I can help ease the blow to Carl considerably, and perhaps ...

Hanigan: I understand that.

Strang: Talk to him ... it's just perhaps an emotional thing, but it's something I can't ...

Hanigan: Well, look. You're a human being. He's a human being; so am I. And I'm not enjoying this exactly.

Strang: I can believe it.

Hanigan: I probably put this off a little longer than I should have, but I wanted to give him every chance.[14]

Near the end of the emotional conversation, Hanigan tried to persuade Strang one more time to accept the presidency. Strang was scheduled to attend a charity boat race to benefit mentally retarded children the following morning, which would commence from Chicago's Navy Pier on the shores of Lake Michigan. Hanigan asked Strang to break away from the race when it was feasible, and come to his house in the suburbs, or failing that, come to his house on Sunday. He further told Strang that Kiekhaefer would soon know of his plan, for Walter M. Heymann, a member of the board, was going to call Jack Puelicher, Carl's good friend and fellow board member, and Puelicher would then inform Carl. Fortunately, Carl was to have a forty-eight hour advance notice of Brunswick's secret and treacherous plan.

Carl acted quickly to line up support on Brunswick's board of directors in an effort to foil Hanigan's vindictive scheme. Eighteen members sat on Brunswick's board, including the Bensingers, Hanigan, Carl himself, and Jack Puelicher, the additional director Carl was promised during merger negotiations. Besides Puelicher, however, Carl had methodically developed a great deal of respect and admiration over the years with other directors, especially Heymann, the influential director of the First National Bank of Chicago. The trouble was, nine full members of the board -- exactly half -- were either employees of the Brunswick Corporation or retired Brunswick executives. According to Hanigan, only a simple majority was needed to implement his plan. The voting would be extremely close, and no one doubted that Hanigan and the Bensingers would be able to bring a tremendous amount of pressure to bear on the other directors.

During the race festivities at Navy Pier, Strang walked across Chicago's busy Outer Drive, pushed a dime into the pay telephone at the Grand Hotel, and ended his career with Kiekhaefer Mercury. He phoned Jack Hanigan and told him that he wouldn't be coming over to his house, and that there was nothing left to talk about. "I told him I appreciated what he was offering, thanked him greatly, but there's no way I was going to do that to Carl. Therefore, I'm leaving the company."[15] Still, Hanigan persisted.

"I can understand that," Hanigan said. "Suppose we make Carl Chairman Emeritus?"

"That's not going to do any good," Strang pointed out. "Let's just forget the whole thing.

"Well, I'm telling you, I'm going to fire him anyway, and I'm going to put Erb in as president on Tuesday." Strang was deeply troubled and saddened by the whole affair. "I said, 'Good-Bye,' and that was the end of it." That evening, Ann Strang, Charlie's strongest and most vocal booster, "passed out, she was so upset, because of the whole turmoil." She, too, had reluctantly arrived at the same conclusion, that Charlie now had no alternative but to leave the company.

Hanigan began to vacillate. Now convinced that Strang would not accept the presidency, and in fact would leave the company, he phoned Strang on Monday morning, June 1, 1964 -- the day before the board meeting -- to tell him that he was modifying his plans. Carl wouldn't be locked-out. In fact, Hanigan had decided to offer Carl a generous consulting contract, "but make it clear that Carl could consult *only when asked*." Aware of Carl's passion for construction, Hanigan also decided that he would let Carl bid on Brunswick buildings, including a new Zebco division plant. And, most surprising,

Hanigan informed Strang that he had decided not to put Charles Erb in as president, but would take the job himself! "That would be quite a trick," Strang was quick to point out. Hanigan said he would fly up to Fond du Lac one or two afternoons a week from Chicago, as long as Strang would stay on to actually run the operation for him. "In a year or so," Hanigan suggested, "You might feel differently about Carl, and then we'll make you president."[16] In the mean time, he promised Strang that he would treat him like the president. Strang steadfastly refused. It was to be Hanigan's last attempt to maneuver Strang into the presidency of Kiekhaefer Mercury.

That evening, Strang was at home, preoccupied with the prospect of the bloodletting at the board meeting the following morning. Then, Hanigan called. "He had changed his mind. The whole thing was off. There would be no firing. He would lay down the law and try to come to a workable agreement. 'Tell King about it.' He expected us to maintain strict confidence about all that had transpired. 'Good-Bye.'" It was as if a crippling weight had been lifted from Strang's shoulders. Even though Strang was determined to leave, perhaps now more than ever, Carl was to be spared. Hanigan had been perhaps bluffing all along. Strang would never know for certain.

Hanigan decided to focus his attention on keeping Strang within the Brunswick organization at all cost. On Wednesday, June 4, 1964 -- Carl's 58th birthday -- Hanigan informed Strang that he was to be his right arm, and that the next time he moved against Carl, Strang would move with him. Hanigan offered Strang the position of director of research for all of Brunswick Corporation. It was a combination of compliment and insult, for at the same time he said, "You're really happy about the way things turned out aren't you?"

"Yes, at least for Carl's sake," Strang responded.

"You taught me one thing," Hanigan continued. "You are too soft-hearted to be my kind of operating man. You'll never head up a Brunswick division."

"I agree. I'm not *your* kind of operating man," Strang rejoined with only thinly-veiled sarcasm.

Hanigan was desperate to entice Strang to remain. He outlined the glorified research position in the most seductive terms, telling Strang how a new research center would be built in Long Grove, Illinois, and that he would report directly to Hanigan. "But, if you take the job," Hanigan warned Strang, "remember it's much bigger than the one you've had. Understand that you're *my boy*, and the loyalty that went to Carl now goes to me no matter what the circumstances, whether it's a conflict with Carl or anyone else. If you agree to that, and want the job, just tell Carl to send you down to Chicago."[17]

Strang knew that Hanigan was going out of his way to create the best possible engineering job in the corporation just to keep him. Hanigan had even sweetened the deal by offering a very lucrative package of compensation and a generous stock option. "It was a big job, probably the best I'll ever be offered, but after that warning, I could only see it as a political nightmare under Jack Hanigan's thumb. There was no real choice by my standards but to leave Brunswick, even without a [consulting] contract." Both Strang and King would leave the company within forty-eight hours.

Hanigan was furious that he was losing two of the three brightest minds in the only division with the profit potential to pull the ailing Brunswick Corporation out of their continuing financial crisis. He had gambled and lost, owing to Strang's deep personal regard for Carl Kiekhaefer, and for Strang's deep convictions and firm moral principles. There was perhaps only one way to save face for Hanigan, now that Strang and King were leaving, for he would have to deal directly with Carl once again. Strang received a call not long after he had turned down Hanigan's final offer. His source informed him that "Mr. Kiekhaefer had come back from wherever he was, and was tremendously upset ... because Mr. Hanigan had told him that *I had come down there and asked them to get rid of Carl and make me president!*" Hanigan had saved Carl by dismissing the conspirators -- Strang and King -- according to the ruthless accusation. Strang was absolutely shocked. "Rose [Smiljanic] said that Carl was going mad, was really upset. At the same time he had all the locks on all the buildings changed so my keys wouldn't work. Instead of getting the keys back, he changed all the locks *on all the buildings!*" Strang's reward for secretly saving Carl's honor and position -- including saving Carl from being locked out of his own company -- was to be locked out himself and branded a traitor by his best friend. It was with the heaviest of hearts that Charlie Strang left his offices at the Kiekhaefer Corporation for good, near the end of the first week in June 1964.

"I was a real son-of-a-bitch for a while," Strang remembered of Carl's tragic error of judgment. Carl was quite vocal about his feelings of betrayal by his friend and top executive, spawning vicious and stealthily creeping rumors of Strang's duplicity throughout the industry. Rose Smiljanic finally called Strang and pleaded, "Mr. Kiekhaefer is in such a state over this thing, please come and talk to him." It was then, after Carl had already told many of his closest associates that both Brunswick and Strang had deceived him, that Strang told Rose, "I recorded all the telephone conversations related to this matter. I'm going to bring them up. Have a player handy."

Strang brought his box of dictabelt tapes up to Carl's sprawling home above Taycheedah, overlooking Lake Winnebago, eager to confront Carl and to vindicate himself in this deepening morass of confused loyalties and broken friendships. In the presence of Rose Smiljanic, Armand Hauser and a stoic Carl Kiekhaefer, Strang played tape after vindicating tape. "Suddenly, I was a *golden boy*, even though I wasn't working for him."[18] Yet, even though Strang would play the same conversation that appeared on these pages between himself and Hanigan, proving his rigid moral stand and ultimate rejection of the presidency over the corporate corpse of Carl Kiekhaefer, Carl continued to doubt.

Jimmy Jost was still working for Carl, when it became clear to him that Carl hadn't realized how Strang had sacrificed himself for his friend. It was at a press conference in Canada, when Carl summoned Jost for a talk. Carl was sitting on a bench, alone, halfway down a long path leading to the water's edge.

"You sent for me?" Jost asked the brooding Kiekhaefer. Carl ignored him, and remained fixated at the dirt on the ground between his feet, saying finally, "Won't you sit down?" Jost sat down obediently, and waited for Carl to speak. "It seemed like an hour. I didn't know what the hell was up at all. All of a sudden he says, "Why did Charlie do this to me?"

"Do *what* to you Carl?"

"Why did he and Tom try to take my company away from me?"

"Do you really think Charlie would do something like that?"

"Well, Hanigan said he did."[19] Jost could feel Carl's agony, and it seemed like Carl would never forgive himself for having lost Charlie, and never forgive Charlie for having secret negotiations with Hanigan, regardless of the outcome.

Even in the face of Charlie Strang's thirteen years of intimate friendship, and given the irrefutable evidence of the recorded conversations between Strang and Hanigan, Carl continued to judge his friend harshly, and within a year, would denounce both Strang and Hanigan in long and blunt letters to fellow Brunswick directors. In a letter to Simpson E. Meyers, Carl explained his ongoing confusion in the matter.

> "I am still completely perplexed as to what brought about the narrow squeeze I myself had and my escape from being fired, when for years I have been operating almost the only profit center and the highest sales center of the entire Brunswick Corporation. At times you have criticized me for feeling that someone was trying to injure me. I am sure you will agree with me that there have been reasons, a time or two, for this sentiment. How would you and the rest of the Board have explained my exodus to the 18,000 stockholders? Indeed, there may be more surprises ahead. ... I have sustained losses in prestige and position; and I am going to be extremely interested in Corporate management from now on."[20]

When Strang and King left the Kiekhaefer Corporation, their plan was to work together as a high-caliber marine industry engineering and marketing consulting firm, available to anyone who might need such a specialized service. They formed United States Executives, Inc., and one of their first clients was Rover, the small turbine engine manufacturer in England with whom Carl had unsuccessfully negotiated for a manufacturing license. Strang and King would drift apart, though, and King began a search for a suitable corporate position for his talents. One of the firms to which King applied was the Hughes Tool Company in Santa Monica, California. When Carl received a letter of inquiry from the company about King, he wrote a scathing letter which he never mailed, hoping to insure that King would never be hired.

> "As you know, it is often typical of men who are dismissed from top executive positions to claim the credit for all progress, and Mr. King's claims are no exception. ...
>
> "Perhaps our most serious criticism is that in the final analysis any offers of friendship we made were completely aborted in the next move he made; namely, to form the U.S.

Executives, Inc., in which he took in as partners one of our former attorneys and a former chief engineer to act primarily as a professional management group. As such, there was an attempt made to the parent corporation to have me removed as ... President of the Kiekhaefer Corporation.

"This indicates some immature thinking since I was still the second largest stockholder in Brunswick at that time and was an officer elected by the stockholders. Regardless of the impressions that were made with the present Brunswick Corporation president, this move also aborted since I happened to be running the No. 1 profit and No. 1 sales centers of the Corporation -- with sales having gone up better than 100%, and profits better than 300%, since the merger in 1961.

"Of course, I don't wish this fellow any harm. I have always felt that we are not only building products in this organization, but also building men. I have a feeling that somewhere along the way I failed in this latter duty to develop a young man who looked so good in 1957 when I first hired him. Perhaps you can do a better job with him."[21]

Carl's efforts to the contrary, King was named vice president of advertising for Braniff International Airways only a week later, on May 3, 1965.

As the talented and well-schooled group of ex-Kiekhaefer employees grew, and as they took on responsible positions both within and outside the marine industry, they continued to look back on their years at the Kiekhaefer Corporation with a combination of disbelief and affection. So strong was their sentiment, that they eventually formed the Alumni of Kiekhaefer Club, or A.O.K. Club, even issuing membership cards to those fellow travelers who had once worked with one of the most interesting industrial leaders of all time. The official looking card gave members the right to be "banded together to exchange anecdotes of their experiences ... and to congratulate each other on the successes enjoyed since severing their connections. ..." On the back of the card, the honor of membership becomes more apparent.

"This card also attests that said member has fought in the various campaigns and battles around Fond du Lac, Oshkosh ... Cedarburg ... and innumerable minor skirmishes throughout the land.

"That said member conducted himself in an outstanding manner despite wounds (both physical and mental) and fought a brilliant but losing battle against midnight phone calls, long trips and the slings and arrows of outrageous misfortune before retiring from 'the service' and returning to normal ... life."[22]

Near the end of 1965, Charlie Strang was contacted by C.W. "Doc" Jones, a one-time Kiekhaefer Mercury dealer, who had become an aggressive and dedicated OMC distributor, and who maintained a high-performance machine shop in Phoenix, Arizona. Would Strang join him in an attempt to establish a new world's outboard record? Strang thought of his old friend Carl, who had forsaken him and already labeled him a traitor. He knew that if he accepted Doc Jones' challenge, and helped OMC to beat the best that Kiekhaefer could build, there would be no turning back. Doc Jones, intimately familiar with the products of both the Kiekhaefer Corporation and OMC, "knew that OMC had to have Charlie."

"We [OMC] were in *all kinds of trouble* with our outboard engines," Doc Jones admitted. "I got Charlie to come out there and stay with me for six months. I personally talked to Ralph Evinrude ... and asked him to come out to Phoenix. The three of us went to dinner. And then they stayed up all night and talked. And Charlie saw me the next morning and said, 'They've offered me a job, but they won't tell me what it is.'"[23]

One of the most difficult considerations in bringing a formidable talent like Charlie Strang into OMC, would be in providing him with the optimum position. OMC needed to carefully juggle its corporate rosters, and make the necessary advancements and promotions, in order to properly take advantage of Strang's unique credentials and potential. This process can take time, and Ralph Evinrude realized that the only way to both have Strang and to make the necessary adjustments at Waukegan, would be to put Strang "on ice," that is, retain his services somehow, with a "signing bonus" of sorts to lock him in at the proper time. Doc Jones' high performance shop and world record attempt provided the perfect vehicle, and Strang was provided with a significant retainer of $50,000 to remain with Jones for the few months it would be required to prepare for his arrival at OMC.[24] Clay Conover, OMC's celebrated chief engineer for the previous sixteen years, was elected a vice president of OMC, and general manager of Johnson Motors. On June 1, 1966, Charlie Strang joined OMC as director of marine engineering. From the exalted position of being Carl Kiekhaefer's closest friend and most trusted confidant, Charlie Strang was about to become *the enemy*.

Carl flashes his signature look of determination and intimidation while testing a new *MerCruiser* stern drive on a chilly morning at Lake X.

Joe Swift (left) and Frank Scalpone were vital components of the Mercury public relations and publicity brain trust. In San Francisco, 1969, they were admonished by Carl (center) for not having invited the mayor of San Francisco to see the *Mona Lou* win the race from Los Angeles to San Francisco.

Below: Publicity in 1958 included setting a new world's record for pulling 31 skiers at Ralston Beach, near Tampa, Florida with a pair of 70-horsepower Mercury Mark 78s. Total weight of the pull, including skiers, motors, boat and its four occupants, was 7,531 lbs. The skiers took off simultaneously from a wide dock.

Left: Mercury engineered this water-skiing elephant stunt to disprove the "fast but no power" slogan circulated by Johnson and Evinrude dealers. Pulled by a pair of 70-horsepower Mercury Mark 78s, this elephant, who normally skis during a show at Ponce de Leon Springs, Florida, was brought to the Hudson River in New York for this dramatic photo. The elephant, his companion Marge Rusing, and his special water skis weigh a total of 3,000 pounds.

Joe Swift, Kiekhaefer's publicist, couldn't resist drawing the rope between the 125-horsepower Mercury and the *Queen Elizabeth.* Hundreds of publications carried the photo, again proving for Carl that Mercurys *really* had pulling power.

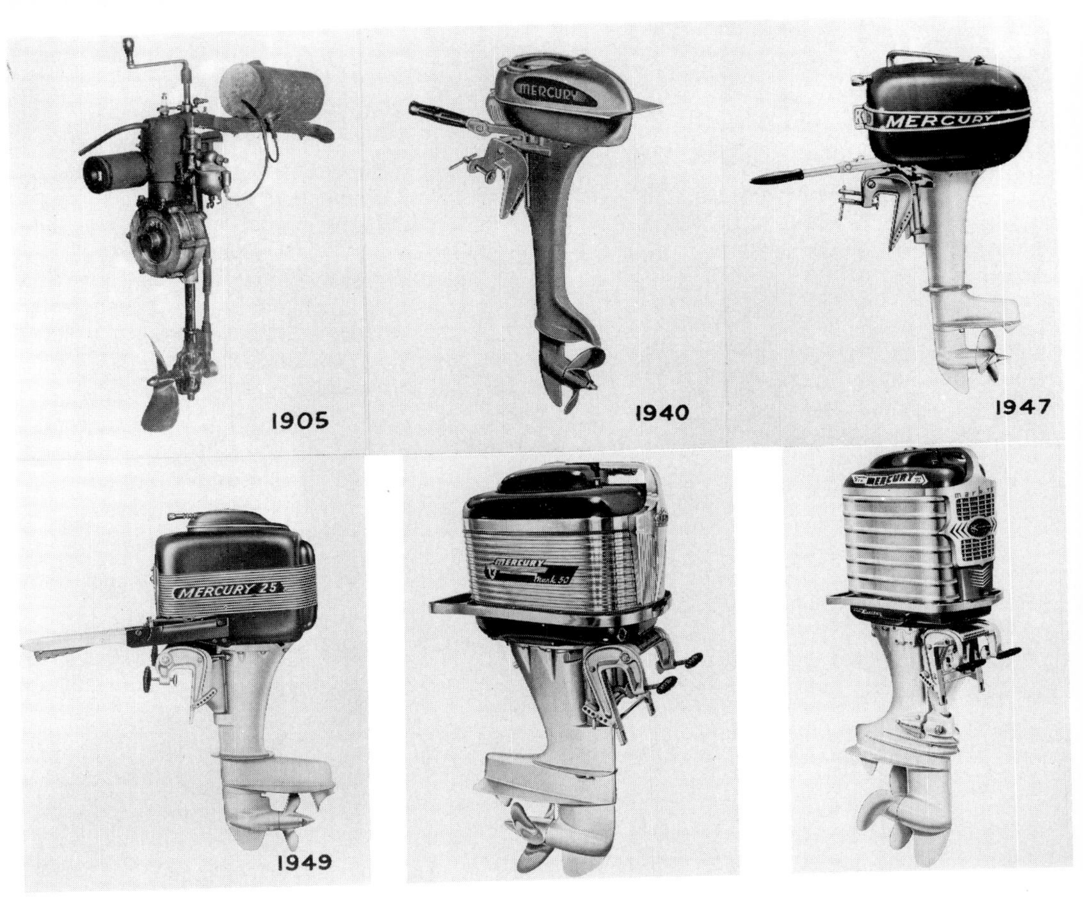

1905: Waterman *Porto*, the first mass produced outboard motor.

1940: First Mercury, single-cyl, first one-piece streamlined drive shaft.

1947: Mercury *Lightning*, 10-horsepower, twin-cylinder.

1949: Mercury *Thunderbolt*, 25-horsepower, world's first 4-cyl-in-line.

1954: Mercury Mark 50, 4-cyl, 40-horsepower, gear shift, electr. start.

1956: Mercury Mark 75, 60-horsepower, world's first 6-cyl-in-line.

1961: Mercury Merc 1000, world's first 100-horsepower engine.

Charlie Strang and Carl Kiekhaefer discuss the design of a new *MerCruiser* powerplant at the engineering lab in Oshkosh. Downstairs is Strang's *Empire Room*.

DIRECT DRIVE

Opposite page and below: Carl delighted in sawing down trees to make chips for his legendary barbecues. The steaks on the grill are "Kiekhaefer Cuts," up to two-inches thick and the finest grade of aged beef available. His obsession with thickly marbled steaks and a penchant for hand-rolled Cuban cigars (left), led to a severe heart condition resulting in five bypasses.

Opposite page top: Don Aronow, founder of Cigarette, Magnum and Donzi boat companies, rode Mercury power to two world championships and three U.S. national championships. Here, driving without a helmet, Aronow bucks a pair of 475-horsepower *MerCruiser* stern drives aboard a 32-foot Cary hull to win the Sam Griffith Memorial trophy for 1969.

Opposite page bottom: Dick Genth of Miami, Florida, rode this 28-foot Thunderbird powered by a single 427-cubic-inch *MerCruiser* stern drive to win the 1967 253-mile Hennessy Long Island Marathon. Genth, Thunderbird president, would become president of Chris-Craft and chairman of Donzi boat companies.

Nothing could distract Carl when he was watching the boat races. Here, he combines his two favorite passions: offshore powerboat racing and cigar smoking.

CIGARETTE

99
Thunderbird IV

Left: On October 27, 1967 Carl drove the firs Mercury 150 E snowmo bile. Critics claimed th sled was so top-heavy i would fall over sitting i the garage.

Opposite page top: Jin Bede's famous *Bede-5*, build-it-yourself was powered by Carl's speci ly modified snowmobile engines in 1972, until problems separated the two stubborn designers

Opposite page bottom: On May 15, 1973, Fred Kiekhaefer (standing) a Carl signed merger agre ments with Bombardier Limited President Laure Beaudoin. The tired m were up most of the nig ironing out final details the merger, which woul never be carried out.

Below: The first shipment of KAM snowmob engines to Bolens, division of FMC in the summer of 1971.

5
N500BD

Top left: Dr. Robert Magoon, five-time APBA offshore national champion, three for KAM, is Miami eye surgeon. Top right: Carlo Bonomi, 1973 U.I.M. world champion with Aeromarine engines. Below: Sandy Satullo (holding cup), owner of Copper Kettle restaurants, wins the 1973 Hennessy Hurricane Classic in St. Petersburg, FL, powered by fuel-injected Aeromarine 468 c.i. engines.

Above: *Kiekhaefer Aeromarine I*, piloted by Dr. Robert Magoon, along with throttleman/mechanic Gene Lanham, was first in APBA history to win six national championship races in a single season, and winner of 1971 U.S. National Championship. Below: *Kiekhaefer Aeromarine IX*, driven by Carlo Bonomi of Italy, missed 1972 U.I.M. world crown by single race.

Above: *Dry Martini*, the renamed *Aeromarine IX*, driven by Carlo Bonomi, wins the 1973 U.I.M. World Championship.

Below: *Kiekhaefer Aeromarine III*, with Dr. Robert Magoon and Gene Lanham aboard, are 1972 APBA national champions, powered by a pair of Aeromarine 468 c.i. "Champion Maker" engines.

Right: Four-time world champion throttleman Richie Powers of Miami, Florida.

Below: Tom Gentry's *American Eagle*, 1976 U.I.M. World Champion, throttled by Richie Powers in left hand seat, powered by Carl's "Champion Maker" Aeromarine engines.

Above: Kiekhaefer Aeromarine Motors, showing part of the snowmobile test track.

Below: The "Champion Maker" Aeromarine engine final assembly line on the "Ledge."

CHAPTER TWENTY-NINE

Fast, Faster, Fastest

"It's a reality of life that men are competitive and the most competitive games draw the most competitive men. That's why they're there -- to compete. They know the rules and the objectives when they get in the game. The objective is to win -- fairly, squarely, decently, by the rules -- but *to win!* "[1]

Vince Lombardi

When Carl Kiekhaefer sanctioned Charlie Strang's design for a six-cylinder, in-line 60-horsepower engine, he signaled the start of a race for speed and outboard horsepower that would endure for decades to follow. When the first Mark 75 engine rolled off the assembly lines in Fond du Lac in 1957, Kiekhaefer had, for the first time, taken the lead in outboard horsepower over rival OMC and their 50-horsepower V-4 "Fat Fifty." Seldom again, in Carl Kiekhaefer's lifetime, and only for brief periods of time, would Mercury fall behind *the enemy* in the race for outboard horsepower.

When government contracts for production of target drone aircraft engines were canceled following the Korean War, Ted Jones left the Kiekhaefer organization and returned to work building the fastest boats in the world. His three-point hydroplane designs were to capture world record after world record, powered by the most powerful inboard engines in existence. The so-called "three-point" design was coined because the boats, basically of a catamaran design, would speed across the water supported only by a small point on each sponson, and by the thrust of the propeller which was only about halfway in the water. The wind, rushing through the center of the two hulls, helped to lift the boat out of the water and stabilize it at high speeds. In the nine years between 1950 and 1958, Ted Jones designs captured an incredible eight American Power Boat Association Gold Cups, the most coveted prize in boat racing history. His *Slo-Mo-Shun IV* and *V* unlimited hydroplanes monopolized the coveted trophy from 1950 to 1954. His *Miss Thriftway* design won it again in 1956 and 1957, followed by his co-design of *Hawaii Kai III* which captured the cup in 1958. Carl was eager to use a smaller version of this basic Ted Jones configuration to establish a new world speed record for outboards.

The 100 mile per hour barrier had just been broken by Italian Massimo Di Priolo near Milan in 1956, propelled by a specially designed, one of a kind, supercharged 160-horsepower outboard engine. Even with this enormous engine, the Italian barely passed the century mark at 100.3 miles per hour, and the record was not universally trusted. The world's record hadn't been held by an American since 1937.

Carl contracted with Hubert Entrop and Jack Leek from the state of Washington on March 14, 1958 to spearhead the attempt. Entrop, a model builder for Boeing Aircraft wind tunnel tests in Seattle, would build the boat based on a Ted Jones design, while Leek would serve as engine mechanic to specially prepare a 60-horsepower Mercury Mark 75 for the record setting effort. In his agreement with the men, Kiekhaefer insisted that "Every precaution shall be taken to maintain secrecy during boat construction and tests preliminary to the official run in order that information shall not leak out to the press in advance of the successful conclusion of the project."[2] Further, Kiekhaefer forbade them to make an official run at the record without Kiekhaefer Corporation staff on hand to coordinate publicity, and statements were prohibited to the press following the attempts without clearance through Charlie Strang, Tom King or Kiekhaefer personally.

Carl insisted that the record attempts be made on Lake X, so that he could control not only security, but publicity. A stock engine was modified with special carburetors built by Charlie Strang to allow the engine to burn pure alcohol, and a special racing lower unit was used with a very thin, high speed propeller. Normally, the engine would develop 60-horsepower at 5500 rpm, but with this arrangement, the engine developed 83-horsepower at 7500 rpm. Entrop's first trials on Lake X fell short of expectations and the record, reaching only 96.134 mph.[3] Carl, with his usual bulldog diplomacy, let Entrop and Leek understand in no uncertain terms that they had failed, and that they had better figure out a solution. Entrop felt insulted by Carl, and, according to Charlie Strang, "finally just pulled up stakes and left." Strang would later calm Entrop down, and convince him to continue the attempt for the record back in Entrop's home state of Washington.

Ted Jones came to Lake Washington near Seattle to personally assist in modifying the balance and motion of the boat for the trials. Kiekhaefer, though convinced that a secret romance had blossomed between Jones and Rose Smiljanic while Jones was working on the drone aircraft engines, described Jones' work on the record boat in glowing terms.

> "The [record attempt] is the result of a tremendous amount of preparation and at great risk of limb, if not life. The entire crew feels physically responsible to the driver who undertakes the actual run. I personally know about the anxiety experienced by Ted Jones, the original designer of the boat, who also made the final refinements. Refinements that were almost continuous as he would, with practiced eye, study the action of the boat with binoculars as it whizzed by at something like 150 feet a second."[4]

High speed photography studies were also made of the boat's motion over the water, and Jones, along with his son, Ron, would fill-in or remove small portions of the lifting sponsons of the tiny 13.5 foot, 250 pound hydroplane to get the exact action needed. "By actual count," Kiekhaefer would later confirm, "some 68 runs over the official course were made at the 100 mile an hour speeds. Not so much to gain additional speed but to gain stability of the boat, to develop more efficient propellers, to develop electronic timing equipment, to determine the best spark plugs for the job, to develop better aerodynamics of the boat as well as hydrodynamics of the boat."[5]

Then, on the morning of June 7, 1958, with Carl Kiekhaefer personally directing each step, Entrop made two runs in opposite directions across Lake Washington to challenge the record. His first run was timed by official observers of the APBA at slightly over 109 mph, while the second was clocked at just over 106 mph, producing an average run of 107.821 mph to establish the new world's record. Following the spontaneous celebrations that took place on the docks, a promotional strategy meeting was planned for the evening, and a most bizarre conversation developed over dinner.

Carl was very excited about having established the new record. Not only had Entrop brought the record back to the United States after a lapse of some 21 years, but he had smashed the standing United States record by almost exactly 30 miles per hour. Most new records beat existing speeds by a few tenths of a mile per hour, or by a few miles per hour. Since the Italian record of 100.3 mph was always somewhat suspect, it was not only the first time that a production outboard had exceeded 100 miles per hour, but the new Kiekhaefer Mercury record annihilated the old records by the widest margin in history. It was big news, and Carl knew it. However, he was still mad at Entrop for "deserting the ship" at Lake X. Carl decided that they would widely publicize the new world and American record, but *wouldn't name the driver!* Strang remembers the strange incident.

> "I think the funniest thing I ever heard was that night. Carl said, 'Well, we're going to get this thing publicized, but we're not going to tell anyone who the driver is. That son-of-a-bitch is a friend of [*the enemy*], and there is no sense in mentioning his name. So, here is the poor press guy scratching his head. 'How the hell do I tell the world that an outboard has exceeded a hundred miles an hour for the first time and I can't tell who the driver is! It took most of the night to resolve that one."[6]

Carl finally came to his senses, and allowed Entrop to be proclaimed the world's outboard speed record holder. Privately, Carl had a great deal of respect for Entrop, and wrote then APBA Vice President and National Safety Chairman Jimmy Jost, that "Hugh Entrop is a bear for punishment. At the time of this run, his knees were infected with boils; perhaps one might consider them the necessary irritant to have him get the run over with. What a jet pilot this guy would have made."[7] After news of the record became headline news all around the world, Carl decided to be even more humble, go with the flow, and could be found praising Entrop in the most flowering terms, as in this excerpt of a letter to the editor of the *Houston Chronicle*.

> "Obviously, the greatest share of the glory goes to the driver. In this case, he deserved it. Hugh Entrop ... is one of those rare combinations of engineering talent who has the intelligence to report his observations as he reads them from instruments while pioneering century mark speeds. ... Now please, let there be no mistake. The credit for the record must go to the Entrop-Jones team for setting the mark that will be the target for many other teams, and it takes a team to make

a star, whether it's football, baseball, sports car racing or boat racing. And that's yours for the record."[8]

Carl's acquisition of the new world speed record for Mercury in 1958 signaled the start of a never-ending race for higher and higher speeds between Mercury and OMC, and the regular exchange of world records. In the fall of 1959, C.W. "Doc" Jones, the Evinrude distributor for the states west of the Rocky Mountains, was attending a sales meeting in Milwaukee, and approached Ralph Evinrude about getting an OMC engine over 100 mph. Jones had been a Mercury dealer in Northern California between 1946 and 1949, when he was awarded the distributorship of Mercury for Washington and Alaska. In 1955, Jones switched to Evinrude, and gained a considerably larger territory in the bargain. This, of course, immediately made Carl's old friend "Doc" Jones *the enemy.*

Jones had set up a special machine shop in Phoenix, Arizona to build high-performance racing engines, and was confident that with the right combination of boat and engine, he could surpass the Mercury record. For many years "Doc" Jones had been close friends with Hubert Entrop, the boatbuilder who had set the world's record for Carl. This was another reason that Carl initially didn't want to use Entrop's name as the driver of the boat. Friends of *the enemy*, as far as Carl was concerned, were also *the enemy*. Jones suggested to Ralph Evinrude that with Entrop's help, the record could swing back to OMC. "How much will this cost?" Ralph asked Jones. "I don't know," Jones replied, "but somewhere around $25,000." Evinrude then surprised Jones by saying, "Just go do it, I'll underwrite it *personally*." The next day at the continuing Evinrude sales meetings, "Doc" Jones got a call from OMC President William C. Scott. At lunch, in the company of L.D. "Denny" Watkins, head of OMC engineering research, Scott told Jones, "I want to tell you something. This speed attempt you have in mind is yours and Ralph's alone, and *no part of this* is company policy." Later in the day, Jones received a call from OMC vice chairman and director of manufacturing research, Joseph G. Rayniak, who was also anxious to distance himself from the proposed speed attempt. "This project is you and Ralph alone," Rayniak warned Jones, "and you are not to mention Outboard Marine Corporation in any sense."[9]

Six months later, in March 1960, Entrop and "Doc" Jones met at Lake Berryessa near Woodland Hills, California. Entrop's new 14 foot hydroplane, *Starflite Too*, was named after the new 75-horsepower, V-4 Evinrude *Starflite II* that "Doc" Jones had modified for the speed attempts. The new engine had replaced the "Fat Fifty", and even though it packed fifty percent more power, it weighed five pounds less. After preliminary trials in California, a jubilant "Doc" Jones phoned Evinrude vice president and general manager W.J. "Jim" Webb to tell him that the boat and motor were flying well over the old record. "You never saw something become company policy so fast in all your life," Jones later recalled. "Airplanes started to fly and everything happened then." On March 29, Hubert Entrop dashed through the timing traps across the still waters at Lake Havasu, Arizona at a sizzling 114.65 miles per hour, smashing the Kiekhaefer record by over 7 mph, even though the engine burned regular gasoline and never rose above 4900 rpm. In Evinrude announcements of the new record, "Doc" Jones infuriated Carl by saying, "This should prove for all time that you don't need high engine speed or special racing fuels to get high boat speed *IF* you have the *right kind* of power."[10]

Kiekhaefer responded at Lake Washington two months later, on May 5, 1960, with a new world's record of 115.547 mph, with another Ted Jones designed and built hydroplane, *RX17X*, piloted by Burt Ross, Jr, 32, a Seattle salesman. Again Kiekhaefer and Strang used a modified 60 cubic inch Mercury Mark 75H, beating the one-third larger Evinrude engine by less than one mile per hour. Carl was careful to point out that he used this "smallest of the Merc six cylinder engines because the Union of International Motorboating [the world sanctioning organization for speed records] does not recognize records set with engines over 1000 cc (61 cubic inches)." This meant that the OMC record "wasn't recognized", and that his was. Ross, however, said that he could have gone much faster, as the weather "was far from ideal, it was an overcast and cold day, with a drizzling rain and relative humidity as high as 96%."[11]

Carl was growing weary of this seesaw battle between OMC and Mercury, and decided to increase the odds in his favor. Following his new record, he dispatched an entire team of private detectives to shadow OMC driver Hubert Entrop and "Doc" Jones' chief mechanic for the attempts, Jack Leek. Entrop was never aware of the sophisticated surveillance that Kiekhaefer established first in March, and again in May 1960. Kiekhaefer received daily reports of Entrop's movements by his squad of sleuths as the young boatbuilder drove to the super market, picked up his dry cleaning, and was caught red handed in such conspiratorial activities as "sweeping the garage, mowed and trimmed the lawn." Carl was convinced that Entrop, by assisting *the enemy*, had somehow violated the agreement that he had signed before Entrop set the first record for Mercury. Carl wrote to his patent attorneys, declaring, "It is unfortunate Mr. Entrop is attempting to take liberties with the truth as he has done on a number of past occasions."[12] Even Carl's key vendors were fanning Carl's flaming anger, with comments such as "I don't think you were any more disgusted than we were to hear that Evinrude set a new record using the driver and boat combination that owed much for its development to Mercury. If they think they are such pioneers in everything, we wonder why they didn't pioneer in this." [13] Carl also suspected Ted Jones of secretly helping Entrop design and build the OMC record-setting boat, and so added Ted Jones to the ambitious surveillance effort. During many of the reports, it seems Carl was anxious to find Jack Leek's car on Ted Jones' property, but he could never locate it. When Entrop was in line at the ticket counter at the Tacoma-Seattle airport, one of Kiekhaefer's detectives was in line right behind him, reading the ticket flight information, and relaying it back to Fond du Lac before Entrop could even get on the plane. Carl had found the use of private detectives to be quite beneficial during his increasingly frequent storms of paranoia, and would regularly hire undercover agents to monitor all manner of people, including his own closest staff, like his secretary Rose Smiljanic and his vice president of engineering, Charles "Alex" Alexander.

The race for speed was getting quite personal for Carl, and any encroachment on what he considered to be his rarefied territory was taken as a direct affront. James H. "Jimmy" Jost was vice president of the APBA, and head of the Stock Outboard Racing Commission during these record attempts, and would soon join the Kiekhaefer team. When Jost wrote Carl a short letter -- "Congratulations! What took you so long?" -- after the new record, Carl wrote a letter he never mailed to Jost to organize and vent his feelings on the whole matter.

> "Please realize, Jimmy, that when Mr. Entrop absconded, he did it on the sly, lying on repeated occasions to Mr. Strang and myself about the matter and even after he had newspaper reports on their practice runs, of course. The minute Mr.

Entrop was confronted with the newspaper article that Mr. Strang read to him, he defiantly admitted it. This was the beginning of the so-called 'falling out' we had with him since he was under contract with Ted Jones, who in turn, had some developmental contracts with us. The minute Mr. Entrop knew that we knew about his 'turn coat' activities, his group accelerated the project and almost immediately ran for the record. ...

"Please understand, Jimmy, we are not making excuses, but when Outboard Marine enticed Mr. Entrop, they got, for nothing, six months of testing preparations and experience on boats, propellers, power plants and driver experience. No doubt, Outboard Marine thought they had us completely blocked. Like many other Mercury firsts, they stole this one. I need not remind you that the American record of 79.04 m.p.h. lay dormant for some twenty years until Mercury changed all that. Mercury points the way! When Mercury does something, competition can, of course, always buy it, steal it, copy it or imitate it. Should Outboard Marine get orchids for taking a 50 percent larger engine, for taking our driver, our boats, our propellers, our experience and only come up with five per cent more performance? Really, I think it is Charlie and his boys who should get the orchids as well as Ted Jones for slapping together a boat in a matter of six working days.

"Apparently this event shook competition enough and embarrassed them enough that the "syndicate" [Entrop, "Doc" Jones, Jack Leek and Evinrude] is meeting this morning in Milwaukee for their next move while 'licking their wounds."' [according to Carl's private detectives]

"The next time you send me a needle, please include a tube of vaseline."[14]

As Carl learned, "Doc" Jones and Hubert Entrop were already hard at work preparing for the OMC response to the recent Kiekhaefer record. On September 17, 1960 Entrop once again flew across Lake Havasu, this time in a new boat christened *Starflite III*, boosting the stakes in the outboard record race to a blazing 122.979 miles per hour, topping the Kiekhaefer mark by over seven mph. Entrop was quoted as saying his rig still had speed in reserve. This claim of "only part throttle" infuriated Kiekhaefer. As Carl mulled over his response, he received a report from Charlie Strang that the Kiekhaefer driver, Burt Ross, had publicly "complained that Entrop had received $18,000 for breaking the initial record, $6,000 for the second record, and received a retainer of $600 per month while he, Ross, got nothing but promises." In the same report, as if Kiekhaefer wasn't upset enough over being embarrassed by Ross, he learned that Entrop had entered the Kiekhaefer Mercury booth at the Seattle Boat Show and "said something to the effect that we had a wonderful product, but he 'couldn't stand the guy that made it.'"[15]

In 1966, another attempt was being mounted by OMC to establish a new world's record. Charlie Strang was "on ice," -- soon to join OMC as director of marine engineering -- thanks to "Doc" Jones and Ralph Evinrude, and was helping to coordinate the technical aspects of the attempt. Through his network of spies, Carl Kiekhaefer knew exactly when OMC would be testing their new 100-horsepower *Starflite 100-S* engine for an assault on the record. "Carl called me into his office one day," Jimmy Jost remembered, "and said, 'I want you to go out to California -- I want to make darn sure that they're doing everything legal and that they're not cheating.'"[16] Carl flew a photographer up from Florida to accompany Jost on the espionage assignment, and instructed him to practice taking pictures of very fast moving objects. Jost took him out on the highway in front of the plant at Fond du Lac, "drove a 100 miles per hour and had him shoot telephone poles." The film, quickly developed in Kiekhaefer's own labs, showed a problem with the camera settings, which was corrected. The following morning, Jost and the photographer were on their way to California.

Carl liked to organize his espionage missions against *the enemy* much like a well executed military maneuver, and he always assumed that OMC was equally secretive and paranoid about being observed during record speed attempts. Jost, who was president of the APBA at the time, was also Charlie Strang's closest friend, and Carl knew it. As a precaution, he ordered Jost not to make any contact with Strang whatsoever during the mission, and not to be seen by any of *the enemy*. He instructed Jost and the photographer to split-up when they reached the vicinity of Lake Havasu, with Jost on the Arizona side of the lake and the photographer on the California side. Jost would observe the preparations and runs through powerful binoculars, and the photographer was to capture everything he could through his long lenses.

Jost had arranged for the photographer to phone him once every evening at an assigned time, at a phone booth outside a rural restaurant. This way, they could exchange intelligence, and coordinate their spying efforts without fear of being observed or overheard. The first day of a predicted five or six days on assignment, everything went off without a hitch, and the phone rang in Jost's booth on schedule. But the next night, the phone never rang and Jost was concerned. The California side of the Colorado River above Parker Dam was a rugged, barren, almost lunar landscape, uninhabited and without roads. The photographer had rented a four-wheel drive dune buggy to navigate the dangerous terrain, and during the second day of observations, misjudged his speed, and the brand-new machine began to plunge down a sharp ravine. In fear of his life, he sprang from the open car, grabbing only his cameras before the dune buggy slid over the precipice and was totaled in the boulder-strewn void below.

Meanwhile, Jost, who couldn't help but contact his best friend Strang, was invited to witness the record attempts *right along side the OMC crew.* Jost, not sure whether it would be better to accept and get first hand information for Carl, or to decline, fearing Carl would never believe him and be accused of "giving aid and comfort to *the enemy,*" decided to accept and take his chances. "They allowed me to take *any pictures whatsoever that I wanted,* even to the point of shooting the propeller and the set-up, and pictures of the lower unit," Jost said.[17] It was like he was one of the group, watching the activity along side Charlie Strang, "Doc" Jones, Hubert Entrop, Jack Leek and even Ralph Evinrude himself. To make this suspicious activity look even worse for Jost, OMC arranged to have Jost moved to their own motel for convenience, and then picked up the tab when the record attempt was completed. It was Strang's idea to have Jost observe the attempt, because he wanted Carl to be confident that it was a legitimate effort, and preclude any possibility of a subsequent protest by Kiekhaefer.

On March 16, 1966, Gerry Walin drove the *Starflite IV* to a new world's record of 130.929 miles per hour. Jost, of course, recorded the entire event, and was satisfied with the timing methods and equipment to confirm the record to Carl. When Jost returned to Fond du Lac, he was again called to Carl's office. Carl was in the middle of a conversation with Ron Jones, Ted Jones' son and an able boatbuilder and designer in his own right. "All of a sudden it dawned on me that they're talking about another record attempt," Jost remembered. "I asked them, 'How fast are you running now?'"

"'About 131 or 132,'" Jones replied.

"Is this by speedometer or what?" Jost asked.

"By speedometer."

"Clocks make a liar out of speedometers," Jost suggested, adding, "Carl, I told you in my report that OMC is *sandbagging* you. They're waiting for you to run a mile or two faster -- and then look out!"

Jost was upset that Carl was going to risk life and limb just to gain another mile or two per hour on the OMC record. He asked to be excused from the meeting, but Kiekhaefer ignored him, and just kept on talking to Jones. Finally, out of frustration with Carl, Jost left the room and soon resigned.

As the emotional race for speed supremacy continued, tragedy was inevitable. Often the balance between the light-weight and nearly flying boat and the highly modified engines was such that safety became the speed limiting factor, rather than engine power. To establish a new world's record had become a supreme test of courage as well as technical skill, as time after time drivers held their breaths while accelerating alone between the speed traps of the record book and the glory beyond.

Eight years after he had set the Evinrude record, Gerry Walin was back at Lake Havasu to move the mark up another notch. Jimmy Jost, by then an OMC employee, was on location coordinating an effort to produce a promotional motion picture of the record runs, assisted by five cameras set up at strategic points along the course. The weather was dreadful, and postponed the attempt for five straight days. "The lake is so large," Jost would explain, "that it can be calm on one end of the run, but kicking waves up in the other end." Finally, on the morning of September 12, 1974, the waters calmed. Everything was in readiness, except the driver was nowhere to be found. Calls to the motel in Parker, 17 miles from the record preparations, were to no avail, and Jost was dispatched to find Walin. It turns out that they had been calling the wrong room, and when Jost knocked on the door, he awakened Walin and his wife. When Jost told him that today was the day, Walin exclaimed, "Oh my God! They forgot to wake me up."

Jost waited for Gerry Walin and his wife at the secured gate to the Parker Dam pumping station, so he could lock the gate behind them. "When he came by," Jost said, "I noticed that he didn't look good at all, and his wife looked the same way. It looked as if they had been fuedin' -- let's put it that way."[18]

Because of the critical design of the *Starflite IV*, Walin was carefully instructed by boatbuilder Hubert Entrop and "Doc" Jones not to exceed 135 miles per hour, because the danger of a "blow-over" was imminent beyond this speed. A "blow-over" occurs when too much air is rammed underneath the tunnel hull of the catamaran design at extreme speeds, causing the front end to lift uncontrollably out of the water, meeting tremendous air resistance, and is "blown over" backwards, usually with terrifying results. "Entrop knew that the boat would only be safe up to about 140 miles per hour," Jones would say. "And we wanted to run around 135, and we told him so."[19] Entrop had installed two different speedometers inside the *Starflite IV*, so that there could be no mistake on Walin's part when to back off on the throttle. "We knew that [Walin] was on the raw-edge of being uncontrollable," Jones said. "Just like a basketball player that's not coachable, you know. And we had warned him and warned him not to run it at over 140 mph, because we knew it would take off for the sky -- and that's exactly what it did."[20]

With five motion picture cameras rolling, Walin began his historic attempt. Because buoys couldn't be placed in the water to guide Walin, he picked out the top of a mountain peak in the distance and started to push the throttle forward. "He idled at the other end, and then he started coming into the traps," Jost remembered. "And we could hear that motor go and go and go, and we knew this was it." The first run across the lake was clocked at slightly over 135 miles per hour, exactly the speed they were looking for. To qualify for a world's record, another run in the opposite direction is required, and the average speed for both directions is recorded. Walin idled again, waiting for the waves to roll away from the course, and then started off in the opposite direction. According to tracking cameras, Walin was screaming down the lake at 147 miles per hour when the front of the *Starflite IV* rose towards the sky. "That thing went up into the air," Jost explained, "and did some shenanigans that are not visible by eye." Walin was ejected from the fragile craft during the first seconds of the blow-over. "There was a Japanese photographer from *Sports Illustrated* there, and with his motor drive camera he got off 18 shots while the boat was in the air. And Gerry was coming out of it and skipping across the water, and one leg was bent all out of proportion. He looked like he was doing a cartwheel, only his one leg was still in the water, dragging. ... We also heard a noise across the lake -- it sounded like a bullet whistling -- and then the splash."[21] *Sports Illustrated* reported that as Walin was thrown from the craft, his wife, Lynne, cried, "'Oh, Gerry,' and her voice seemed to hang in the air."[22]

Rescue boats reached Walin within a minute of the crash. He had broken his back, and was paralyzed from the waist down. He was rushed by the ambulance which is always standing by during record attempts, to the Parker Hospital, and soon thereafter ferried by OMC corporate aircraft to a major hospital in Phoenix, where he remained for well over a month in intensive care in critical condition from spinal, thoracic and leg injuries, and underwent heart surgery. "We tried to do everything we could for Gerry when he went back home," Jost explains. "He had beautiful three-story redwood home up in the mountains outside of Seattle." Because of the configuration of his home, OMC sponsored the installation of a special elevator that would allow him to ride his wheel chair up the side of the steep stairs. "I was ordered to go out there," Jost remembered, "to let them know that there would be some fellows out there soon to put it in." Jost spent a pleasant evening at the Walin home, "just like a group should, talking about boat racing until we got blue in the face. And Gerry was in the best of spirits." A few weeks later, Gerry Walin connected a hose to the exhaust of his Volkswagen, sealing it inside the car and killed himself. "His wife had left him," Jost said quietly, "she went to Hawaii with somebody. ... That was the final straw."[23]

On November 30, 1989, Robert Wartinger of Seattle established a new world outboard, straight-away speed record at 176.556 mph. The record was set on a three mile stretch of the Colorado River near Parker, Arizona, this time aboard a 20-foot hydroplane powered by an Evinrude 3.0 liter, V-8 outboard rated at over 500-horsepower. Wartinger, a veteran record-breaker, set his first world record in 1971, and since then has established 47 separate records in various outboard classes.[24] The race goes on.

CHAPTER THIRTY

The Limits of Patience

"My confidence in Brunswick, of course, has never recovered since the time I was supposed to have been 'fired.' How such a situation could have developed to the point it did, will someday have to be explained to me. ... I can assure you, ... that had this move come to pass it would have been the most horrible mistake of all on the part of Brunswick. The action would have been so drastic and the investigation so intense that any survival would have been questionable. ..."[1]

Carl Kiekhaefer

When Charlie Strang and Tom King left Kiekhaefer Mercury in the late summer of 1964, Carl promoted Charles Alexander to vice president of engineering, and Armand Hauser to vice president of marketing. Though the loss of Strang and King was strongly felt throughout the organization, Carl was confident that Hauser, a 20-year Kiekhaefer veteran, and Alexander, most recently chief engineer for *Mer-Cruiser* stern drives, could meet the challenge. In 1965, nearly 43,000 engines rolled down Kiekhaefer's lines in Wisconsin, and by 1966, over 6 million outboard motors were in use throughout the world.

"Let's be realistic and recognize that these are wonderful years for Mercury," Carl announced to his dealers on the 25th anniversary of Mercury in 1964. "We are associated with a product that is enjoying a degree of popularity and public acceptance that is almost unmatched in history and certainly is without parallel in the industry."[2] Things were going Carl's way in the marketplace, and he was rapidly closing the gap on *the enemy*.

"While Outboard Marine has made its mistakes and has lost some of its dominance and stature in the marine industry, they will be like a wounded bear, ready to fight back with new weapons. ... The boys that built our competitor's business, the old heads, have already retired and have left in their place a lot of youngsters who are fumbling around at the moment. ...

"We've had our share of good luck, too ..., we were very fortunate that the competition got fat, dumb and happy in their own business - got so self-centered that they didn't pay any attention to anyone."[3]

Carl was awash in the changing currents of emotion he felt about his company. He was undeniably proud of the progress that Mercury had made against OMC, as well as the thirty-three companies that had come and gone in the industry since he had opened shop over a quarter century before. His bitterness, though, over the near dismissal by Brunswick president Jack Hanigan had left him even more paranoid and insecure about his future, both corporate and private. Even his best friend, Charlie Strang, had become *the enemy*. Who could he trust?

On the corporate side, his suspicion swelled to such a degree, that he imagined virtually everyone in his organization was secretly passing disparaging information along to Hanigan or other Brunswick officials, and that a trap was carefully being laid for his embarrassment and eventual dismissal. On the personal side, he began to suspect that Rose Smiljanic, his secretary and personal assistant for fifteen years, was once again secretly seeing someone else within the organization. He must have thought back over the years, remembering his embarrassment and anger over her clandestine rendezvous with Jim Wynne, and her secret flirtations with race boat designer Ted Jones. A fresh target of his paranoia was the new vice president of engineering, Charles "Alex" Alexander.

Carl leaned hard on Alexander almost from the moment he became head of engineering. Alex had been working for Kiekhaefer Mercury for eleven years, ever since Jim Wynne and Charlie Strang had recruited him to leave the U.S. Naval Air Missile Test Center at Oxnard, California. He was accustomed to the long and irregular pulse of work within Carl's remarkable company, but he had always had the buffer of Charlie Strang to insulate him from Carl. "One thing I remember," Alex said, "is he used to tell people, 'I pay my engineers twice what they're worth, and then I make them earn it.' And he did that. And you see by paying you the extra money, he had you on the hook."[4]
As Charlie Strang was preparing to leave the company, he recommended to Carl that Alexander take over as head of

engineering. "Charlie told me one day, 'I've got to get out of here while I still have my sanity.'"

Though Carl traveled extensively with Rose Smiljanic, often she would remain behind on his forays throughout Mercury's expanding empire. Carl began to suspect that Alexander and Rose were having an affair. No one knows for sure what suspicious behavior first made Carl aware of their subterfuge, but as his suspicions grew he undertook a complex and sophisticated program to trap them in a compromising situation.

Alexander was perfectly aware of the ongoing relationship between Rose and Carl. "Well, you know that from my own position, it was quite obvious what was going on all the way along. It didn't bother me personally," Alexander admitted. "Kiekhaefer's relationship with Rose and his wife didn't bother me or anybody else within the company that I knew of. In other words, I never heard anybody say, 'Well, God damn, I'm a little upset with him because he's doing this. ... Hell, I figured he could do what the hell he wants to do. It annoyed the hell out of my wife. Because she viewed that kind of behavior as a threat to her own security. ... She wasn't all that thrilled about Rose.'"[5]

Enough employees were aware of something going on that word began to spread, and before long, Kiekhaefer began to suspect just about everyone who had access to Rose. Carl once accused Owen L. "Billy" Steele, a long-time Kiekhaefer Mercury employee of "going to bed with Rose." Steele was ordered out of town unexpectedly by Kiekhaefer one day, and asked Rose if she could loan him money for traveling expenses. Rose obligingly gave him a $100 bill. Sometime later, on a subsequent trip to a race at Lake George in upstate New York, Carl inadvertently observed Steele returning the loan in the hotel in which they were all staying. "I met Rose in the hall," Steele recalled, "and reached in my wallet and I handed her a $100 bill. [Carl] walked out the door just as I was handing it to her, and he thought I was bribing her to go to bed with me. He didn't say anything in front of her, but later ... he said, 'Here you are trying to proposition my secretary! Have you been screwing her?'"

"Come on [Carl], you know better than that," Steele replied. "I realize she's your girl, and I've never touched her." He was jolted by the accusation, though, and wondered if Carl had heard the same rumors. "I always heard that Alexander was taking Rose out. ..."[6]

In order to prove his suspicions about Alexander and Rose, Carl had to enlist the support of several trustworthy employees, especially chief of maintenance operations Wilson Snyder, and assistant to the president Fred "Fritz" Shoenfeldt. "He trusted me, I think, more than anybody cause he knew I was honest," Fritz explained. "I never lied to him, and mainly because he could trust me around Rose, I think basically that was why he put his trust in me."[7] Carl also trusted Wilson Snyder greatly, a fellow farmer and rough and tumble employee and confidant who was often the first to be called when a personal favor was needed by the boss. "I think the old man found out himself," Snyder reflected on Carl's knowledge of Alex and Rose. For some time, Rose Smiljanic maintained an efficiency apartment at the Dartmore Motel in Fond du Lac, though spent a great deal of her free time at Carl's home overlooking Lake Winnebago. "She was staying at the motel," Snyder explained, "and Alex's car -- I'd go by the motel and I'd see it. Fuck yeah, he didn't give a shit. It wasn't no secret. ... I knew she was running around, but shit if I would say so."[8] Though both Fritz and Snyder agree that they eventually were coerced into telling Carl details of the clandestine meetings between Rose and Alex, both are equally convinced that the other "spilled the beans" first.

Fritz Shoenfeldt had been working for Carl for nearly twenty years when Carl phoned him from Lake X one evening with a startling question. "I just got home for supper. The phone rings. 'Fritz, what the hell do you know about Alex and Rose?'" Carl demanded. "Wilson [Snyder] tells me that you know all about it." Fritz was shocked to be so openly asked about such an intimate and dangerous subject. He was also afraid of losing his friendship with Rose, with whom he worked nearly every day, and through whom Carl communicated to just about everybody in the company. Further, Fritz was living on a farm, and was speaking on a party-line telephone that had extensions in ten other farms in the area. "I didn't want to get involved in it," Fritz recalled. "*No way* did I want to get involved in it. ... But, you know, he didn't want to believe it at all. ... Even though it's obvious. And that's what made Wilson so mad, that everybody was talking about it. And the boss didn't know, and he felt that he ought to know about it. And I said, 'I wasn't ever going to be able to tell him.' Then, of course, Rose had it in for me for a long time."[9]

"Carl, I really would not like to discuss it," Fritz replied. "Enough people know about it already. Let's leave it 'till you get back."

"I'm leaving *right now!*" Carl snapped. "I'll be up tonight." At one o'clock in the morning, Fritz Shoenfeldt and Wilson Snyder parked on the apron of the taxiway at the Fond du Lac airport, nervously awaiting the approaching sounds of the corporate aircraft. "What have we done?" Fritz said desperately to Snyder in the car. Kiekhaefer sternly grilled his two shell-shocked employees after his arrival, and, according to Fritz, "We filled him in." Foremost on Carl's mind were methods to catch Rose and Alex together, and a number of ideas were discussed. "If you don't want to believe us," Fritz told Carl, "just make off like you're going to Chicago for a trip. Fly out and turnaround and land in Oshkosh and find out." Carl was visibly upset. "Then, around ... five o'clock in the morning I get a phone call at home. 'Fritz, it's Wilson. Get your ass up here, the old man is chasing Rose around in the woods!' So that night, I don't know what happened. She didn't show up for work for [days]. When I got up there they were gone and he had her in the car, and they went down to the motel where she lived."[10]

Carl began an intensive campaign to confirm his suspicions, and decided to try the "doubling back" ploy from a pseudo-business trip with the company airplane. For this he needed the help of one of his pilots, Neil Oleson. Oleson began working for Kiekhaefer in 1965, following 15 years of crop dusting in North Carolina. Oleson explained to Brunswick chief pilot Wayne Predmore how one of Carl's secret missions took place. Oleson had flown Kiekhaefer to Chicago's O'Hare airport in one of the company twin-engine Beechcraft model-18 aircraft, so that he could transfer to a commercial airline that would start him on his way to Australia.

> "And the instructions from Rose were, that he was to help Mr. Kiekhaefer with his bags and actually put him on the airplane. So, Neil did that, and he got him on the airplane, and Neil saw the door close on the airplane, and he saw the tug pushing this 707 back away from the gate. So, then he went over to the phone and he called Rose and said, 'He's on his way. ...'
>
> "Neil got back in his airplane, the Twin-Beech, and he dead-headed back to Fond du Lac, put the airplane away, and he was just ready to go home. ... The phone rang. It was Carl.

He says, 'Is my car in the hangar?' and Neil says, 'Yeah.' He said, 'Well, bring it up to Oshkosh -- come up here and get me.

"Now, we don't know to this day how he did it. That airplane had to come back to the gate. He had to insist. Got off of that 707 and got on [another] flight to Oshkosh, and had Neil drive up to Oshkosh and brought him back into town so nobody would know. ... Rose was living at the Dartmore, and he drove through the parking lot. ... I believe Alex's car was there."[11]

Carl would repeat his mysterious "false starts" on many occasions, and it kept a number of employees perpetually nervous. "He would go to Florida," Neil Oleson remembers, "he'd say, 'Take me to Chicago.' We'd take him to Chicago. I'll put him on an airplane. He'd be heading for Florida. ... I'd be at home and he'd call me up and say, 'Pick me up in Oshkosh.' I picked him up in Oshkosh, drive him back down here [to Fond du Lac], and he was just coming down here to check on people. ... Might be her [Rose], it might be Alex, it might be any of them."[12]

It became so nerve-wracking that people were afraid to even be *seen* with Rose, for fear that Carl might misinterpret the circumstances. One night Oleson got a call from Rose, who had some papers that needed to be on the plane with Carl in the morning. "She said, 'Can you come over and get them?' I said, 'Yeah, I'll be right over.' I hung the phone up, and I started thinking about this. 'This isn't going to be right', I said. 'Just as I'm coming out of that room, Kiekhaefer might be coming in.' I don't want none of that, so I get back on the phone and I call Miss Rose and I said, 'Miss Rose, I changed my mind here.' I said, 'You better bring them over to the hangar.' I said, 'I don't want to get caught coming out of that room when Mr. Kiekhaefer might be coming.' She said, 'That's a good idea.' So she brought them over to me."[13] Another night, after Oleson had flown Carl back to Fond du Lac at two in the morning following meetings in Detroit, Carl "got in that car and burned out of the hangar just like he was going to a fire. I don't know where he went to, or what happened. ..."[14]

As the evidence mounted against Rose and Alex, so did Carl's frustration and level of commitment to expose them. Through Armand Hauser, he hired a full-time private detective to shadow Rose, and to report her movements when Carl was out of town.[15] The trouble with this scheme was that the private detective drove around town in a fire-engine-red Mustang convertible, and almost from the moment of his arrival, many people knew who he was, and why he was in town. "We knew he had a guy watching her," Wilson Snyder says, "but I don't know what the hell he watched. Because when Rose wasn't in the motel, she was up to the God damn ledge [at Carl's house]."[16] Once the detective lost his usefulness as an undercover agent, Carl hired him as a security guard at one of the plants.

Carl then approached Fritz Shoenfeldt with a special, dark and highly secret assignment. He asked him to acquire a secret bugging device, and to have it installed above Alexander's office, to discover when Alex and Rose were to meet so he could catch them red-handed. In the 1960s, remote listening devices were not only illegal, but difficult to obtain, and to actually plant one in the office of a high-ranking executive was a dangerous request. But Fritz was steadfastly loyal to Kiekhaefer, and obeyed without question. Fritz immediately contacted Wilson Snyder, whose responsibility it would be to actually obtain and plant the device above Alexander's office. "I'm the one that got the bug and planted the son-of-a-bitch in his office," Snyder admitted.

> "He [Carl] asked Fritz, as much as I know, to have me put it in because I had keys for even his office up there. I had keys for the whole fucking plant. The old man trusted me, really. I had keys for all his fucking cars, I had keys for everything. ...
>
> "The only ones that knew about the bug was me and Fritz. I went down to Milwaukee and got that from someone. You know, they're illegal to have. They are illegal to own."[17]

Wilson Snyder crawled above Alexander's office late one night, and carefully planted the bug over the vice president of engineering's desk. It was quite small by the standards of the day, "like a tie clip," Snyder remembered. It wasn't, however, the most effective installation, for the ceiling tiles were quite thick, and Snyder was afraid to expose the device. "I know the wire should have been through the ceiling, but, shit, you could see it then. ... We could have changed some ceiling tiles there, with some with holes in them." The device was designed to be received by an ordinary FM radio, which Fritz had placed in his office. From the receiver, recordings were made on a tape recorder or Dictabelt machine. "We had a little radio," Snyder remembered. "You turn the radio on and listen, and tape it like anything." The signal from the device was strong enough, that it could even be heard on a carefully tuned car radio if it was parked close enough to the building.

Carl Kiekhaefer and Fritz Shoenfeldt would meet in Fritz' office early in the morning, shortly after Alexander got to his desk. They would turn on the equipment and wait for Alex to phone Rose at the motel. "See, Alex lived in Oshkosh," Snyder said, "and he'd come in early and then he'd call her on the phone. And she'd be at the hotel, yet. And that's when they were listening in." The bug was in place and monitored regularly for quite some time. "God, it must have been a couple of months," Snyder recalled. "He [Carl] just wanted to hear him [Alex] make a rendezvous with Rose so he could catch them." Snyder was uncomfortable with the whole concept, however, and would rather that Carl had left Rose in peace. "Yeah, that's where, you know, I often thought, 'What the fuck does he give a shit about whether somebody is fucking Rose or not. Christ, he was out fucking a couple of other old bimbos around town anyway."[18]

Actually, Carl was always on the lookout for sexual outlets. From the house of negotiable affection in Chicago where he used to entertain military brass, to regular visits to a house of ill-repute in Milwaukee, Carl was known to his intimate employee-companions as a man of great sexual appetites. When the weather was bad and company pilots couldn't pick Carl and his party up in Chicago, sometimes Wilson Snyder would drive down from Fond du Lac to shuttle them home.

> "There were a couple of places in Milwaukee where we would stop. I don't remember where they were. I know one place had a doorman. Kiekhaefer would always give that guy about fifty bucks, and he'd leave us parked right in front of the fucking door, you know. They stopped and there was some old bags come fucking around. ...
>
> "But the old man he'd spend money and, Christ, he'd tip them and everything, you know. ... It was a whore house, that's what it was. ... Christ, they wanted me to go in the back room with them. They was too expensive for me."[19]

When Carl would attend the New York Motor Boat Show in New York, he would ask Billy Steele to line up some feminine distractions in the evening. "He'd say, 'Steele, where's all the women?' You know, after he's got a half load on him."

"Hell, I don't know," Steele would answer.

"God damn it! Find some! You've got to be good for *something!*" was Carl's urgent reply. "The paying kind."[20]

On many trips, Rose would accompany Carl, and it would make for some rather comical schemes to get Rose out of the way to clear the decks for Carl's indiscretions. "She was in the same hotel, usually in an adjoining room," Steele said. In fact, according to Joe Swift, the young and aggressive public relations coordinator who joined the staff in 1967, Carl once paid thousands of dollars to have a connecting door built between his and Rose's suite at Hilton Head, South Carolina.

Perhaps the classic episode of Carl playing 'hide and seek' with Rose Smiljanic occurred in Miami, on the occasion of the Miami-Nassau powerboat race in the early 60s. Then Director of Public Relations Frank Scalpone, who today is a top executive with the National Marine Manufacturers Association (NMMA), remembers how Carl pulled off the minor miracle of an evening alone in the middle of race-weekend chaos.

> "We all came down to Miami for the race. We would bring down maybe twenty-four drivers, two to a boat, and twelve boats, and we put them at Dave Craig's Marina on 79th Street. ... And right around the corner on Biscayne is the motel we had, a pretty seedy motel, and [Carl] puts everybody in there.
>
> "Well, Rose and he go down to what was then the best hotel, the DuPont Plaza. They had adjoining rooms, they always did. Everyone is getting ready for the race, but one of these storms comes up. And the Gulf Stream starts boiling and boiling, and you can't race. So it's postponed. Once the Gulf Stream starts acting up it takes three or four days. So, the second day comes and we get up in the morning and no chance of a race, so Red Crise [the race organizer] calls it off. Everybody is getting a little restless. We were playing cards, what else could you do. All the guys, we had a crap game going in this motel. ...
>
> "Carl would come to Miami alone, once in a while, and Dave [Craig] would fix him up with a girl. You know that's quite true. He did, and it was a regular thing. Carl was getting restless, you know, he had been with Rose for a long time, and he's stuck with her down there and he's getting restless. The race is not going to come off for another couple of days. The third day he gets Rose on the phone. He's up with us and pacing the floor thinking, and Rose is at the DuPont Plaza and he gets Rose on the phone and says, 'Rose, these poor guys are going to run tomorrow for sure, and they're going to go over there [to Nassau] and they're going to get there and they're

going to be tired and dirty and they're not going to have any clothes, whatever, as they put them in the boat they're going to get wet. I want you to go to those guys, collect all their shaving kits and a set of clothing, or whatever they need, put it in the Beechcraft, go over to the Pilot House [hotel in Nassau], check them in, because sure enough we're going to go tomorrow.'

"Poor Rose goes around to all these guys, gets their shaving kit, gets a change of clothing, gets the pilot, loads up a couple of cabs, gets to the Beechcraft, flies over there, checks all these poor guys in -- twenty-four guys -- into the Pilot House and stays overnight -- so Carl and Dave Craig can fool around!

"The next morning, and sure enough this is the truth, he wakes up. There's no race. 'Where's Rose? Get Rose on the phone!'

'Rose! What are you doing over there? These poor guys don't have any clothes, they can't shave!'

"She loads up; she checks them all out, loads up the Beech; brings all the stuff back again. Honest to God."[21]

Paradoxically, as Rose Smiljanic recalled the episode, when the pilot showed up in Nassau to bring her back, she thought Carl "must have sent them over to get us cause he was probably worrying that we might meet some guys and be having a great time with them. ... [H]e always checked up on me."[22]

As crafty and perfect as Carl must have imagined his various schemes to evade Rose's watchful eye, she kept pretty well informed of his antics anyway. "I always found out," she explained. "The fellows always told me, I don't know why. But anytime anything would happen, somehow I would always -- I didn't go out of my way to find out -- but somehow you hear a little from here, a little from there, and I could place the whole thing. So what I did was to trick and I'd say, 'Oh!' I didn't want to get the other fellows in trouble because, you know, he would fire a person."[23]

When Carl had achieved his purpose in secretly monitoring Charles Alexander's phone calls to Rose, Snyder removed the bug and stored it at his house in Fond du Lac for many years, eventually giving it to Carl. "We thought Alex was done," Snyder said of the period following Carl's discovery. "Fuck, we thought he was done. ... Yeah, because he'd fire you because there was mud on the field, and then Alex stays. ... That's what me and Fritz used to say, 'Geez, I don't understand it. She [Rose] didn't have nothing on him [Carl] that we didn't know, you know, like they'd say she had something on him."

Much speculation occurred as to why Carl, who was capable of firing people on the smallest of pretense, spared Alexander. One theory is that because of the embroiled and delicate nature of his position with Brunswick, and his soured relationship with Jack Hanigan in particular, that he dared not risk the public disclosure of his reasons for dismissing Alexander. Certainly the airing of Carl's dirty laundry in the halls of Brunswick in Chicago would have destroyed any remaining influence Carl enjoyed with Hanigan or the other, quite conservative members of the board of directors. And worse yet, should Alexander or an injured and vengeful Rose Smiljanic learn of the office bugging, Kiekhaefer would have been open to at best a financially crippling personal law suit, and at worst a criminal prosecution and conviction for illegal wire-tapping. Alexander would have to stay. Besides, Alex was a most capable engineer and administrator. So soon after the loss of Charlie Strang, Carl could ill-afford to squander additional key engineering talent.

Whether due to Carl's constant pressure, or due to Brunswick's own carefully measured resources, money became available for critical expansion at Kiekhaefer Mercury in 1964. For the three years following the merger, Carl made relatively minor additions to facilities, and argued vociferously that the golden opportunity to overtake OMC and become the world's leader in marine propulsion was slipping through their collective fingers. In 1964, a major expansion was finally authorized, and Carl embarked on the most ambitious building program in the history of the Kiekhaefer Corporation. But as Carl prepared for construction of a mammoth, 180,000 square foot distribution building in Fond du Lac, and finalized plans for an even larger, 372,000 square foot assembly building, a political blunder by the City of Fond du Lac threatened the future of the company in Wisconsin.

The proposed $4.5 million investment, part of a forecasted $10 million expansion program, screeched to a halt amidst a blaze of Kiekhaefer anger in late February 1964. The City of Fond du Lac voted to annex the properties and buildings of the Kiekhaefer Corporation, adding substantially to the taxable industry available to the city treasury. Carl was furious. He told the city that he would not only halt his proposed expansion in the community, but actually pull up stakes and move away, perhaps even to another state. Carl published a terse statement of his convictions in the Fond du Lac *Commonwealth Reporter*.

> "The representatives of Fond du Lac made us a promise when we came to build plants and create a local industry, and that promise to a struggling small company was that the city would never annex the property upon which we built. That promise was broken ... when your city government did in fact annex our property. We consider that a breech of faith. ...
>
> "We had a plan to build our modern consolidated facilities right here. We think the multimillion dollar construction program could also be a turning point in the history of Fond du Lac, for it would mean literally thousands of new jobs and thousands of new families coming here. ...
>
> "However, because your city government and ours (for we live here also) broke the promise made to us many years ago, we are not willing to begin on this project in the Fond du Lac area until we receive from the city a simple statement that the city will never again annex our plants and lands unless we ask them to."[24]

The City of Fond du Lac erupted in a shower of support for keeping the Kiekhaefer Corporation and its internationally recognized product line in the community. Dozens of prominent businesses and individuals lent their names and lobbying efforts to the cause. Kiekhaefer and project coordinator Armand Hauser sat stoically with their arms crossed at public hearings that drew the largest spectator crowds in local governmental history. Carl would not budge. Ultimately, in the face of nearly universal community support and Carl Kiekhaefer's poker faced bluff of moving the Kiekhaefer Corporation, the City Commission relented, and rescinded their plan of annexation. On April 26, 1964, local residents were relieved to see the eight-column banner headline in the morning paper which read, "Kiekhaefer Will Build Plant."[25]

The Brunswick Corporation, not Carl Kiekhaefer, would have had the final word on any decision to move all nine Kiekhaefer Corporation plants out of the city, or even more unlikely, out of the state. Fortunately for Carl, the city decided to reverse their order of annexation, as very likely the Brunswick Corporation, in the absence of Carl Kiekhaefer, would have allowed the annexation to proceed unimpeded. Carl, too, never wanted to leave, and admitted feeling a common bond with the predominantly German ancestry of the community and the dedicated work force which he had carefully trained for over a quarter century.

> "... We believe in the direct, practical, shirt-sleeve approach to a problem and not in the ivy tower, abstract, theoretical attempted problem-solving. ...
>
> "It is true that we have many disadvantages with a plant location in this area; however, if we count our assets, we can counteract some of the disadvantages -- for example, pride and skill of the craftsmen are the most important factors in building a quality product. In this area, the skill and productivity of the Teutonic races are a big asset."[26]

Carl broke ground for his giant assembly building in spectacular, and typical Kiekhaefer fashion, on July 10, 1964. Where some executives would be content with the simple ceremonial silver shovel ground breaking tradition, Carl took the opportunity to deliver a command performance. Weeks before the ground breaking, Carl decided to organize a parade of earth moving equipment to parade to the site of the proposed "mile long factory" building on the corner of Hickory Street and Highway 41 on the outskirts of Fond du Lac.

Carl, though, wasn't content with the giant Caterpillar bulldozers which he already had in the Kiekhaefer construction arsenal. He decided to acquire a Euclid earth moving tractor, one of the largest and most complicated -- and most expensive -- pieces of construction equipment available in the world. With an articulated body, and *two* enormous diesel engines, the behemoth earthmover bristles with levers and controls, and its enormous bucket can rip a deep and wide swath across an excavation site. Carl ordered one delivered to him in Fond du Lac, and didn't even bother to negotiate its $75,000 price tag until it was delivered. When he saw the gigantically impressive and freshly painted leviathan of heavy machinery roll off the special railroad car, he was transfixed. He could see the gleaming monster rolling before his assembled executives and Brunswick visitors, and he could see it leading the parade on the morning of the ground breaking. Well, since it looked so good, why not buy *two*? He shocked the delivery personnel by asking how much he would save by buying a pair of the machines, like he was bartering with a Tijuana street vendor for a pair of straw hats. The home office confirmed it would shave $10,000 from the price of each, and according to Wilson Snyder, "Carl had to have it the next day."

> "We delivered it up on the ledge, there. ... Shit, you know he didn't even know how to start that. But, you know, he'll never say nothing, you know, 'Show me how to run it.'... Christ, they'd think he was a dummy. ...
>
> "And then he brought a couple of guys up from down South, you know, that knew how to run equipment, that run the same kind. They drove them in the parade."[27]

As far as anyone knew, Carl never received instruction in how to operate the complicated earth mover. The audience, including Brunswick Chairman Ted Bensinger and President Jack Hanigan, was indeed impressed on that July morning as the giant machines lumbered down the service road to the ground breaking site. Frank Scalpone and "Ham" Hamburger coordinated festivities for the occasion, including a long string of boats on trailers with *MerCruiser* stern drive engines and Mercury outboards that followed the awe-inspiring Euclids. After Carl turned over the first shovel of dirt with a chrome plated shovel, forty guests were given silver painted shovels and duly turned their few pounds of dirt. Carl, though, quietly slipped away while the guests were rummaging around in the dirt with their shovels, quickly changed into a pair of size 46 farmer overalls, and mounted the giant Euclid. Wilson Snyder was terrified.

> "Then the old man, he comes with his pair of coveralls on and he goes over by that big Euclid ... and he climbs on. I said to Andy [who worked with Snyder], 'Jesus Christ! Get the fuck out of here!' I didn't know he knew anything about it.
>
> "And you know, he slapped them levers on the back, let the bucket down, you know, and he's coming off the ground about one inch -- just like a professional with that fucking thing -- he went right down the line. ... He dumped the son-of-a-bitch, he come back and he cut the next one the same depth. You'd think he was running the fucker for six months!"[28]

Carl began a building frenzy, racing to regain the time lost waiting for Brunswick to authorize expansion. In 1965, the board of directors of Brunswick Corporation met in Fond du Lac, toured the new building, and gave Carl the opportunity to tell them how Kiekhaefer put up buildings.

"It began with an immense mile-long parade down Main Street to this site. Included in this parade were all our construction equipment, many of our semi-trailers, cars, boats -- the works.

"We broke ground not only with silver shovels, but immediately set to work in earnest with this giant Euclid earth mover. Gentlemen, we build plants in a hurry, and we build them for four dollars a square foot. In just over five months from the date of the ground breaking this plant was in operation."[29]

Over the next four years, Carl would supervise the construction of over 1.5 million square feet of additional Kiekhaefer Mercury production, engineering, sales and storage capacity, primarily in Fond du Lac, with major new facilities also built in Cedarburg and Oshkosh. Though Brunswick Corporation was still struggling to regain profitability following the collapse of bowling markets, sales and profits of Kiekhaefer Mercury products continued to soar. *Fortune* magazine put it bluntly: "The leading money loser among the 500 largest industrials in 1965 was Chicago's Brunswick Corp., which racked up a deficit of $76,932,000."[30] *Forbes* was no less kind, reporting in 1965 that, "Nearly $400 million of the company's $650 million in assets consists of accounts receivable for bowling alleys sold long ago. Until this mountain of bowling paper is substantially reduced, it will be hard to tell whether Brunswick is simply on the downgrade of the roller coaster -- or heading into a bottomless pit."[31]

It seemed that the harder Carl worked, the farther behind Brunswick was getting. He was rapidly making up his mind that it was time to get out. It was time to try something new.

CHAPTER THIRTY-ONE

The Stuff of Legends

"One is tempted to reminisce: the many occasions, serious, humorous, trivial, momentous (at the time), hundreds of them come to mind as we go back over the years.[1]

"The stories about me seem to refuse to die and are getting more fantastic with my age. Intended originally to disparage my name and our good company, they seem to have picked up momentum by others until they are completely illogical."[2]

Carl Kiekhaefer

Carl Kiekhaefer's notoriety as a brilliant, if not eccentric empire builder spread throughout industrial America, along with scores of humorous anecdotes attributed to his activities. The classic firing of the loafing non-employee who was actually delivering materials to his plant, and the omnipresent story of Carl burying his Cadillac at Lake X, have been discussed already. The more Carl would become frustrated in his ongoing battles with Brunswick or OMC, the more flamboyant would his occasional escapades become. Though some "Kiekhaefer stories" grew out of proportion to actual incidents, perhaps due to the pressures and frustrations of working for one of the most demanding employers in industry, an equal number are absolutely, and often comically true.

Carl's capacity for work was virtually matchless, and one of the surest and quickest ways to experience his wrath was to shirk responsibility or duty in any way. When Brunswick Corporation authorized the modernization and expansion of Kiekhaefer Mercury facilities in Wisconsin, funds were also released for the construction of badly needed engineering and testing facilities at Lake X in Florida. During the height of expansion activities there in the late 60s, the local population of construction crews grew considerably, and occasionally Carl would take the discipline of individuals into his own hands. Joe Swift, who joined the ranks of Kiekhaefer public relations staff in 1967, remembered when Carl reached his limit of patience one afternoon in nearby St. Cloud.

> "There was a lot of concrete poured in Lake X, and Kiekhaefer had a constant string of ready-mix trucks coming out from St. Cloud to Lake X. There was a guy on the crew at Lake X, a supervisor or something, who almost every morning would have to go back into town to get a finger wrapped up by the doctor, or have his ears examined, or some excuse that Kiekhaefer felt was phony.
>
> "Kiekhaefer followed him in one day and found that he was parking his car in front of a lady's house whose husband had gone off to work. Kiekhaefer drove out to the main road, and diverted one of the transit mixer trucks. This guy had a convertible, and the top was back. [Kiekhaefer ordered it to] put about three yards of concrete in this convertible. It just went down, you know, and the truck drove off. And now, of course, by the time the guy came out the concrete may not have been hard yet, but it was setting up and there was no getting it out of there anyway.
>
> "Kiekhaefer ended up firing the guy, but he bought him a car and he paid the City of St. Cloud to bring a derrick, or whatever, over there and get the car to the dump. So, it cost [Carl] a pretty good tad, but he had the satisfaction of teaching that bastard, 'Don't you ever lie to me again.' Now that's a true story."[3]

Another of the fascinating Lake X stories is the eye-witness account of Charlie Strang to Carl's intrepid skills as a negotiator and buyer of heavy equipment. If there was anything that Carl enjoyed more than operating the largest construction equipment, it was bargaining for its purchase. Strang relates a tale of how Carl determined which of several competitive front-loader tractors to buy.

> "I recall flying into Lake X ... at the time he was carving it out of the wilderness. He wanted to buy a pay-loader or front-loader -- a big earthmoving machine with a giant scoop at the front end. There at the base, arrayed on a large field, were a half-dozen competing brands of earth movers, a like number of great heaps of sand -- and Carl with a batch of stopwatches and data sheets.
>
> "As you may have guessed, he ran a race between the behemoths, timing them as they furiously moved the sand heaps from one end of the field to the other! Then he switched drivers between machines, and raced them all again. When that dusty day was done, he had selected and bought the best machine -- on a competitive basis!"[4]

Career pilot Wayne Predmore secured the confidence of Carl Kiekhaefer on the first day they met in November 1965, when Carl characteristically insisted that he come to work for him immediately. Predmore had been working for Butler Aviation in Illinois for five years, and told Carl that he would have to give his company two weeks notice before starting. Carl glared at Predmore, and the young 34-year-old pilot felt that the alluring job might disappear any moment. But then he said, "Mr. Kiekhaefer, I really have to give Butler Aviation two weeks notice, but I want you to know that if I came to work for your company, I would have the same consideration for your company. I would never walk out on your company either. And he looked me right in the eye, and he said, 'That's fine. You come when you can.'"[5]

Aboard one of the many flights that Predmore made to Lake X in Florida from Fond du Lac, Wisconsin, a young engineer made a fundamental mistake that would prove fatal to his career: he beat the boss at gin rummy. Many Kiekhaefer employees succumbed to Carl's insistence on a game of gin, but most were secure in the knowledge that next to being beaten by *the enemy*, Carl hated losing at cards more than anything else. Frank Scalpone, Carl's head of public relations, remembered the agony of sitting in the noisy twin-engine Beechcraft model 18 aircraft on a long trip with Carl, as he was systematically whipped at gin. The system, however, was Scalpone's.

> "I'm a violent non-smoker. The greatest strain is six and a half hours in the Beech, Florida to Fond du Lac and back, and he smoked his cigar and played gin rummy -- and *you have to lose*. It's *hard* to lose. I'm a pretty good gin rummy player, and it's much harder to lose than win because you can't get caught. You can't obviously get caught. It's a great skill. I was good at it."[6]

An uninitiated young engineer was making the trip with Carl to Lake X, and Wayne Predmore, assisted by a co-pilot, was at the controls. The new employee had been hired less than a week before he sat down in the soft, leather seat of the aircraft across from the legendary industrialist. The Beech 18 had a small, stowaway table that Carl would noisily slam into place when he pulled it up and out from the fuselage wall. Carl always sat facing aft, on the port side of the airplane, virtually back to back from Wayne Predmore in the front left cockpit seat. Separating the cockpit from the cabin was a light, accordion style door to provide pilots and passengers a degree of privacy.

Carl informed the junior engineer that he wanted to play gin, and brought out the deck of cards usually found on the airplane. The flight was proceeding smoothly somewhere over Tennessee, and Wayne Predmore decided to check on the passengers to see if he could get them any coffee or soft drinks. Before long, according to Predmore, the brash young newcomer started to beat the boss.

> "I went back to serve them. He was just a young fellow, and they were playing gin. Mr. Kiekhaefer was not a good loser, and this guy was beating Mr. Kiekhaefer. We were maybe an hour out of Atlanta, and I went back to see if I could get him one last cup of coffee, and also to advise him of our arrival time in Atlanta.
>
> "And they were engrossed in this card game, and so they kind of ignored me there. Mr. Kiekhaefer dealt the cards, and after their hand was dealt, he took the engineer's hand and kept that for himself, and gave the engineer his hand. Now -- he had dealt, and yet he still switched the hands.
>
> "The engineer picked them up and he had a 'no-brainer,' he just had to lay down a gin hand! He hadn't even touched the cards except to pick them up. And Mr. Kiekhaefer said, 'I want you to land in Atlanta.' And we landed in Atlanta, and he paid him off in cash, because he 'didn't want card sharks with our company!'"[7]

When consumer advocate Ralph Nader exploded into prominence following the publication of his book, *Unsafe At Any Speed*, the biting criticism of the design and construction of the Chevrolet Corvair, he gained many enemies in American industry, including Carl Kiekhaefer. Carl was capable of the most vicious, and at the same time humorous, arguments against "Naderisms" in business. In a letter to Senator William Proxmire, Carl said, "Why doesn't Nader just go to some country of his liking? No one is holding him here. We don't need professional pessimists."[8] Carl's classic description of what could happen if Nader were to eliminate automobiles completely, leaves little to the imagination.

> "Relative to just one of his many subjects, would Nader prefer that we go back to the horse and buggy or what? To go back to the horse and buggy would pose some interesting

problems with our expressways. For example, with twenty million cars on our highways with say approximately 150 h.p. apiece, we would assume an effort of three billion horses, and assuming optimistically that the family would have to have at least two one-horse rigs to cover the daily transportation needs, this would assume six billion horses. If we eliminate the automobile, I suppose it would include farm tractors necessary to provide the feed for these six billion passenger horses and to grow the feed, at least another billion horses could be fed, the distribution transportation needs of not only the feed but the by-products, we could well assume another billion for a total of eight billion. Those of us who have been around horses might have some idea of what this might do to the fly population problem of rather extensive proportions, to say nothing of the odor standards which have not been written as yet [by Washington bureaucrats].

"Now we have not yet discussed air transportation and perhaps this is getting beyond all imagination, but could you conceive standing up to your neck in manure while another billion horses (if they could be made to fly) were dropping pollution from the heavens? ... One would need more than an ordinary umbrella to ward off the flying fertilizer. ...

"If Mr. Nader has any other suggestions, constructive, he hasn't voiced them as yet, at least not publicly, and while the automobile may not be bio-degradable, I do think it has been a useful machine and will probably be here to stay in some form or another."[9]

It didn't take much to turn Carl from an affable, delightful and informative breakfast companion, into a terrifying obstacle to eating. Carl had always been overweight, ever since he was a little boy. Throughout the years, he would battle his weight problem, either as a self-imposed improvement plan, or as a request by a physician following his yearly checkup at the Mayo Clinic. When he was on a diet, those around him knew well enough not to order anything that might be considered fattening or extravagant. Occasionally, however, an unsuspecting table companion would reach for the menu.

On this particular occasion, Gene Wagner, manager of the Siesta Key salt water proving grounds, was accompanying Miami Mercury dealer Dave Craig and Rose Smiljanic to breakfast. Carl had just returned from the Mayo Clinic, where, because of juggled schedules, the company plane had to depart before he had completed his examination. Carl was an impulsive automobile buyer, and throughout his career personally purchased hundreds of brand new cars, often presenting them to key employees to use as company cars. So, when he found himself stranded in Rochester, Minnesota without transportation, he walked into the local Cadillac dealership and bought a brand-new, powder blue Cadillac. He dashed off a check for the sticker price, and drove it across Minnesota and back to Fond du Lac.

He liked the car very much, he told Wagner, Rose and Craig on the way to the restaurant on the fateful morning, and pointed out all the latest engineering features and passenger comforts, like automatic door locks, of the new model.

"I never liked Cadillacs," Carl commented, "but this is a pretty nice car."

The trio went into the restaurant in Oshkosh, found a table and waited for the waitress. "I didn't know at the time that Mr. Kiekhaefer was on a strict diet," Wagner admitted.

> "So, we all started ordering, and he [Carl] ordered a poached egg on toast. And I noticed that Rose didn't order much at all, and it didn't strike me, you know, and then it was my turn. I said, 'Yeah, I'll have pancakes and sausages and a vanilla milk shake. I always liked milk shakes. My first boat was called Milk Shake, because I was the only teetotaler on the race crews.
>
> "Well, anyhow, I seen some looks, especially on Rose's face, you know, but I wasn't really thinking. And then Dave orders something, and again it wasn't very much. So, here comes the food. I was pouring on the molasses and everything, and I wasn't paying attention. Mr. Kiekhaefer is sitting right across from me, and all of a sudden he just *blows up!*
>
> "'Why in the hell can't I have some real food? Here I am with a piece of dry toast and a soggy egg and here's this guy right across from me with a *tower* of pancakes, pouring *rivers* of molasses, and sausages, and milk shakes. ...'
>
> "He gets *real mad*, was just ranting and raving. And I had taken the first bite, and it was in my throat when he started yelling, and I just about choked to death. I couldn't get it down, and it wanted to come up, and I just about died right there. And things got very quiet and tense. ... I felt so sorry for Rose. And Dave -- he was just kind of crouching down, you know. *Oh!* I tell you, if there would have been a crack I would have crawled into it.
>
> "So, the rest of the meal went very tense, and we got out to the car, and he had a little problem in getting the door locks to work, and *that did it!* Boy, he started cussing and swearing that car. 'God dammit!' you know. Here it was a pretty nice

car all the way down there, but after that meal and everything he started just ranting and raving. And we got in that car, and he was running like a maniac, you know, and we finally got to where we were going, and he *screeched* up there, and he gets out and he tells Dave, 'I want you to get in this piece of crap and drive it back to Miami, and I never want to see it again in my life!' And so Dave gets in the car and he takes off for Miami, and he had that car for a couple of years. Just because I ordered the wrong thing, it changed his whole outlook on everything."[10]

Food, for Carl, was a most intimate affair. How it was handled, cooked, presented and served would often become as important as how it tasted. He was most opinionated on how food should be barbecued outdoors, and throughout his career his legendary steak barbecues became the stuff of legends. He prided himself as a steak barbecue chef *par excellence*, and imagined that all who dined at his picnic tables on the geologic ledge overlooking Lake Winnebago were swept away with his grilled masterpieces. Actually, no.

He had a custom barbecue grill built right at the plant in Fond du Lac, which was over twelve feet long. It was made of reinforced 55 gallon oil drums cut in half and welded end to end, with heavy steel legs welded throughout the bottom to hold the extreme weight required for Carl's typical barbecue. Across the top were set after set of glistening, heavy-duty stainless steel grills that could tip up to admit enough charcoal briquettes to heat the entire woods behind his house.

Every ingredient connected with the barbecue ritual was important to Kiekhaefer, right down to the oak chips used to smoke and flavor the steaks. Like everything else Carl did, providing oak chips was a major production. He had a giant tree chipper which he would trailer into the woods behind his pick-up truck, searching for just the right oak tree to sacrifice for his nearly ceremonial feasts. Once his selection was made, Carl would pull out his heavy-duty one-man chain-saw that he had designed and manufactured for soldiers to cut their way through armored tank traps in World War II, and reduce the tallest of trees to man-size logs in short order. These he would toss into the churning wood chipper, spewing a fountain of green oak chips onto an industrial-sized mound in his truck bed.

Charcoal, too, was prepared in staggering quantities. "When he had a fuckin' barbecue," Wilson Snyder recalled, "he used more coal than I've used all my life. Christ, he'd buy half a ton of coal -- more than that. ... He'd fill that up with coal and then he'd pour gasoline on it to start it."[11] With many hundreds of pounds of charcoal soaked with gasoline blazing in the barbecue troughs, the barrel iron would quickly change from cold-blue to red to white-hot within fifteen minutes. Carl mounted a high speed fan at a strategic angle to the roaring inferno, turning the whole contraption into a nearly uncontrolled blast furnace. At the proper moment, he would dump a few hundred pounds of moist, green oak chips over the coals, sending giant clouds of steam billowing skyward through the trees on the ledge, sizzling and crackling like the onset of a biblical prophesy.

Preparation of the steaks was no less an art for Carl, and it is said that he could nearly destroy a kitchen just getting the steaks seasoned and ready for the waiting grills. He would buy the very best steaks available in Fond du Lac, specifying the thickest, most aged, prime porterhouse steaks that the butchers could locate. From an inch and a half to two inches thick, and up to a foot in diameter was his standing order. For a large function, he would buy several hundred pounds worth. Even today, Wilson Snyder can go to his local butcher in Fond du Lac and ask for a "Kiekhaefer Cut," and they know exactly what he wants.[12]

Across the floor of the kitchen Carl would scatter the latest edition of the Fond du Lac *Commonwealth Reporter*, which, if a letter to the editor had somehow slighted him, he would refer to as the *Common Filth Repeater*.[13] First, he would wash the steaks in the sink, and then toss them onto a jumbo bread board on the counter. "He's got Adolph meat tenderizer, he's got garlic salt, he's got celery salt," Fritz Shoenfeldt recalled Carl's recipe.

> "He's got every imaginable kind of seasoning you can think of, and he's pouring it all over the steak, and he's taking his big, ham-hock hands and he's kneading it into this meat, and he flips it over and he does the other side. Meanwhile, this shit is getting all over the kitchen floor."[14]

As the moment of truth approaches, the barbecue ritual slowly changes from a leisurely outing in the back yard into a close order military drill. Carl was obsessed with the precision of the grilling process, and insisted that the steaks be seared, scored, flipped, rotated and removed at absolutely precise intervals. The chef became an engineer again. To this end he would have an assistant stand by with a stop watch, or several stop watches, to time the process. If it was a large, more formal company-style barbecue with perhaps up to a hundred steaks to grill, he would have his chief engineer, Charles Alexander, man the timepieces. If it were a smaller, more intimate affair, Rose Smiljanic was appointed timekeeper. Either way, he never listened to the timekeepers anyway.

Those in the know at Carl's famous barbecues would fight in line so they wouldn't get the last steak left on the grill. Since it was impossible for Carl to flip even the first steaks over in time, the last one was indistinguishable from the charcoal beneath. "The last fucker that he'd get to flip would be burnt," Snyder recalled, shaking his head.

> "Billy Steele was always in there trying to help him, and Kiekhaefer hollered, 'Who wants this steak?' Billy says, 'I'll take it.' Jesus Christ, he gave him that fucker -- my dog would have buried it!
>
> "But he'd get them *all* fucked up. He wants to do everything at once, and you can't do it. You can't put a dozen steaks on there and then keep them from burning. Fuck, they'd get away from you. And then, if you try to help him, it's 'Get out of here! I'm doing this!'"[15]

Some people absolutely loved Carl's steaks. "Cooked everyone alike, and right to perfection," long-time friend General Mabry Edwards recalled of his barbecues with Carl. But with as many as a hundred steaks sizzling and flaring up on the neolithic sized fires, more often than not the orderly flipping and turning of the steaks turned into a headlong dash from one end of the oversized grill to the other, with steaks literally flying through the air. Fritz Shoenfeldt has the scene indelibly etched into his mind, of Rose standing nervously next to the indomitable chef moving faster than a hook and ladder crew at the Chicago fire.

> "Rose is there with a clip board and a stop watch. And he says, 'Now, I want them a minute on each side.' And these things are an inch and a half thick, so he'd throw one on and she'd start the stop watch, and by the time the second one is on, he flipped the first one and it falls on the ground.
>
> "'I'll eat that one,' he says.
>
> "'Minute's up,' she says.

"OH NO! It *can't* be a minute yet!' So, pretty soon, by the time he flips them over, they're burnt, and there's about maybe three quarters of an inch in the middle pink. And, of course, everybody says, 'Carl, you make the best steak. ...' And they all say it tastes good, and they don't eat it. They just don't eat. They carve out that little pink part, and then, of course they eat some salad and shoot the shit a little while. Well, I ended up bringing it all home, and we had a bunch of pigs on the farm, and they ate good that night."[16]

Carl was also fastidious about the buns or rolls that would grace his picnic tables, and Sylvan "Ham" Hamberger, Kiekhaefer Mercury's affable ace sales manager, finally found a way to always get the right rolls for the boss. Before one of Carl's cook-outs on the ledge, Ham was dispatched to the local bakery to buy a generous supply of bratwurst rolls. But at the barbecue, an unhappy Kiekhaefer stalked Ham and finally cornered him with complaints about the rolls.

"He walked up to me with a hot dog in his hands, and he's eating this thing -- like it's running down his shirt. He says, 'Hamberger, this food is not fit for human consumption.' Man, he's shoving it in with both hands. 'And you got the wrong rolls twice in a row!'

"So, I went to the bakery and I said, 'Give me every roll that you could possibly have a bratwurst in it.' The guy gave me five rolls. I took them into Carl. I said, 'Now Carl, I want to make sure we get this right.' This was an expensive party [coming up]. We want to do it the best we can. So I said, 'Which is the roll that you like?'

"'Well, God Dammit, that one!'

"I gave him a felt tip marker and made him write ECK on the roll. So for years the story was the *approved rolls*. And I kept these rolls in a baggie and they got stale and hard, and I had them for about two years. And that's the story. That's exactly how it happened. He was a *perfectionist!*[17]

Among the strangest and most humorous of Kiekhaefer's food escapades occurred at a formal banquet hosted by the Governor of Florida, and again illustrates Carl's lightning quick reversals of temperament. It was a meeting of the Florida Council of 100, a prestigious advisory group ostensibly assembled by the Governor of Florida to represent the best industrial, social and intellectual minds in the state. Joe Swift, the bright and innovative newcomer to Carl's public relations staff, viewed the

Governor's Council of 100 somewhat differently. "[It] doesn't mean anything, except that if you've got money and you donate it to the Governor's political fund, you're on the Committee of 100."

Swift, an eye witness to the incident, had built a phenomenal reputation in hydroplane race boat design and construction with his company, Swift Hydroplanes, before being lured to Fond du Lac by Carl Kiekhaefer and Frank Scalpone in 1967. After building over 9,000 race boats from his headquarters in Mt. Dora, Florida, and capturing a long string of national hydroplane championships and speed records, Swift would soon gain a reputation for promotional largess seldom matched in the marine industry.

The Governor's banquet for the group coincided with a meeting of the American Power Boat Association (APBA) at Port of Call, St. Petersburg, Florida. "It was a pretty lush place at the time," Swift recalled, "Guy Lombardo and so forth." Swift had ended up in St. Petersburg after staging one of the more bizarre promotions of his career. He had outfitted a houseboat with Mercury outboards that towed skiers from the Montreal Expo in Canada, all the way down the Mississippi River to New Orleans, across the Gulf of Mexico and into St. Petersburg. In the Gulf, during the dead of night, a violent storm tossed a cinematographer overboard who was already wearing a leg cast from a previous spill on the trip. Sinking from the weight of his cast, he took a key from his pocket and managed, with great difficulty, to saw off the heavy plaster cast before he drowned. He drifted all night, and wasn't located until the following day by Coast Guard search and rescue units. Swift was just about to take his first break after the harrowing voyage and the all-night vigil for the overboard victim from the *Miss Sunshine*, when Kiekhaefer called from his home in Winter Haven. "Why aren't you over here for this meeting?" Carl demanded. Carl then told Swift to pick up a pair of black shoes he had in his closet at Port of Call, where Kiekhaefer had reserved a suite for the duration of the APBA meetings, and bring them to Winter Haven to join his party attending the Governor's banquet. "I'll send the Beech over to pick you up," Kiekhaefer grumbled.

Carl's twin engine Beechcraft 18 managed to land on a small spit of sand near the hotel where Swift was staying in St. Petersburg, in violation of a long list of FAA regulations, picked up Swift and flew to Winter Haven with the shoes. "I went into his closet and there were three pairs of black shoes, so I took them all," Swift remembered.

When Swift arrived at Kiekhaefer's home in Winter Haven, Carl was "sitting there in his 'skivvies,'" in the company of Dick Genth and Robert L. Rodman, chairman of Thunderbird Products Corporation. Swift said, "Well, Carl, here are your shoes. I didn't know which to bring, so I brought them all so I wouldn't make a mistake."

"Are you trying to embarrass me in front of my friends?" Carl snapped as he grabbed the shoes.

"No, Carl, I was just trying to get the right shoes," Swift explained.

Carl had just bought a new Volkswagen sedan to try out, the first new product from the German manufacturer that was different from the traditional VW beetle. Swift, Genth, Rodman and Kiekhaefer piled into the little car, but when Carl slammed his door shut, the overhead dome light wouldn't go out. Impatient with searching for a switch to turn it off, Carl reached up and ripped the entire dome assembly out of the roof, wires and all, and tossed it out the window, and said, "Well, it's out now!" Carl's passengers were already apprehensive about the outcome of the evening, given Carl's black mood.

When Carl and his entourage arrived at the formal banquet being held at the Holiday Inn, the Governor's host was flustered at first, because Kiekhaefer hadn't reserved any seats besides his own, and space was quite strictly limited. "The guy looks at Carl and then this string of people behind him," Swift explained, "there was Rodman and Genth and Swift, and doesn't know what the hell to say because he's got one seat saved for Carl at a table right in front of the rest-room. He saves the day by directing us to a table that is clear across the room up against the wall." Swift and company were greatly relieved, for they feared Carl would blow up if there had been a problem with seating. But then Swift noticed that the host had seated Carl opposite Dick Pope, of the Cypress Gardens Pope family. Speaking to Dick Genth, Swift said, "Dick, it's not going to be long but there's going to be an *explosion*." Originally contracted to promote and use Mercury outboard motors in the world famous Cypress Gardens water ski shows, Dick Pope had recently switched to *the enemy*.

Carl was to be in the spotlight twice during the pre-dinner program. First, Carl was presented with a huge, ornate trophy commemorating his generous sponsorship of many Florida outboard racing events. Later in the program, Carl presented the Governor with a check for $5,000. On the check was the customary corporate time limit for cashing checks of sixty days. The Master of Ceremonies snapped a harmless joke, saying something like, "What's the deal, Carl, is the company going to be broke after sixty days?" It received a laugh from the audience, but Carl nearly turned purple. He abruptly left his table where he had been glaring at Dick Pope, and conspicuously marched between the Master of Ceremonies, the Dixieland Band and the Governor, and over to the table where Swift, Genth and Rodman had been seated at the opposite side of the assembly.

Everyone at the banquet had just been served a steak on a plank, rimmed with mashed potatoes. Swift and his companions were starving, but when Carl joined them they couldn't start eating because Carl didn't have a serving, having left his plank behind at the other table. "In the meantime," Swift recalls, "someone else has brought the trophy over and puts it down in front of me. The waitress does a quick job and brings another steak for Carl, a brand new one with new potatoes and everything. Genth, Rodman and I think, 'My God, we're going to eat now *finally*." But something had snapped in Carl, and he chose this moment to demonstrate his anger and frustration over the whole evening.

> "Carl picks up this plank, pushes his chair back a little bit, puts it on the floor, and with his black shoes, puts both feet in the mashed potatoes and the steak. Right into this, see! And then he turns to us and said, 'Let's go!' and with that he gets up with his shoes rimmed with mashed potatoes and opens the door directly behind us and walks into a china closet! Booiing!!
>
> "Now, we would have enjoyed eating the God damn steaks, see, but -- nothing! We're leaving *right now*! He can't get out that way, so he looks around and here's the aisle we're going to have to follow, which is between the audience and the MC, and in front of the band. Here goes this caravan, and I'm carrying the trophy, followed by Genth and Rodman. And here goes our boss with the potatoes all over his feet and it was a little difficult at that moment to say, 'Hi, I work for him you know.'"[18]

Carl's day had gone from bad to worse, but it wasn't over yet. On the way out of the banquet, the trail of mashed potatoes led into the cocktail lounge where two of the young girls who had rotated water skiing from Montreal to Florida had stopped for a drink. "And Carl had nothing against pretty girls, and these were pretty girls," Swift admitted. "So nothing would do but when we finished the drink the girls had to come with us out to the house." Freda, Carl's, wife wasn't at the Winter Haven home this particular weekend. "Well, there was a girl for Carl, and the there was a girl for Genth, but there was no girl for Swift or for Rodman. So we ended up sitting in the kitchen looking at each other." Hoping to impress the girls, Carl had taken them down to the dock where Mrs. Kiekhaefer kept a little runabout with a small Mercury outboard on it for touring the small lake. "Carl had gotten everybody in the boat," Swift learned, "and he had cranked on that Mercury and it didn't even sputter, it didn't do a damn thing. Finally, he unscrews it and throws it in the damn lake. And then he notices that the fuel tank hadn't been plugged in!"[19]

CHAPTER THIRTY-TWO

A Bitter Reward

"A man in America is free to choose whichever side he wishes to be on. This reminds me of a quotation by Henry Thoreau which I read not long ago, 'If a man does not keep pace with his companions, perhaps he hears a different drummer. Let him step the music which he hears, however measured and far away.'"[1]

Carl Kiekhaefer

Carl continued to struggle with his feelings of betrayal by Brunswick during the waning years of the 60s. Though reluctant to face the inevitable possibility, he cautiously prepared for a future without Mercury. Having managed his company as a form of benevolent dictatorship for nearly 30 years, his frustrations with Brunswick control were often overshadowed by his feelings of loyalty and responsibility to his employees and customers. In 1968, he spoke of his widely rumored discontent with Brunswick at a gathering of Kiekhaefer Mercury elite: his distributors, branch managers and corporate executives.

> "Too numerous to mention, in the past 28 years ... are deeds of loyalty and devotion to this company which I shall never forget. While there have been many times before and after the date we merged with Brunswick when I have felt like walking out, and while I might well have been able to afford it and survive, I would not have been able to live with myself -- and because of my feeling of gratitude and devotion to the people who helped build this business. ...
>
> "If one had only a selfish interest, perhaps things would be easier; but, at the risk of being conceited, I have felt in the past that I and my team would never walk out on the 100,000 shareholders who own this business, on the 7,000 dealers on a worldwide basis who believe in this business to the point of personal investments and long devoted service and dedication, and some 4,000 employees who with all their good points and their failings as humans have nevertheless maintained a product which has been the envy of not only our competition but of businessmen in many other areas. ...

"Last of all, but not least, who could walk out to ease the tensions, compressions and worries by retirement or going fishing when we have this team of employees who are continuing to build ... in face of treason and disloyalty by over 20 people who succumbed to the incentives offered by competition."[2]

Indeed, Carl was most distressed by the defections which had taken place, and agonized over the thought of his former friends and employees in league with *the enemy*. Though the innovative and talented Edgar Rose was a recent OMC convert, Carl continued to focus his exasperation on his once closest friend, Charlie Strang. Since joining OMC in 1966 as director of marine engineering, Strang's talents became as evident in Waukegan as they had been in Fond du Lac. In 1968, Strang was promoted to vice president, and later in the year was elected by OMC's board to the critical position of group vice president for marine products. As such, Strang became responsible for "coordinating and directing the marine product development, manufacture and marketing activities of the company's domestic divisions and subsidiaries."[3] In short, Strang was then able to control both the technology and marketing of OMC marine products. Kiekhaefer lashed out at his old friend in a letter he never mailed to fellow Brunswick board member Walter M. Heymann.

"I do appreciate all you and Jack Puelicher have done for me by keeping me from being fired several years ago in favor of Charles Strang. By first building his hopes with our company and then suddenly dashing them, Strang has become our worst enemy and is being driven on with a zeal that the devil couldn't match to become Outboard Marine's next president.

"This treacherous act, namely Brunswick's negotiating with this man for a year or more behind my back, was a burden difficult to carry after I became aware of it. It was a narrow scrape not only for me but for Brunswick. Brunswick would have had the damnedest proxy fight on their hands, the likes of which they would not have seen before. In fact, there wouldn't have been any Brunswick Corporation -- even if I had to die with it -- had the act been consummated ... the act of putting Charles Strang in over my head as a surprise move."[4]

Carl's paranoia with employees, both past and present, extended to Peter D. Humleker, who at one time was in charge of all Kiekhaefer Corporation personnel and labor relations. In 1967, Carl received feedback from his various "spies" that Humleker, who was recovering from a heart attack, was "denouncing me and other executives," an offense for which he would be terminated.[5] In the exchange of invectives leading up to Humleker's departure, Carl was most disturbed by Humleker's accusation that Carl was "bugging the phones of various executives in the office and in their homes. How ridiculous!" Carl exclaimed. "... [T]he very thought was so abhorrent to me that I could never do it."[6]

Kiekhaefer was justifiably concerned that his own bugging of Alexander's office would become known to Hanigan in Chicago. According to Carl, Humleker told Hanigan that Carl was bugging executives, but stopped short of specific examples. In a letter to fellow board member Jack Puelicher, Carl felt the need to explain his practice of recording his own personal calls, and to shift the blame of secret recording to others. Carl's deepening paranoia about losing his position is evident in his remarks, and serves to show how preoccupied he had become with ferreting out the many imagined threats to his leadership, especially those who he thought were secretly reporting to Brunswick.

> "So many things have happened in the past are now beginning to fall into a pattern. One of these is Pete's [Humleker] inordinate interest in these electronic 'bugging' devices. ... My office group reports that Pete took a great delight in testing this type of equipment ... he would wear a secret lapel or tie clasp microphone and walk about the office and engage people in provocative conversations that would be picked up by an FM radio and listened to by another group in our Board Room. ...
>
> "In a way, it is easy to understand how Hanigan might put some stock in what Pete told him about phone bugging. He claimed that King and Strang had told him before that all of his (Hanigan's) calls were recorded regardless of who he was talking to. So Pete's story fit into the pattern. Yet when the Strang-King episode exploded, who had the damaging recordings of Hanigan's phone conversations and was eager to make vindictive use of them? None other than Strang and King.
>
> "Of course I have a Dictaphone recorder in my office, along with all my other executives, but it is used solely as a convenience to record conversations involving service difficulties, engineering reports and other calls so that I am able to pass the tape on to the proper department for handling. The idea of using it to actually 'invade one's privacy' is preposterous. I

have no tapes or recordings that are anything other than simple 'convenience' records for our business.

"As our Vice President in charge of Industrial Relations, Pete was in charge of our plant security, our employee investigations, our ordinary surveillance of new employees and all of the routine precautionary measures an operation such as ours should take. Looking back at this activity, it is easy to see how easy it would be to extend this effort to cover employees or executives that might stand in the way of his future progress. You must bear in mind, Jack, that Pete was a vain and ambitious man who truly believed that he would some day be President of Kiekhaefer Corporation. This is not an idle jest ... it has been reported to me many times and, as a matter of fact, his kids have said it around the neighborhood to other kids, which proves that it has been discussed openly in his home.

"It wouldn't surprise me one whit if Pete has been reporting to Chicago for years. His lust for power would be a motivating factor and with the departure of Strang he believed the path clearer than ever ... to the point where Pete actually fancied himself as the only man who could protect Brunswick from Kiekhaefer's machinations, whatever they might be. ...

"What choice do I have but to discharge a man who blatantly leads such an insurrection? While Hanigan talks about being 'humane' to a 21-year employee, where were the 'humane' characteristics of Pete who for the same 21 years was one of my most trusted employees? All else is academic.

"Now the whole ugly mess confronts me in the form of veiled threats, innuendo and criticism from Chicago. How can I even answer them without humiliating myself?"[7]

Concerned that someone might be tapping his own telephone lines, or listening to his private office conversations, Carl hired an elite electronic investigation firm to "de-bug" his house and offices. On October 30, 1967, the firm completed their exhaustive search and were able to report that everything in his environment was "clean."

"We checked out the offices of Mr. Kiekhaefer and found everything in order. The telephones were clear at the time we checked them. ... There was no RF radiation from the main office and the adjacent offices. We traced out every available route for wires and found nothing in the way of a hidden microphone. We checked over the walls, ceiling, furniture, floor and fixtures for possible hidden transmitters or wires and found none. We walked the entire roof ... the furniture was checked over ... the electrical system had nothing attached to it that was transmitting information. The speakers are not being used as microphones. The outboard motors do not have anything that is strange or out of place."[8]

Carl had become so paranoid that he actually thought some of the outboard engines he built could have bugs planted in them. As Carl's anxiety of being separated from Kiekhaefer Mercury grew, he began to take steps that might insure a quick start following any irreconcilable differences with Jack Hanigan. His insecurity led him into negotiations with Hanigan as early as 1966, to guarantee a fruitful and smooth transition out of the company. Hanigan offered Carl the possibility of being a third partner in a proposed new company, with Brunswick and the British Rover Company as the third. Carl, though, had hoped that he would still be able to stay as president of Kiekhaefer Corporation if the new company were formed. "Well, this wouldn't be too bad," Carl told Hanigan in a telephone conversation recorded by Carl, "this wouldn't be too bad. But this means that I sacrifice something else in the process -- that puts a new dimension on the thing ... this means sacrificing my presidency here."[9]

Hanigan began to take Carl to task for his frequent comments about leaving the company, and slowly contoured conversations around to what Carl would expect should they ever part ways.

"Carl ... you wanted to get loose and be on your own, and get your status back by making something good. And this was your suggestion - not mine."

"This feeling of insecurity, of course," Carl replied, "probably ought to be solved -- resolved one day. I hate to bring up the bloody past, but you know once before my situation here was quite insecure. We didn't know whether Strang was going to run it or whether I was going to run it."

"I reiterate that you're the one who brought it up," Hanigan said.

"Is there something that could be done about resolving the insecurity?" Carl asked. "Because I never know from one day to the next when we have a disagreement but -- what the hell -- that I'm going to be cut loose - not that that would fry me - but I just hate to think of the impact on the organization, if I can be so proud of myself."

When Carl asked Hanigan if he could get a two-year contract, he was surprised when the Brunswick president offered one for more.

"I'd like to make it five years instead of two, if that would make you feel any better. ... I want to be very clear with you that as far as I am concerned, we can do this in any way that you physically can handle it, but you said to me the other day that you'd like to get out and have something that was your own, and take pride in, and so forth. And I said, 'It was alright with me.' And I was trying to suggest a way that you could continue to be at Mercury and still be free to spend as much time as you wanted on [something else]. ... I don't want you out of there."[10]

It was time for Carl to decide how he could start something new while at the same time stay on as president of Kiekhaefer Corporation. He quickly replied that "there would be one other thing in this contract. ..."

> "Would there be any objection to my having a little tool and die shop where I could do some model planning on my own, away from the regular Engineering Department? You know very well that Brunswick is protected on anything and they will have first refusal on anything I develop. I feel that I've got to have a group that I can work with evenings, Saturdays, Sundays. ..."

To this novel suggestion, Hanigan replied, "I'll do my best to fix it the way you want it fixed. ... But for goodness sake, stop worrying."[11] When nothing came of his suggestion, Carl enlisted the support of Armand Hauser to fabricate a memo as if the establishment of this special facility was a serious company necessity. In an internal memo, Hauser first discussed the overloaded condition of engineering facilities before supporting Carl's subterfuge.

> "One course appears open to us. A separate facility manned and staffed by a group not under our regular engineering supervision but reporting only to Mr. Kiekhaefer should be built and equipped as soon as possible. Only by doing this can we provide him with the flexibility and freedom required for our long-range planning to succeed.
>
> "A drafting room, tool room, test cells and an assembly area and perhaps one or two small offices would be required. ... A suggested location is on the ledge, with the exact site to be picked by Mr. Kiekhaefer."[12]

Carl didn't have to pick a site, for he had already purchased the sprawling farm adjoining his house on the natural limestone ledge overlooking Lake Winnebago. He had resurrected the name Kiekhaefer Aeromarine Motors which he had consolidated with the Kiekhaefer Corporation in the years before the merger with Brunswick, and had already drawn up plans for building the facility that Hauser was independently "suggesting" in his memo. It was planned to be Carl's safety net; his escape valve to the future -- when the final blow up inevitably came.

Offshore powerboat racing, or racing in the ocean, became more of a political item on Carl's marketing agenda than a sporting one during most of the decade of the 60s. Carl was almost constantly embroiled in heated arguments with race promoters, APBA committee chairmen, and occasionally, competitors. In 1962, Johnny Bakos drove *AOKone* to victory in the Miami-Nassau race at a record average of 48.6 mph, propelled by a pair of 327 cubic inch *MerCruiser* stern drives. In this, one of the toughest and most demanding of all open-passage ocean races, six of the first ten finishers were powered by Carl's engines. The following year, *Mona Lou*, driven by Odell Lewis, set a new speed record for the same race at 55.4 mph. Again, seven of the first ten finishers were powered by *Mer-Cruiser* stern drives. Then, for a period of several years, Carl withdrew most of his factory participation in racing, being disgruntled over racing venues, rules changes and official personalities. Typical of his feelings at the time, Carl minimized the consequences of victory by saying, "We frown on racing in all of its forms. It throws manufacturers at each other's throat ... and is an item of considerable expense. I have yet to see where a race has sold an outboard motor. ... While Mercury has done well in all types of competitive events, we, nevertheless, are discouraging racing. ..."

Jim Wynne, campaigning a wide variety of boats, engines and races, became one of the most consistent winners in the sport. He built up an unbeatable lead in driver points in 1965, and when the season was over, was declared the World Champion offshore driver. Near the end of the 1966 season, Wynne created a furor by winning the celebrated Miami-Nassau race with a turbine powered 32 foot aluminum boat, the *Thunderbird.* Wynne's unique boat, powered by a pair of 450-horsepower Pratt & Whitney jet aircraft engines driving twin propeller shafts, won the race in wild seas rather handily, but was ruled illegal following the victory. Wynne, though, was awarded the driver's points for the race nonetheless, which helped the Kiekhaefer alumnus to become World Champion for a second year in a row.

Carl had long been enamored with the potential for turbine power in the marine industry, and purchased the *Thunderbird* turbine boat. Renamed the *Mona Lou II*, he unsuccessfully lobbied the APBA and race promoters to allow the new technology to compete with traditional piston powered engines. Arguments for and against the turbines were loud and many, and confusion over the unfamiliar powerplants left the fate of the boat unresolved. Wynne had gone through similar chaos before entering the 1966 race. As Carl faced obstacles to entering *Mona Lou II* in the 1967 Sam Griffith Memorial Race from Miami to Bimini and back, he was furious that Wynne was allowed to enter the same boat in the same race a year earlier. "For a period of three weeks before the race, turbine power was publicly declared legal, declared illegal two weeks before the race, [and] declared legal again one week before the race. ..."[13]

For a while, Carl seriously considered pulling out of racing altogether, feeling that maybe Mercury needed to "change our image." He asked Armand Hauser for a policy statement on the company's reasons for racing. Hauser was quick to chastise Kiekhaefer for even considering giving up their world-renowned high-performance image, saying, "What would we change it to? What, if anything, would be better than what we have now? ... In the minds of our dealers, we are a dynamic, progressive, fast acting and hard-hitting company ... With 'esprit de corps' like this, can you imagine how their enthusiasm would wave if we tried to change our image into that of a stodgy old manufacturer who built products ideal for school teachers and little old ladies?"[14]

Bluster as he might, Carl was privately determined to get back into racing with a flourish. In 1967, the year of Mercury's return to racing, Carl entered his products and teams in a total of 27 races. Calculating the various divisions within the races, his remarkable record shows the level of his determination and the durability of his products: 43 first places out of a possible 50, and 93 places (1st, 2nd or 3rd) out of a possible 113.

Kiekhaefer Mercury dominated the race courses throughout the year, with Mercury outboards and *MerCruiser* stern drives collecting more prize money and trophies than all other competitors combined. Certainly, a volume of imposing proportions could be assembled which only enumerated Carl's racing successes. Yet, when controversy arose, Carl attacked the source as if there was only one race a year.

On February 9, 1967, Carl's turbine powered *Mona Lou II* driven by Odell Lewis, and his 27 foot Magnum *Old Yeller*, driven by Bill Sirois and powered by a pair of powerful 450-horsepower *MerCruiser* stern drives, finished within seconds of each other to win the grueling 172 mile Sam Griffith Memorial Race. The piston-engine *MerCruiser* stern drives barely edged out the controversial turbine engines, which lead Jim Wynne to speculate that Carl had rigged the results so that *MerCruiser* products would win. The *New York Times* liberally quoted Wynne, who didn't race, but was on the docks for the finish.

> "Jim Wynne, the world offshore driver champion, described ... the gas-turbine boat that finished second in ... [the] race as a 'clay pigeon,' doomed by its own people.
>
> "They put it into the race to be shot down by their own engines. ... It was a logical thing to do. If I had been in their position, I probably would have done the same thing. But nobody should get the idea from this result that turbines aren't all they were cracked up to be. ...
>
> "Look, I think *MerCruisers* are wonderful engines. ... But I don't think the turbine ever ran the way she's capable of running. At the finish, she was just idling along, staying comfortably ahead of the third boat but not challenging the leader."[15]

Carl nearly went into orbit when he read Wynne's comments. Since Wynne himself had crossed the finish line an astounding *50 miles* ahead of the second place finisher in the same boat a year before, people were quick to believe him that Kiekhaefer had instructed the driver to finish second like "crooked horserace jockeys" behind the *MerCruiser* powered *Old Yeller.*[16] Sportsmanship was fundamental to Carl's competitive character, and to be accused of rigging a race for publicity's sake by his old friend Wynne was almost more than he could bear. He blistered back in passionate scorn over the accusing comments, and tried so hard to counter the damaging remarks that the *Boston Globe* headlines declared, "Kiekhaefer Has to 'Defend' Even His Race Victories."[17]

> "The dockside opinions of a disgruntled former employee who was neither a spectator nor a participant in the recent ... race should be given little credence, and is disturbing to the whole industry that his innuendo about the finish of the race should be given such wide publicity. ...
>
> "Obviously the poor sportsmanship of Wynne in disparaging the victory of the stern drives over the turbine has hurt us badly. You can well imagine how Odell Lewis (who drove it) feels when someone questions his integrity and ability as a driver and infers that he 'threw' a race. How can he ever hope to defend himself against these vile accusations? Certainly Wynne accomplished nothing constructive by this atrocious behavior. ...
>
> "All of the above was due to the fact that Wynne was on the beach, without a boat and out of the limelight in which he loves to bask. ... Perhaps Wynne's attitude is understandable when you view the over-all results of the race. Maybe he was 'snake bit' with nostalgia as he realized 9 out of the first 10 boats were powered by Kiekhaefer equipment."[18]

More important than his occasional racing controversies, Carl was determined to thrust Mercury into new and rarefied financial territory. "For the first time in the history of the Kiekhaefer Corporation since its inception," Carl advised his loyal friend and Australian plant manager Herman Stieg, "We have a fighting chance to produce a hundred million dollars in sales in a calendar year ... believe me, we will need every nickel, dime and dollar we can get."[19] The steady gains in sales of Kiekhaefer Mercury products around the world was bringing Carl closer to his number one goal, to become the world's leading manufacturer of marine propulsion. He held an open house in Fond du Lac on a Sunday in September 1967, and was astounded by the crowds that flocked to tour his facilities.

> "Over 21,000 came through the front lobby and toured through the offices, engineering, distribution center and assembly plants. ...
>
> "Our parking lots were filled to capacity, the big fields adjacent were jammed out to Pioneer Road and the police estimated that thousands turned away when they saw the crowds. They were here lined up a half hour ahead of the 10

a.m. opening, and the stragglers were still starting the tour at 6:30 p.m.. What a mob!"[20]

As Kiekhaefer Mercury grew in both stature and size, Carl's ongoing disputes with Brunswick's home office in Chicago continued to overshadow his joys of achievement. He kept up a heated campaign of open rebellion against corporate programs, and perhaps thought himself immune from attack because of his glowing sales and profit records. Slowly, Brunswick president Jack Hanigan and chairman Ted Bensinger were reaching their limits of forbearance and patience. In a sort of *management manifesto*, Carl attempted to explain the fundamental differences in management philosophy between entrepreneurial leadership and the anonymous faces of corporate interference.

"Perhaps the greatest danger in exposing the Kiekhaefer staff group to the Brunswick staff concept of management by committees, study groups, seminars and conferences lies in the retarding effect it would have on those directly exposed and the danger of the infection spreading within the Kiekhaefer group.

"Within the framework of the Kiekhaefer concept of division management, the very word 'management' provides the clue to its success. One man assumes over-all responsibility for its progress. Under him, another man assumes responsibility for engineering, another man for accounting and finance, another for marketing, etc. Under these men are other men who assume responsibilities for the multiplicity of departments that must exist in a corporation. Management is men ... with responsibility and authority, a specific chain of command and clearly outlined duties. It is not an evanescent, nebulous and anonymous committee.

"The Kiekhaefer group are hard chargers, sometimes unorthodox but thoroughly schooled by experience in getting the job done. In its scramble for success in a highly competitive field dominated by a powerfully financed, well entrenched 'monopoly,' it gained this 'can do' know-how in its fight for survival.

"To subject this group to the delays and procrastinations of 'group planning meetings' and 'study seminars' would put them completely out of character and thoroughly cripple their effectiveness.

"To subject them to the paper tiger of staff planning is like having the surgical chief of staff prescribe an enema for every patient in a hospital regardless of ailment. What may be beneficial to one could be the cause of death to another.

"Companies are successful because a man made them that way. While this progress is continuing, what an economic crime it is for staff theorists and paper planners to tinker with smooth running machinery. Doctors are called in for the ailing ... the healthy man goes on about his business. We recognize the functions of 'staff' but maintain they should cure the sick, not sicken the well.

"In an organization such as Brunswick, there is such diversification that we find ourselves totally apart from any other division with no common base in materials, manufacturing methods, products, marketing methods or customers. Staff plans and programs, obviously geared to the slowest horse in the field, are nothing but hobbles on a fast horse.

"We do not question the need for nor the value of staff. Without a doubt, there are divisions which by their size and product and problems need the care and feeding that staff can give them from every standpoint. A division not large enough to support its own research and development group or its own market research department will certainly benefit from this corporate facility.

"What we do object to is the misapplication of staff functions as outlined here. We like to think of ourselves as a lean and rugged race horse running smoothly in a long race ... if we have to stop and add the weight of staff dictates to our saddle, we could be in for trouble.[21]

Carl had eloquently and precisely explained why it had become imperative that Brunswick deal with the growing rift between the Kiekhaefer Corporation and themselves. Even though the sterling results of Kiekhaefer growth and profitability were continuing sources of pride and financial comfort to Brunswick, Carl pushed too deeply into the dark and private recesses of personal pride, making their leadership feel impotent, unwanted and meddlesome. The success of the Kiekhaefer Corporation was always Carl's, seldom shared publicly with Brunswick. Carl was the first to point out blemishes in Brunswick management, and the last to seek ways to benefit from nearly 125 years of business experience. Though Carl had done little to prepare an effective succession of management at Kiekhaefer Mercury, Brunswick was becoming confident that he had built his organization with sufficient balance, and with such superbly engineered products, that the company could now risk his departure.

As both Carl and Brunswick skidded closer to the point of no return, Carl summarized his feelings in a letter he never mailed to his friend Jack Puelicher.

"... I can no longer take the insults, beating and violations of the fundamentals of good leadership. While from time to time good service has been given to improve at least the surface relationship, the built-in and deep resentment against the principles of this business just doesn't seem to dissipate. The principles in this case transcend all financial benefits of continuing, which in direct remuneration means almost $1 million in the next 5 years. What's another million after you have already given up 30?

'Termination is not without its compensations, however. It will permit me for once to do what I want to do ... like take a vacation ... like setting my financial management in order and doing a bit of technical writing and to do a bit of inventing in other fields. I also have ambitions to write a book that might be called 'Sequel to the Organization Man,' or 'The Rise and Fall of Brunswick,' or we might even call it 'S.O.B.' (which means 'Sons of Brunswick.'). ...

"I am sorry to trouble you with all this, Jack, but after having tried just about everything it is difficult to have faith in a company that has tried to fire you for being the most outstanding sales and profit center manager in the corporation. While it didn't materialize, some real damages did in that Mr. Strang left with ... our complete file and roster of technical and scientific personnel ... a complete knowledge of our patent successes and failures ... a complete knowledge of our design successes and failures ... a complete knowledge of our philosophies and policies and a complete knowledge of our forward plans. What an aid and comfort to the enemy! What a bungling bunch of fools!

"I have tried to protect the interests of worthwhile Mercury distributors, dealers and branch managers that have been acquired through ... years of sales and promotion efforts. I have tried my best to protect the best interests of the Kiekhaefer employees who helped make possible the great penetration into a market that was already a monopoly.

"Above all, I believe that I have to the best of my strength and ability protected the best interests of the 100,000 Brunswick shareholders. With humble apologies for such errors as I may have committed, I think it is time I tell my shareholders, distributors, dealers and employees why I can no longer continue."[22]

Jack Hanigan didn't need to be hit over the head to understand the heavy hints emanating from Fond du Lac that Carl was discontent and wanted out. Hanigan finally reached the end of his long-suffering patience with Carl Kiekhaefer, and decided to take action. In a letter to memorialize a meeting that took place between the two stubborn men on October 8, 1969, Hanigan held nothing back, and put his true feelings about Carl out in the open.

> "I cannot wait until your retirement date to replace you in charge of the Kiekhaefer Mercury Division. It is my opinion that your personality is so strong that a man strong enough to run this large a division would find it impossible to work for you. It is my hope, however, that a way can be found to have him work with you, not only until your retirement date but for many years after.
>
> "In listening to your comments over the past few years, I have come to have what I think to be a pretty good idea of what you want to do when you do retire. I am suggesting that this program be started now instead of waiting until 1971."[23]

Both men agreed to meet again in Chicago on November 6, 1969 to finalize plans for a smooth transition of control at Kiekhaefer Mercury. Following the meeting, Carl began to draft the terms of separation between himself and his company. Basically, the plan called for the establishment of a new division known as the "Propulsion and Planning Division," with Carl as president at his existing annual salary of $100,000. Carl would then be appointed "Chairman Founder" of the Kiekhaefer Corporation, and basically step aside. His new division would have research and development authority in the fields of marine propulsion and snow vehicles, along with responsibility for testing operations, including Lake X and other testing sites. He would also be in charge of plant construction for Brunswick, air transport and racing activities. This new division would lease space from Carl, at his modest facilities on "the ledge," at a generous price, and give him the option to buy out any equipment installed at the facility whenever the more or less open-ended agreement might lapse. Lastly, the agreement called for Carl to "provide reasonable advice and assistance" to a new general manager of Kiekhaefer Mercury, allow Carl access to main plant facilities at all times, and have suitable office space at the main plant in addition to office space at "the ledge."[24]

Brunswick ultimately objected to the formation of this special division just to appease Carl, and decided instead to allow Carl to build his own company, Kiekhaefer Aeromarine Motors, and appoint the new company as a consultant to Brunswick for research into marine propulsion and snow vehicles. This precarious relationship could then be terminated with only 60 days notice from either side. Under the new scheme, Kiekhaefer Aeromarine Motors would be "free to develop its own interests outside of the marine propulsion field, but will not compete with Brunswick in that field, at least prior to August 1, 1971."[25] Both Carl and Jack Hanigan agreed to the loosely specified conditions on November 21, 1969, and joint announcements were prepared for release to the press, distributors, dealers and employees.

In order to create an atmosphere of cooperation and a controlled transition, it was agreed that a meeting would be held in the Kiekhaefer Mercury auditorium four days later, on November 25, for Carl to announce his "semi-retirement," and to introduce the new president of Mercury. Hanigan selected K.B. "Brooks" Abernathy, the bookish hero of Brunswick's campaign to collect the massive delinquent receivables due from the bowling division disaster, to be Carl's successor. "We were owed $500 million dollars at the time," Hanigan said of Abernathy, "which is quite a piece of change. He then took over as treasurer of the company and did a very good job of gathering all our money up, and sweeping it up from the nooks and crannies, and putting it where we could use it."[26] Hanigan rewarded Abernathy by promoting him to head Brunswick's International Division, which also thrived as a result of his skills in money management.

One of Carl's trademarks was being a little late to meetings. Unfortunately, when Carl hadn't arrived at the auditorium within a few minutes past the 9:30 a.m. time for the dramatic announcement, Hanigan refused to wait and started the meeting without him. Assembled in the auditorium were about 100 Kiekhaefer Mercury executives and key employees. Hanigan strode up to the podium and read the press release which would be sent out following the meeting. As Hanigan announced the appointment of Abernathy as the new president, Carl arrived at the level above the auditorium. He was absolutely stunned that Hanigan and Abernathy had started the meeting and made the announcement without waiting for him. The realization of what was happening just down the familiar flight of stairs seemed to shock him for a moment. Then, Rose Smiljanic started to cry. He slowly turned around, his heart breaking that he wasn't permitted, after over 30 years of leadership, to tell the assembled employees the news himself. He was crushed, and those who saw him slowly walk to his car and drive away were silent, as if something great had suddenly died and vanished in their midst.

As Carl drove through the gates for the last time, Hanigan explained to the astonished group that, "This is a step to bring younger management into a position of top executive responsibility in this division. Mr. Kiekhaefer, who is one of the great pioneers in marine propulsion in this country, will be on hand to aid in this transition and simultaneously will be working on special projects where his considerable engineering talents will be available to the corporation."[27] The meeting was recorded, and a transcription of the comments Hanigan made following the announcement reveal the fallacy of the announced transition period.

> "I want to say to all of you ... that this is a very difficult thing for Carl to do. This place is his baby, and we want to make it as easy for him as possible by being just as polite and gentle as we possibly can be with him. However, there ought to be no question in anybody's mind that you are working for this man [Abernathy] now, not Carl. But don't make it too obvious if you can avoid it because he's contributed a hell of a lot to this business. The unfortunate part of Carl's style of management is that he, I think, would be here when he'd be 95 if left to his own devices. And you can't take that risk if you have stockholders to whom you're responsible. I've been worried for the last four or five years that a truck would hit him, or something, and I wouldn't know what the hell was going on up here."[28]

Carl was devastated by the ultimate snub of being coldly and summarily replaced without Hanigan deeming his presence even necessary. After storming back to his home on "the ledge," Carl tried to collect his thoughts and retire his feelings of rage. He was especially stinging from the realization that some of his most loyal and favorite employees appeared to immediately turn their affections and attention to the new boss. "He had heard some stories," Rose Smiljanic said, "[about the meeting] that Abernathy asked something be done and Joe Swift put his arm around Mr. Abernathy and said, 'It shall be done,' and then Armand Hauser put his arm around the guy and walked down the hall with him. And some of the guys felt they were betraying him and they told him. Instead of being quiet about that, they added insult to injury. They told Kiekhaefer that, and that just upset him so badly, 'cause he thought these were his loyalist, and here they were turning around."[29]

The evening following the announcement, Carl called an intimate meeting at his house. Among those in attendance were Charles Alexander, Bob Anderegg and Rose Smiljanic. "I told you that they were going to replace me," Carl said to the pensive and somber group. He had often cautioned them in the months before the announcement, saying, "You guys, you know you're going to have to worry after they bring in someone else. You never know when they're going to get rid of me."[30] Indeed, Abernathy had been showing up occasionally at the plants in Fond du Lac, ostensibly on routine Brunswick errands. Bob Anderegg, Kiekhaefer's controller, recalled in retrospect how transparent the visits seemed. "They sent Abernathy up here a few times on the pretense of doing some credit investigation and looking over our credit procedures and so on," he explained. "It was kind of a thinly veiled attempt to get Abernathy in here and get familiar with the place. There was so much speculation, and just his appearance here so often made everybody convinced that this would happen. ... He was obviously one of Hanigan's rising stars."[31]

Not everyone shared Carl's horror at his perceived mistreatment at the hands of Hanigan and the Chicago home office. Though Carl demanded, and was generally given, the highest measure of loyalty among his executives and management staff, his volatility and unpredictability alienated a certain minority as well. "People who imagined themselves as more dominated and abused by Carl," Charles Alexander reflected, "imagined that it would be a brighter day with the enlightened big corporation. That's the kind of 'grass is greener' syndrome. None of us knew exactly what was going to happen."[32]

Among those who must have wondered how Carl would *really* fit into the new order of things was Brooks Abernathy himself. Within a few days of the stunning announcement, Abernathy risked a meeting with Carl at his house on "the ledge." Carl had bought land across the road from his first house above Taycheedah, and built a spacious home, with a beautiful view of forest lands and Lake Winnebago below. He invited Rose Smiljanic to move into his first house, a stone's throw distant. While Carl was still living in the house now occupied by Rose, he had rigged it with listening devices and recorders. "He rigged up a tape recorder, and then I had the bugs in my house, so he wanted me to listen to this," Rose confirmed. Carl had been bestowed the somewhat nebulous title of *Chairman Founder* of the Kiekhaefer Corporation, a position for which there was no actual purpose, save perhaps the occasional ceremonial function. Naturally, Abernathy was most concerned that Carl understand that he had no real authority anymore, and that if Carl attempted to exert authority it would only serve to undermine Abernathy's position and ability to do his job effectively. Rose has a vivid recollection of the tense and vitriolic conversation that ensued.

> "E.C. [Carl] really got upset and really told Abernathy off, and this is why I think Abernathy never liked him. E.C. was trying to tell him what to do. Here he is, the new president now, you know, and Kiekhaefer's telling him they're wrong what they're doing, and on and on and on. And after that time they never saw each other. And Abernathy didn't try to come back to see him. ...
>
> "[Carl] said, 'You're going to ruin the company. It's going to take you five years to [figure it out]', and 'What does a financial man know about running the operation?'
>
> "And Abernathy said, 'Well, here you climbed your mountain. You made your success, and you'd think you'd want to retire and take life easy.' But they were like two enemies after that. And anytime that they had a function which he was invited to, he just got really upset because here was Abernathy, you know, taking over his company and his people. So I said, 'Let's just stay away and not continue running into those functions."[33]

Rose Smiljanic was still technically employed by Kiekhaefer Mercury and Brunswick, but hadn't reported back to the offices in Fond du Lac since the fateful announcement meeting. "He told me not to go back," Rose remembered of Carl's dilemma. "I was so deeply involved in handling all his personal work, and his [fiduciary] trust, so I couldn't have left him. ... There he sat with no one but me. ... They treated him like a jerk!" Once when Carl couldn't locate Rose for a few hours, he worried she had gone to a Brunswick function without telling him. When she arrived at home later that evening, she called him, "... and he was really feeling down in the dumps. I went over there, and it was like ten-thirty and he was in bed, and 'Oh,' he said, 'Rose, I thought you too had forsaken me. ...' And I always felt so bad for him. ... I kept saying, 'Hey, If they hurt you that much, just turn your back away and don't think about them.'"[34]

For a few days, some of his oldest friends from the plant would call and stop by, but then Carl told Rose, 'I just don't want to see them,' because the conversations quickly turned to new Brunswick procedures and policies, and Carl would end up being more depressed than ever. Rose did much to comfort him during the first confusing and demoralizing weeks following his replacement. "I said, 'The only way we're going to have piece of mind is by turning our backs to Mercury and then going forward.' So, that was my cheer leading speech whenever something like that would come up."[35]

Life without Carl at Kiekhaefer Mercury was very frustrating for some who tried to orient themselves to the new regime. Fritz Shoenfeldt, who for nearly a quarter-century had assisted Carl with a wide range of demanding and critical assignments, found himself without a viable position. When Brooks Abernathy took over, he called all the top executives into his office and said he "wanted to get acquainted with everybody and see what we did." In front of Shoenfeldt's friends and colleagues of many years, Abernathy asked him, "Now, Fritz. According to this [list of duties] you're 'Assistant to the President.' What does that mean? If I can't walk across the street, you're going to assist me?' I just about felt like hitting him right in the head." Shoenfeldt was shuffled from position to position until Abernathy had him reporting to one of his old friends from General Electric, and Fritz quit.[36]

After not receiving any instruction or communication from Hanigan, Abernathy or Brunswick for several weeks, Carl decided to act. He felt betrayed by Hanigan and Brunswick, and became convinced that the only way he could regain his honor and pride would be to renounce any agreements he had made with Hanigan and formally and completely *resign.* On December 30, 1969, Carl sent a short and terse letter to Jack Hanigan, Brooks Abernathy and to each member of Brunswick's board.

> 'The purpose of this letter is to inform you and the board of directors of Brunswick Corporation that effective on my thirty-first anniversary with Kiekhaefer Mercury, namely, on January 31, 1970, I am resigning as an officer and director of Brunswick Corporation, as a member of its executive committee

and also as Chairman Founder of the Kiekhaefer Mercury Division."[37]

Carl's recollection of the events surrounding his replacement by Abernathy would shift somewhat conveniently when he explained his departure to long-time friends and colleagues. To one of the men who had recently completed 25 years with the company, Carl revealed his own interpretation of his departure.

"The implication that I 'retired' or that I had 'stepped down' is not quite factual. The move didn't come from me contrary to the impression that Brunswick is trying to effect. It irked me from the news release that stated, 'Mr. Kiekhaefer will be on hand to aid in this transition and simultaneously will be working on special projects where his engineering talents will be available to the corporation.' Hanigan told the organization and I quote: 'Carl will be up on the Ledge. ... He will be working on assignments that will be assigned to him by me and also things that Brooks will ask him to do. Some of these will have to do with Kiekhaefer Mercury Division; some of them may have to do with other divisions of the corporation.'

"Apparently they were insincere about this since there were no assignments and virtually no communications for a month ... and as a result our attorneys recommended that we might as well terminate for good.

"After a number of meetings it was decided that I was to introduce Mr. Abernathy to my top management group and give the usual information about background, history and future duties, and also explain to my organization what my future duties comprised to which I agreed. However, when I appeared at the meeting for the presentation I was told I was to remain away from the meeting and not attend and instead Mr. Hanigan gave the ceremonies which made my entire organization wonder as to what had happened to me, particularly as I had told certain members of the group of our plans and made meeting arrangements. To not be there after this and being forced to return home I feel was the ultimate insult which Hanigan was trying to brush off with the statement, 'It was hard for Carl and he is very shook up about it and he was unable to be here, and words to that effect. ..."[38]

From that ice-cold morning in the early weeks of 1939, Carl had worked an industrial miracle from the ruins and wreckage of an abandoned building in Cedarburg, Wisconsin. In the intervening 31 years, Carl's vision and skills were to change the entire nature and face of the world's marine industry, and change the lives of countless men and women whom he nurtured and trained in the process. The record is remarkable. The net worth of his company grew from $25,000 to over $65 million. The floor space of his operations grew from 16,000 to well over 2.5 million. His annual sales went from $0 to over $175 million, and the value of the products he would sell would top $1.25 billion. He built over two million engines of all varieties, while his payroll reached over $30 million a year. The good name of Mercury was distributed in 118 countries around the world by nearly 8,000 dealers, and assembled and shipped by nearly 5,000 employees.

Carl had every right to feel proud of his accomplishments, and every right to expect Brunswick Corporation to be proud of their association with him. Sadly, Carl soon felt discarded and abandoned by the company he had built, and betrayed by the organization that he had nourished and to which he had given every measure of his strength and vigor. "Apparently they are attempting to erase all my contributions to the company I founded thirty-one years ago," Carl admitted forlornly. "I guess this is my reward for hard work and sincere effort."[39] Of his decades of labor, Carl was most concerned that somehow his remarkable contributions would be lost to the passage of time, the erosion of memory, and to history itself. With great emotion and melancholy, Carl Kiekhaefer resigned himself to the disparaging possibility that it was all done in vain.

"I just cannot understand what horrible crime I have committed to warrant a role like Nathan Hale's 'Man Without A Country.' I hate to see the time come when my only reminder of the past in the marine industry is by looking through the patent records."[40]

CHAPTER THIRTY-THREE

The Second Lifetime

"After having built a company that has produced over $1.5 billion worth of product with my name on it, I don't know whether any man is capable of two ventures of this type in his lifetime. ... It's always easier, I suppose, to liquidate and buy tax-exempt bonds, but I wouldn't know what to do with myself since I don't like fishing or traveling."[1]

Carl Kiekhaefer

Carl Kiekhaefer was determined to prove to the world that he still had what it takes to build a successful company of his own. He was approaching his 64th birthday as he brushed aside his anger and bruised pride, and set out to renew confidence in himself.

He was soon jolted by the many perplexing and exasperating challenges of starting completely over in business. Perhaps the first shock was just how alone and isolated he really was. The doors of the re-born Kiekhaefer Aeromarine Motors officially opened on February 1, 1970, and not since 1939 had Carl been without the generous assistance of a large and well developed organization. At Mercury, his smallest whims had been translated into quick action, and his boldest schemes were diligently executed by bulging ranks of executive talent. Now, starting over with a skeleton crew of thirteen wide-eyed and somewhat anxious individuals, Carl faced the same daunting, time consuming and frustrating challenges that beset the greenest of start-up enterprises. "It is the team that I shall miss the most," Carl admitted forlornly, "and their silence is the most stunning -- something like the silence one experiences after suddenly shutting off a riveting hammer."[2]

Carl undeniably felt deserted by his many former associates at Mercury, and though bitter and resentful of their "silence" following his departure, he also understood the heavy political pressures brought to bear on those who remained behind. The widening distance between himself and Mercury, however, allowed him the opportunity to feel-out his eroded friendship with Charlie Strang at OMC, following nearly five years of often vocal protest over what Carl wrongly imagined as Strang's treachery in league with Jack Hanigan. "My reactions to date indicate that I have more friends than enemies," Carl reasoned philosophically. "I suppose in a transition of this type when the heat of battle is over, communication with the enemy begins."[3] A few days before opening his doors at Kiekhaefer Aeromarine Motors, Carl wrote a letter to Strang in an effort to break the ice.

> "[Quite amazing are] some of the things that happened during B.B. 'Brunswick Bondage.' You can imagine it wasn't easy working under those circumstances, doing your very best while enduring frustration after frustration, but I am sorry now only for those who remain. I did my best for them and it was my sense of loyalty to the organization I built, perhaps more than my sense of loyalty to the company that was paying my salary that I continued as long as I did."[4]

Of his future business prospects, Carl was most familiar with the marine industry, and with the products he had invented, tooled, manufactured and marketed for the previous 31 years. He was prohibited from competing with Brunswick and Mercury products, though, due to a ten year non-compete restriction attached to the Brunswick merger in the fall of 1961. So, until September 1, 1971 -- a distant twenty months -- Carl wouldn't even be able to draw upon his widespread friendships in the industry to help him get started. The iron-clad restriction required Carl to establish a fresh direction for his new company, while at the same time plan for the day when he would be free to compete against his old associates.

Over several years, Carl had carefully equipped what Mercury referred to as the Tech 1 and Tech 2 research and development complex. Located on the sprawling Kiekhaefer properties adjoining his homes on the ledge above Fond du Lac, the facilities were originally envisioned by Carl and Brunswick to test Mercury snowmobile products. When Carl negotiated the purchase of the equipment located there for the bargain price of $250,000, he acquired nearly all the hardware he needed to design, tool and build prototype engines of virtually any description. Though the specialized equipment and Carl's own expertise would dictate that two-cycle engine development would naturally follow, Carl considered a wide variety of enterprises, all broadly connected with the general recreational industry. He attacked the exploration of new opportunities with abandon, feeling out the potential for the manufacture of such diverse product lines as motor home waste incinerators, portable generators, chain-saws, motorcycle engines, drone aircraft engines, lawn mower engines, engines for home-built aircraft, hovercraft engines, helicopter and gyrocopter engines and snowmobile engines. By addressing so many potential projects at once, Carl hoped that at least one two-cycle application would bear fruit in terms of sufficient advance orders to merit mass assembly.

Carl's marketing theory was simple. He surmised that his own spectacular record of innovation and production at Mercury would be sufficient to attract orders from even the most cautious organizations. His new facilities, valued at well over a million dollars, would provide the proof that he was serious, while his well-known name and heralded career should demonstrate that he was certainly able. He quickly arrived at the conclusion that he couldn't manage both the production and distribution of products, owing to the overwhelming investment required in both money and lead time. His answer was to concentrate on the design of a small and versatile two-cycle engine, targeted to meet the special needs of the growing snowmobile market. By manufacturing a major component of a growth industry, he reasoned, he could avoid the pitfalls and financial risk associated with developing the wholesale distributorships and retail dealerships required of products for the public at large.

Carl was no stranger to the snowmobile market. Living in the heart of the American snowbelt, he had watched the introduction of snow vehicles and their growing popularity since the late fifties. By the early sixties, demand for snowmobiles had begun to steadily rise, and Carl became enamored by the possibility of providing this unique product to northern Mercury dealers to keep them active and profitable throughout the long winter season. The snowmobile market began to nearly double each year, with 10,000 units produced in 1963, 18,000 in 1964, 30,000 in 1965 and 60,000 in 1966. Carl became convinced that Mercury expertise in two-cycle engine design and manufacture, in combination with engineering innovation in chassis and track design, would be sufficient to capture a large share of the burgeoning market. Spurring him on was the fact that in 1964, OMC had introduced Johnson *Skee-horse* and Evinrude *Skeeter* snowmobiles, powered by relatively under-powered 14-horsepower engines. By 1966, OMC introduced their second generation of snowmobiles, including such refinements as reverse gear, lighter sled construction and 16-horsepower engines that would carry two passengers to nearly 40 mph. Carl grew increasingly anxious that this potentially lucrative business was slipping through his fingers.

Billy Steele remembered the Saturday morning meeting in Fond du Lac when Carl, as president of Kiekhaefer Mercury, muscled his way into the snowmobile business. Carl had recently visited a group of Canadian dealers, and was impressed by their success in marketing snowmobiles in the winter months, "and all he would talk about was, '*We'll* build snowmobiles.'"

> "And so at a staff meeting on Saturday, he goes around the table, 'I want to know what you think about us going into the snowmobile business.' He went around the table, and everybody had to give their views.
>
> "Everybody said, 'Well, we think we ought to stay in the business we know best -- the marine business.' Well, Armand Hauser would always say, 'Yes.' Armand was a 'yes man.' So, it gets back down to [Carl] and he says, 'We're in the snowmobile business!' God! Why did we go through all of this dissertation all the way around the table? Thirty-five people there, you know."[5]

Carl selected John C. Hull to spearhead Mercury's marketing analysis of the snowmobile industry, and Hull was soon pulling his hair out by the roots with both hands. "He was a marine engineer," Hull said, "and he didn't know didilly about snowmobiles and it frustrated him. We went in eighteen different directions on that snowmobile. ... It was a product that he didn't understand." Many companies were rushing into the fray, and by early 1965 it looked like Mercury would have to start at no better than sixth place. Bombardier Snowmobile, Ltd. of Canada (Ski-Doo) led the field with a generous 32% of the market, followed by OMC with 23%, Polaris Industries, Inc. with 17%, Fox Body Corporation with 6% and Arctic Enterprises, Inc. (Arctic Cat) with 5%. The other 17% of the market was shared by no less than 40 manufacturers, all scrambling along in the flurry of snow behind the leaders and fighting desperately for a share of the exploding sport. Hull reported to Carl that OMC "had sold every snowmobile they built [for the 1964 season], and planned on tripling their production in 1965. This would roughly mean 15,000 units in Canada and the united States."[6]

By early 1967 Carl was nearly desperate to get into the market. He had received a glowing forecast from Armand Hauser and Bob Anderegg that predicted that the snowmobile industry "could expand in the next five years to $100,000,000." They recommended that Carl purchase Polaris Industries, Inc. of Roseau, Minnesota, rather than attempt a costly and time consuming start from scratch development program. "Our entry into the field would knock out some of the 'fly by night' operators. ... The field could rapidly narrow down to Mercury, OMC and Ski Doo [Bombardier] ... especially so if we bought Polaris."[7] If the glowing report from Anderegg and Hauser wasn't enough, two days later Carl learned that his old nemesis, Bob McCulloch, already the clear leader in the chain-saw industry, was ready to manufacture a new snowmobile of his own.

Carl was anxious, however, to look beyond his association with Brunswick, and considered buying Polaris himself. His attorneys, notably Tom Fifield, cautioned him that he could only attempt such a transaction after a specific agreement was reached with Brunswick. In view of his deteriorating relations with Jack Hanigan, Carl decided to scrap the idea. Instead, he argued vociferously with Hanigan that Mercury should enter the market immediately, and that Mercury should produce the entire snow vehicle, not just the power plant.

> "We need to supply our dealers around the world with a line of snowmobiles to protect them from the encroachment of our competitors. ... OMC will not be the only threat. Chrysler and McCulloch are reported ready to introduce machines for the '67-'68 season. Our flanks will be further exposed and vulnerable when this occurs. The manufacturer with an 'instant' marketing organization such as our distributor-dealer network can quickly move into a position of leadership. Witness OMC who did $5,000,000 their first year and $10,000,000 the second year to become the second largest producer in the industry."[8]

Only OMC and McCulloch were producing engines for their snowmobile lines. The rest of the industry purchased engines built in Germany (Hirth, Sachs, JLO), in Austria (Rotax) and Japan (Honda, Yamaha, Fuji, Kawasaki, Suzuki). Engine buyers were continually frustrated by long delivery times, marginal horsepower, and the feeling of being held hostage by a major supplier that snatched considerable profit from each snowmobile produced. Dozens of under-financed companies were springing up to enter the market, assembling snowmobiles totally from components manufactured by others. Profit margins on these assembly operations were dangerously low, and increased competition served to keep prices relatively depressed, favoring those builders that were strong enough to manufacture the most components. Carl was quite critical of the state-of-the-art of most snowmobile builders, and in a letter to a young nephew, explained the compromises required to build a good machine.

> "Many manufacturers today hardly know how to build a snowmobile. If you build it so the weight is on the front skis so they steer well, then they complain they don't climb hills. If you build them so they climb hills, then they lose their steerability. If you make them nice and wide and stable so they don't tip, then the people complain that they can't go up the side of a hill diagonally, so you have to build them a little bit 'tippy' so you can half run them like a motorcycle when you want to and like a motorcycle, you must lean in order to turn."[9]

Carl led a contingent of Mercury executives to Roseau, Minnesota on January 10, 1967, to evaluate a possible merger with Polaris Industries, a division of Textron, Inc. Though little interest was shown by Polaris due to the quickly rising market, their chief engineer and plant manager both ended up working for Kiekhaefer, leading to a charge that "your attempts to hire selected key employees of Polaris to work for you are unlawful," and quickly soured any possibly of cooperation or joint enterprise between the two companies.[10]

By June of 1967, Mercury had finished their first prototype snowmobile, and flew out of the hot and humid Fond du Lac airport for Aspen, Colorado. At Independence Pass, 13,000 feet high in the Rocky Mountains, test crews encountered difficult conditions of warm air and icy trails, and began to uncover serious design deficiencies. New ideas in the system of wheels (bogies) that support the continuous track of the prototype were discarded, as they tended to load up with snow and slush. The prototype was also slow and heavy, due to a 15-horsepower engine overwhelmed with Carl's insistence on such standard features as electric starting, heavy duty transmission and nearly bullet proof chassis and cowling. Carl accompanied the prototype evaluation crew to Alaska, where he asked John Hull to go into town and buy the smallest snowmobile he could find, hoping to impress the local Mercury dealer by having the new Mercury snowmobile beat it in comparative tests. "We were up there in cold weather, and he gets up there with little thin shoes on. He was cold, but he wouldn't admit it. He says, 'Well, go down town and buy one of the competitors. ... So I went and bought one that just ran the crap out of ours."

"You know this is our dealer here!" Kiekhaefer screamed at Hull. "You embarrassed me in front of our dealer!"

"Well, what did you want me to do?" Hull answered cautiously. "I bought the smallest sled I could find and it beat ours."[11]

On Friday, October 27, 1967, Carl proudly escorted the first production model of the newly announced Mercury 150E snowmobile to Fond du Lac's Lakeside Park. Auspiciously, it began to snow. Carl and a small group of Fond du Lac city officials took turns plowing through more leaves than snow throughout the cool autumn morning.

Carl announced that only 1,000 lucky owners, mostly dealers, would receive this first generation Mercury snowmobile, and that full production was scheduled for the following summer. As it turned out, few mourned the shortage, for the first product had a whole host of problems. One of the problems concerned Carl's methods of testing. He had built a 1.33 mile banked oval testing track on his ledge property adjoining his house, where he put the prototypes through their paces. John Hull explained the problems that surfaced as a result of the closely controlled testing environment.

> "It was a product that he [Carl] really didn't understand. He didn't understand that the terrain it had to operate on. ... I'll never forget -- he had a track up there, and the sleds would go around the track, around the track. And he tested those machines, week in and week out, turning to the left. But when you turned to the right, [it broke]. They never turned it right!"[12]

The first snowmobile was so bad that one customer wrote to Hull and said, "You know, this thing is so top-heavy and unstable that it fell over just sitting in my garage."[13] Word of this cleverly phrased customer complaint spread like an unfortunate wildfire, and soon everyone was repeating the half-serious joke. Then, dealers started getting snowmobiles back from unhappy customers. "We told the dealers after they got them back from the customers to set them on fire, and send us a picture for credit," Hull admitted. "We didn't want to pay for the freight to ship them back." It would be many years before Mercury finally ironed out the wrinkles in their snowmobile line, in fact, not until after Carl left Mercury, but eventually they would produce some of the top performers in the marketplace. Interestingly, for their next generation of snowmobiles, Mercury pulled the names *Lightning*, *Rocket* and *Hurricane* off the dusty shelves of years gone by, names Carl had used to change the history of outboard motors.

By the time Carl Kiekhaefer began to re-evaluate the snowmobile market in 1970, over sixty companies were assembling some version of a snowmobile, and yet only three American manufacturers were producing their own engines. Mercury, OMC and McCulloch powered their own models, and of these, only OMC could claim even modest success. Carl's own informal surveys of the other leading builders convinced him that if he could provide a reliable, compact, efficient and suitably powerful snowmobile engine, his new company could rapidly erode sales of the German and Japanese suppliers. Moreover, because most snowmobiles were sold from "winterized" outboard motor franchises, Carl was confident that the well respected Kiekhaefer name would benefit the buyers of his engines. Industry estimates for 1970 revealed over one million snowmobiles were owned in the United States, and over 100,000 were registered in Wisconsin alone.

Carl decided to blueprint three basic engine sizes: 433cc 35- to 38-horsepower, 649cc 56- to 76-horsepower, and a 787cc model that could deliver nearly 100-horsepower. Each of the three basic models would benefit from Carl's experience with capacitance discharge ignition systems, which Carl would name the *K-Tron* ignition. The alternate firing two- and three-cylinder engines could be cooled by free-air or integral fans to blast fresh air over the cylinder heads. Though slightly heavier and more expensive than the German and Japanese competition, the designs were compact, exceptionally rugged, quick-starting and to a high degree benefited from interchangeable parts.

With characteristic Kiekhaefer efficiency, the designs were finalized within 90 days of opening shop. Assisting Carl in the designs was Herman Meir, whom Carl described as "one of the world's foremost two cycle experts who was a consultant for most of the European motorcycle manufacturers at one time or another."[14]

Though Carl was financially secure, both from the large blocks of Brunswick stock he still held, and from his extensive land holdings of Lake X, Midnight Pass, and Fond du Lac properties, he couldn't risk completing tooling for production models until sizable orders and financial guarantees could be landed. His very first support arrived in the form of a $1 million line of credit from his long-time friend and trusted business advisor Jack Puelicher and the Marshall & Ilsley Bank of Milwaukee. Since the scale of his potential engine sales were large enough, Carl also let prospective snowmobile companies know that he would consider merging Kiekhaefer Aeromarine Motors with the right organization to close the right deal.

In the fall of 1970, Carl was approached by a number of concerned citizens living in and around Taycheedah, including the affable Jack Cotter, Carl's friend and the popular bartender at *Petries* restaurant and lounge in downtown Fond du Lac.[15] Cotter was a long-time friend of a local judge in the area, and over the years when Carl would receive tickets for minor driving infractions, Cotter would see to it that the tickets were dismissed on a busy day in traffic court. It seemed that the City of Fond du Lac was negotiating a contract with the City of Taycheeda to accept refuse and garbage at a site within a mile of Carl's new complex at Kiekhaefer Aeromarine Motors. A loud and boisterous verbal exchange ensued between city lawmakers and those opposed to making the City of Taycheedah a dump site. Carl led the fight against the proposal, and eventually, in order to win the day, offered to buy the 150 acre dump site, convert it into a park and snowmobile competition track, and donate the whole works to the City of Taycheedah. When the City suspected ulterior motives, Carl agreed to build a city hall for the town across from the park, donating the earth moving and construction himself. Carl explained how he managed to get involved.

> "It might interest you to know how I got involved was that we heard that the City of Fond du Lac was about to make a deal with the Town of Taycheedah to be allowed to fill a huge, abandoned gravel pit with garbage for $10,000 a year, the proceeds of which after five years would have been sufficient to buy the park. Owning at that time some 800 acres of land in this community and having had experience drilling several wells, and having two gravel pits on my own land, I realized

how absorbent the subsoil was in this area and I could foresee damage to hundreds of fine wells due to seepage from the City's garbage. I fought the movement at that time but could make no headway until I conceived purchasing the land and giving it to the Town with a resolution that the Town of Taycheedah would never be a garbage dump for anyone.

'The humorous side was that almost of a man standing on Hollywood and Vine giving away $20.00 bills. People skirted the proposition suspiciously but after they finally realized there was nothing in it for me and I was merely trying to save the ecology of this community, the matter was resolved by my donation and as a result, it was named 'Kiekhaefer Park'.[16]

No sooner were the donation papers filed, when Carl and a contingent of giant earth moving equipment swarmed over the site, to begin construction of a half-mile oval competition snowmobile track. Under Carl's watchful eye, the unique fifty-foot wide clay track took shape, featuring constant radius and banked turns, so that the angle of entry into the turns would be equal to the angle of exit, thus minimizing the common danger of spin-outs at turns. Carl rushed against the clock, for the First Annual Midwest Championship Snowmobile Race was scheduled at new Kiekhaefer Park for early February 1971, and the site was rugged terrain covered with high weeds and brush. During the following summer of 1971, Carl completed finishing touches on the park, including pit facilities for over 300 competitors, parking on premises for 5,000 cars, and spectator areas to accommodate over 30,000 people. Carl also provided fencing, retaining walls, a judges' stand, a public address system and a test track area for competitors to tryout repairs. In addition, Carl provided a system of snowmobile and motorcycle trails, bridle paths and general recreation areas. When completed, Carl declared the *Aeromarine Sno-Bol* "the fastest, safest, finest snowmobile speedway in North America."[17] The entire project, including the land and the adjacent city hall, cost Carl and Kiekhaefer Aeromarine an estimated $300,000.

Carl's continued visits to various manufacturers like Featherweight (Bangor Punta), Rupp and Arctic Cat generated great enthusiasm for Carl's new engines, but most companies had to contract between two and three years in advance to secure engines from Germany and Japan. Frustrated in his initial efforts to penetrate the American markets, he considered having his engines built by Aspera Motors in Torino, Italy, as he began to fear he couldn't compete with the low cost of overseas labor.

> "Considerable discussion and changing of plans have occurred in the past month because of the Japanese invasion with their air cooled engines to the snowmobile industry with rash prices and delivery promises, which have upset German suppliers and have caused them to lose business. A highly competitive situation apparently is appearing, which makes it more interesting that you rather than we in America be the producers. It could well be that we in America shall become distributors and service outlets for engines produced in countries that have lower manufacturing costs and manufacturing costs competitive to the Japanese costs."[18]

On one of his many sales trips, Carl flew on the newly inaugurated Boeing 747 jumbo jet and was infuriated by the experience. Over the years, Carl wrote many long and bitter letters to the presidents of the various airlines that he patronized, but this brief excerpt of a long letter to the *New York Times* gives a fair representation of his unique frankness when expressing his displeasure.

> "A customer is not particularly impressed with being plied with liquor, beer, bad food and bad musical tapes amidst a milling throng of hundreds of people who often cannot even find seating space and must spread out on the floor.
>
> "I am sure that the health inspectors have approved the sanitation situation aboard the 747's, but I don't feel comfortable having the food dished out in cabinets that are less than six feet away from toilets that are conspicuous as in the form of flowing juices evidenced in the aircraft carpeting. The toilets are certainly evident by their odor. If forced to fly on one of the 'elephants' again, I am inclined to bring my own lunch."[19]

Carl's many overtures to snowmobile manufacturers were disappointing. Anxious to overcome the lack of orders to get KAM into engine production, Carl widened his search for business into other industries. He tried to land the design and prototype contract for a proposed two-cycle automobile engine for Ford Motor Company of Australia, to power a "small, inexpensive automobile for sale in the Asian countries."[20] He lobbied Lee Iacocca bravely for the chance to prove himself in the automotive world, but the newly appointed president of Ford assigned the project to more traditional sources. Carl also consulted briefly with the owner of an Indianapolis 500 contender to provide a *K-Tron* ignition for their turbocharged Offenhauser race engines. When no other business surfaced, Carl consented to evaluate the potential North American and international market for the Swedish-built Monark-Crescent outboard motor under a consulting contract with Bangor Punta Corporation President Nicolas Salgo. Unfortunately, his $5,000 fee and travel expenses to Sweden went unpaid for many months, leading Carl to speculate that just about everything seemed to be going wrong.

Carl had received enough encouragement from the major snowmobile manufacturers, however, to give him confidence in his plans for the new engines. To secure large scale orders, he needed to demonstrate the ability to manufacture engines as well as build prototypes, and so on November 13, 1970, a familiar sight greeted over 50 business and community leaders assembled in the stubble field behind Kiekhaefer Aeromarine Motors. It was a grinning Carl Kiekhaefer, adroitly at the controls of a giant front-loader tractor, scooping out the first bucket of earth in dramatic ground breaking ceremonies to expand his plant. The press was quick to make comparisons with Carl's surprise appearance on an earthmover at the 1964 ground breaking for Mercury, and reported that he hadn't lost his touch. "Again Kiekhaefer mounted an earthmover. At the age of 64, he piloted the vehicle with the same skill, and ground was broken for another industrial adventure by a genuinely remarkable man."[21] The expansion would double the size of Carl's existing facilities to 60,000 square feet, situated on his 650 acre tract of land at 1970 Aeromarine Drive on the ledge overlooking Lake Winnebago.

Carl confidently told those assembled at the ground breaking that employment would quadruple to 200 workers at the plant during the following year, and that "I feel a tremendous responsibility and workload, and the trials and tribulations that go with this responsibility. But I have no interest in going fishing; I might as well do what I like."[22] And so, with the addition of nearly $1 million in additional plant and equipment, Carl was preparing to start the next phase of his operations.

As he continued to work long hours and invest heavily in his facilities and machinery, many wondered just what Carl was trying to accomplish, and Carl felt the need to explain why he even bothered to start all over again.

> "Not because I am trying to prove anything in particular and not because I think I am thirty-five again, as I have been accused by some of the late hirelings at [Kiekhaefer Mercury], but because I take a genuine interest in developing and producing a product. ...
>
> "Sometime in January [of 1971] there will be an announcement of great interest and importance -- an announcement that will shut up forever the Doubting Thomases and the vicious tongues who, in want of anything else to do, take delight in disparagement of something which I thought was an effort of some note and appreciated by this community. But then it will be ever thus and it doesn't particularly bother me. ..."[23]

Of Carl's unflagging struggle to succeed, a reporter for the Fond du Lac *Commonwealth Reporter* perhaps expressed it best. "E.C. Kiekhaefer is a doer. He's also a brilliant exponent of creative engineering. Combining exceptional ability with optimism, he is always looking ahead to new horizons."[24] In his headlong dash to those new horizons, Carl would surely confront the most difficult challenges of his life, and experience triumph and defeat in nearly equal proportions.

CHAPTER THIRTY-FOUR

The Invisible Family

"Happy families are all alike; every unhappy family is unhappy in its own way."[1]

Leo Nikolaveich Tolstoi
in *Anna Karenina*

Little has been said of Carl Kiekhaefer's family, for in a nearly literal sense they were largely invisible to him and to his busy career. Freda and the three children, Helen, Anita and Fred Kiekhaefer, weren't physically neglected in the sense that their primary needs went unmet, but rather they were somewhat emotionally abandoned by Carl along the road to his destiny of fame and success. Almost before their brief camping honeymoon was over in the spring of 1932, Carl had disappeared into his struggle for promotion and financial advancement at Stearns Magnetic in Milwaukee, and Freda quickly came to realize that his work would always be paramount in the organization of his life.

When Freda gave birth to Helen Jean Kiekhaefer on November 8, 1933, Carl made it clear that she would never return to work as a nurse, and that her rightful place would be in the home. The first clear signal to Freda that the family would always represent a distant second priority in Carl's life was almost the exact moment she was giving birth to Anita Rae Kiekhaefer, Carl opened the doors to the Cedarburg Manufacturing Company for the first time on January 22, 1939. Though it was Sunday, and though Carl wouldn't legally take possession of the property for several days, he chose to roam around the frigid and abandoned ruins of the bankrupt Thor outboard facility rather than be at his wife's side in Milwaukee. And finally, as Freda gave birth to Frederick Carl Kiekhaefer on August 23, 1947, so was Carl giving birth to the *Lightning*, a machine that he would ultimately consider his most important engineering accomplishment. And so it would continue throughout Carl's frenzied industrial existence, work and discipline first

-- family last, and then only if there was absolutely nothing else to do.

The arrival of Carl's children was spread out over a period of some fourteen years, and Carl would often joke that his children "came like the locusts, every seven years." Their disparity in ages seemed to handicap Carl somewhat in adjusting to the role of father. No sooner had he grown accustomed to dealing with one of his children at some critical age of life, than a new round of infancy would challenge his gruff patience. Freda became an island of sanity and refuge for the children, insulating each carefully and lovingly from the sharp words that Carl could hurl at the innocent like shards of broken glass. Each of the children, nevertheless, would be deeply affected by the stranger they called father, and in later years each would bare scars of resentment and confusion, wishing that their youth had not been spent in the near total darkness of the giant shadow he cast.

Others, too, could see the impact that Carl was having on the children, and were quietly sympathetic. Once, when Charlie Strang was at the Kiekhaefer home at Cedarburg, Wisconsin in the early fifties, he could see that Carl never made a connection between the development of his children and his own somewhat eccentric behavior.

> "I think one of the funniest things I saw [concerned] his two daughters. They were gorgeous, gorgeous young girls. Helen was very much like her mother, temperamental, calm and quiet. Anita was exactly like Carl. They had a boxer dog, Pokey. Anita was a very cute, little blond girl ... maybe ten or thereabouts. Pokey was running around or something, and Carl just kind of kicked the dog with the side of his leg, and Anita *blew her cork*, and she screamed, and she jumped up and down, and she howled and howled because he had kicked the dog. Carl turned to me and says, 'I don't know *where* this girl gets her temper!'"[2]

Anita's earliest recollection of her father is one of fear. "I was a little girl, about five years old, and he used to take me for a ride in the car -- a *very fast* ride. He said, 'Hold on! Why don't you hold on *more*! Don't you cry, Don't you dare cry! We don't want any sissies around here!' He tried to toughen us up."[3] Freda Kiekhaefer remembered how Anita took advantage of her father's position once when she was still a little girl. The school she was attending was raising money by having the kids sell candy bars. Carl had brought her down to the plant in Cedarburg on many occasions, so just about everybody knew her. She positioned herself at the exit to the plant as the men were filing out after work, selling these candy bars. And, of course, just about everyone that came out had to buy a candy bar from the boss' daughter, and she won the prize at school for selling the most. Her unique method was, of course, kept secret from Carl.

It was nearly impossible for Carl to display affection and tenderness with the children. "He didn't want to pick me up or anything," Helen remembered. "I think he was just really kind of uncomfortable with that kind of picking up and cuddling. ... I know that a couple of times I tried to tell him I loved him, and he almost got angry he was so embarrassed. He just didn't know what to do with it." He was also all thumbs when it came to any kind of playing with the little girls, and his roughhousing often sent them crying to their mother for asylum. "Yeah, I mean he would do things like carry us over broom handles and he'd wrap us up in a blanket and sit on us," Helen continued. "I honestly felt like we were going to smother. I mean it was awful. But it was so painful the way he'd play that we couldn't help but cry."[4] Freda regretted the fear the children had of their father, but felt helpless in his presence. "The kids, when he'd come in, they'd scoot up to their room," she said. "And, you know, that hurts when there's that feeling. And then, I was the buffer zone, always."[5]

Though Carl insisted that the girls could benefit from the hard knocks and tough discipline of the farm life he endured to help mold their self-assertiveness, he pressured them to conform with his own vision of proper young ladies in high heels and flowing dresses. Both daughters rebelled somewhat against this pressure, leaving Carl almost perpetually at odds with them and with Freda. When the girls were old enough, they worked briefly during the summer in the Kiekhaefer Mercury offices in Cedarburg, where Rose Smiljanic remembered how he would needle the girls about their appearance. "He didn't like the way their hair was done, it should be done like the other girls in the office. And the mother would get mad about that." He picked on Anita especially, Rose recalled. "The poor thing. He liked her but he was always picking on her. He wanted her to be -- you know, he had the wealth -- he wanted her to be a lady. She was like him, and then he couldn't understand that, and there would be friction, and the mother would get in between."[6]

The triangle of friction between Carl, Freda, and the children, created a perpetual environment of tension and anxiety around the Kiekhaefer household. "He always criticized us," Helen said. "How we looked, how we ate. We used to all fight to see how far away from him we could sit. Really, and you know he was like a kid. I mean the way he acted, too. I mean he would be upset if he didn't get the chicken liver when everybody else wanted it."[7]

Carl spent so many hours at the plant he was like a surprise guest when he showed up at the house, and would often shock Freda by bringing a large group of people home with him for dinner without notice. "I know that my dad was always very abusive towards my mother," Helen said quietly. "Not *always*. Once in a while there were some tender moments. But most of the time, you know, he had really expected tremendous things of her. He would pop up at the last minute with a whole bunch of people and expect my mother to whip out a meal and then embarrass her terribly about how terrible the meal was. I remember him throwing dishes, you know, against the refrigerator because there was a crack in one of the plates. And, you know, he flung it against the refrigerator and all of it broke, food everywhere. Then I remember one time he went out the door, slamming it, saying, 'You all would eat buttered door knobs if they were served. He was always critical -- always critical."

Conflict wasn't confined to the immediate family either, and the oldest daughter, Helen, recalled a disturbing scene in prelude to Carl's firing of all of his sisters and their husbands in the early days at Cedarburg. "I remember one time ... he was *screaming* and *screaming* at my aunt Ruth. And my aunt Ruth was standing there crying, and it was so devastating to me that when I went to school that day, I just couldn't do any of my work and I went and told the teacher what had happened. And then I went home and told my family what had happened. Oh, my dad got *so mad* at me for having said something to the school. It was like I had to go somewhere with it. It upset me so badly. ... He never praised me for anything. He always downgraded us so all the time. ... I grew up, in my life, I had absolute terror of him. I was so afraid of him that it was devastating. It was emotionally, mentally. He never laid a hand on me. He never had to. All he had to do was look at me or say a word and I was *terrified.* If there was a phone call from him I immediately got nauseated and my palms started sweating and I'm shaking and I'm in absolute terror."[8]

A portion of the anxiety that permeated the adolescent years of Carl's two daughters, was due to the small community in which they were raised. Carl's company was the largest employer in the area, and as a consequence, the children of the homes that relied on a Kiekhaefer paycheck often kept the girls at arms length. "Everything was ordinary about the way we lived," Helen said. "And that's why I never could understand -- all my friend's fathers worked for him, because he had basically the only business in town. And I'd hear my cousins, everybody would say, 'Your daddy's rich,' and I'd say, 'What are you talking about?' 'Cause I had no concept of that. I mean, not that way. I had a heck of a time growing up. I was ostracized, rejected, gossiped about, and was so easily hurt."[9]

Helen and Anita both developed into very pretty young ladies, trim, athletic and excellent water skiers. Through Carl's friendship and business relations with the Polk family, for a time they worked as Aqua Maids and ski performers in the popular shows at Cypress Gardens in Winter Haven, Florida. By the time Anita reached teen-age, however, Carl had already put her to work with the photography department of Kiekhaefer Mercury. "I was only about thirteen years old and he said, 'Well, I believe that you have to learn the value of a dollar.'" She was sent out with the advertising and public relations crews during her summer vacations from high school, assisting them in getting catalog or brochure shots of Mercury products in action. As her experience grew, she operated the tiny darkroom on the second floor of the old Corium Farms barn. The trouble was, the men's rest-room was directly underneath her enlarger, and each time the men slammed the door to the stalls, her pictures came out blurry. "Finally, when I was twenty," Anita recalled, "I was doing most of the photography, the processing and the developing ... we felt like history was in the making."[10]

Anita became frustrated with the limitations of still photography, and graduated into promotional 16mm motion picture films. She worked on a number of projects, with such titles as *Water Ski Antics*, traveling throughout the country on behalf of the company. She would climb the scaffold of a specially designed photo boat speeding through the water, and capture the performance of her father's engines on film. One day, with the camera pressed to her eye, the boat hit a large wave, throwing Anita twenty feet to the hard deck below and breaking her shoulder. She was game, though, and many compared her to her father. She was back in the photo platform in short order, filming such demanding events as the Bahamas 500 marathon, and other offshore powerboat races in difficult conditions. She was also required to shoot from helicopters when the speed and distances of races were too great for the photo boat. "I was hanging out of them half of the time, trying to get a camera in one hand and my leg tied up with the other. ... A lot of times I'd forget where I was, and I'd just be hanging in there and I thought, 'This is crazy. ...' I'm hanging out of a helicopter and I'm wondering, 'Am I doing sky stunts or am I doing photography?'... Of course my father wanted me to be more of a son."

After receiving a masters degree in film from New York University's Institute of Film & Television in January 1968, Anita embarked on a career of film making, specializing in religious documentaries, and scouring locations amidst the ancient cities of the bible. Before she left the United States, she wrote Carl a letter on her birthday -- the thirtieth anniversary of Carl's founding the company -- expressing her gratitude for his assistance in helping her to learn her craft, and her regret that the family was in turmoil.

> "My diploma ... reminds me of all your trouble to provide for my education, when you endured hardships to obtain yours and to get to where you are now. ...
>
> "Unfortunately, circumstances of late under which we were together like last Christmas, was not conducive for discussing what is on both our minds. I am sure you are well aware of the tension of the family breaking up in more ways than one.
>
> "I know your schedules and responsibilities are wearisome, your tasks unbelievable [and] hard to bear. But I want to tell you another thing before I leave. ... However estranged we've been during these thirty years of company expansion, I want to reassure you that you are a father, not just another employer or corporation boss ... and your help to get me started will not go unheeded. I feel this more strongly since you have had to endure 'the worst' whether it was for yourself or for the family. Certainly you have done your best to provide for us at the cost of peace of mind and heart as well."[11]

When Helen was elected "Miss University of Florida" at the Jacksonville campus in 1953, Carl honored her by placing her name on the side of one of his race cars competing in the Mexican Road Race that year. Typical of Carl's private reactions to news of his children's accomplishments, though, was contained in a letter he sent in response to receiving a news clipping about Helen's honor from the local Jacksonville Mercury dealer.

> "Frankly, I do not approve of all the publicity which has been more than kind and which has been flattering. It is easy to spoil children in their most impressionable years and I would have been more proud of Helen if she had grown up as a farmer's daughter knowing the meaning of work, thrift and a few rough knocks."
>
> "Needless to say, her mother disagrees and when a man is outvoted three to one and is exposed to the public relations produced by two grandmothers, one sometimes begins to doubt one's own judgment and the easy way out is to say nothing."[12]

She was Queen of the campus, and he was a four-year starting veteran at left end on the "Gators" varsity football team. Like the plot of a fifties campus movie, she was the pretty pom-pom wielding cheerleader on the grid-iron sidelines as he dashed for the end zone. They fell in love, and the year after graduation, on August 24, 1957, Helen Jean Kiekhaefer and Robert Ray Burford were married in Winter Haven. Carl was on hand to slowly walk his daughter down the isle, in many respects a stranger to both bride and groom.

Carl and Freda showered the newlyweds with kindness, cars, cash, a house and an all expenses paid honeymoon to the Bahamas. When Carl sold the Kiekhaefer Corporation to Brunswick in 1961, he established a trust fund for Freda and each of the children, enough to provide for reasonable, yet far from boundless, financial security. Into the trust for the newlywed Burfords, Carl placed a generous portfolio of Brunswick stock. When young Bob Burford finished a short hitch with the armed services, Carl surprised him with a position on the board of directors of the Kiekhaefer Corporation, and a generous job. The Burfords adopted an infant son, Derek, and to the casual onlooker, their lives may have seemed like the American dream come true. Unfortunately, the bubble soon burst.

On April 17, 1961, Burford was fired after working for the company a little more than three years at various positions.

Carl accused Burford of a long list (22 items) of infractions, including misrepresentation, insubordination, "telling president of Kiekhaefer Corporation what to do," and "Worrying and upsetting his family and relatives in such a manner as to adversely affect the operations of the family-owned company, who at this time, cannot afford such additional stress."[13] A deep antagonism developed between Carl and Burford following the son-in-law's dismissal from the company. Five months later, Helen wrote a most difficult letter to her father, pleading for understanding, and revealing the very tenuous nature of Carl's relationship with his children in general.

> "This is a very difficult letter for me to write. I have always been afraid of you, Dad, and whenever I've been in your presence I've always been too nervous and ill-at-ease to express myself in terms of how I felt or what I thought about various matters. So, I am attempting to organize my thoughts in a form of a letter which is written in complete earnestness by a daughter seeking approval, sympathy and understanding from her father. ..."[14]

The strain of a deteriorating marriage and a fearsome father completely exhausted Helen, who was hospitalized in September 1963, in need of total rest and relaxation. Carl reacted sharply and awkwardly to her hospitalization, and if anything, helped to increase Helen's anxiety. He flew down to Sarasota to visit her in the hospital, but Helen avoided seeing him as long as she could. Carl grew more and more agitated with each passing hour, until to Helen and the hospital staff it seemed like Carl would explode. "The [doctor] said, 'Please go out there and hug your sad father,'" Helen said. "And when he came in to see me, he just went by the door and screamed and screamed and *screamed* at me. He screamed so loud that all the way down the corridor and all the way out at the waiting room behind closed doors people could hear him. He just could not handle illness of any kind. And I think it might have stirred up his own guilt, his own ... helplessness or whatever, because it's not in the area of nuts or bolts, and he just could not handle it."[15] When Helen was transferred to a clinic in Massachusetts, Carl secretly hired a private detective to send reports back to Fond du Lac on her continuing improvement.

When she was discharged from the clinic, Helen returned to Florida, and with Carl's assistance, began divorce proceedings against Burford. Carl hired more detectives, and started round-the-clock surveillance of his son-in-law. Though Carl was convinced that his daughter's fatigue was due to her broken marriage, transcripts of the divorce proceedings often pointed the finger of blame squarely at Carl himself.

"Are you aware of pressure in your home that caused your difficulty?" Helen was asked under oath.

"Well, certainly. ... Inability or not allowed to make decisions," Helen answered.

"Not allowed to make decisions? ... By whom?" she was asked.

"Well, my father. He's a very domineering sort of person ... and this is a German trait," Helen replied.[16] Later in her testimony, Helen was asked more about her father's personality.

"He has harangued [your mother] though, and domineered?"

"Yes."

"He's a tyrant, isn't he?"

"If you want to describe it that way," Helen answered. At this point, her attorney jumped up and offered, "*Tycoon* might be a better word." To which Helen's examiner replied, "Both. Sometimes one follows the other."

"Let me ask you this. Do you feel that a child under the influence of your father is in a wholesome environment from your own life and your own experience?"

"Well, I have thought about that a lot. ... My grandfather was the same way, and a lot of German people have this, you know. I mean this has been their culture, and the man is Lord and master of the household and this sort of thing."[17] Helen and Bob Burford were granted a divorce on November 21, 1965.

By 1957, Freda, along with ten-year old Frederick, had moved away from the chaos and daily calamity of Carl's growing industrial empire, choosing instead the warmth and unruffled calm of Winter Haven, Florida. Carl had purchased a pleasant, conservative home across from fabled Cypress Gardens, when Mercury's relationship with the Pope family, owners of the popular attraction, seemed unimpeachable. It had also become painfully obvious to Freda, that her husband's intimate relationship with his secretary, Rose Smiljanic, had somehow become rather common knowledge, and she was moribund to endure the clumsy knowing glances and nervously chosen words that issued from Carl's many understanding employees.

In spite of the exigencies of his unique family life, Carl's only son, Frederick Carl Kiekhaefer, managed a rather remarkably normal childhood and adolescence. In a rare philosophical mood, Carl once wrote a letter he never mailed to Fred's future wife, Dr. Carol Stafford, and described from his own perspective his impression of the growth and development of his son, who had at the time already finished college with advanced degrees in both engineering and business administration.

> "Fred has been exposed to older people all his life, not only in his own family, being the youngest, but also by association with company personnel. As soon as he was able to drive a boat, he got a job at Lake X as a test driver and naturally he was exposed primarily to technical and professional people in the engineering fields. After several years of that, he became a mechanic at the test center and after several summer vacations of that, I brought him back to the plant during summer vacation to work in the Engineering Department where he began as a detailer and ended up a designer and layout draftsman.
>
> "Then, of course, he was exposed to the whole picture of our growth from a very small company when he still lived in Cedarburg through the time that we grew into a $160 million company. During this period I tried not to preach too much because boys must learn how to grow up in their own element and reach manhood by easy but steady stages.
>
> "When he reached the reckless age when young manhood is beginning to express the aggressiveness of the male animal, it is easy to become impatient with their compulsiveness and I daresay there were times when I could not control my temper.
>
> "To my knowledge, he stayed out of trouble pretty well by showing him fine equipment and during his latter teens, fine automobiles with enough horsepower that he need not prove anything amongst his fellow 'rodders', with the warning that if he ever got into accidents due to misuse of the most powerful automobile on the road, I would have to put him back in a V-W or something like that -- or worse yet, into a Chevrolet or Pontiac.
>
> "We went the route on motorcycles too. Because of his knowledge of the limitations of equipment, he never got into serious trouble there. I was happy to see that he survived the drinking and drug age without becoming an addict, and one of my important reasons that I started back in business ... was to offer him the challenge he needs after getting the finest educational opportunities.

"He inherited the tempers of the Dutch and Irish, which causes him at times to say things he doesn't really mean and I am sure later regrets.

"He has heeded my advice to not become too seriously involved with young ladies until he had completed his education, a mistake that I made and have dearly paid for. His opportunities are obviously so much greater after having completed his education and has reached a higher degree of maturity and with it, a better sense of values that apply to nearly all things. A mature judgment, therefore, is much more sound and spells much more happiness in later life and enables him to get a foundation in his life's work before undertaking the responsibility of marriage and parenthood where earning ability also makes him a better provider.

"My other two children are earning their own livings, but it took sending them back to college twice, once to get an education and second time to teach them how to make a living, should anything ever happen to the savings of their parents, which may not be there to pass on, which is certainly greatly reduced with the present trend in our taxation philosophies. Our nation is on a socialistic trend, which is nothing but a constant process of giving to the 'have-nots' and taking from the 'have-gots' without working for them, which in the extreme degree probably spells communism. ...

"This is the point I have been trying to make with Fred, who has professed many times that he didn't intend to work as hard as I did, and neither do I want him to. I hope life will be easier for him and future family, but he must earn this privilege and get in his licks when he has physical health and mental alertness and energy, and this usually means you have got to have it made by the time you are fifty, which seems a quarter of a century away perhaps. There will be many new problems because the rules change, the government changes, people themselves change, human nature may change somewhat, but it will not change to the better. It is for this that we have the energy, the optimism and the enthusiasm of youth, and thank God for that, that have for generations managed to cope with life and its problems.

"One other important factor as to why I went back into business was to give the young people of this community opportunities when their elders have become so complacent and so steeped in their day to day and comparatively good affluence that they have lost their serious-mindedness as to their duties to their families, their employees, and their community. ..."[18]

Of his family, Carl would in later life admit, "I have had grief-stricken times with my two girls and for a while, it didn't look like I would be able to communicate with my son, Fred. All this is past and I am reasonably happy to say my family has turned out fine, even though the girls are both going to be executives rather than grandmothers from all appearances."[19]

Holidays represented a particularly stressful time for the Kiekhaefer family, for in most cases the family had no idea if Carl would be home or not. "He'd say, 'I'm not coming for Thanksgiving,'" Freda said, "so I'd make [other] plans. The church always had a program for people that were alone, so I signed up for it ... and of course the night before, here he comes. But he was always there for Christmas. He'd be in a lousy mood. We would just wish that he stayed home. We had gifts for him and he wouldn't even open them. ... Helen had painted him a picture one time, and he never opened it. So, we put it behind the piano. Next year, we put it out again and he *still* didn't open it! Three Christmases before he opened that package. We would give it to him and he'd pile it up there by the chair and crack nuts. He loved to crack nuts at Christmas time. Sit there and eat nuts."[20]

Even Carl's Christmas cards betrayed his lack of family identity. When most families were sending out photo-cards of mom and dad kneeling on the front lawn with the kids all around, Carl would send out cards with himself and his three dogs, Duna, Bonnie & Clyde in the photo. Inside, regardless of the implications of the photo on the front of the card, the inscription would read, "Carl Kiekhaefer and family."

When Freda realized that Carl had completely forgot their twenty-fifth wedding anniversary in 1958, she was crushed. "I said, 'Okay, I'll forgive you for that. But if you forget the fiftieth, I'm going to divorce you!'" She kept reminding him every ten years or so, but when it finally came around, he forgot anyway.

Very soon after he opened the doors of the Cedarburg Manufacturing company, Carl stopped telling Freda anything about his work or his business. It made her feel somewhat abandoned, and often anxious about what to say whenever he was around.

> "That's when it all started. Just the tension of the business that he took out on the family. And that's why we stayed out of the way as much as we could. Just let him go. Just have to realize that [his work] was the main thing. It took

> a lot of blood and tears for that. Yes, he would be hurt, but he didn't want to discuss it. See, I thought that I could help him if he would just talk. Then, I'd ask him about it, 'No, no,' he didn't want to discuss it. But I really think it would have helped him a lot if he had been open and discussed his problems. To talk to someone that could understand, but I guess he thought I just didn't know enough about the business to understand the problems. And I didn't because he wouldn't tell me anything about it."[21]

Freda, over the years, also became satisfied to keep Carl at an emotional distance, and to keep out of his way as best she could. When she was asked whether she had ever discovered Carl's soft spot, she replied, "Well, I'll tell you. To be truthful, I didn't even look for one. I thought it was safer just staying out of him. If I had started an argument, I don't know how it would have come out."[22]

Whenever friction developed between one of the children and Carl, or between Freda and Carl, he was quick to threaten them with being cut out of the will. "Everybody was threatened," Freda admitted. "I don't know how many times that [the kids] were going to be cut out of the will ... so many times. Finally, they got so they said, 'Let him!'"[23]

Though Carl had established a significant trust fund for Freda following the sale of the Kiekhaefer Corporation to Brunswick in 1961, he controlled it with an iron grip. Once, after Freda had donated some money to her local church in Winter Haven, Carl cut-off her income completely, leaving her to fend for herself with only a small social security check to pay for groceries. Even minor expenses like paying for a tankful of gas for her car with Carl's credit card, would draw an attack from Carl. After one such incident, Freda dashed off a note in anger to her "keeper" in Fond du Lac. "Since I dislike being accused of stealing, I'm writing a check to you to cover the gasoline bill ... also including the credit card. I didn't steal!"[24]

When Anita was filming in Israel, Freda had to sell her coin collection to raise plane fare and travel expenses to go visit her. "Oh, I was so sick, I could have gotten three times what I got out of that coin collection," Freda recalled. Carl was furious whenever he found that she had done something without his permission, even though they lived nearly 2,000 miles apart. Fred Kiekhaefer described the emotional dilemma that confronted his mother.

"My dad would expect her to always be there when he called. And if she wasn't there, 'Where were you? Why weren't you at home?' You know, that kind of thing.

"There was definitely an old world type of relationship, and that's why I don't try to impose too much of my value structure and attitude on it. They're of a different generation, and people in relationships tended to make things permanent in those days. For better or for worse. And more often than not it would have been intolerable by today's standards. *Intolerable.*

"So, it's very easy to sit here as a third party and say what should have been or what they should have done. We're talking about two grown up people with the ability to make their own decisions."[25]

Once, when Freda and Rose Smiljanic were driving to Milwaukee together, the two rivals for Carl's affections spoke philosophically about Freda's awkward situation. "Why don't you leave him? You would have an income, you know," Rose asked Freda directly. "God, she would have taken half of what he owned and turned around and sold it and he couldn't have done anything," Rose would later say. "She said, 'Well, I don't know why I didn't do it."[26] Curiously, Rose and Freda would eventually become friends, as both seemed to understand that Carl's indomitable presence and spirit was in control of events in both of their lives, leaving little room for self-determination.

The awkward triangle affected Freda and the family profoundly, though, and they suffered in quiet anguish with the knowledge that so very many people were aware of their predicament. "I can tell you that an awful lot of people from time to time have told my mother that things were going on," Fred Kiekhaefer disclosed. "The crush and hurt often went so deep that after a while it sort of stopped hurting. ... I dislike dishonesty in relationships. There will never be any in mine. 'Cause people get hurt and good people don't hurt good people. ... I think my mother and my father -- and that's my personal opinion -- I think they made a mistake. ... I wouldn't have tolerated it, let me put it that way. I wouldn't have put up with it."[27]

Carl's invisible family loved him, feared him, challenged him and eventually grew somewhat impassive and numb to the passionate man who would storm in and out of their lives. Carl wrote to other people about his family, more than he wrote to them. He understood the damage his unyielding drive for success had done to his battle-weary family, and was saddened to see the pain reflected in their lives. In later years later, Carl would confess in a letter to Helen that he felt himself responsible for the family's sadness, but was quick to argue and rationalize that his neglect was an unavoidable consequence of his career.

> "Indirectly, or perhaps more directly, I blame myself for everything that has happened. It is a penalty of the drive to build a large corporation in the modern, competitive world. This tremendous concentration of meeting the great, wicked competition left little time for the family and gave your mother almost the complete responsibility of raising the children, which has its penalties.
>
> "The reason for the tremendous drive, almost beyond human endurance, was to lift, for once and for all, my family out of the poverty experienced not only by myself but my parents and relatives around me. This business of going to work after dinner in the evening for twenty years and working till late, the wee hours of the morning, was not exactly a joyful procedure in spite of my ambitions and the challenges."[28]

Soon after Carl left Brunswick, and was wrestling with the start-up problems of Kiekhaefer Aeromarine Motors, Fred Kiekhaefer would become a vital resource to both his father and to the new company. Recently graduated with a masters degree in engineering, Fred felt strongly the bonds of kinship with his father. But the seeds of rivalry had been sown liberally throughout Fred's youth, and the road ahead, working beside his father, was strewn with the emotional boulders of a lifetime.

CHAPTER THIRTY-FIVE

Dashing Through The Snow

"I am a born optimist and a fighter -- not a quitter, but I know the problems that go with developing any new going concern! I don't want to be compared with people who know how to run a business which is already there and established. I think my track record will stand on both counts. ... It takes stamina, health, wealth, experience, and talent to build such an 'empire' with many sleepless nights -- and no salary, often an added penalty. Is it worth risking bankruptcy?"[1]

Carl Kiekhaefer

The pent-up frustrations of not competing in *anything* must have finally taken its toll on Carl Kiekhaefer, for starting in 1971, he was once again off to the races -- this time campaigning on both land and sea. Ever since the *Lightning* hit the water in 1947, Carl had been involved in racing and competition in one form or another. Without a competition-ready product for nearly a year, Carl had been forced to abstain from the adrenalin-enriched struggles for victory that had shaped his personality for the preceding quarter century.

In January 1971, Carl installed one of his prototype 76-horsepower, 649cc engines into a Polaris snowmobile chassis, setting his sights on the World Championship Derby at Eagle River, Wisconsin. The nearly untested combination was entered in the Open Class for non-production machines. The driver, Mike Rychlock of Three Lakes, Wisconsin, had never driven Carl's entry, and was "sternly cautioned ... to take it easy and bring her home in one piece."[2] When the snow flurries settled back to the frozen ground, Rychlock and Carl's new engine had won a double championship, winning both the oval and obstacle course events. The new engine never fluttered during the formidable competition, even beating 13 out of 19 entries in the larger 800cc class.

Rychlock continued to roar to victory after victory in the KAM/Polaris snowmobile, while a second machine and driver were prepared to join the team. On February 20, 1971, Rick Scholwin, also from Three Lakes, mounted an identical racing sled, and thereafter, the Kiekhaefer team completely dominated United States Snowmobile Association sanctioned events, and was nearly always first and second over the finish line in national competition. Starting in 25 events, they captured an incredible 19 first-place wins, four second-place finishes and two third-place. To cap off the remarkable season, on March 21, 1971, Rychlock won the World Series U.S. Snowmobile championships in the Open Class before a crowd of 47,800 spectators in Booneville, New York. One year later, Rich Scholwin would clinch a second consecutive World Series Championship for Carl by taking first place in the 1972 Open Finals at Ironwood, Michigan.

Carl was ecstatic. Three national championships on his first foray into snowmobile racing with his new engine design. "Quite a record for a brand new engine designed less than a year ago, eh?" Carl quipped following the final race of the season.[3]

The impressive string of racing victories were like a shot in the arm for Carl's effort to attract new business. Following the successful racing season, the new Kiekhaefer engine quickly became the dark-horse darling of the industry. Requests for evaluation engines from major manufacturers came across Carl's desk in a steady stream, and the prototype shop was soon swamped. Don Thompson, Carl's director of marketing, described the blaze of activity that ensued throughout the plant.

> "In the meantime, Kiekhaefer's engineering team was busy doing the final design work on the new engine line. Dynamometer checks, endurance running and field test continued at an accelerated pace. While the engine was being readied for production, construction crews were working feverishly on the plant expansion in fair weather and foul. The objective was a common goal: the first line of two-cycle, air cooled engines produced exclusively in North America for the recreational vehicle trade."[4]

One of the companies that was impressed with the new engine was the Bolens Division of Food Machinery Chemical (FMC) Corporation, Port Washington, Wisconsin, who began to consider an exclusive arrangement with Kiekhaefer. Once it looked as if Bolens were serious, Carl and Don Thompson decided to suspend their efforts to lure additional clients. "The letters and phone calls started to roll in, however," Thompson said, "and from all segments of the recreational vehicle industry.

Aircraft, all terrain vehicles, air drive boats, dune buggies, hovercraft, race cars, motorcycles, amphibious vehicles, snowmobiles, -- the inquiries came from every class of manufacturer, designer and engineer and from all over North America and Europe as well."[5]

Among the unique products that Carl designed during this period, was a wheel-conversion kit for snowmobiles. Although not an original concept, Carl's version was perhaps the most practical and durable of those offered, and allowed snowmobiles to hurl around his test track on rubber tires at over sixty miles per hour on a hot June afternoon. Though they performed wonderfully, Carl would ultimately fear the liability of death or injury resulting from the fast, marginally stable contraptions, and no more than a few prototypes were ever built.

When the Bolens order finally came through in April 1971 it was for only 1,600 of Carl's 433cc engines. It was an awkward size order, too small to take advantage of mass quantity discounts from suppliers, but large enough to shock Carl into full-scale production. An additional order for 400 similar engines was placed by Northway Snowmobiles of Pointe Claire, Quebec. Almost immediately, Carl fell behind schedule. "I certainly appreciate your patience and understanding in living with our production start-up set-backs," Carl explained to his impatient clients. "I fully realize how hard this has been and certainly am aware of the position which we have placed you with your distributors and dealers."[6] Being helplessly late with commitments was a new experience for Carl, and he became more and more agitated as his many start-up problems caused delay after delay. "There are days when the process of starting a new company seems almost like it should be the last. Perhaps I am overly sensitive. Perhaps I wouldn't be if I didn't have a deluge [of problems]. Sometimes it's like standing chained on an ant hill."[7]

In July 1971, FMC had taken Carl's many overtures to heart, and submitted an option to acquire up to 80 percent of the stock of Kiekhaefer Aeromarine Motors by January 1, 1972, in exchange for $2.6 million in FMC stock and cash. Mindful of Carl's spectacular track record in the marine industry, FMC was sagely looking ahead to the expiration of his non-compete clause with Brunswick, and the possibility of a new Kiekhaefer/FMC outboard engine. Carl gushed with anticipation over the prospect of an FMC merger, and having gone through two sets of merger discussions with FMC in his days with Kiekhaefer Mercury, Carl thought he had them pretty well figured out.

> "We have been engaged in many talks on arrangements with partnerships or mergers, and it looks like one is being firmed up with [FMC]. They have now completed the marine and recreational market studies, and are 'hot to trot.' We expect a closing in January [1972]. They are huge operators, well established with the military in the Ordnance Division, and also make amphibious tanks. ... It's an engineering oriented rather than a merchandising oriented company who seem to have unlimited funds for R&D. ..."[8]

The option was never exercised. It would be the first of many failed merger attempts, each more enticingly certain than the last, and each viewed by Carl as a major set-back and personal embarrassment.

The Bolens Division was delighted, nevertheless, with Carl's new snowmobile engine, and on November 8, 1970 increased their order to 10,000 units, with an option for 5,000 more. Response from dealers and customers was making the Bolens/-Kiekhaefer snowmobile among the most popular in their history. Carl geared up for the substantial production run, negotiating contracts with his many suppliers to supply not only the components needed for the 10,000 units, but an additional inventory of service parts. It was a massive build-up for a small company, and Carl's rosters soon quadrupled to nearly 200 employees. Enormous machinery was ordered and delivered, including a pair of 65,000 pound multiple spindle machines that were so large, a 50 ton crane had to be driven from Milwaukee to unload them from oversize trailers.[9] Though each of these machines was valued at over $250,000, they represented only a small portion of the giant effort underway to prepare for full scale production. Like the earliest days of production at Cedarburg, it was "carloads Kiekhaefer" once again, rushing to produce the avalanche of orders. And then, the worst happened.

With a little over one-third of their order built, FMC, the parent corporation of Bolens, shocked Carl on February 24, 1972, by announcing their decision to discontinue snowmobile operations. Carl was absolutely stunned by the news. Carl had actually attended a marketing meeting with a number of top Bolens personnel the very morning of the crushing news. Not only was Carl overnight ensnared with the financially devastating problems relating to enormous vendor over-production of castings, forgings and hundreds of other components, but he was also stranded with nearly 200 employees and not a single pending order. Worse still, it was far too late to solicit production orders from the other snowmobile manufacturers for the 1972-73 season, and the horrible reality sank in that he wouldn't even have a chance at the market for at least another full year. Carl eventually settled with Bolens and received over $1 million for his troubles, representing less than half of his actual losses. "It took us quite some time to get over the state of shock the Bolens affair caused us," Carl wrote to a friend a year later. "It was a dirty business and a situation that would have broken many another businessman. We had the choice of settling down to a 3-year law suit to collect the $3 million they owed us, or accepting roughly half that amount in the interest of having it over with and getting into something new. We chose the latter."[10] Not since the spectacular fall of the value of his Brunswick shares in 1962 had Carl suffered such an overwhelming sense of defeat.

Fred Kiekhaefer completed his masters degree in engineering from the University of Wisconsin at Madison in the summer of 1972. He had already been working part-time at Kiekhaefer Aeromarine for several months while he completed preparations for graduation. Carl eagerly awaited his arrival, and on many occasions wrote to his friends and associates with glowing reports of Fred's impressive progress, and hinted that his hard work to establish the new company would be rewarded when he could turn it over at last to his son. In January 1972, as Fred's arrival became imminent, Carl wrote, "I have been anxiously awaiting my son to complete his thesis and I believe that by February 1 he will be with me, hopefully to take some of the load."[11]

Fred began to work his way up and through the Aeromarine organization, completing assignments as a mechanic, detailer, draftsman, designer and engineer. His master's thesis explored the acoustics of two-cycle engines, and among his many projects was the silencing of intake and exhaust systems on Aeromarine engines. "My son, Fred, is taking hold and is turning out to be a pretty good engineer," Carl boasted, "which gives me some new incentives and some new encouragement."[12] When it became obvious that a second generation of snowmobile engine would be necessary for the changing marketplace, Fred was placed in charge of the inverted-V, two-cylinder, liquid-cooled snowmobile engine that would be considered among the most innovative two-cycle engines ever designed. Carl, at least on the surface, seemed willing to share the creative spotlight with another engineer.

> "In the meantime, my son, Fred, took hold of the two-cycle development and, primarily due to his leadership, we have developed a very fine liquid-cooled, extremely quiet snowmobile engine. Suddenly we are becoming the 'darlings' of the American snowmobile industry with the honeymoon in Japan having lost its romance."[13]

Occasionally, when Carl was feeling his age, he would find solace in the certainty that Fred would be at his side to build the company. "... The Horatio Alger story in America is fast becoming a myth," Carl wrote with some melancholy. "With that, plus my age perhaps, I have lost some of my touch in attracting and developing a top organization, which is so necessary in starting a new business. Nevertheless, we are making constant progress and I am certain that with the help of my son, who is now active in the business ... we will eventually succeed in making a mark in the U.S.A."[14]

Less than six months after taking charge of the inverted-V (I/V) snowmobile engine, Fred and his team of eight designers, draftsmen and detailers, had a prototype running in the lab. It was a unique design in a number of significant ways. All other two-cylinder snowmobile engines had their pistons in-line, like Carl's early Thor outboard engines. Fred's design resulted in a 90-degree V-2 configuration, which he inverted so that it rested on the two cylinder heads like two widespread legs. As a result, the centerline distances between the cylinders was unusually short, allowing for a short and more rigid crankshaft, reducing the amount of vibration usually felt in two-cylinder engines. Further, the unique design forced any left-over unburned fuel/oil mixture down into the combustion chambers to be exploded by the 40,000 volt *K-Tron* ignition, making the I/V a cleaner burning engine. Furthermore, since Fred designed the engine to be liquid-cooled, it was much quieter than any other snowmobile engine on the market, and made it more readily adaptable as a future outboard motor power source. "Water cooling is the best way to make an engine quiet," Carl said. "The biggest rap on snowmobiles is the unearthly noise their air cooled two-cycle engines make. Snowmobiles are banned in many areas simply because of this horrible racket." And lastly, Fred engineered the I/V to take advantage of recent advances in die-casting complex shapes, by designing a two-piece engine block with identical halves, minimizing the number of joint faces, gaskets, studs, nuts, bolts and washers needed to hold traditional engines together.

Both Carl and Fred were proud of the new I/V engine, and made the rounds of major snowmobile builders with the prototype, offering production in 20-, 30- and 40-horsepower sizes. Unfortunately, the industry was experiencing a severe recession, as a combination of over production and two years of minimal snowfall left producers and dealers with an enormous inventory of left over models. The Kiekhaefers began to worry that their window of opportunity to supply engines for the snowmobile market was about to close for good.

During the early months of 1973, they began discussions with Bombardier Limited of Canada, the worlds largest manufacturer of snowmobiles. The Bombardier *Ski-Doo* line of snowmobiles were the world's most popular and best-selling brand, generating revenues of $150 million in 1972. Laurent Beaudoin, the president of Bombardier Limited in Valcourt, Quebec, was attracted to the I/V engine for his snowmobiles. But he was also interested in expanding into the marine industry, and with Carl's track-record it began to look like a perfect match.

On May 8, 1973, Carl and Fred escorted Beaudoin and a trio of Bombardier executives through the Aeromarine facilities, and presented a proposal to merge the two companies for $3.5 million in bonds, Bombardier stock and cash. The Canadian manufacturer was receptive, and another meeting was set to take place on May 14 in Toronto, the day before Carl and Fred were scheduled to introduce the new I/V engine at the International Snowmobile Trade Show there. Bombardier requested, however, that KAM withdraw from the show to avoid enticing potential competitors to a Bombardier-KAM merger, and join them instead in a press conference to announce that negotiations were in progress to merge the companies.

Between 1970 and the May 1973 meetings with Bombardier, Carl had approached 32 different companies with varying degrees of interest in joining forces with Kiekhaefer Aeromarine Motors. This time, Carl felt it was going all the way. On the eve of the International Snowmobile Trade Show, a most serious group assembled in Carl's small suite in Toronto. Carl, Fred, KAM treasurer and advisor Willis Blank and attorney Wayne Roper represented Kiekhaefer Aeromarine Motors, while Laurent Beaudoin, Charles LeBlanc, Warren Daust and Jean Rivard negotiated for Bombardier. The meeting didn't get started until ten-thirty in the evening, and didn't adjourn until two o'clock in the morning, but when the cigar smoke cleared, an agreement to merge KAM into Bombardier was accomplished. A formal signing was scheduled a few minutes before a joint press announcement later in the morning. Clauses in the agreement were initialed to protect both parties who were eager to avoid a repeat of the Bolens type of breach issue. The final language of the agreement stated that a closing of the merger would take place on July 1, 1973.

At the press conference, Laurent Beaudoin announced that an agreement in principle to acquire the assets of Kiekhaefer Aeromarine Motors had been signed, and that both Carl and Fred Kiekhaefer would be retained in "top management positions," and that "We are looking to them and their research and development team to help us reinforce our position in the engine field."[15] Beaudoin joked about how long Fred Kiekhaefer has been associated with two-cycle engines. "There are those who say he was not born, he was *die cast!*" The following morning, headlines in *The Toronto Star* proclaimed that "Bombardier agrees to buy U.S. company."[16] But the same day, when another newspaper announced the agreement, they also disclosed the fragile condition of the snowmobile market, and of the potential vulnerability of Bombardier. "The Snowmobile industry is a troubled one, with more than 340,000 unsold machines of all makes in dealer and distributor networks."[17] The article also ominously stated that Bombardier's profit had fallen from $11.6 million the year before, to less than $400,000 for the 1972-73 fiscal year. It was a dangerous time for investment in the fragile snowmobile industry, but the Kiekhaefers at least had the satisfaction of concluding an agreement with the largest player in the field.

As the closing date of July 1, 1973 approached, Carl began to worry that Bombardier was having difficulties with financing. A number of minor obstacles, including formal searches of Carl's over 200 patents were mentioned as reasons to delay the closing. Another problem arose when Bombardier's wholly-owned subsidiary, Bombardier-Rotax in Gunskirchen, Austria, one of the world's largest two-cycle engine manufacturing plants, was suddenly able to offer the home office a new liquid-cooled engine design and reduced cost per engine. And finally, Bombardier was unable to secure favorable financing with European banks because of unstable international exchange rates. On July 7, 1973, Bombardier and KAM agreed on the language of a joint press release which said in part, "the timing for merger of the two companies did not appear to be feasible at the present time ... negotiations have been suspended. ..."[18]

Carl, though, was certain that some hidden reason was behind the collapsed merger, and he wrote to Laurent Beaudoin to try and salvage an agreement. "... We can understand the attitude of a captive engine plant who perhaps has been shocked into a little greater activity. Some time when we meet privately, perhaps you will tell me the real reason for the turn-down. Naturally we were quite disappointed."[19] Ever paranoid, Carl even imagined the possibility that his old associates at Mercury had spoiled the deal by communicating with Bombardier.

> "As you may surmise, we are not quite sure as to the real reasons since conditions hadn't changed from the beginning of our negotiations. Of course, if you have been influenced by my former associates in any way, we should like to know about it since we all know that employees of your organization have been approached in an aim to disparage us, and to discourage you.
>
> "To say we are disappointed is the understatement of the day."[20]

Secretly, Carl and Bombardier had agreed to introduce a new design of a marine stern drive in 1974. Part of the reason that Bombardier was having difficulty in acquiring financing for the proposed merger was because over $9 million was required, nearly $6 million of which was for Carl's secret stern drive which was still on the drawing boards. "We thought you and we were buying time in starting a new marine venture," Carl said to Beaudoin. "We thought we were saving you money in this new venture as compared to having to start absolutely from scratch. Do you really believe that you could make up this time and money already invested by starting from scratch today? How could you possibly match it?" Beaudoin replied by asking Carl, "The stern drive that you have shown to us, still in the drawing stage, will it be competitive in price and performance with what is available on the market presently? Will we be able to compete successfully with other stern drive manufacturers such as Volvo, Mercury, OMC? It was pointed out to me, with good reasons, that the financial requirements to merge with your company was involving a huge sum of money, some $6,000,000 to market a successful stern drive. Frankly, we are afraid ... we might find out that we do not have the right product in the stern drive business. ... Please understand that this is being said, as I stated before, very frankly, and being aware of your marvelous track record in the marine field and of the great potential of your son Fred, as a man and as a technician."[21]

Months later, as Carl unsuccessfully attempted to generate merger talks with others, he wrote a sharply worded letter to Beaudoin, explaining how Kiekhaefer Aeromarine Motors had gotten a reputation of being "damaged goods," hampering negotiations with others.

> "Your rejection of our engine, of course, and the resultant worldwide publicity has cause us to be considered as tarnished merchandise, thus hindering our negotiations with others. In addition, it appears that someone in your organization may have leaked information to interested prospects that our engine costs were too high. Subsequent to your decision, we resumed negotiations with one company ... and the 'high cost' issue was raised several times and eventually soured those negotiations also.
>
> "We furnished personnel to work with your engineering group who, incidentally, loved our little engine -- and admitted it was at least three years ahead of anything else on the market. I guess it was this atmosphere of combined confidence and cooperation that left me shocked after receiving your July 30 call."[22]

The double failures of the FMC-Bolens and Bombardier ventures were the death knell for Carl's dreams of building the best two-cycle engines for snowmobiles. The Bolens cancellation caused him to miss an entire year of marketing opportunities, while the collapse of merger talks with Bombardier in combination with the depressed snowmobile market, spelled the end of his hopes for Fred's remarkable little engine. Even as of today's writing, no other engine has been designed or manufactured that has so successfully conquered the special problems and requirements for snowmobile applications. Bad timing and bad luck haunted Carl's snowmobile engine efforts almost from the start, and the failed merger with Bombardier had been the last, great hope for sustained production. The engine, along with Carl's dreams of building upwards of a hundred thousand engines a year, were permanently shelved.

Trouble, just like success, seemed to arrive in oversize portions for Carl Kiekhaefer. His crippling experience with the snowmobile industry would have been sufficient to demoralize even the most optimistic entrepreneurs. Unfortunately, Carl had to endure even more disappointment and failure during the same depressing period.

During the celebrated Experimental Aircraft Association (EAA) annual fly-in and air show at Oshkosh in August 1971, home-built aircraft designer James R. Bede heard glowing reports about the new snowmobile engines being built by Carl Kiekhaefer. Carl's marketing director, Don Thompson, invited Bede to tour KAM facilities and inspect the prototype engines. Bede was at once struck with the solid engineering of Carl's designs, and announced to Thompson that he would like to use the engines for his new *BD-5* airplane kit. The beautiful little airplane was reminiscent of a miniature, one-place fighter aircraft, mostly aluminum with a single, pusher-style engine and propeller in the rear. The response to Bede's new design was very positive. In fact, by the time Bede toured Carl's facilities, he had accepted deposits from nearly 1,000 buyers, eager to begin building the over 200 mile per hour dream plane in their garages.

Within a week of his tour, Bede informed Thompson that he was considering an order of over 6,000 engines, with deliveries to commence in January 1972, coinciding with his own projected manufacture of ready-to-build *BD-5* airframe kits. Even before Bede received an evaluation engine, he was preparing news releases about using Kiekhaefer engines in the aircraft, and had prepared a logo for the side of the aircraft which proclaimed "Powered by Kiekhaefer Aeromarine."[23] Arrangements were even made to have a completed *BD-5* on static display at the Aeromarine Sno-Bol during race activities the following February. By November 1971, Bede reported that the Kiekhaefer evaluation engine "... is running beautifully in the aircraft with no heating problem, muffler problem, or other difficulties."[24] When the order finally arrived in December 1972, it was for only 1,150 engines, scattered throughout three different models, and Bede wanted the first deliveries in six weeks. With aircraft kit orders then topping 2,000, Bede was jubilant with the success of his marketing efforts for the *BD-5*. He told Don Thompson that his new airplane could end up to be the number one selling aircraft in the world.

> "I truly believe that a new industry is being developed with a personal or leisure time aircraft. It could not only be a major sport but a transportation vehicle as well.
>
> "Right now we are the only one in this business and in our first nine months of sales, before we have even delivered one aircraft, we have sold more units than Piper did in their last twelve months. They are second largest in the world. In no time we will have sold more units than Cessna, the leader. But we have just scratched the surface. ...
>
> "If the production can fulfill the needs, you will become the leading manufacturer in this new exciting market."[25]

When Carl had to inform Bede that KAM needed more time to deliver the specially modified engines, Bede increased his order to 5,000 engines to be delivered over a five month period.[26] Trouble began when Carl received a deposit for only $30,000, which would have covered only a fraction of the first month's order. He began to suspect Bede's ability to deliver all the kits he had sold, and at the optimistic delivery schedules he was quoting customers. Within two weeks, Carl was worried enough to return Bede's deposit, telling him, "Perhaps, when the situation is clarified, we will be able to reopen negotiations with you."[27]

What Carl really wanted was a letter of credit that would guarantee payment by Bede of the entire run of 5,000 engines, fearing that Bede's precarious situation could collapse on all involved. In a transcribed telephone conversation with an officer of Bede's bank, Carl made his reservations perfectly clear.

> "And the man is a wild man. He's a real way-out -- he ought to be pitching snake oil at a county fair rather than selling this sort of thing, and we're getting very concerned because it's got our good name in it, and he's maligning a product, and if we don't deliver, he'll malign the product and our reputation. ...

"And he's giving -- what I don't like is he's giving customers a lot of misinformation and he's getting twenty times as much for an engine after we sell it to him ... and this is just not right and he is defrauding the public -- that's what it amounts to in my estimation right now. That's just plain fraud."[28]

Later the same day, Jim Bede called Carl to try and get his order back on track. Carl was absolutely stubborn, refusing to consider even a $1 million letter of credit. Since Bede had already committed to delivery of Aeromarine engines to his customers, the aircraft designer compared his predicament to how Bolens had canceled their contract with Carl, leaving him in a most delicate situation.

"So, what we're saying is that we think it would be better if you got your engines elsewhere," Carl informed Bede.

"Well, this is going to put us in a heck of a bind ...," Bede replied, "when we have to come along and say, 'Gentlemen, after one year of negotiating with Mr. Kiekhaefer and so on and so forth, he is unable to build the engine for us and deliver the engine for us,' I think this is, of course, going to hurt our reputation. ... Again, I can come up with a good million and a quarter line of credit if that's what you would like. ..."

"Well, at this stage of the game we'd feel a lot better if you just didn't give us an order," Carl concluded later in the conversation. "We'd like to return this and forget it. ... we just feel that we're going to get caught in something that is going to be a real consumerism item. All we have to do is kill one person and our name is going to be dragged down with yours."

"Well, I don't intend on having my name dragged down any place ...," Bede retorted. "Now what kind of fool could I possibly be to be paying salaries to 42 people only to go ahead and have the whole thing flop? Baloney! I've got some of the damn top flight test people in there now that are truly absolute experts on this thing, and I can't say that no one will ever have an accident in this airplane just the same as people can get hurt on roller skates, but I can tell you one thing, it'll be a helluva lot better than anybody else's!"

"Sure, sure," Kiekhaefer responded eventually to Bede's arguments. "But, Jim, as it is, I'd rather you get your engines somewhere else."

"Boy!" Bede finally exclaimed. "Here you go and get apprehensive about your name and what might happen in the future and so on, because -- Sir, you have a helluva good reputation and no one can ever take that away from you. I'm still young and I'm just starting to make mine and I'm 'gonna watch my name about five times as much as I'm watching anybody else's name, and it's not gonna go bad, it's gonna be exactly the other way around."

"I just can't see this," Carl said stubbornly. "I'm sorry, but I'd rather not sell engines for this application."[29]

Carl solicited the help of his old friend Dr. Alexander M. Lippisch, the noted aircraft designer credited with the German Heinkel bomber and the concept of the high-performance delta wing. Dr. Lippisch also designed the world's first successful rocket-propelled airplane, and created the Messerschmitt ME 163, the first combat rocket propelled aircraft in 1944. In his report back to Kiekhaefer, Dr. Lippisch said in part, "... I must say that this project is not very sound and I would not engage in any definite agreement until you have seen the aircraft being flown and tested by a trustworthy pilot. The discussions with Mr. Bede and his assistant showed me that their philosophy of design is more in the direction of optical impression than engineering effectiveness."[30]

Carl finally told Bede in June 1972, that he refused to power the *BD-5* for various reasons. "For example," Carl wrote, "the engine has apparently now gone through three different positions: right side up, laying down on the side with downdraft carburetion, laying down on the side with updraft carburetion, and now with a complete upside down configuration and we wonder what next. ... Frankly, we feel that the aircraft is being designed by sales rather than engineering people. ..."[31] Bede, frustrated in his efforts to negotiate with Carl, wrote back, saying, "I am well aware of the problems you are having with your company, aside from our aircraft engine application. ... I must say, however, that it is an exceedingly difficult thing to be a customer of yours."[32]

Bede's dream of supplying more *BD-5*'s than Piper or Cessna could deliver conventional aircraft would slowly grind to a halt. In 1976, *Air Progress* editor Keith Connes commented on the ambitious entrepreneur's plans. "If there are any of you out there wondering who the devil is this Jim Bede, let me say that he is regarded by some as an engineering genius, by others as a snake oil huckster, and may very well be, in truth, an engineering-genius-snake-oil-huckster. To put it another way, Bede has designed some unique aircraft that promise to do an awful lot for relatively little money, when, as, and if they are delivered."[33]

Bede continued to have trouble securing the proper engines for his proposed kits, starting with Polaris snowmobile engines, Carl's Aeromarine engines, Hirth of Germany, Xenoah of Japan, and ultimately turbocharged Honda motorcycle engines. According to *Popular Mechanics*, of the 3,000 kits that were sold, 50 or 60 were completed by home builders, and of those an estimated 30 were flying by the end of 1989. The National Transportation Safety Board, according to the same source, revealed "25 accidents for BD-5's in total, nine of which were fatal."[34] As of December 1989, Bede was still at it, this time with a supersonic home-built jet aircraft, the *BD-10J* that you can assemble in your garage for a mere $160,000 -- less engine and instrumentation. Perhaps most revealing about Bede's ambitious plans for the ill-fated *BD-5*, were among the first words in his 1971 sales kit, "For years now we have been dreaming. ..."[35]

In the space of a year, Carl had lost all of the large orders that would have put his new engines into production. Together, the Bolens, Bede Aircraft and Bombardier orders would have totaled over 60,000 engines, justifying the great investments Carl had made in property, equipment and staff. Combined, the cancellation of these orders had placed Carl in the miserable position of having lost the largest orders in both the snowmobile and recreational aircraft markets in one bleak and disparaging twelve month period.

The lack of substantial, volume orders for any of Kiekhaefer Aeromarine Motors products led to industry rumors of imminent financial collapse for Carl's new company. Nothing could upset Carl quicker than to intercept rumors of this sort, but even as early as November 1974, word had begun to spread. "There is a 'rumor' circulating in the marine trade that you are troubled by deep financial problems," one of Carl's consultants on the west coast wrote. "Sometimes your best friends won't tell you what is being said exactly, so I can't say I know anymore than that."[36]

Among Carl's greatest challenges in keeping the doors of the plant open was collection of debt. Whether individuals or giant organizations, the problem of getting the money to the bank was often exasperating. "The problem is on the part of some customers who take advantage of their political situation and their banking organization," Carl complained. "No tickee, no washee," he announced as the central idea of his new financial strategy.[37]

Carl certainly had sufficient reserve funds to continue losing money for a dozen years or more, but more and more he was forced to tighten his corporate belt. Following the failure of the Bombardier merger, Carl's controller, LeRoy E. Kirchstein, wrote a memo entitled, "Urgent Cash Requirements," in which he chastised Carl's continued unprofitability by saying, "We have come to that point where a decision has to be made regarding our cash position. ... If there is a negative response ... [to our need for cash] I suggest that we make some arrangements to discontinue operations."[38] By the end of the month, as he did many months, Carl had to inject the company with more capital to meet payrolls and satisfy payroll taxes. The following summer, Carl sent another memo to Kirchstein, threatening to close up shop.

> "... The important items that must be taken care of as priority one are:
>
> American Express
> Airline travel cards
> Fuel credit cards
> Skyport [aircraft operations]
>
> "Without this, we are shutting down the business for good because management needs this to travel. I realize that you are getting accounts payable calls but I think you can understand my situation -- that we cannot afford to shut off all travel and communications.
>
> "We must make an earnest reduction in our payroll. I have never seen the yard so full of automobiles since the day we were in production on snowmobile engines. Hiring has been done loosely, haphazardly and superfluously."[39]

Undaunted, Kirchstein fired a reply back to the boss the same day.

> "I am totally in agreement with your need to travel. ... As it is now, we are having problems meeting the payroll. ...

"If cash is not brought into the company within the very immediate future, you will not have much choice in shutting the plant down. If we are unable to pay our COD's (as our credit is now shot) we will not be able to continue to process billings to enable us to at least meet payroll If you can not meet payroll, I am sure that the employees will not work.

"If we do not have work in the house, we should not continue to have the employees on the premises; but, at the present time, I have no control over this matter.

"I will be happy to follow your instructions to the tee but I must have instructions as to how I am going to pay items you specify, accounts payable and payroll without cash."[40]

Carl was greatly depressed by the steady losses he was experiencing in his new business. Nothing seemed to be as easy as it was the first time around with Mercury. To a friend he admitted the pain of being chief executive of a losing operation.

"It is somewhat embarrassing to one who has operated in the black for thirty out of thirty-one years to find one's plans not materializing as rapidly as they had in the past. The net result is that I spent roughly $8 million in the acquisition of the present Aeromarine plant, its land, its machinery, its mistakes and also its cost of development, prototyping, testing and proving of designs, patent searches and expenditures all in preparedness for the marine marketplace efforts ahead."[41]

Carl did what he always did when things were going wrong. He invented something new. One of the most important enticements that Carl had offered Bombardier during their failed merger talks was a new generation of stern drive. Though the design, credited in the patent applications to one of his engineers, Larry Lohse, was then only on paper, Carl completed working prototypes of the *K-Drive* within 90 days of those collapsed negotiations. In February 1974, Carl startled the marine industry by unveiling this very innovative stern drive at the Miami International Boat Show. When Carl stepped up to the microphone at the Miami Convention Center, the marine press and boatbuilder audience was filled with expectation, knowing that if Carl had gathered them together, there must be a pretty good reason.

"Much water has gone under the bridge since I was active in the marine industry -- much more than we can tell about here. But I remain firm in the belief that the most important thing in any new endeavor is the product itself. Without the product, all else is meaningless.

"This product has been designed to live up to a track record that I am sure you are all familiar with and of which we are very proud. There were reasons for the four-year delay, of course. First, we had to wait out a non-compete clause and for the want of anything else to do, we dabbled in the snowmobile field. Unfortunately, that industry ran into difficulties."[42]

The new design created great excitement. It was 30 percent smaller and had 30 percent fewer parts than stern drives by *MerCruiser*, OMC or Volvo. But even though it was smaller, it could handle nearly twice the horsepower of comparable models, up to 600-horsepower for offshore racing applications. He circumvented existing patents by designing a unique trunnion tilt-up mechanism, much how a cannon is tilted up and down on its carriage. To further avoid patent infringement, the bottom half of the lower-unit of Carl's design pivoted back and forth, rather than the whole outdrive unit. It was truly an engineering marvel, and even the *New York Times* gushed with platitudes about Carl, "... the outstanding marine mechanical engineer of his time ..." and his welcome new invention, "... an outdrive unit (propeller and steering in one package on the transom) that is likely to make all others obsolete."

"If anyone but Kiekhaefer made these claims, they could be dismissed. But the 67-year-old industrialist from Fond du Lac, Wis. has a track record so outstanding in boating that he demands attention and credibility. ...

"There is nothing this one-time Wisconsin farmboy would rather do than fuss around in his laboratory, letting his remarkable mind roam over the solution of mechanical problems. That is the life he would prefer to the end of his days. ...

"The marine engine world is waiting for the next move of the master, Kiekhaefer, and so are the consumers who stand to benefit the most."[43]

Carl delighted in watching wave after wave of demonstration drivers take the new *K-Drive* through its paces, dashing across Biscayne Bay aboard a 24 foot runabout. Everyone agreed: Carl had done it again. The only problem was, the drive wasn't available. Though Carl speculated that at least one version of the *K-Drive* would be produced during 1975, it was contingent on a number of factors. Most importantly, Carl would have to locate and persuade a large enough company to undertake the tooling, dies and assembly operations along with a sizable marketing investment. Estimates for properly getting the new design to market ranged from $10 million to $80 million, depending on how many models might be introduced initially. Once again, Carl was in the position of having a novel and formidable prototype, but no buyers.

Many organizations showed interest in the new drive, particularly those who had substantial investments in heavy manufacturing, like John Deere and Company or Chrysler, and were attracted to the relative high volumes of marine industry production. Certainly, the new prototype was most impressive, and Carl's forecast of capturing a third of the industry's sales was alluring, but Carl knew it was his company's "hidden assets" that would have to make the final sale.

> "The value of good will and reputation is always difficult to assess -- especially when two billion dollars worth of product, one quarter billion dollars worth of paid advertising and thousands of pounds of 'free ink' have projected the name, Kiekhaefer, before the public for 35 years."[44]

After a string of disappointing presentations, Carl sidelined the *K-Drive*, to concentrate on other KAM opportunities. Fred Kiekhaefer is convinced that the whole idea of the new stern drive was developed only to worry and irritate Brunswick. "It was all smoke and 'hocus-pocus', and a strategy to keep Mercury nervous about what he was doing. He liked his revenge from the way they treated him, and I think he deserved to get every ounce of it."[45] In a fit of despair, Carl blamed U.S. policy makers and obtrusive government intervention for his woes. He even suggested that Fred might be better off working for the Government than for private enterprise.

> "All industry and producers of product are vulnerable from the standpoint of taxes and welfare costs, and will be the target of all sorts of police and punitive action to the point where there soon will be no more free enterprise, except in the form of very large conglomerates and multinational companies. Since 20% of our income is used to support some sort of government policeman, and with the increase in police action due to the environmental and welfare programs, it may be that soon 30% of our income will be used to support some surveillance or

regulation activity on the part of the government. My advice to my own son is not to try to follow in his father's footsteps, but to get on the safe side of existence, and nothing is safer than civil service or working for some government agency. We are going to face a problem of survival in the face of the world population growth, and it won't be military -- it will be human survival."[46]

Carl had experienced a long and disillusioning string of failures, from snowmobile engines, aircraft engines, to foiled mergers and a shelved stern drive. Not since he was fired from Evinrude Motors in 1927 had Carl experienced personal failure to such a harsh degree. And yet, even as he calculated the millions of parts he would need to mass assemble the products he would never manufacture, Carl was carefully building individual offshore racing engines one at a time, with a degree of precision and performance unmatched in the history of marine engineering. He still had yet another surprise in store for the marine industry.

CHAPTER THIRTY-SIX

Return To Glory

"They which run in a race run all, but one receiveth the prize. ... So run, that ye may obtain."[1]

1 Corinthians 9:24

Even as Carl was preparing in 1970 for his triumphant debut in snowmobile racing competition, he was carefully planning his return to offshore racing. Impatiently waiting for his non-compete clause to expire with Brunswick, Carl reasoned that the best way to keep his name alive within the marine industry was to sponsor and campaign an offshore powerboat, and compete against his old associates at Mercury head-on in the only area legally left to him: racing. He bought one of the first 36 foot Cigarette hulls from Don Aronow in North Miami, Florida, and installed a pair of 496 cubic inch *MerCruiser* stern drives carefully reinforced and meticulously modified by himself and the KAM engineering staff. Mercury had dominated offshore racing almost from the moment that Carl had introduced the *MerCruiser* stern drive, but Carl was certain that much more could be done with the same basic equipment if preparation, accessories, fine tuning and absolute attention to detail were combined with the right driver. These were the same basic tenets that had won Carl three national stock car championships, and a greater percentage of victories in American Power Boat Association (APBA) competition than all other marine manufacturers combined. "Consistency," Carl was fond of saying, "thou art a jewel."

Selected to pilot the *Aeromarine I* was two time APBA offshore outboard racing champion Dr. Robert Magoon of Miami Beach, Florida. Magoon, a 6-foot 3-inch, trim and athletic eye surgeon, had won the coveted championship in both 1968 and 1970, assisted by throttleman/mechanic Gene Lanham, a City of Miami fireman and ten-year veteran offshore racer. Though Magoon had never competed in an inboard boat, Kiekhaefer recruited him for his superb driving and navigating skills, and knew he would be up to the rigorous challenge of a sustained year-long campaign.

The first offshore test for the *Aeromarine I* and Magoon was the 173-mile Hennessy Key West Race from Key West, Florida around historic Fort Jefferson in the Dry Tortugas and return. Although two other entries, Tommy Sopwith of England and Vincenzo Balestrieri of Italy, were locked in a year-long struggle for the world offshore driving championship, Magoon and Lanham crossed the finish line first after three hours and 17 minutes of hopping and skipping and slamming across the Gulf of Mexico. The November 1970 race was actually the first race of the 1971 offshore season, and it served notice that Carl Kiekhaefer was back on the water, an event that drew almost as much attention as the race itself. "It marks his return to racing," one commentator proclaimed, "and you will be reading a lot about this boat in the 1971 offshore racing season."[2]

In January 1971, Bob Magoon wrote Carl to discuss the racing schedule for the coming year. In his letter, he outlined the strengths and weaknesses of the competition, and then matter of factly told Carl what the competition was already beginning to suspect. "I am not bragging but I think I am the best driver around today. I think I have the most experience, running most of these races three or four times. I am the best navigator in the fleet. I can out-think them and I feel I am physically better."[3] To prove his point, Magoon, assisted by Lanham at the throttles, pushed his Kiekhaefer prepared engines and drives to six first place finishes out of nine during the year, along with a second and third. Carl's *Aeromarine I* was the first boat in APBA racing annals to win six national championship races in a single season, and Magoon became the first driver in history to win the Hennessy "Triple Crown", with victories at all three of the famous Jas. Hennessy & Co. of France cognac distiller's sponsored races in New York, Long Beach and Key West. Kiekhaefer Aeromarine and the team of Magoon and Lanham were declared the 1971 American Power Boat Association U.S. national offshore racing champions. Combined with his three championships in snowmobile racing, Carl had managed a phenomenal *four* national championships in his first year of business.

Even while *Aeromarine I* was racking up victory after victory on the way to the U.S. championship, Carl grumbled about the costs of campaigning, about the distraction to his business, and the politics inherent in ocean racing.

> "Right now, I am obligated and committed to devote all my resources, my time, my energy, and my brains to building Kiekhaefer Aeromarine Motors. Initially, I became involved in ocean racing because it represented a challenge to my technical capabilities and because I believed in its value as a means of testing and improving our products. I subsequently learned to love it as a thrilling and exciting sport -- to the extent, some say, of addiction. I participate to the extent that circumstances (and my conscience) permit and I don't need anyone to spur me on."[4]

Of the costs involved, Carl would comment, "No need to mention that offshore racing is perhaps the most expensive, exceeding the costs by far of an Indianapolis campaign or an unlimited boat campaign."[5] Of the publicity value of racing, Carl was certain. "Outside of the little bit of fun that I can get out of ocean racing, there is another and important reason for participating and that obviously is to keep my name and my company's name before the public, and for this I pay the bill, which amounts to some two hundred grand a year [plus the cost of equipment]. When I don't get value received, I naturally get irritated at those who are hogging the show at my expense."[6] And finally, of the distractions to his business, Carl expressed his feelings quite openly in a memo to his employees. "I know how interesting racing is and how 90% of the people would rather yak about racing than do a lick of a day's work, and while no harm is intended, any conversation [about racing to outsiders] unless handled by [me] is apt to give aid and comfort to the enemy. ... Violation of this in the future will be followed by dismissal. Our investments are too great."[7]

Of the excitement involved in offshore racing, there can be little doubt. For sponsors and spectators, there is probably no more thrilling sight that a fleet of high performance race boats crashing through heavy seas and watching the brave pilots endure the violent pounding of airborne launch and thundering landing, wave after wave. To the drivers and throttlemen aboard the boats, it is a most serious affair, requiring the keenest concentration and effort to keep the boats above the water and on course. *Saga* magazine once described the wild melee of offshore ocean racing.

> "Of all the wild and woolly sports enjoyed today, this has to be the toughest--pushing five tons of super-powered fiber glass hull to speeds approaching 90 miles an hour in rough seas, caroming from crest to crest like a flying fish, leaping 15 feet clear of the surface and soaring 20 yards before crashing down again in bone-jarring explosions of spray, time after time for hours on end. And when you add the unpredictable elements--blinding thunderstorms, rain squalls, treacherous reefs, and dangerous shoals--plus the gut wrenching punishment every driver suffers in the stand-up cockpit of a porpoising boat, you have all the action-packed, thrilling ingredients of deep sea powerboat racing."[8]

Of the abilities of Bob Magoon, Carl would wax euphoric, and in a letter he never mailed to *Yachting* editor Mel Crook, he described the paradox of this delicate and yet rugged individual.

> "I won't belabor you again with his record ... but the man is a terrific, lion-hearted competitor and a real credit to one of the toughest sports in the world. You might imagine a man of his type to be a burly, rough and tumble, hard drinking and hard playing sort of individual. Not so. First of all, he is an eye surgeon who does delicate sutures under a magnifying glass, is an outstanding athlete with tennis and swimming his major sports, but also has time for several small children, a charming wife and mother and pets -- truly a gentleman, a family man, very refined, sensitive and soft-spoken by nature. He seems completely out of cast when he wears the boat. Not only a tremendous driver, but I believe his is probably the world's outstanding navigator because you know how tough it is to read a compass in a boat that is making little Kitty Hawk flights every few seconds."[9]

Throughout Magoon's victorious season, Carl began to manufacture more and more strategic components for the *Aeromarine I*, and to make them available for others. By the time his marine products non-compete clause had expired with Brunswick on August 31, 1971, Carl was manufacturing complete assemblies for fuel injection, water cooled exhaust headers, oil pumps, power steering, after planes (trim tabs for boat pitch and plane stability and control), propellers, optional gear ratios and heavy-duty vertical drive shafts and mounting plates for *MerCruiser* stern drives, heavy-duty transmissions, oil coolers, C.D. "K-Tron" ignitions, among other critical engine and drive accessories. His engines, though starting out as stock *MerCrui-*

ser engines evolved from Chevrolet blocks, were considerably more powerful and dependable than the originals as delivered by Kiekhaefer Mercury, and yet were completely legal by APBA standards. They were *Aeromarine* engines, and other powerboat racers were soon lining up to buy their own.

Among Carl's first clients was Dr. Carlo C. Bonomi, an Italian entrepreneur from Milan who ordered a fully rigged boat from Kiekhaefer Aeromarine in the early spring of 1972. Bonomi, 32, was a principal officer in his family's business, Gruppo Bonomi, one of Italy's largest land developers and builders. Bonomi received a doctorate in economics from the Bocconi University of Milan, and taught for a time at the University of Pavia.

For Bonomi, Carl selected an identical 36-foot Cigarette like the *Aeromarine I*, and equipped it with beefed up *MerCruiser* drives and a pair of new, improved Aeromarine 482 cubic inch engines. Even though Bonomi's boat, christened *Aeromarine IX*, wasn't ready to race until six months of the U.I.M. (Union Internationale Motonautique) international season was over, he managed to catch up with the leader by September, and the plucky European missed capturing the world title by a single race. The power and reliability of Carl's Aeromarine engines had become evident to racing competitors around the world. For his own racing team, Carl outfitted another pair of boats for testing and back-up chores. *Aeromarine III* was a sister ship to the 36 foot Cigarette, *Aeromarine I*, and *Aeromarine V* was a 36-foot Memco-Bertram design.

One of the things that continued to annoy Carl was the fact that Brunswick-owned Kiekhaefer Mercury was still using his name. Occasionally, Carl would pick up the Fond du Lac *Commonwealth Reporter* or the Milwaukee *Sentinel*, and the name Kiekhaefer would leap out at him from the headlines, even though reporters were referring to Kiekhaefer Mercury. As the slights, both imagined and otherwise from his previous company continued to grate on his sensibilities, Carl became increasingly possessive about the use of his name. When Carl obtained permission from the Ford Motor Company in 1940 to use their registered trade name "Mercury", it was with the understanding that as long as the word "Kiekhaefer" preceded and qualified the name, Ford would allow the usage.

In June 1971, Carl noticed several ads prepared by Kiekhaefer Mercury that referred to the company only as Mercury.

Anxious to undertake whatever he thought would needle Brunswick the most, Carl wrote to Henry Ford II to tattle on his former associates, and attempted to denigrate the image of his previous company.

> "We certainly are not proud of having my name used with Mercury any more since the product has greatly deteriorated in quality, but we believe that you might not object if it were called 'Brunswick Mercury'. Since it was decided in the courts that you own the name 'Mercury', I thought this might be of interest to you."[10]

Carl had been so successful, however, in promoting the good name of Mercury throughout the world, that even the Ford Motor Company realized that the name had been established on its own merit. When Mr. Ford's counsel replied, he had to wonder why in the world Carl would want his name removed from such a universally praised product.

> "As matters now stand, I doubt that we could successfully challenge the use of 'Mercury' alone when applied to outboards because of the public acceptance it has achieved. ... As a purely personal matter, however, I might be concerned about your reasons for disinterest: I bought by first Mercury outboard twenty-five years ago, have had no other makes in the interim and my ninth is now on order to be delivered next week.
>
> "I should be reluctant to see the use of the name 'Kiekhaefer' diminish in an industry in which it has played so substantial a role. ..."[11]

Brunswick had been taking note of Carl's bluster over the use of his name, and on November 23, 1971 Brooks Abernathy announced a new name, which is the current usage. "Effective immediately the name of this Division is changed to Mercury Marine Division of Brunswick Corporation. ... Out of respect for Mr. Kiekhaefer, there will be no company initiated fanfare over this change."[12] Since the 1972 models of outboards, stern drives, inboards and snowmobiles were already in production, "Mercury Marine" wouldn't appear until the 1973 model line, breaking a 33-year tradition of having the Kiekhaefer name appear on their products.

The change presented a new problem for Carl, though, as he angrily explained to Brooks Abernathy. "Since your company is no longer listed in the telephone directory under the initial 'K', we are being deluged with calls for 'Kiekhaefer Mercury', averaging at least twenty-five calls per day. This not only places an added burden on our switchboard operators, but delays your calls."[13]

Carl received a shock in early 1972 when placing a routine order for *MerCruiser* 496 cubic inch engine crankshafts from Mercury Marine. Gary Garbrecht, in charge of Mercury high-performance products at the time, informed Carl that he would have to order *the whole engine block* in order to get the crankshaft. Carl was furious to be asked to pay for an expensive block just to receive a critical component, and was certain that Mercury Marine, embarrassed by his recent success, was simply trying to push him out of racing. In a transcribed telephone conversation with Garbrecht, Carl got right to the point. "It's the silliest thing in the world. We can get more cooperation from Outboard Marine than I can from the company that I founded and the company whose neck I saved and kept from going bankrupt. I was officially thanked by the Board of Directors at one time for saving Brunswick and that's a matter of record. ... Are they mad because I quit?"[14] Carl also vented his anger in a letter to the *Cincinnati Enquirer*. "It's an interesting thing about this unusual and unconventional competition we have encountered in the form of my former company, who have done their damndest to prevent publicity and also the actual winning. I guess it was all right for Carl to win when he was part of the company, but now that I am no longer with the company, they expect me to lose."[15] Carl responded by switching his emphasis to a different primary engine block, the Chevrolet 454 which he would enlarge to 468 cubic inches during race modification. Parts for the popular eight-cylinder automobile engine were plentiful, and Carl went through a long and exasperating period of documenting the engine with both the APBA and U.I.M. race sanctioning authorities to make sure the engine was legal for competition.

To introduce a new racing engine is an enormous undertaking. Carl described how complicated it can be, and at the same time, assumed right from the start that he would lose money for his efforts.

> "A manufacturer, as ourselves, must do some soul-searching when he realizes how small a part the engine really comprises, after he has complied with the bore, the stroke, the

> displacement ... just to qualify for the rule[s]. The real work and cost begins with the marinizing, or marine dress, of the power plant containing some 500-odd additional pieces, which have to be designed, proved, raw materials purchased, machining fixtures made, machine tools provided, the labor force to manufacture, a service parts stock provided, parts lists, price lists and sales literature provided, backed up by suitable marketing personnel, promotion and advertising planned and executed, and the whole procedure has to be cost calculated to see how much money is lost on the project."[16]

The first appearance of Carl's new engines was on August 31, 1972 at offshore powerboat speed trials sponsored in part by *Powerboat* magazine on the Pacific Ocean off of Marina Del Rey, California. Dr. Robert Magoon flew the *Aeromarine III* across the ocean swells and through the timing traps at 82.499 miles per hour to set a new world's record in his class. Unfortunately, a controversial speed run of a completely different kind of boat in England, across glassy smooth lake water, served to rob Kiekhaefer and Magoon of the official record. It was a wonderful introduction of Carl's new Aeromarine engines, however, and by the time Magoon captured first place in the 165-mile Catalina Challenge Trophy Race the following day, Carl had received orders for fifty engines at nearly $12,000 apiece.

Unfortunately, the politics of offshore racing soon soured Carl's outlook, as one of the more vociferous race promoters, Capt. Sherman F. "Red" Crise of Florida, ruled his engines illegal. Crise, who seemed to enjoy the controversies that erupted before, during and following the races he promoted, was accused by Carl of interpreting the rules in his own peculiar fashion. Tom Pratt once described Crise for the *Miami Pictorial*.

> "The man held the microphone as though it were something he was being forced to eat. He was a big and ruddy man with a battered sea captain's hat pulled hard down on his salt and rust red hair. The man opened his mouth and his voice filled Freeport Harbor.
>
> "'Damn it, I wrote the rules and you're going to follow them!' the man roared. ... 'Fish or cut bait. ... I don't care if a driver sinks his boat, or blows it out of the water. He's going to wear a tie when he shows up after the race.'"[17]

Carl was incensed that anyone would think he would cheat to win. "Not only just in power boat racing and outboard racing," Carl fumed, "but in stock car racing and sports car racing, never once as a contestant have we been disqualified for illegality and I am hardly tempted to start now. We get our kicks out of winning with legal, rather than illegal equipment since anybody could win by cheating."[18] To Bill Wishnick, chairman of the Offshore Racing Commission of the APBA and a fellow competitor, Carl wrote, "Why should the eligibility of engines be subject to the whims of promoters whose principal investment is in pencils, paper, printing and postage?"[19]

By the end of the 1972 racing season, Magoon and *Aeromarine III* had competed in eleven races, taking three first places, three second places and two third places to capture the U.S. offshore championship for a second straight year. Though Carl was elated, he was preoccupied with the many problems of his money-losing business. He was struggling with contradictory feelings, basking in the pride of engineering accomplishment and languishing in the disappointment of another unprofitable year.

> "If I appeared to be lost, with a vacant look on my face, it is because I am concerned about intimate details of the birth of a new engine, very much like an anxious father, perhaps, who has a hostile doctor and midwife, if you know what I mean. One is torn between opposite emotions -- whether to have an abortion or to keep the product in longer gestation."[20]

Almost from the moment Carl began offshore racing again, he threatened to get out. In fact, between November 1970 and November 1972, Carl had written to others to say he was going to quit no less than twenty-seven times. Like the boy who cried "Wolf!", though, after the first few outbursts nobody really paid any attention. He certainly had sufficient cause, for not only was he spending nearly $500,000 a year to campaign on the wide-flung circuit, but he was also losing money on the racing engines and accessories he sold to others. Perhaps most troubling of all, he was enjoying it less and less as the political struggles with Mercury Marine and race promoter Red Crise rubbed salt into his widening financial wounds. "There is little fun in racing," Carl admitted sadly in late 1972. "It's a deadly serious business to the point where we never even stay for the victory dinner, we don't attend cocktail parties, and no one brings his wife or girl friend because when we go racing, it's all business."[21]

When Mercury Marine assisted in a challenge of legality over the racing of Aeromarine's new engines in Europe, Carl discovered that not only were his engines perfectly legal, but that on a number of technical points, most of Mercury Marine's *MerCruiser* engines could be declared illegal. He was tempted to blow the whistle to the international governing authority in Geneva, but he admitted in a letter to Dr. Bonomi what a trauma it would be to the entire sport.

> "We have enough information to blow and destroy the legality of almost every MerCruiser engine running in the U.I.M. circuit. If we use this detailed information to its fullest extent, we will be parties of the greatest dissension movement that has ever occurred involving many innocent people, honorable officials and raise confusion and havoc in two national authorities. It seems we have no alternative except to say nothing, bow out quietly, discontinue in a sport that we have loved for thirty years, refuse to build and repair existing ... engines and return all the orders for the 1973 engines. We would do this in a minute if we knew that it would stop the self-appointed rabble-rouser [promoter "Red" Crise] and his disparaging tactics. ...
>
> "We have a feeling that the drivers want our equipment, but I must be given some immunity, respect, cooperation and encouragement for making [the] heavy investments required."[22]

Of all his doubts about continuing to race, the harsh treatment he felt he was receiving at the hands of Mercury Marine angered him the most. To Bill Wishnick, Carl even admitted he was ashamed that his son, Fred, would see his great father in such dire straits.

> "You have no idea of the nastiness of the situation where Mercury is refusing to sell us stern drive parts. Nevertheless, they send their customers to us to buy gears, drive lines, propellers, transmissions, etc. without which their equipment would not be running today. ...
>
> "I don't think I'm going fishing or playing in the sun, but I am sure as hell that I'm going to build something else! I cannot admit failure to my young son, who is just out of college with his Master's Degree, and who is yet to see his first example of 'failure'.
>
> "If I didn't commit murder, perhaps I have committed embezzlement or a fraud of some kind with my old company,

for which you cannot, of course, be blamed or even bothered. It's my problem, but I'm giving you all this to show you why I must quit the offshore racing activity."[23]

At the last race of the year, the first race for the 1973 season, Carl surprised Bob Magoon and the entire racing community by placing a giant red ribbon on the front of *Aeromarine III* before the start of the Hennessy Key West race on November 11, 1972. Carl had finally stopped bluffing and decided to quit sponsoring his own race entries, to concentrate on building high-performance engines and accessories for others. Apart from the frustrating political battle scars of offshore racing, it had always been his philosophy not to compete with his customers. As a parting tribute to the driver who had worked so hard to win two national championships for Kiekhaefer Aeromarine, Carl made a gift of the $65,000 *Aeromarine III* to Dr. Robert Magoon. As if to prove his gratitude, Magoon went out the next day and won the grueling race, narrowly defeating Tom Gentry of Honolulu, another true racing sportsman, who miscalculated the home stretch while in the lead, and ground to an expensive halt on a shallow reef.

True to his word, Carl indeed concentrated on refining and building among the best powerboat racing engines ever produced. In the U.I.M. international circuit during the 1973 season, Kiekhaefer Aeromarine engines started in 35 races, capturing 29 firsts and 4 seconds. Aeromarine engines, gaining a reputation as the engines of choice for those serious about winning championships, were selected by Carlo Bonomi, Vincenzo Balestrieri, Giorgio Mondadori, Sandy Satullo (owner of Copper Kettle Restaurants) and Colonel Ronald Hoare throughout the season. During the race at Palma de Mallorca, Spain, Bonomi set a new world speed mark of 83.2 miles per hour in his 36-foot Cigarette, *Dry Martini*, which put Carl's engines in the *Guinness Book of World Records* for 1973. A month later, Balestrieri increased the record to 87.8 miles per hour at the Pescara (Italy) - Markarska (Yugoslavia) race in the 36-foot Cigarette, *Black Tornado III*. Even more satisfying to Carl was the fact that Mercury Marine's *MerCruiser* engines failed to win a single race in Europe during the season. Controversy, though, once again soured Carl's victories.

Two-time U.I.M. world champion Vincenzo Balestrieri of Rome (1968, 1970), was campaigning Carl's 600-horsepower Aeromarine engines aboard *Black Tornado III* on the international circuit, and was frustratingly close to an unprecedented third world crown. The owner of three racing boats, the diminutive five-foot, four-inch Balestrieri ordered ten Aeromarine engines from Carl during the summer of 1972, two for each boat plus spares.[24] During the 1973 season the Italian ran the engines in 13 races, and only once did he experience a single engine failure which cost him a race. He had driven to victory at such wide-spread racing venues as Buenos Aires, Argentina and in Yugoslavia, but victory was snatched from him almost everywhere else due to problems unrelated to the engines. He was leading in the Miami-Nassau race when he withdrew due to illness. He missed a buoy at the Punta del Este, Uruguay race. In another Yugoslavian race a stern drive pump failed. In Naples, Italy, the power steering broke. Even with all of these seemingly unrelated problems, Balestrieri ended up missing third place in the international standings by a single point, and missed second place by only five points.

Balestrieri, however, blamed Carl and his engines for his relatively poor showing in an article written for *TUTTO MOTORI*, a leading Italian racing publication. To make matters worse, Balestrieri switched to and paid compliments to *MerCruiser* engines. Further, Aeromarine was still owed some $22,000 by the Italian racing team, and it looked as if it wouldn't be paid. Carl exploded with anger, feeling he had bent over backward to help Balestrieri in his ill-fated campaign for the world title. When Balestrieri threatened to sue Carl because of his lost opportunity to win an unprecedented third world championship, Carl promptly filed two separate actions in U.S. Federal Court, suing the Italian for $1 million dollars in revenues lost due to the damaging *TUTTO MOTORI* article, and an additional $5 million in punitive damages, claiming Balestrieri's actions were "maliciously motivated in order to damage Kiekhaefer's business reputation."[25]

Racing engines seldom come with any kind of warranty, because of the extreme conditions of use. It appeared that Balestrieri was simply blowing off steam in the ill-fated magazine article after a disappointing season, when he triggered the violent eruption in Carl Kiekhaefer. Carl characterized the growing feud to his attorney, Wayne Roper. "It is unfortunate that Balestrieri, in a fit of rage, fired his mechanics and went to the press with a completely distorted and erroneous story on the integrity of our products and our corporation, at the same time extolling the virtues of our competition."[26]

Less than two months after Carl filed suit, Balestrieri called him to attempt a reconciliation. Carl demanded that Balestrieri write another article for *TUTTO MOTORI*, setting the facts straight as Carl saw them, and retracting his earlier statements.

"You tell me a retraction," Balestrieri said to Carl in a transcribed telephone conversation. "I like to say it's not retraction because I like to be your friend. If I can't say to everybody that we are friends again ... I don't make retraction."

"In fact, you might tell them that *MerCruiser* didn't win a single race in Europe this year," Carl later added. "I'll tell you, this thing has got me so disgusted that I don't even know if I want to build engines next year."[27]

The matter was eventually settled amicably, but it was a deeply disturbing episode for Carl. Even though the success record of his Aeromarine engines provided overwhelming evidence of their quality, doubts were raised throughout the racing community because of Balestrieri's strong comments. Carl explained to Balestrieri how the matter made him feel.

> "Well, I'll tell you. I lost so much money on this racing that I don't know whether I would want to build any race engines this year. I got a few of them left but I have no appetite to race any more. It is so controversial; there are so many angles and I don't like that. I'm getting too old to take this kind of a beating. I've been a gentleman all my life and I'm not used to this kind of attrition, this kind of anger and all this sort of thing. I just want nothing more to do with racing."[28]

Despite the unfortunate controversy during the 1973 season, Carl managed to acquire the two best reasons to keep his engines racing: championships in both the U.S. and international offshore circuits. Dr. Robert Magoon, driving *Aeromarine III* with Kiekhaefer engines won an unprecedented fifth national title, winning five APBA offshore races, and placing for points in five others. Dr. Carlo Bonomi, also campaigning with Aeromarine engines, passed Balestrieri and everyone else on the circuit to win the coveted Sam Griffith Trophy, symbolic of the U.I.M. world offshore drivers championship, for 1973. Bonomi had almost won the previous, 1972 championship as well, settling for second place, after starting six months late in the season. Carl's engines finished a resounding 1-2-3 for the 1973 international circuit, with Bonomi, Don Shead and Giorgio Mondadori all racing with Aeromarine power. The story behind Bonomi's success provides a revealing look into the fragile relationship between Carl and his son, Fred, and to the extraordinary lengths that Carl would go to protect a friend.

When Fred graduated from The University of Wisconsin and arrived from Madison in the early summer of 1972, he shared an apartment with Richie Powers, a friend of many years who had worked, like Fred, as a test driver at the Kiekhaefer Mercury Siesta Key proving grounds years ago. Richie had always been interested in boating and mechanics, and from the day he started working with the company, he viewed Carl Kiekhaefer as a living legend in the history of marine engineering.

> "I just always remember the magnitude of Carl Kiekhaefer, when he would come to the base down there. ... 'The old man's coming!' It was like everybody was running in circles - 'The old man's coming! The old man's coming!' The gate opens and you see some hot car come driving in - he'd come driving in a Dodge with a hemi-engine in it. You know, all of us kids were hot rodders and ... here comes the 'old man' in a car that does 110 miles an hour in a quarter of a mile."[29]

Carl was so much bigger than life to the young, impressionable test drivers, that to actually meet and speak with him was a genuine thrill. For Powers, sometime in 1966, it was almost rapture. He had the honor of taking Mrs. Kiekhaefer out "shelling" along the deserted beaches of small islands near Midnight Pass. "She was a riot," Powers explained, "because she'd take us out ... buy lunch and we always had a couple of beers at lunch, which was *absolutely* prohibited. 'Oh, don't worry about it,' she'd say. 'Have a beer.'"[30] After one such outing, Richie was summoned to the Kiekhaefer residence at the base to see Mrs. Kiekhaefer.

> "I walked up and E.C. opened the door, and he had just a pair of shorts on and no shirt or anything. I was impressed because he was a huge man, but he was all muscle. I looked at this guy, and I said, 'Oh my God! It's the 'Old Man!' I was dumbfounded, because he was like a *god* to me.
>
> "And he said, 'Oh, you're the guy that took Mrs. K out on the boat today.' ... And he thanked me and invited me in for a beer. And I thought, 'Geez, this is not the 'Old Man' that everybody is talking about. This guy, you know, he's a nice

guy. ... When I left, it was about the greatest thing that happened to me."[31]

After Powers graduated to co-driving one of the Kiekhaefer Mercury team race boats with senior driver Gene Wagner, he also earned the privilege of preparing team boats at Lake X. One day, Carl was at the lake and saw Powers' Pontiac GTO, considered a pretty hot car at the time, and said, "Who owns the GTO?" When Powers replied, Carl said, "Do you think it's fast?"

"Well, I doubt if it's as fast as yours," Powers answered prudently. Carl took on everybody at Lake X one hot Saturday afternoon, screaming, "Who thinks they got a fast car?" One by one Carl took on the Corvettes, GTO's and anyone else who dared, screeching off from a standing start on the one-mile long concrete runway at Lake X. "He beat everybody," Powers said. "He'd have that cigar in his mouth, and there he'd go. ... He'd get a flagman and everything else. But, I mean, he still had machinery that we had never dreamed of."[32]

As Richie continued racing for the Kiekhaefer Mercury team, he was worried that Carl would overlook him, for he was racing with the well-seasoned, headline names that won so many races for him. "I felt like the lowest, littlest guy on the team," Richie confessed, "because there's these *heroes*, you know: Odel Lewis and Mel Riggs, and Bill Sirois, and all these guys have ridden the big boats and the fast boats, and here I was struggling to get going. But I knew that eventually I was going to get there."

When the informal agreement between OMC and Mercury eliminated the factory racing teams, Carl invited the team drivers to Fond du Lac to train for new jobs as factory representatives or work relating to high-performance. Richie became an engine re-building mechanic, and was eventually responsible for preparing engines that would win a string of national and world records for Kiekhaefer Mercury. He was part of a crew headed for the hotly contested Lake Havasu, Arizona marathon races when they heard that "Kiekhaefer just retired." The crew was dazed. "It came as such a shock. We had no idea what was going on."[33] They didn't even know if they should continue to Arizona. It was like the day when President John F. Kennedy was shot, nobody knew what to do, or what was proper to do.

Richie let Carl know that he would like to work with him at the new company, knowing it meant losing eight years of seniority at Mercury. When Carl finally called, Richie was excited. "I thought I was going to be a throttle man on one of the boats or something trick." Unfortunately, all Carl wanted was a good mechanic to put his new snowmobiles on the dynamometer. "I was so disappointed," Richie remembered. "I still had the feeling that he didn't know who Richie Powers was."[34] But Richie gradually worked his way into the offshore racing side of the business, working on the Aeromarine engines that would propel Dr. Robert Magoon to a string of national championships. Soon, he was helping to configure Carl's first fuel injection systems, boosting Magoon's engines from the stock 475-horsepower *MerCruiser* range, into the winner's circle at over 550-horsepower.

One Christmas Eve, as Richie and Carl worked to boost the power even further, Richie assumed that he could go home to Sarasota if they managed to get 600-horsepower from the tortured engine screaming in the test cell. When the gauges passed 605-horsepower, Richie said, "Merry Christmas Mr. K! Think it'd be okay if I went home?" Carl blew up. "*Goddamit!* Tell Magoon he's not going to race this year, Richie's got to go home for vacation! Shut everything down! Close the plant!" After a while, Carl cooled down, saying, "Ah, hell. Go on home - take a couple of days off." "So I got home," Richie said. "The next day he's calling on the phone, 'Where are you? We got to run this motor.'"[35] So, on Christmas Day, Richie flew back to Fond du Lac, after having spent less than a day at home in Sarasota.

As the engines became more complex, Richie began racing as the third man aboard *Aeromarine I* with Magoon and Gene Lanham. As the racing team discovered problems with various systems, Richie would recommend changes or new products to Carl, often to the dismay of the regular engineering staff. When he wasn't off to the races, Richie was working long hours at Aeromarine. After five months without a break, Richie was becoming discouraged with his future. One afternoon he left work with another Kiekhaefer employee, went drinking, and never made it back to the plant. Speeding along the narrow Wisconsin highway between Fond du Lac and the plant in his convertible 1969 Corvette, Richie and his companion hit another car head-on. In the other car were three couples returning from a wedding reception. Only four people survived. Richie was thrown completely through the windshield of the Corvette, taking the frame with him. He didn't remember a single thing from the time he left the plant until he woke up in the hospital.

"I opened up my eyes and the first person I saw was Kiekhaefer," Richie said quietly. "He was standing there holding my arm ... and he just looked at the doctor and he said, 'He's okay now.' He just kind of hit my arm and walked away. And I said, 'My God - that really struck home."[36] Richie's companion received a severe laceration to his leg.

"It was summertime," Richie said, "I was young, I had a Corvette, it was Wisconsin, go out, have some beers, have some fun. ..." Richie received severe injuries to his head and face, smashing his cheekbone, injured an eye and fractured his skull. When Carl visited him, he was in intensive care, and wasn't discharged from the hospital for ten days. Meanwhile, even though evidence was present that the occupants of the other car were similarly guilty of driving under the influence, a case for vehicular manslaughter began to form. Carl decided to take the matter into his own hands.

Carl knew that Dr. Carlo Bonomi was in need of a first-class mechanic and co-driver who understood the nuances and complexities of Aeromarine engines. Mel Riggs had returned from campaigning with Bonomi, and was considered Carl's first choice to return. But as the heat of a potential manslaughter case grew, Carl quickly and secretly shipped Richie overseas to escape the mounting inquiries. The manhunt for Richie continued for well over a year, as Carl feigned ignorance of his whereabouts. After Bonomi discovered the reasons for the switch, Carl confirmed the dangerous and unlawful scheme to the Italian.

> "As you know, Richie got involved in an automobile accident when, instead of coming back to work as scheduled, he decided 'to have a few beers.' The result of the accident was a complete totaling of Richie's car and another car, in which a manslaughter case resulted. It was, therefore, quite timely that Richie went to Europe to evade authorities who are still looking for him. ..."[37]

Bonomi, far from upset over the ruse, was delighted, for Richie was an exceptionally gifted mechanic and throttleman. Bonomi's success, taking second place in the 1972 U.I.M. international circuit, and winning the world crown in 1973, was in no small measure a consequence of a horrible accident along a stretch of Wisconsin highway, and a bold and imaginative getaway planned and executed by Carl Kiekhaefer to keep a friend from prison. Fred Kiekhaefer, however, was shocked by his father's action, and rightly feared that everything his father had worked for might be shattered by this one, reckless act. "It just was sleazy," Fred grimaced, "pure and simple and I really detested it. I [thought less of] ... the individual for being a part of it, and I certainly didn't respect my father for doing it."[38] Worse still for Fred, who was struggling to assist his father to develop a creative and talented staff at the time, was the fact that at least two exceptional people resigned over the incident. "It hurt the company because some very, very good people left. And, in fact, I have tried to rehire ... every one of those people. But unfortunately for me they were doing very well where they are."[39] It wasn't the last time that a conflict between father and son would erupt over Richie Powers.

When Richie left for Europe, he was far from fully recovered. Though extremely uncomfortable, Richie was able to prepare Bonomi's *Aeromarine IX* for the first race, the Pescara (Italy) - Markarska (Yugoslavia) race. Disaster almost struck, as Richie attempted to race so soon after nearly being killed in Wisconsin.

> "It was rough and I still wasn't right. I remember about half way through the race, I started seeing stars. I thought, 'Oh my God! -- I'm going to pass out. Oh, this is going to be so embarrassing. Kiekhaefer is going to *kill* me! I'll never hear the end of it.' And - BANG! - The prop broke. I went, 'Oh my God!' I jumped right out of the boat into the water and it was freezing cold. It was in the Mediterranean. I said, 'Hand me a prop.' They got up and got a prop and we changed the prop and after that I was fresh as a daisy."[40]

Even though Richie was supposed to be on temporary loan to Bonomi, he never returned permanently to Aeromarine. After winning the world championship with Bonomi for a second straight season in 1974, this time aboard *Dry Martini II*, a 35-foot Don Aronow *Cigarette*, Bonomi surprised Richie by saying, "Rich, we've won everything we wanted to -- we won the Italian Championship, the European Championship, the South American Championship, the World Championship -- twice. ... I don't want to race anymore. I'm done."[41]

During the Miami International Boat Show in February 1974, Carl received the Gulf Marine Hall of Fame Award. Sponsored by Gulf Oil, the award and induction was given in recognition of Carl's many contributions to the marine industry and to the sport of offshore racing. Jim Martenhoff, the respected boating columnist for the *Miami Herald* said, "Kiekhaefer's achievements in engineering and administration have made him a wealthy man. But you may still find him at a race, sleeves rolled up, cigar clenched in his teeth, supervising mechanics as they labor to exact the last ounce of power out of a racing engine."[42]

In 1975, Brazilian auto parts distributor Wally Franz won the world championship, campaigning the 38-foot Bertram, *Pangare Gringo*, equipped with Aeromarine engines. The 47-year old father of five from Sao Paulo jumped out in front of the international circuit by winning the first three races in South America: the Mar del Plata Race in Argentina, the Punta del Este Race in Uruguay, and the 245-mile Santos to Rio de Janiero Race in Brazil. It marked the third consecutive year that Carl's engines had won the world crown.

With Richie Powers available following Bonomi's retirement, Hawaiian real estate entrepreneur Tom Gentry recruited him for an attempt at the 1976 world title. Since the title had been in foreign hands for several years, it was Gentry's plan to return the Sam Griffith Trophy to America during the Bicentennial. "I want to bring the cup back," Gentry told him. Richie agreed to race with Gentry only if he would install a pair of Carl's new 625-horsepower Aeromarine "Champion Maker" engines aboard Gentry's *American Eagle*. With navigator Bobby Beich pointing the way and Richie manning the throttles, Gentry drove to win the European Championship, the South American Championship and finally the World Championship for 1976, by winning six races in Brazil, Argentina, the U.S., Sweden and Italy.

Carl had won seven championships in six years with five different boats and four different drivers. The only common denominator to this remarkable combination of elements were Carl's Aeromarine "Champion Maker" engines and accessories. APBA national titles were won in 1971, 1972 and 1973, while U.I.M. world titles were won in 1973, 1974, 1975 and 1976. Offshore racing historian John Crouse summed up the spectacular success of Carl's engines in his book, *Searace - A History of Offshore Powerboat Racing*.

> "By the termination of the '76 offshore season, Aeromarine engines had powered the winning hulls in no less than 59 major events, second in all the world only to engine maker Carl

> Kiekhaefer's former firm Mercury Marine whose vaunted *MerCruiser's* and Mercury outboards [combined] had collected 99."[43]

Once, during the summer of 1975, Richie was in Fond du Lac working with Aeromarine to test some new engines. Carl had passed the word that he didn't want the boat to run unless he was there to watch. When the boat was ready to run, a search was started to find Carl. He was finally located driving a tractor pulling a hay baler with some of his farmhands. The farmhands were quite grateful when Carl finally got down off the tractor and left.

> "... The 'Old Man's' out there, and he's got his tractor in high gear as fast as it's going, and the baling machine -- he must have had it pumped out to about 400-horsepower or something like this, but it's just taking these bales and it's just going, 'Whoooomphh' and it's *shooting* the bales over the cart. ... And he's got three guys out there just picking them up and putting them in a pickup truck. About every third one was hitting the trailer where it was supposed to go.
>
> "And the guys said, 'Please -- *Please* take him down to your race boat. Get him off the tractor. We can't get anything done. He wants this place hayed by the end of the day."[44]

Richie's respect for Carl continued to grow over the years. "He became the Vince Lombardi of my life," he said. "He really did. And we even discussed that, and my nickname for him was 'Coach.' My nickname was 'Champ.' He used to call me 'Champ' all the time. So we really got just very, very close." So close, that occasionally it seemed like Richie Powers was closer to Carl than Fred Kiekhaefer, although certainly the foundation for the relationship was completely different.

Fred Kiekhaefer had been promoted to vice president and director of engineering at Aeromarine. As such, he was working directly for his father, and when they disagreed over a point of design or styling, the two engineers collided with an uproar that could be heard throughout the large plant, even over the clamor of machinery. "You'd be standing out in the hall," Rose Smiljanic remembered, "and the whole plant would hear!" The two Kiekhaefers alternated between hard working associates and hard hitting rivals. At first, Fred was no match for the intimidating style of his father.

> "At the age these things were occurring I was a much less mature adult [mid-20s], and therefore I was not as in control of my emotions as I would have liked to be. Secondly, there was a sensitivity there being in the family, and being very, very concerned about my own future -- about establishing my own value. I think, therefore, I was a lot more sensitive and I had never had that kind of experience before in my life, so I had no preparation for the emotions to expect.
>
> "I probably also felt, and this is the first time I've ever really thought about it, that I had very little to lose. It wasn't like your boss you were arguing with, it was like your dad you were arguing with. ...
>
> "But, yeah, I'll agree he could find my hot button, and he'd keep pushing it and I'd get better and better at restraining myself. But he'd just push long enough and hard enough until he'd get the right way -- until I went off -- and then he'd say, 'See! You're not qualified ...' to do whatever the argument was about. And you know, by God, he was right. I demonstrated once again that he could manipulate me. But he'd push too hard, and ultimately I said, 'I don't need this,' too."[45]

Some of the disagreements between Carl and Fred concerned issues of administration and finance. Fred realized two things in debating these questions with his father. First, Carl never gave him enough financial information to understand the problems the company was experiencing, and secondly, as an engineer, Fred felt he hadn't enough experience in the real world of business to offer useful solutions. "My father was not a person who shared business information with anybody, including me," Fred observed. "[He] would talk marketing strategy and business philosophy and product development until he was blue in the face. But when you asked to see the numbers he wouldn't share them with anybody. And in those days, I didn't realize that I really needed to know that. I was not a marketing oriented person. I was trained as an engineer. And you've got to remember, I was in my mid-twenties and people in their mid-twenties just don't have enough life experience, in most cases ... just don't have enough experience to know the questions they should ask. ... I didn't know what a loan amortization was. ... I didn't know how to talk to accountants. I didn't know how to talk to anyone outside of technical management."[46]

As the confrontations continued in the summer of 1975, Fred decided on a strategy that would both remove him from the increasingly charged emotional environment, and round out his business education. He applied and was accepted at Northwestern University at Evanston, Illinois near Chicago, and began graduate studies toward a Master's degree in Business Administration (M.B.A.). For Fred, it was a great relief from the pressure and confusion surrounding his overbearing father. To Carl, it was viewed as a strategic opportunity for Fred to obtain advanced business skills that could be of great benefit to the company when he returned.

Almost from the first day that Carl entered the outboard business in 1939, the focus of his competitive indignation and wrath became Outboard Marine Corporation, manufacturers of Johnson and Evinrude products. Ole Evinrude, whom Carl had worked diligently in the 1950s to refute as the "inventor" of the outboard motor, was arguably the most influential spark for the development of a modern marine industry. It came as somewhat of a shock to Carl, then, when he learned in early 1976 that he was to be awarded the prestigious Ole Evinrude Award for "immeasurable contributions to the sport of boating." Carl was absolutely thrilled, and perhaps more than any other honor he would receive in his lifetime, the Ole Evinrude Award represented for him the pinnacle of marine industry achievement.

In the six years since Carl had left Mercury, he had completely discarded the competitive passion that he had demonstrated towards Evinrude in three decades of engineering and marketplace warfare. Thanks to the many key executives which had migrated from Kiekhaefer Mercury to OMC over the years -- particularly president Charlie Strang, chief engineer Edgar Rose and Evinrude public relations manager Jim Jost -- Carl had received welcome contracts at KAM for the machining of stern drive lower units and other components. Even Ralph Evinrude wrote to say, "So nice to hear that we are working together. ..."[47] He had also renewed his great friendship with Charlie Strang, who in 1974, was appointed president and chief executive officer of OMC. Among the very first outside of the OMC boardroom to hear the news of Strang's great honor was Carl, in a phone call from the new president.

> "Just calling to say thanks," Strang began the conversation mysteriously.
> "For what," Carl asked.
> "Well, at our board meeting this afternoon they made me President and Chief Executive of OMC and I wanted to thank the guy who made it possible."

"Why, you're very kind. Congratulations! ... Well, God bless you. I think that's wonderful. I'm proud," Carl said.
"You should be. ... Thank *you* very much," Strang replied.
"How's this going to affect our relationship now?" Carl asked.
"Oh, to the better, I hope," Strang rejoined.
"Are we going to take a common stand against Mercury?" Carl wondered.
"Haven't we been?" Strang replied not entirely in jest.[48]

Not surprisingly, Charlie Strang was selected by OMC Chairman Ralph Evinrude to deliver the speech on the occasion of Carl's receipt of the Ole Evinrude Award on February 22, 1976. The OMC Gulfstream Jet picked Carl up at the Fond du Lac airport a few days earlier, and Strang joked that Carl should wait at the Mercury hanger for the flight. Carl thought better of the idea and remarked, "With their underground and monitoring system, [Mercury] will know well enough what took place when the OMC Gulfstream lands in Fond du Lac -- and I don't want to do anything to lower the value of their stock any more!"[49]

At the Key Biscayne luncheon, held during the Miami International Boat Show, Strang and Evinrude showered praise on Carl's career and enormous contributions to the industry and to the pleasure of boating enthusiasts around the world. Carl listened, dabbing his wet eyes with a handkerchief, as "a distinguished crowd of boat manufacturers, corporate and company personnel and awed media members watched the history making event of one marine giant honoring another."[50]

Ralph Evinrude said, "Carl Kiekhaefer galvanized the marine business into action. His gift for 'making things go' will long remain memorable in the marine field," and "I can't think of a finer man to represent the marine industry than that great catalyst for the trade. ..." Charlie Strang spoke at glowing length of his friend and mentor's accomplishments, and attempted in conclusion to circumvent Carl's professed modesty. "If I know him, in accepting this award, he will probably ascribe most of this success to a 'team'. True, any achievement as great as Carl's calls for the help of others. ... But I was there ... saw the team change through the years ... saw the team members come and go ... with the one solid, stable factor always Carl himself. *He did it his way*, and he can't take that away from his legend, his innate modesty notwithstanding!"[51]

The unique circumstance of Carl receiving this honor from the Evinrude Foundation was picked up and printed by hundreds of newspapers and magazines throughout the country. The award served to remind Carl of how little recognition he had received from his own company for the work of a lifetime. "The most valuable player on my team ... was Charlie Strang. When the team breaks up, and the 'greatness' turns into disparagement, it's almost hard to understand how former 'rivals' in the marketplace have a higher regard than my former colleagues. ... Somehow this seems like a great mistake, -- awarding the prize to me."[52] When only one representative from Mercury attended the well publicized awards ceremony, Carl remarked, "Are they under such duress that they dare not express friendship? ..."[53] But to Ralph Evinrude, Carl gushed with sincere appreciation.

> "I am sorry only that we didn't get to know each other better during the heat of the marketplace over the past three decades. ... [The press and my friends] feel, as I do, that it must have taken considerable courage to select a former competitor, but usually with the final thought that some high principles remain in the minds of our industry leaders such as yourself.
>
> "As Charlie says, we were competitive. We respected your competitiveness; in fact, you set a pattern of deportment businesswise that we understood and followed, such as your marketing structures, product integrity and image in the ever changing marketplace. ..."[54]

A year later, when Ralph Evinrude was being honored for his fiftieth year of service to the marine industry, Carl and Fred Kiekhaefer were invited to Ralph's table of honor during the gala banquet in Milwaukee. For all of Carl's contributions and success in the industry, he still felt out of place seated next to Ralph, and confessed as much to Jim Jost. "I could well believe that there were others who deserved a spot at Ralph's table in view of the tremendous history of [Evinrude Motors] and I felt a little out of place in all this distinguished company. I am just wondering if Ralph Evinrude himself didn't think I was somewhat of an impostor in view of some of his old friends like Jim Webb, Hugo Biersach, the Briggs boys and Clay Conover, etc."[55] Again, the festivities arranged for recognition of Evinrude made Carl feel sad for never having received recognition from his own company, Mercury. Of the occasion, Carl wrote to Charlie Strang, "I am only sorry that my former company couldn't have done the same for you and me when it came our time to depart. When it was over, I left the table with a lump in my throat."

Not long after, as Fred Kiekhaefer was well into his second year of graduate business studies at Northwestern University, Carl phoned with disturbing news. He had just returned from his annual physical at the Mayo Clinic, where doctors had recommended that he undergo open-heart, coronary bypass surgery to increase the blood supply to his heart. "He was worried that he would die," Fred recalled, "and he asked me to come back and run the business for him."[56]

CHAPTER THIRTY-SEVEN

Changes of Heart

"The pressure and frustration (mostly the latter) of my ten years with Brunswick accelerated a condition that might well be the penalty of ambition."[1]

Carl Kiekhaefer

It was the winter of 1960 when Carl first felt the pain deep in his chest. On one of those breathtaking and bitterly cold Wisconsin mornings, when the first deep breath out of doors shocks the lungs, Carl felt something else. It was an acute and stabbing pain, like an ice-pick buried in his breast. Though the ache went away after a few minutes of rest, the anxiety of waiting for the next episode of angina would haunt him for the next seventeen years. "It was not easy to live with this fearful anxiety," Carl suggested to a friend, "particularly when one could not confide in anyone but one's doctor or have the whole thing collapse around his head."[2] During those long, stress-filled years, his condition worsened. The colder the winters and the shorter the deadlines imposed by his rigorous schedules at Kiekhaefer Mercury, the worse the symptoms became.

Rose Smiljanic remembers the trip to the New York Motor Boat Show in 1961, the year 54-year old Carl sold Kiekhaefer Mercury to Brunswick.

> "He was ... entertaining constantly, not to have fun, but for customers. I remember several instances we took people out ... and there'd be pressure, pressure, pressure at the show all day long, and then big dinners at night -- a couple of drinks before dinner and then a big steak and ... baked potatoes and a salad and deserts. And then we'd be walking back to our hotel or a cab and he'd say, 'Stop ... look into the windows ... I'm just going to let them get ahead for a few minutes.' He didn't want them to know, so we'd kind of slow down, and then he said, 'Okay,' and we'd head on again. And it was getting worse all the time."[3]

Carl had been a heavy eater all of his life, and would often attribute his more mild attacks of angina to gas or indigestion. Angina pectoris, the severe pain in the heart, is caused by the restricted flow of blood to the heart muscle itself, due to coronary arteries clogged with the consequences of a life of rich and fattening foods. Ever since childhood, when Carl gorged on the abundant fare of the homestead dining table, he learned to love the greasy sausages, thickly marbled steaks and heavily buttered pastries so common to Wisconsin diets.

> "Knowing what the medics know now, it would be easier to live the life over again without all the fine dairy products we were taught were necessary for good health -- the cheese, the butter, the wholesome milk, the fresh farm eggs, etc., which we now find are heavy in cholesterol, as are the well-marbled steaks with potatoes and gravy. As a child on the farm, we were chastised if we left the fatty part of the meat. Sandwich spreads consisted of pork fat and gravy from a roast. The bacon and fried sausage products that still make my mouth water are also on the taboo list as is highly seasoned German cooking in general."[4]

Each year, Carl ventured to the Mayo Clinic in Rochester, Minnesota for a general checkup. He developed diabetes in his forties, and learned to inject himself with insulin in his arms and legs. He kept this diabetic condition almost a complete secret, fearing that the discovery of his condition would somehow smudge his patina of invincibility and reputation for legendary strength and vigor. During his exams, his physicians would take care to check that he wasn't developing any problems which can be associated with diabetes, such as poor circulation to the extremities and the possibility of encroaching blindness. He passed the usual tests with each visit, and was almost ritually instructed to monitor his blood sugar, lose weight and quit smoking. Of these, he would only pay loose attention to checking his need for insulin, using a urine test strip at his home rather than the more accurate and much preferred finger-pricking blood test.

Carl also feared that his father's premature death due to a heart attack, was the progenitor of hereditary heart disease for himself, again adding to his anxiety. To Frank Scalpone, his former director of public relations who had risen to a key position in the National Association of Engine and Boat Manufacturers, he admitted struggling with his fears and anxieties during the hectic and pressure-filled years of Kiekhaefer Mercury. It would be to Scalpone in 1977 that Carl would first suggest, perhaps more for effect than reality, that more than financial distress caused him to merge with Brunswick.

> "I can now confess that one of my reasons for merging with Brunswick in 1961 was my concern for what seemed to be an impending coronary problem. With Dad dying at 67 from a coronary, I wasn't sure whether it was hereditary or just what it was so naturally I had apprehensions. Subsequently, I learned to live with the problem after studying up on it and without pushing my luck, and waited for an opportune time to deal with the situation.
>
> "Severe angina pains often cramped my style at a time of pressure such as press receptions and other important functions, that made my behavior suspect to many of my crew, including you. Can you imagine the confusion and consternation any explanation would have caused trying to explain it? A man in a position other than mine need not have suffered keeping a stiff upper lip."[5]

As Carl considered his own mortality, he opened up about the mental and physical torment he endured throughout his Brunswick days, and began to blame both the merger and his steadily mounting angina on Jack Hanigan and the whole corporation.

> "The pressure and frustration (mostly the latter) of my ten years with Brunswick accelerated a condition that might well be the penalty of ambition. No one will ever know the amount of effort it took to build Mercury and it's a pretty lonely task. It gets even lonelier when you merge. Few people wish to extend themselves to the magnitude that it requires but it seems that everything has its price. If one were to live a life the second time, it could be a lot easier."[6]

By 1975, as Kiekhaefer Aeromarine engines were winning their fifth straight offshore championships, Carl quietly had oxygen installed in his home and office. When he returned from a long walk, or when he mounted the many steps to his offices at Aeromarine, he could put the refreshing mask over his nose and mouth and catch his breath. The bottles, like his need for insulin, Carl kept hidden from nearly everyone. He kept them so hidden that one day when he needed it, it couldn't be found. Ruth Backus, who had started at Mercury in 1953, and then followed Carl onto the ledge in 1970, remembers how Carl struggled against his increasing illness.

> "I remember one time I was working ... and I had an appointment with the hairdresser. I went into Mr. Kiekhaefer's office and he looked just awful. He was white as a sheet and he asked me to get him the oxygen, and I wasn't sure it had been moved and I was scrounging around trying to find it. And he was just gasping and I thought the man was having a heart attack. So I went out and ... I was going to cancel my appointment and he heard me and he said, 'No! Keep it.' And so I stayed with him and his color seemed to return. He seemed to be a little bit better."[7]

Two years later, his condition had become critical. When his annual Christmas flight from Milwaukee to Tampa made a stop to change planes in Atlanta, he confided to his friend, General Mabry Edwards, that he could hardly carry himself around anymore. "He told me, 'I had to get a wheel chair, I couldn't walk across that big Atlanta terminal.' And that's when I realized he really had some problems there with his heart."[8] A month later, in January 1977 the brutally cold Wisconsin winter aggravated an already deteriorating condition. His embarrassment over being discovered ailing by the small town communications underground, upset him as much as the medical emergency.

> "After a severe siege of cold weather for almost two months, when people stayed out of the outdoors and, consequently, lacked exercise, I too reached a situation that required emergency action. Consequently, on the evening of January 20, while working at the office at my desk, I got severe chest pains inducing nausea and extreme difficulty. ... I called an ambulance and alerted my local family doctor ... and was brought into the Intensive Care Unit ... after a newsworthy C.B. radio broadcast by the ambulance driver that the great Kiekhaefer was heading to the hospital -- no doubt a choice bit of news for a change. Also, at least one nurse from this hospital talked to her neighbor and it was spread around the plant and as usual, they got everything all crossed up."[9]

Carl was discharged from the hospital within twenty-four hours, and was told that "my excessive fluid got in my system, which is what caused the chest pains. ..."[10] His recovery was very rapid, and he said he "felt fit as a fiddle within a very short time." But within days the symptoms returned. He couldn't sleep lying down, forcing him instead to doze uncomfortably and intermittently in his recliner. His anxiety continued to rise, and it seemed the more acute his anxiety, the more acute the pain would become. Carl described one of his many episodes. "Oxygen starvation on the part of the heart muscle brought on angina with enough severity that all action ceased and I, almost like a robot, froze on the spot until the pain disappeared. ..."[11] Within two weeks he had checked into Mt. Sinai Medical Center in Milwaukee, to undergo preliminary tests to determine his suitability for coronary bypass surgery, known also as myocardial revascularization surgery. Freda was in Florida and Fred was attending graduate school at Northwestern University, so when the hospital called to confer with the family it was Rose Smiljanic whose name Carl had given.

> "They called me one day and they said, 'Hey, can you come down here, he's having severe pains -- we're going to have to talk.' So, you know, ignorantly I drive down to Milwaukee to the hospital. They had him real sedated because he was having such severe pain and they thought that he might have to have surgery, that his blood vessels to the heart were clogged. I said, 'I don't think it's that.' I said, 'He's a big eater and he's always talked about gas.' And then this cardiologist ... said, 'If I could tell you how many people keep talking about gas, and it always turns out to be that they need surgery. And I mean this is a *shocker*."[12]

The pace of events began to overwhelm Rose. Though Carl perhaps trusted her more than any other person, she felt out of place discussing his fate with doctors who were recommending procedures that could kill him. He had been sedated with what she thought was morphine, and could tell he wasn't totally coherent. Carl had already signed a surgical consent form, but she wasn't sure he knew what he was doing. About twenty after seven in the evening, Carl was rolled into the operating room for a heart catheterization to determine what procedures should be performed. Rose began to panic. "I sat there an hour, and two hours went by. My legs began to shake. I thought, 'I bet they're going into surgery and I haven't called the family yet!"[13] She called Fred during the evening, but due to a mix-up of who should call who, wasn't able to inform Freda in Florida until the following morning. Carl didn't reappear until eleven o'clock that evening, and Rose learned to her great relief that an operation hadn't been performed.

Freda flew up to Wisconsin as the physicians were deciding that a coronary artery bypass operation was necessary to restore the flow of blood to his heart muscle. Fred's fiance, the future Dr. Carol Stafford, was completing her medical training at the time, and was able to steer the family to one of the most highly qualified heart surgeons in the country, Dr. W. Dudley Johnson of Milwaukee. After further evaluation, all agreed that a coronary artery bypass operation was essential to save Carl's life. Though generally successful, a significant number of patients succumbed to the relatively new open-heart procedure. Carl was justifiably frightened when he wrote, "Time gets a bit hectic when you suddenly realize the important, last minute decisions that must be made before such an operation since you are not certain that you will be there to make them afterwards."[14] He prepared for the worst and summoned his son, Fred.

> "He was seriously terrified of the operation and what it meant. He was worried that he would die, and he asked me to come back and run the business for him. And he promised me that if he did come out of this operation in good health that he would systematically turn the business over to me -- including the ownership. He would gradually withdraw from the day-to-day operations. He would like to have an office where he could tinker, and he'd like to continue to help us develop new products and that kind of thing -- but I would run the business."[15]

Taking his father on his word, Fred quickly rearranged his life to meet the sudden challenge and new responsibilities. He was in the middle of the second quarter of his final year of graduate business studies, and switched to an intensive commuting program between Fond du Lac and the outskirts of Chicago, driving over three hours in each direction. He was dictating business letters into a tape recorder on the way down to classes, and dictating his homework assignments and research drafts on the way back. It soon became obvious to him that he couldn't maintain a full-time curriculum and also dedicate himself full-time to the many exigencies of running the business. "I got special consideration from the professor to come down and miss essentially a couple of classes each week." When he finished the quarter, some six weeks and over 6,000 miles later, Fred transferred to night classes, even though it would mean driving to Evanston for another full year to complete requirements for the M.B.A.. "I had a bet with the admissions director when I switched into the program," Fred remembers. "He said, 'People *never* finish.' I said, 'I'd finish, because I'm a finisher."[16]

Two days after Valentines Day, February 16, 1977, Carl was wheeled into the operating theatre at Mt. Sinai Medical Center in Milwaukee. Four anxious and fidgeting people waited impatiently in the family lounge: Freda, Fred, Carol Stafford and Rose Smiljanic. For almost eleven hours they thumbed through old magazines, made nervous small talk and thought about the 71-year-old warrior whose heart was laid open under the bright lights only a 100 feet away.

Over three feet of one large leg vein had been stripped from Carl, and the surgeons were grafting it to many locations on his struggling heart. The blood flow through his major coronary arteries was found to be severely restricted, in some cases nearly totally. Graft after graft was performed as the hours went by, and the circulation was slowly restored to Carl's heart. Before they finished, Carl received five complete bypasses, among the most extensive and demanding of such procedures that had been accomplished to date. "Oh God!" Rose Smiljanic exclaimed. "I can still see him laying on that slab of the table. They brought him out there, tubes in his nose, tubes in the body and all that. Ohhhh, it was just *terrible*. And his eyes -- he'd look at us and he had those big eyes, you know, and he looked to me -- it looked like an animal that you would have slaughtered. He didn't know what was going on, and of course they had to kind of keep him awake to get him out of that."[17]

Carl's strong constitution and fighting spirit overcame the feared handicap of age, and after three weeks in the hospital he was ready to return home. But March in Wisconsin can be among the coldest and most unpleasant of months, and it was recommended that Carl spend some time at Freda's home in Florida to recuperate. Though it had been seven years since Carl had left Mercury, the news of his operation and his need to go to Florida became a catalyst for a thaw in the icy relations between him and the company he had founded.

In May 1971, Jack F. Reichert had become president of Mercury Marine, replacing K.B. "Brooks" Abernathy who was promoted to president and chief operating officer of Brunswick, while Jack Hanigan was promoted to chairman and chief executive officer. Reichert had first encountered Carl a few years after he joined the Brunswick Bowling Division in 1957. As a young sales engineer, Reichert's territory included Fond du Lac, and when a new Brunswick Bowling Center was proposed for the area, he pitched Carl on the possibility of becoming an investor. "I remember he was very nice to me," Reichert said. "This was early [following the merger] and he said, 'No,' he didn't want to make an investment of that kind. But he thanked me for calling and said they would support it."[18] After fourteen years of steadily working his way through the ranks, the intrepid Reichert was promoted to vice president of the Bowling Division, and in 1971 he was appointed director of marketing for Mercury Marine, replacing the retiring Armand Hauser.

Reichert had been on the job only a year when Abernathy selected him as his successor at Mercury. Charles Alexander, vice president of engineering for the previous thirteen years, felt that he had been passed over. "... [T]hat aggravated me some," Alexander said, "because I figured he [Reichert] didn't know a damn thing about the business. So I complained a little about it, and ended up negotiating a little better deal for myself, and expressed my unhappiness in being passed over. ... The main thing it did, it told them that I was interested in that job, enough to raise hell about it. ..."[19]

Rose Smiljanic had remained friends with Sharon L. Trescott, in 1977 the executive secretary to Jack Reichert, and who had been Carl's dictation secretary for many years. Rose spoke with Trescott, who in turn mentioned Carl's need to go to Florida to her boss, and the quick-thinking Reichert recognized a golden opportunity to begin to soothe the injured feelings Carl had for his old company. "I said to Sharon," Reichert remembered, "I want Mr. Kiekhaefer to fly down on our airplane. And she said, 'Well, Mr. Reichert, there's no way to do that unless you ask him.'" Within minutes, Reichert and Trescott were in the car on their way to Carl's house on the ledge.

It was Reichert's first trip to Carl's house, and as they drove up, Carl and Freda were returning from a short exercise walk.

"He asked me very gruntly, what did I want? I asked him how he was feeling, and, you know, grunt, 'I'm fine, I'm feeling alright, very cold. Well, what do you care, however?'" Reichert explained that he had driven up there to offer the Brunswick jet to fly him down to Florida, and how much easier it would be for him than to attempt to fly commercially.

> "And he said, 'NO!' And I said something charming like, 'Goddammit Carl! Stop being such a stubborn German! For your information, I'm a German from Wisconsin too, and I can be just as stubborn as you can.' Well, I guess that kind of shocked him, because Carl was used to people caving in, I guess. So Mrs. Kiekhaefer said, 'Carl, stop being such a jerk,' or worse, and 'Why don't you see what this young man is all about?' So, because of her, he said, 'Okay.'
>
> "We went into his home and we sat down, and he says, 'I don't understand why you want to do this for me. I don't work for Brunswick. I don't work for Mercury. Nobody has ever treated me right at Brunswick! Why do you want to do this?'
>
> "'Because you are the founder of Mercury,' I said, 'And I don't want to lose the founder, and you are very, *very* important to our company, and I want you to come back. And I don't want you to die, and that's why I want you on that airplane.'"[20]

Reichert explained to Carl how last year, when he was only 40 years old, he had nearly died himself from a somewhat freak gastrointestinal ailment that required the emergency removal of a large part of his colon. "Don't tell me how sick you are," Reichert admonished Carl. "'At least you can afford to lose a foot of your colon, [but] you can't afford to lose your heart. Your body doesn't work too well when that stops. I just want to do everything I can for you.' Well, I think he was kind of moved and he said, 'Okay, I'll think about it.'"[21]

> "And then he proceeded, for about an hour, to just spew out all the poison about Brunswick, and how he hated Brunswick and how he hated Ted [Bensinger] and how he didn't want to sell, and then [talked about] Jack Hanigan, who he hated intensely. ... So, he talked and talked, and went on and on and I just let him get it out of his system. And I said, 'Carl, there is nothing I can do about the past, *nothing*. But I can do *everything* about the future. And I want you to come back, but the first thing I want you to do is get well. So, I want you to take my plane and I want you to go to Florida.

"'Well,' he said to me, 'Will you let me land at Lake X?' Well, I had just spent $35,000 repaving the runway. We leased Lake X from Carl -- Carl *owned* it. And I just shut the runway about two months earlier, so that we didn't have to fool around fixing it. He said, 'I want to land at Lake X.' I said, 'You want to land at Lake X, we'll land at Lake X.' So, he said to me, 'Well, I'll think about it. ...'

"So, the next morning my telephone rings and Carl's on the phone and he says, 'I'll think about going, but I've got to have Wayne Predmore [Carl's chief pilot at Mercury] as my pilot. I won't let anybody else fly me.' I said, 'Okay.' He said, 'I must leave at' such and such a time. I said, 'Okay.'

"And so, to make a long story short, Carl finally reluctantly agreed, and it was the damnedest thing you'd ever see. I was giving him a free airplane ride and the conditions he thought of were *unbelievable*."[22]

Reichert had met each of Carl's conditions for the flight, and on March 12, 1977, Carl flew to Lake X, and then was driven to Freda's home in Winter Haven for six weeks of rest and recuperation. Freda, whom Carl had retired as a nurse so many years before, now had her solitary opportunity to exert some small measure of authority over her tough but wounded husband.

Carl was confused over Reichert's generosity and apparent genuine concern. Less than a year earlier, Carl had attempted to meet with Reichert at the Miami International Boat Show, but their schedules didn't match, prompting Carl to say, "I guess they are up to their old tricks of playing games with me and as such, I do not feel they are serious about talking about anything."[23] Even shortly after his flight to Florida on the Brunswick Sabreliner, Carl said, "The most confounding experience in the whole thing was Reichert driving up to my house and insisting on flying me to Lake X in a company jet. This happened within three hours after I made reservations on a commercial flight, which would have been just as practical. Of course, it was a nice gesture but I can't for the life of me see them doing any favors suddenly, seven years after my resignation."[24] Carl eventually realized, somewhat to his embarrassment, that Reichert was sincere, and wrote, "You are the first officer of Brunswick who has shown any such compassion or kindness since my resignation in 1970 when I thought I had left in good standing."[25]

Reichert continued to convince Carl of his sincerity. When Carl returned from Florida, Reichert shocked him with a phone call that would solidify their trust and friendship. Mercury had prepared a special pictorial montage of their history, which included an image of Carl and his early engines and accomplishments, and Reichert wanted him to have a special copy.

> "He said, 'Fine, I'll send somebody down,' I said, 'No you *won't!* Nobody is going to be a middle person between you and me anymore.' I said, 'Those days are over. Either you come down to my office and pick it up or I'll come up to your office and give it to you.
>
> "And he said to me, and it was really very sad, he said, 'You mean to tell me you'll let me inside my plant? You'll let me in there?'
>
> "I said, 'Of course I will. Anytime you want to come you're welcome.' He said, 'I'll be down in five minutes.'
>
> "Well, I called the guard house and I said, 'Open the gate because Mr. Kiekhaefer is coming, and don't ask. And if the gate's closed, he'll go right through it. Just open them up and let him in!'"[26]

Carl met with Reichert in his office at Mercury for over an hour, and "probably the first 30 or 40 minutes was again a tirade on Brunswick," but from that day on, the two men became good friends. Within months of that meeting, Reichert was appointed president of Brunswick and Charles Alexander finally got his wish in May 1977, and became president of Mercury Marine. Carl was proud that Reichert had asked his advice about who should be the next president of Mercury. "I naturally recommended Charlie Alexander," Carl said.[27] Reichert was also sympathetic to Carl's business situation, and recognizing a good opportunity, made sure that KAM would get some component manufacturing business from Mercury. "He gave instructions to Charlie Alexander," Carl said, "that he 'is to keep us so busy' that we will never again be able to accept Outboard Marine contracts any more."[28] As Reichert was leaving to assume his new responsibilities in Skokie, Illinois, Carl wrote him to say, "Most of all, I want to thank you for the change of attitude at Mercury. I hope this will continue at the Brunswick level as well. I could never understand why I was so completely ostracized. It caused a bad reflection on both companies, particularly in the absence of any wrongdoing or reported wrongdoing on my part."[29]

The following year, when Mercury was hosting an open house, Reichert and Alexander surprised Carl by inviting him for a special guided tour of the plant. "We put him on a golf cart," Reichert said, "and we took him through the plant and it was like he had died and gone to heaven. And the people were so happy to see Carl, it was *unbelievable*."

> "I don't want to sound like I'm some kind of a brilliant guy," Reichert continued, but it was clear the company came together. And I've always said, of all the contributions I made at Mercury, and I think we did some pretty decent things while I was president, the one thing that I did better than anything else, is I brought Mercury back together again.
>
> "I brought Carl back to the company, and the people felt good about it. ... He might have been a tough guy to work for, but they absolutely admired the guy, and when he was in disfavor, people at Mercury didn't like it. And when he became a part of the company again, the people at Mercury liked it and the family came together again."[30]

Later, Carl would admit, "Now that the company has decided to 'kiss and make up' with me, I am certainly much relieved. I could never understand the campaign against me for the first few years. Hanigan made it absolutely miserable for me and I just had to leave in order to live with myself. The turnaround came with Reichert."[31] [Reichert became chief executive officer of multi-billion dollar Brunswick Corporation in April 1982, and chairman in the fall of 1983.]

Fred assumed his responsibilities as the new executive vice-president of Kiekhaefer Aeromarine Motors with the gusto and enthusiasm of youth. Soon to graduate with an M.B.A., and already possessed of a graduate degree in engineering, he was eager to prove himself equal to the task. He began to cut wasteful engineering programs, streamline accounting procedures and develop and refine new products. All the while, he was courting the long-term buyers of high-performance marine accessories, and opened markets with both large and independent boat-builders throughout the country. Foremost among his new approach to race-proven products for the sport boat market were the *K-Plane*, a line of hydraulic trim-tabs for medium to large sportboats, and a family of high-performance throttle and shifting controls which would eventually be known as *Zero Effort* controls. While his father convalesced, Fred also introduced a new line of high-efficiency propellers, based on the successful designs evolved from competition.

Within little more than a year's time, Fred had managed to turn Kiekhaefer Aeromarine Motors around from a perpetual loss to a modest profit of nearly twelve percent of sales, and at the same time doubling revenues of the company. More importantly, Fred had structured the company to take advantage of their strong image and good reputation over the long term, rather than building a few very expensive offshore engines for a very small and volatile market. He was justly proud of his achievements, but as Carl continued to regain his strength, it became evident that there was trouble brewing.

Physically, Carl was indeed recovering nicely, and more quickly than anyone had imagined. For the first several months following the operation, though, his mental faculties were somewhat impaired, and it was feared that he had suffered a mild stroke during the long operation and deep anesthesia. Carl was frightened that he couldn't accomplish some simple tasks soon after the operation, but fortunately most of the effects eventually wore off. "With an operation of this magnitude," Carl wrote to a friend, "apparently some time is required to re-establish mental alertness, my memory and even habits like tying a necktie. I have simply forgotten how to tie a tie and I never seem to tie it the same way twice. And little quirks almost humorous, as are simple problems in arithmetic."[32] To another he wrote, "Silly little disturbances that are humorous are forgetting habits like forgetting how to dial a telephone number you have dialed for thirty years."[33] Unless Carl mentioned the problem, hardly anyone would notice. Rose Smiljanic, though, was quick to sense the subtle changes. "You know, he would try to remember things," Rose said, "and he'd say ...'I think when you're under [anesthesia] too long that something happens to the mind.' He was sharp after that, but he was never back to being the old sharp person he was."[34]

His angina pains disappeared immediately following the bypass, and he started a regimen of walking one to two miles every day to regain his strength. He had dropped some fifty pounds, and was delighted to see his waist shrink from a size 44 to a size 38.[35] Soon, his humor returned, and he entertained many friends with his engineering analogies of his heart surgery through his letters. Particularly graphic, is this excerpt written exactly one year to the day after his operation.

> "In a more serious vein, they did remove one vein from my right leg from the ankle to mid-section, which was then snipped into short lengths and fitted up for five bypasses. I was worried whether the five-foot wound would heal at 70 years of age, but it did. I tried to get the brand of the little chain saw they used for splitting my chest, but the name escaped me. They truss you in an open position like popping open a clam shell or a steer, while there was a five-hour duty on a heart-lung pump which took over the heart functions while they stitched and fitted the vein from the leg, making sure that the check valves were all pointed in the right direction. But I am glad I made it! It might have been that I was rejected by 'both parties' [heaven and hell] as undesirable, in which event there was nothing more to do than to come back and go to work."[36]

Carl delighted in comparing his surgery to an engine overhaul, and generally making light of the lifesaving procedure. "It might be of interest to you that I had a new manifold installed on my heart with five bypasses -- like the manifold on a five-cylinder Mercedes diesel. My motor (the heart) apparently was in fine shape but all the passages were partially clogged and two were completely clogged, one 99% and the other 98%."[37] To another friend he wrote, "I am feeling much, much better -- something like an engine that has just had a new oil change and a new filter installed."[38]

Carl gained an even greater sense of his good fortune when Carl F. Knuth, his friend and loyal employee of 36 years, passed away on the operating table during a similar bypass procedure in late 1977. "He was a typical German male," Carl wrote in a moment of despair, "who refused to believe there was anything wrong with him."[39] Carl was grief stricken over the loss of his valuable ally and friend, ghost writer and in-house historian, and in association with Boating Writers International, dedicated a new "Carl Knuth Award" for excellence in marine industry communications.

As quickly as he was able, Carl began to inject himself further and further back into the business. "I am quite happy with my son, Fred," Carl wrote a friend, "who at the time just before my surgery was in for some quite serious talks, not knowing whether I would be back, involving the operation of my company. ... Now that I am again fat and sassy, he is holding me to it, and is retaining command, as it should be, to where I don't touch, but watch, and when it comes to the more important connections, such as finance and banking, I still am a 'welcome member of the club.'" Carl was at first careful not to disturb Fred's administrative procedures, but the frustration was proving to be too much for him.

> "... When I went in for the operation, I had to make certain plans with my son, Fred. ... This meant turning over the reins of the business to Fred and he is holding me to it so I feel a little useless and like I had been put out to pasture. Nevertheless, I am trying to keep busy and am at work every day. In fact, I was a bill collector yesterday. I made a trip to Florida and back to collect some $40,000 cash, which I carried in my Samsonite suitcase with nothing more than a fiberglass tape seal over the two latches. It was one way to get the money although I admit not the best way. Sometimes to get the job done, you have to punt or blast."[40]

Before Fred knew what hit him, Carl was back at work virtually full-time and full-throttle. "He came back to strength very quickly," Fred says, "and that's when I learned my most important lesson in negotiating. Because as he regained his strength he reinserted himself in the business. He conveniently forgot our entire conversation about me operating the business."[41] Again the two Kiekhaefers clashed. Fred was doubly frustrated because he had seen the successful implementation of his new business plan, and the company was already experiencing positive results on the balance sheet. Fred wanted to continue to expand the business into the high-performance pleasure craft marketplace, while Carl wanted to continue his domination of powerboat racing. "When my dad came back, the offshore racing became the ultimate priority again, because that's what pumped him up."[42]

When Carl was gingerly asked by his friends or family about "retirement," he was ready with a favorite reply. "People don't wear out, they just rust out like any other idle machinery. And besides, I believe I am fortunate in being able to do what one likes and/or likes what one does."

Once again, Fred felt trapped. He was in the uncomfortable position of having been given authority for the operation of the company, but his father began to systematically regain control of employees and the engineering programs at the heart of new product development. With nowhere to turn, and with virtually no one to share his sentiments with, one day he summoned secretary Ruth Backus into his office and could contain himself no longer.

> "I'll never forget. I walked into his office one day and he said, 'Ruth, close the door.' And he said, 'I've got to blow off some steam.' And he started, and he was practically yelling and I was standing there laughing I couldn't help it. It struck me so funny. He was pacing the floor, and he was screaming at me, and he was yelling at me and pounding the desk, and when he got all through he looked at me, laughed and said, 'Thanks a lot Ruth, thanks for listening.' Something had teed him off and he just had to get it out. I said, 'Okay Freddie,' and out I went.
>
> "I realized that all the people in the other offices had stopped work and I think they thought that he was reading me the riot act. Because they looked at me ... and I said, 'Oh, it was nothing.' I had to go through engineering and I just laughed as hard as I could because they all thought that he was really telling me off. And he was just blowing off steam."[43]

As Carl continued to back away from his promise to Fred, it was apparent that he simply missed the thrilling but unprofitable excitement he had known in racing. And yet to justify his broken promise, to Rose Smiljanic he said, "I can't *give* the place to Fred, it's just not fair to the other two girls." Rose wanted Carl to relax, get away from business and enjoy what was left of his life. "I was always after E.C. to get rid of that company -- give it to Fred, Rose said. "And you won't have any more worries, and then give the equivalent of money to the girls. Well, he didn't quite want to do that. I guess, you know, he didn't want to think that Fred had won that way."[44] In a more philosophical mood, Carl admitted that he wished Fred had been born earlier, and he would have been able to give him Mercury instead of merging with Brunswick. "I have always said, I'll retire, God willing, when I am ninety. I might never have merged Kiekhaefer Corporation if my son had been my first rather than my lastborn. My children came like the locusts, one every seven years, so this makes it almost a two-generation gap between father and son, which has been ... tough on the old man."[45]

The question of authority finally came to a boil in the early weeks of 1978. Richie Powers had been asked by Rocky Aoki, the owner and founder of the famous Benihana chain of Japanese restaurants, to rig an offshore racing boat to compete in the 1978 APBA national circuit. Richie accepted the job on the condition that Aoki allow him to select Aeromarine engines, drives and propellers. Richie was accustomed to the routine of negotiating with Carl for a package price for the exotic hardware, especially since Carl was partial to both Richie and the dream of another championship.

When Richie presented Carl with a major order for eight engines, eight drives and accessories, Carl surprised him by saying, "Rich, I've turned everything over to Fred. Go and talk with him. He'll give you the prices and everything else."[46]

> "So, I went to Fred and said, 'What's the price for the motors?' And Fred was of a different school than [Carl] Kiekhaefer was. Fred came from making money -- 'We've got to make money' -- the accountant's view. 'We're not making money selling these motors at $18,000 apiece like we were last year.' So, he charged me $25,000 apiece for the engines that I had, just a year prior to that, paid $18,000 apiece. ...
>
> "I said, 'Fred, wait a minute.' I'm here to buy 8 engines, 8 stern drives. I'm going to give you all of the rebuilding business, which over the period of a year, is well over a million dollars worth of business.' Because the rebuilds on those engines - one engine alone - was averaging about 8,000 bucks, because after every race, they go back. I said, 'Can't you make me a package price?' He said, 'I can't. I got to charge what I can charge.'"[47]

Fred's answer surprised Richie and put him in an uncomfortable spot. Rocky Aoki had been campaigning his boat, *Benihana*, with *MerCruiser* engines and drives for the past two seasons, and was reasonably satisfied with their performance. Richie knew that Aoki would save a small fortune by selecting Mercury again, but he also knew that the restauranteur's best chance for a championship season was with Carl's *Champion Maker* equipment.

"Fred, you're way out of the ballpark," Richie protested. "Maybe your stuff is good, but ... you still have to have the people behind it to make it work. I mean, let's be reasonable here. ... Don't let me walk away from here and not go with Kiekhaefer Aeromarine."

"Richie, I'm sorry," Fred replied quietly. "That's the best I can do." It was the first time that Richie had seen his friend assert himself, or really put his foot down, and it made Richie feel awkward and equally stubborn.

"Well, I'm sorry then," Richie said with finality. "I have no alternative then but to go to Mercury and buy these packages, because I can't pass this on to Rocky. He'll kill me. He's raced Mercury before and he's had good luck. So, how am I going to tell him he's got to pay $200,000 more for your equipment as opposed to Mercury? I can't do that."

"Well, Rich," Fred said without emotion, "whatever you've got to do."

Richie Powers left Fred's office and headed down the ledge toward Oshkosh and Mercury's Hi-Performance facility. But on the way he stopped at fellow offshore racer Mel Riggs' home to discuss this turn of events. "We used to call E.C. the 'Buck' and we used to call Freddie 'Velvet,'" Richie said, " It's just when the deer starts sprouting its little velvet up there [before the horns]." Before he continued on his way to Mercury, Richie made one more phone call to Aeromarine. Lucy Lehner, Aeromarine's never flustered receptionist, said, "Richie, we've been trying all over to get a hold of you. Where are you? What's going on? Have you been to Mercury yet? E.C.'s *got* to talk to you -- he's *got* to talk to you!"

"Richie," Carl said when he took the phone, "please come back. Let's talk."

"E.C., we talked," Richie replied.

"Don't pay any attention to Fred," Carl said. "What do you want?"

"You charged me $18,000 last year for the motors. You're asking $25,000. I'm offering you $20,000."

"Done," Carl replied quickly, "You've got it."

Richie and Carl made similar compromises on the stern drives, propellers and other racing accessories the *Benihana* would need for a full year of active campaigning. On April 16, 1978 Rocky Aoki and Richie Powers were setting a course record in the 204-mile Swift Offshore Classic Race near St. Petersburg, Florida, when the boat hooked violently and ejected Richie into the Gulf of Mexico, breaking his elbow and six ribs. Meanwhile, back at Aeromarine, the injuries to the relationship between Carl and Fred over the Aoki/Powers account were even more damaging.

Soon, another confrontation ensued between the Kiekhaefers with Richie Powers again in the middle. Richie had convinced

Carl that a certain type of exhaust header system would increase the horsepower of Aeromarine engines enough to provide an edge against certain competition. It made perfectly good engineering sense, but Fred wasn't sure it would be profitable. Privately, Fred did some calculations, knowing that Carl would soon press for the project. He discovered that by selling only thirteen sets of the $4,000 systems they could actually start making money. Fred's plan was to ask Richie's racing partner to fund the prototypes, but then share in profits during production. Fred was quite proud of his educated approach, and when Carl summoned him for an impromptu meeting with Richie in his office, he was ready to surprise his father with support for the project.

> "So, I came in and had all this back-up, which, whenever my dad saw me with back-up, and knew I had done my homework, that set him off. Before I could even open my mouth he said, 'Richie, we can't do it. We're not the same Aeromarine we used to be. Now we're *profitable!*' you know, a big slur word. I thought to myself -- in fact I probably said it out loud, 'Well, if that's not the objective, what the hell are we doing here?' Which, you know, set him off.
>
> "And by that time, he had hit my hot button, and I was probably in a higher pitched louder voice ... [and] by that time his ears had *slammed* shut and one of the problems was in that stage of my life I was not good at recognizing when people were hearing what I was saying. I was not particularly good at choosing my timing."[48]

The argument continued to deteriorate, and as Carl's passion rose, Fred's would rise to meet it. "He would stand up behind his desk," Fred said, "and he'd look down his nose at you, and he'd pound the desk in front of you, and shout at you and do all kinds of things." Though the day was carried by Carl, Fred was slowly learning something that would take him nearly thirty-years to realize about his father's temper. "What wasn't obvious was that he was *playing a game,*" Fred said. "He could turn it off and on at will. I didn't realize this was a *game.* I thought we were doing things in real life and I wasn't mature enough to know that, hey, this is *bull shit.* ... I wish, I wish I could -- it's funny, I almost made the mistake of saying 'I wish I could go back to those years and relive them with what I know now.' I'm not sure I'd choose to do that. I would be better equipped to understand, but I think what I would ultimately do is prolong the agony. I'm not sure I would change the outcome."[49]

When Fred finally confronted Carl with his broken pledge about turning over the company to him, Carl responded by saying, "Well, If you want this business, why don't you *buy* it from me?" Fred said quickly, "Alright, I *will!*" Fred was brimming with the exuberant business confidence generated by nearing completion of his M.B.A. at Northwestern University, "and like all good M.B.A. types, I figured, sure, I can do *anything*." Fred acted fast to start lining up potential sources of financing. Carl wasn't certain if Fred was serious, but, "in an attempt to call my bluff, he agreed to negotiate." Fred was preparing to make an offer based solely on the physical assets of the company, ignoring Carl's overtures to include the many engine and stern drive designs for which he had no immediate ability to manufacture.

"I felt that under no circumstances would I be able to get away for less than a million, a million two," Fred recalled, "partly because of the scrutiny the IRS would put on a sale of a business to a family member." The negotiations were frustrating for Fred, because he was certain that a complete outsider might be able to buy the company for even less, as "It had a very, very long track record of red ink, and a negative net worth of in excess of $5 million." Fred, in his usual methodical way, had meticulously prepared a proposal for his father's consideration which met all of the points they had discussed. He called for a formal session, in which Fred's attorneys and Carl's attorneys were prepared to evaluate the proposal and perhaps close the deal. And then, Carl tossed an emotional grenade into the meeting.

> "As I was pulling my proposal out of my brief case to share with everyone present, he produced a letter out of his brief case to share with everyone present. And that letter expounded upon how I was unfit to buy the business, how I was trying to blackmail him out of his business, and how I had no respect for him and how a small business like ours could never, ever pay the salary that a person with my education would expect and would command.
>
> "I read this thing and I was shocked. I was insulted. I was deeply hurt. I put my proposal back into my brief case ... I turned to my attorneys and I said, 'Thank you gentlemen,

> please pull your billing together and send me a final bill. I see no basis whatsoever for further discussions here.'...
>
> "I went into the office that afternoon and gave my resignation and donated the keys to my corporate car and I rode away on a motorcycle. ... That was probably the single most devastating moment in my life. It hurt. It hurts now just to talk about it."[50]

Carl conveniently blamed Fred's departure on the Richie Powers incident, and wrote an unflattering letter to Tom Gentry, another of Richie's sponsors, in an attempt to explain the problem. "It was [Fred] who destroyed our long term friendly relations with Richie over pricing of racing engines and drives. We are certainly very sorry to have lost this business to Mercury Marine. Sorry about my son, Fred."[51]

Fred was out of work, had just purchased a house less than sixty days before the explosive, final meeting, and had no way to even make the mortgage payments. But in the midst of this stunning defeat, Fred had a pleasant surprise later that night.

> "I came home that evening and every employee in the business was at my house. Ready to leave with me. To see if we could start up something on our own. I remember that meeting, because that was a pretty touching event. I mean, there you had fifty some employees and they were all there saying, 'We want to do this with you. We understand you want to leave and we think we understand why. We know it's been really tough, you've been going through a lot of hell and we want to do something.' I mean these guys were ready to chip in everything they had to go out and buy a building someplace and try to do the same thing we've been doing, and I said, 'We don't have the name. We don't have any equipment. We don't have any tooling, and while I appreciate this gesture, in about a month we'd be out of money with all of your life savings gone, and I can't in good conscience ask you to do that.'"[52]

Fred would never return to Kiekhaefer Aeromarine Motors for more than a brief visit in his father's lifetime. He was quickly accepted for a management position at Fisher Imaging, a diagnostic x-ray and ultrasound equipment manufacturer, where he had worked as a business intern while still at Northwestern University. "I learned what it was like to work in an organization where your father wasn't the boss," Fred said of his new job. While Fred worked slowly toward an uncertain future, his sister,

Anita, expressed her frustration over Carl's harsh treatment of Fred, and of her father's overall lack of sensitivity. Afraid to confront him in person, she handed him a three-page handwritten letter as he left for Wisconsin after a trip to Lake X on October 29, 1978.

> "I want you to read this on your way back because I know you would not or could not sit down together to talk on your last trip. But I think the time has come where I will have to convey one way or another some thoughts that have been disturbing me & mom.
>
> "First, I am sorry what has happened to Fred. I know I cannot speak for himself but his reactions to frustrations you had caused him are only similar to mine. ... Fred is frustrated. It seems like with me, whenever we become enthusiastic about our life plan, you become to see our plans as 'undesirable or unfitting' because they are different methods from yours. Fred loves the business, but your insistence on ocean racing, etc. etc. takes a decided turn, a confusing turn in your life plan as well as ours. So many times we don't see eye to what you may want to do because we feel so 'left' out, unconsidered, not a part of your plans, ideas. Secondly, when we need your help, it seems like we don't find it -- and if we do get moving in projects, of our own, we don't get encouragement, only frustration and discouragement. I feel, like Fred, that we need to know the business of getting things going -- the space for working out our problems -- not conflict, interference & opposition. ... I know you are in the business of 'making a living' -- but we also have to be in the business of making a living. But, so many times, when we do seem to get our feet on the ground, the encouragement, energy & motivation to go on becomes discharged because of your opposition. ...
>
> "Fred likes the business but is not interested in ocean racing. O.K., maybe Fred could still run the business but with another emphasis -- allowing him to use another approach to business management so he can learn from his own mistakes and not from yours. You cannot teach him this. But he must learn on his own with all of your desired, sincere hope that he will succeed. He needs your love -- not your resistance. ... After all, what's in money or business if there is NO FAMILY.
>
> "Think it over, Dad -- You have done what you could -- even if there is selfishness ... but we all love you --
>
> Helen, Fred, Mom & Anita

"P.S. I think Fred needs to hear from you that he is needed. Perhaps it would be nice to call him."[53]

The call would never come. Fred was recruited by Price Waterhouse, the nationally acclaimed business consulting firm, who recognized a great potential in his unique background and education. Even though his starting salary wasn't up to his expectations, he realized that the new challenge, located in Boston, would be "in the very worst case, a funded graduate school where I could pull away a lot of the things that I felt I still needed to know." He was a hard worker, and yet in some ways still had to prove himself because of his famous father.

> "Everybody supposes we're a rich family because of Mercury. They don't understand the economic realities of the post-Mercury days. I'm glad I had an independent mind and a desire to prove my individuality, and to live financially independent of my father, because today, I need it. But everybody, you know, they find out I'm Fred Kiekhaefer, especially if they don't know me, and they say, 'Well, why are you working?' When I was at Price Waterhouse, 'Why do you work this hard?' Because, I'll have to or I'll get fired. And then who pays my mortgage?"[54]

When Fred drove away from the company in the summer of 1978, it marked the beginning of a long and slow decline in the health and fortunes of his father. For Carl, not having a son to rely on in times of business stress, or to share the dreams and vision of better days ahead for the business, meant a feeling of working towards no special purpose. The objective all along, or so he would say to just about everyone except Fred, was to create a strong and durable legacy for Fred, to carry on the industrial tradition of Kiekhaefer innovation and excellence for another generation. Years later, in a transcribed telephone conversation with Richie Powers, Carl admitted his agony.

> "Yes, I had a battle. But I guess it was frustrations more than anything else, my son leaving me and I could have saved about $9 million if I hadn't planned this damn business.
>
> "And I was thinking Freddie would follow in my footsteps. What an ungrateful lad. Well, he's just immature and he'll probably grow up some day and everything will be all right."[55]

But without this last, prime purpose to carry him, Carl began to wander without design in the affairs of his business, and started a slow process of liquidating his many properties.

Carl's greatest, and final challenge was still yet to come.

Carl was justifiably proud of the Kiekhaefer Aeromarine 468 cubic inch, 600-horsepower fuel-injected V-8 he developed. In its first appearance, on August 31, 1972, Dr. Robert Magoon established a new U.S. offshore speed record of 82.449 mph aboard *Aeromarine III*.

Above: It is noteworthy that almost every member of the family is looking in a different direction in 1956. Left to Right: Helen, Anita, Freda, Carl, Fred, Carl's mother Clara, Freda's mother Mrs. Greenfield.

Below: Carl and son Fred share a happy moment together in 1950.

Top left: Daughter Helen was elected "Miss University of Florida" at the Jacksonville campus in 1953, and Carl dedicated one of his Mexican Road Race cars in her honor.

Top Right: Fred and his sisters each learned to water ski expertly, pulled by a steady stream of Mercury outboards through the years.

Below: Carl indoctrinated Fred into the mysteries of internal combustion at such an early age, that some said he was die-cast, not born.

Carl and Fred Kiekhaefer pose with the new *K-Drive 400*, (left) which featured a unique tilting strategy, integral power steering and "bullet-proof" durability. Though the innovative design was introduced at the 1974 Miami International Boat Show (below), it was never put into production.

Right: Carl grimaces at some well-meaning joke while receiving the 1976 Ole Evinrude Award from OMC Chairman Ralph Evinrude and President Charlie Strang.

Below: Carl visits with Mercury President Charles Alexander and OMC President Charlie Strang at the Mercury Marine booth during the Miami International Boat Show in 1977.

Carl receives the congratulations of his son, Fred, on receiving an honorary doctorate of laws from Marian College in Fond du Lac, May 15, 1982. "I wish it could have been a real one," Carl said.

CARL KIEKHAEFER

1906 - 1983

Chairman Jack F. Reichert of Brunswick Corporation, engineered the emotional return of Carl Kiekhaefer to his factory in Fond du Lac. Reichert, who was president of Mercury Marine from 1972-1977, effectively dissolved years of Carl's exile through diplomacy and genuine affection for the founder.

Fred Kiekhaefer in 1991, president of Brunswick Marine Hi-performance. When Kiekhaefer Aeromarine was sold to Brunswick Corporation in July 1990, the Kiekhaefer legacy returned full-circle to the Mercury family.

CHAPTER THIRTY-EIGHT

Twilight

"Wherever I am going in the future, heaven or hell, I do hope they have an engineering department because I might have some fun being reduced back to a plain, old engineer. The simplest and sweetest things in life are complex engineering; it's the people problem that's responsible for today's ills. I hope some day I can still get a job as a draftsman or as an engineer with a boss who can hand me the work on a platter -- with no other responsibilities except that it works."[1]

Carl Kiekhaefer

The first signs of trouble came in early 1980. Ever since his bypass operation in 1977, Carl Kiekhaefer had regular checkups at the Mayo Clinic, and each time the doctors had given him cause for optimism and peace of mind. In January, tests at the Clinic indicated an obstruction near his aorta, the main artery of the body which carries blood to all organs except the lungs. The surgeons wanted to do exploratory surgery to determine the extent of the blockage, but Carl refused. "I am reluctant to undergo any further anesthesia unless absolutely necessary. Some of my loss of memory has been blamed as a side effect of total anesthesia."[2]

He looked for almost any other possible cause for his feelings of chest congestion. An avid reader, he began to attribute to himself virtually every newly discovered illness, such as the possibility that he developed "ozone sickness" from flying so often, basing his diagnosis on an article he read in the *Wall Street Journal*. "Because of my business flight frequency in the last two years, I am wondering if I did not have a touch of this ozone sickness and if that didn't cause my peculiar sensations in my chest which I noticed during my walking exercises."[3]

Later that month, Carl proudly accepted the 1979 Charles F. Chapman Memorial Award for "outstanding contributions to boating." The award, named for the much admired and longtime editor of *Motorboating* magazine, was co-sponsored by the National Marine Manufacturers Association and *Motorboating and Sailing* magazine. Of the award, Carl remarked, "For me, the boating industry and my work over 40 years in designing and producing better engines and drives has been great fun and a constant challenge."[4]

The following month, Carl's cardiologist ordered a CAT scan at Milwaukee General Hospital to search for this same "chest obstruction," and Carl's level of anxiety began to rise quickly as a "growth" was indicated on the film. Carl then went to Mt. Sinai Medical Center across town for bone marrow tests, drawn from his pelvis, and a lymph gland biopsy was performed from underneath his right arm. The shocking news hit Carl like a sledgehammer: lymphoma, a cancerous condition of the lymphatic system and the real possibility of spreading illness.[5] Succeeding tests revealed that the disease was somewhat stabilized, although one specialist told Rose Smiljanic that once it begins to progress the second time, it would "gallop" through his system.

One of Carl's doctors told him that with luck, he would have five years to live. Even coming at 73 years of age, it was a shock. Carl had literally never had time for church or religion. The few times that he had attended during the years he was struggling to get Mercury off the ground in Cedarburg, the preacher seemed to pick on him. "I don't know if this was just what he imagined, or just his guilt," Rose Smiljanic said, "that the minister would direct sermons toward him. 'People who work during the holy days, just to pile up piles of money,' and things like that." The pronouncement by his doctor of an actual time limit to life, though, spurred him into a religious investigation, "just in case." He located a small church in Fond du Lac that gave sermons in German, and he compared the practices and beliefs of various churches much like he was buying aluminum castings. "And then, all of a sudden," Rose recalled, "he said he was going to join this church. He said, I feel I really need it. ... He was kind of preparing for his death, and he knew that to get into [heaven] that he better, not repent, but start making amends so he could get up there instead of down below."[6]

His weight, normally a robust 210 pounds, had slipped to 185 following his open heart surgery, now continued to fall to a new low of 170 pounds. In December 1980, Carl was at the Marshfield Clinic in Marshfield, Wisconsin for a long and rigorous day of tests. When the results were reviewed, lymphoma was again discovered, and an ominous spot on the lung was detected for the first time. Carl's anxiety began to peak, and he started a long slide into a deep depression. "He still traveled," Rose Smiljanic remembered. "He still went to races. And he'd go off by himself, too, like he went to California to attend a race. I said, 'You can't go alone,' and he'd say, 'Oh, don't worry about me.'" But his driving was deteriorating, and he was picked up for speeding between Los Angeles and San Diego. "And then ... he'd be driving and I'd say, 'Hey! You're driving on the line,' and he'd say, 'No I'm not!' and he'd stop the car and he'd say, 'If you're so smart, you go ahead and drive.'"[7]

The following month, January 1981, Carl underwent prostate surgery to alleviate his many bouts with bladder infections and the urinary discomfort he had suffered for years. During the operation, the surgeons discovered a tumor growth obstructing his kidney function, blocking his ureter and enlarging his left kidney. He began an ambitious program of chemotherapy, which seemed to weaken him even more. Following the surgery, he attended a function with Ralph Evinrude, who, according to Carl, was experiencing similar urinary discomfort. "An interesting point of my recent meeting with Ralph Evinrude was that he called me aside for a brief discussion of how things went with my 'root canal'. He was apparently concerned about a similar problem that he was having."[8] Carl's surgery boosted his attitude considerably, as he reported "I am now sleeping again and can get my rest at night, which has greatly improved my physical well being."[9]

Carl developed pneumonia in May 1981, raising fears that the lymphoma had traveled to his lungs. His condition was recorded as "Very weak, disoriented, ill, despondent."[10] Again, Carl tried to find some mysterious source for his illness. "I am still not sure whether my pneumonia was affected by crop spraying in the middle of the farming country, whether I might have gotten a touch of agent orange or whether I had Legionnaire's disease."[11] Fueling his suspicions, one of his beautiful German Shepherds, "who had never been sick a day in her life, came down with the same shortness of breath during this time and died." By June, Carl's own shortness of breath had returned for the first time since his heart surgery four years previously. In July, Carl's struggle for breath increased, and oxygen was again installed in his home, which he used frequently, often most of the day.

The condition made his hosting of a Kiekhaefer family reunion at his home most difficult. In his first display of any interest in the Kiekhaefer lineage outside of his own immediate family in over fifty years, Carl brought together at his home over 200 Kiekhaefers from throughout America on July 12, 1981. "Some two hundred guests are expected which means accompanying obligations of tents, tables, chairs, music and what have you. The family tree has now been dated back to sometime in 1600, but I am sure there won't be any guests from those generations."[12]

Three months later, a double heart catheterization indicated that Carl's aortic "valve [was] not any worse than in 1977 when ... surgery was completed, and not bad enough to have valve operation."[13] His weight had fallen to 150 pounds. "A total of 105 pills per week were being taken which made me feel somewhat like a small traveling concrete mixer. The effects of these pills caused impaired vision, decreased appetite and general loss of energy and ambition. ..."[14]

On Friday, September 4, 1981, Carl was startled to read the headlines of the *New York Times*: "20 Tons of Marijuana and 33 Seized in Long Island Raids."[15] Listed among the Colombians, New Jersey and Long Island residents captured in the raid on September 3, was one of Kiekhaefer Aeromarine's best clients, offshore racer Joseph Ippolito, Jr., and Carl's close friend, Richie Powers. According to the *Times*, the capture represented a three-year investigation by Federal drug-enforcement agents in coordination with local police. In a three-pronged attack, the agents first stormed an off-loading site on Gardiners Bay on Eastern Long Island, arresting 21 individuals and confiscating three tons of marijuana. Two hours later, another task force led by a 41-foot Coast Guard cruiser and a 95-foot cutter, captured a Columbian fishing trawler that had delivered the contraband to the coastline, along with nine men and an additional 17 tons of marijuana. And lastly, a 40-foot sport fisherman, *Unapplied Time*, laden with sophisticated electronics and night vision equipment, was boarded by Federal officers and Coast Guard troops bristling with shotguns and M-16 rifles, yielding three additional men, including Richie Powers.

Richie Powers had been in a slump. Major contracts for rigging and racing with such offshore luminaries as Dr. Carlo Bonomi, Tom Gentry and Rocky Aoki were behind him, and he had opened an independent rigging and engine shop in Miami.

"I just started my own business," Richie explained, "racing here and there, and basically trying to find some trouble to get into -- all the excitement of the drug situation -- people were coming to me to have their boats fixed -- 'Put new engines in my boat tonight and I'll pay you 20,000 bucks,' that type of thing."[16] The offers were plentiful and always lucrative. "'Drive my boat over to Bimini ... leave it and pick it up the next day and here's $25,000,' Richie continued, "and that type of thing."

Richie had been working with Joey Ippolito aboard the *Michelob Light* offshore racer, throttling Aeromarine engines, and sponsored by the venerable backer of champion unlimited hydroplane boats, Bernie Little. Ippolito had shown promise early in his offshore racing career, and was even named Offshore Powerboat Rookie of the Year in 1977. "Ippolito told me when I went to work with him with Anheuser-Busch that he was a record executive," Richie said, "that he owned K-Tel Records. I really didn't know the difference. God's honest truth. Then eventually, as I got to know him and what was going on, then I knew that he was in the wrong business."

Ippolito asked Richie to go to New York. "I need your help," Ippolito told him. "You got to go with me to Long Island. And, all it is, is you go out and you take two guys off a boat and bring them back in a Cigarette." Don Aronow, the flamboyant boat designer and founder of Cigarette once said, "When you hear of boats caught in smuggling -- which our boats are very famous for -- it's often a Cigarette. We don't mind the publicity, but we do have torn emotions when one of our boats gets caught in a high-speed chase. I always hope that ours was caught running out of gas or something."[17] Richie knew right away there was more involved than Ippolito was telling him. "I knew the thing was loaded with shit. I told him three times to forget it." But eventually, Ippolito was able to convince him that he wouldn't have to be around the incriminating product.

On the fateful morning in New York, the 41-foot Cigarette wasn't running. Richie had to use the $300,000 sport fisherman, *Unapplied Time*, instead. The seas were tall and rough, and Richie couldn't get the two Americans off the Columbian trawler. He led them to calmer waters close to Long Island, retrieved the men, and headed for shore. "My job was done," Richie said, "We were sitting there having a beer." But then he got an emergency call from the crew remaining on the trawler. They had run aground. Richie ran through the heavy seas to return to the stricken boat, which was wallowing in the shallows, overburdened with 34,000 pounds of marijuana below decks, stacked six bales high. Confusion reigned aboard the listing trawler, and as soon as Richie maneuvered alongside, anxious hands tied the two boats together. Richie screamed at the panicked smugglers, "Don't tie the boat off! Don't tie the boat off!" When they continued, he tried to break free of the other boat. "I just took both throttles and rammed it and tried to snap the ropes to get away from the thing. I didn't want to be around it. And then they started loading the shit on the sport fishing boat to pull the weight off the [trawler] to get it off the sandbar. I said, 'Get that shit off the boat! I don't want that shit on the boat! Get it off!' I was yelling and screaming, I was going crazy."

In court, a member of the crew testified that Richie had indeed yelled for them to "Get it off," and "I don't want that shit on the boat." His attorney was able to successfully argue, "Why would Powers say to "get it off" if he wanted to make some money off it. He would want it on." Though others were convicted with stiff sentences -- Joey Ippolito received 8 years[18] -- Richie was sentenced to only one year, and served only three months in a county prison in New York. During his ninety days behind bars, he wrote a long letter of appeal to his friend and mentor, Carl Kiekhaefer.

> "As you know quite well, my situation at this time is not a very good one. ... There are many reasons for being involved in what I have done in the past and which I will never do again, such as, naturally, the monetary gain and another is influence by other people -- when one just can't or won't say no to someone. The biggest reason for my actions ... is my sense of adventure.
>
> "Like all of the aspects of my life with which you have been so helpful and influential such as, lead test driver, factory team racer, marine engineer, race crew chief, racing team manager and organizer ... winner of 40 offshore races and 4-time world champion -- and finally smuggler, I have had to conquer and surpass all of these parts of my life. Now they are behind me and I am grateful and am looking forward to a new and more stable challenge.
>
> "As I hope you know ... I have the highest regard and respect for you, and, also a great sense of a 'father image.' Many of your own achievements have affected and molded my own way of thinking. Your constant coaching and support are

directly responsible for a lot of the good things that have always followed me through life. With this in mind I would like to propose the following. ...

"I have always wanted to perpetuate and play a major role in continuing the Kiekhaefer name and image. ... Carl, at this time I would like to submit my application ... for the position of President of Kiekhaefer Aeromarine Motors. ..."[19]

Though Carl had to ultimately inform Richie that his declining health, and the weakened financial condition of the company precluded accepting his proposal, he wrote letters on Richie's behalf to assist in an early release from prison. "To all who know him, he is a fine outstanding character and of good deportment. ... He is a gentleman and was ready to help someone else who needed it at the time."[20]

In the midst of his letter writing campaign to help Richie, Carl would need help of his own. On April 8, 1982 Carl awoke feeling quite strange. He had no sensation in his left arm, the left side of his chest and abdomen, and left leg. Though he could still move his limbs, it was clear that partial paralysis had occurred during the night, and soon his physicians confirmed that he had suffered a mild stroke. He said his foot felt "like a clump of firewood."[21] Later in the month, Carl was still having difficulty walking, even for short distances. In the report of his condition that Rose Smiljanic would help him write, Carl said, "During month, noticed staggering upon rising from sitting position and a listing or dropping of the right shoulder when walking."[22] Tests performed the following month showed that the blood vessels in his neck were up to 50 percent occluded with plaque. "It is really tough to grow old," Carl wrote in May 1982, "and to face the prospect of restrictions, although for the last five years I have not smoked or partaken of alcohol, and I am behaving myself."[23]

Of his limitations, most distressing for Carl was his lack of ability to conduct business properly. "I [have] to carry on the dual role of patient and Chief Executive Officer; and it is the latter that I feel I have neglected. ... I am still putting in a partial day's work at the office because there are too many things that need my attention."[24] In admonishing his physicians at the Mayo Clinic, Carl wrote, "I feel that I am too young at 76-1/2 to 'lay down my tools' and do nothing."[25]

Those in attendance were touched to see Carl struggle to walk straight to the podium as he was awarded an honorary doctorate degree at Marian College in Fond du Lac only weeks later on May 15, 1982. The honorary degree of doctor of laws was given, according to the College, "to individuals in government, public service, business and industry who have distinguished themselves in general service to the state, and to mankind."[26] Fred returned to honor his father on the special occasion, and a photographer captured the essence of their meeting as Carl smiled proudly at his son. Remembering his own halting education, Carl said of the honor, "I feel a little guilty about being awarded such a degree, and I wish it could have been a real one instead of an honorary one."[27] Carl was also proud to have learned earlier in the year, that Charlie Strang had been elected chairman of the board and chief executive officer of Outboard Marine Corporation on January 22, 1982. Ralph Evinrude, 74, had retired after 19 years as chairman.

Carl once again delved into the dangerous science of self-diagnosis, when in July 1982 he read an article in the *Milwaukee Journal* about a mysterious disease. "Doctors to stalk an elusive, microscopic killer," the headline proclaimed.[28] The nation's preliminary cases of AIDS were just beginning to appear, and even though the initial reports showed the disease to be somewhat restricted to homosexuals and intravenous drug users, Carl imagined that the general symptoms were close enough to his own to warrant suspicion nonetheless. Later, Carl marked an article "quite interesting" with the headline, "Mystery Chemical Found in Schizophrenic's Blood." Carl may have wondered, like many who had worked with him over the years, if there was some special reason why his moods could swing so quickly, and his temper could replace an otherwise affable disposition with such startling short notice. Carl marked the paragraph that described the chemical imbalance responsible for "the bizarre psychological changes in perception, thought, mood, and personality so common to the disease."[29]

Carl was honored on September 11, 1982, as the APBA Offshore Division Race in St. Augustine Florida was renamed the *Kiekhaefer - St. Augustine Classic*, and a well attended testimonial dinner was held in his honor. When he first heard of the affair, Carl protested by saying, "Why me? Nobody in offshore racing remembers me anymore."[30] Charlie Strang presided as the affable master of ceremonies, and tributes were paid to Carl for his remarkable achievements and contributions to the marine industry. "Without question," Strang addressed the banquet guests, "Carl Kiekhaefer provided the impetus that transformed offshore racing from a Miami-based hobby to an internationally recognized sport. ... It's time he received full credit. ..."[31] Though weakened by his loss of weight and intensive medication, Carl's spirits were lifted high as he basked in the spotlight of appreciation and applause throughout the memorable evening.

Less than two months later, on November 1, 1982, Carl was driven to Mt. Sinai Medical Center in Milwaukee, and admitted into the intensive care unit. "Disinterested in everything," his report read. "Depressed. Tired. Extreme shortness of breath. Slept constantly. Talked in very weak, almost whisper voice."[32] He remained hospitalized for two weeks before returning to Fond du Lac. Over the next three months, his weight continued to drop, stabilizing at 139 pounds in February 1983. His ankles and feet began to swell in dramatic fashion, making it difficult for Carl to sleep or walk comfortably.

In March 1983, tests conducted at the Fond du Lac Clinic confirmed that "my lymphoma needed immediate treatment,"[33] but Carl insisted on one more trip to the Mayo Clinic before proceeding with the difficult treatments. He spent an entire week at the Clinic, the results of which deflated nearly any hope Carl may have had about a recovery. "It was Mayo's opinion that my heart was the main source of my problem and that the aortic valve must be replaced; my diabetes was a No. 2 problem and that the lymphoma No. 3. Naturally, this was quite a shock, and I must make an intelligent decision as to whether or not physically I can withstand such an operation."[34] To make his decision even more difficult, his Mayo Clinic physicians told him that his previous bypasses could very well complicate the already risky operation.

Carl's entire family, Freda, Helen, Anita and Fred, were reunited over the Christmas holidays in 1982, and remained with him in Fond du Lac to bring in the new year of 1983. Both before and after their visit, though, it was Rose Smiljanic that kept the family informed of Carl's condition. Throughout his illness, Rose was the rock that Carl leaned on to administer the hundreds of details concerning his medical appointments, tests, prescriptions, and the reams of insurance forms required for his treatments. As the director of corporate administration for KAM, she also coordinated information on Carl's progress, or lack of progress to the Kiekhaefer family, and was often the bearer of bad news. In March 1983, she wrote to both Helen and Fred with the distressing news that Carl's condition was even worse than previously imagined.

> "Your father completed his tests at St. Agnes last week and one of the tests given was an abdominal cat scan. Diffused tumors were noted with the largest across the front. The scan was compared with the last one he had at Mt. Sinai [six months ago] and he learned yesterday that the tumors have indeed grown quite extensively in that period of time. In fact, I'm wondering now if that wasn't a great deal of his problem the last several months rather than the heart; but what good does it do to speculate at this time. ...
>
> "Just the last two days, he has been experiencing a fullness in the stomach wherein yesterday he could not close those brown trousers that he always wears, and yesterday, too, we noticed his ankles had begun to swell again.
>
> "In his favor, the doctor says this is a slow-growing tumor or tumors and they hope that they can contain it for a while."[35]

Over the years, Rose had threatened to leave Carl on a number of occasions, but as his illness progressed, she put aside any feelings of independence she may have retained. Typical of the occasional spats between Carl and Rose, was the exchange of notes that took place in late August 1974. First, Rose announced her resignation:

> "Since it is almost impossible to talk with you, I had intended to tell you in person that I definitely plan to leave the company on September 15, 1974, and I will move out of my house by September 30.
>
> "Twenty-four years is enough years to spend on one cause and I feel a change for me would be best at this time."[36]

Carl scribbled a return message over her neatly typed resignation:

> "Combining, Corn Harvesting, Lake X Sale & Trip to Fla, Daytona, etc. Trip to Europe, and now I'm ready to talk to you! Will meet you ... at Daytona for labor day."[37]

Tiring of the countless tests and diagnostic procedures he had endured for nearly six years, Carl said, "I am sure we have long exceeded the maximum permissive x-rays a human body can tolerate. Not counting all the sonic scans, 'echos', etc. Now I would like to put a limit on them, whatever the price."[38] With each hospital admission since May 1981, to all of the hospital staff he encountered, Carl complained of being cold. "He was always freezing, and the nurse would have to pile blanket upon blanket over him and he would also resort to wearing long underwear as well as pajamas."

By May of 1983, Carl was virtually prohibited from travel. When he was requested to appear in New Orleans to testify as a witness in a trial, his physician at the Mayo Clinic protested, writing "... I do not feel that medically I can allow him to travel to New Orleans at this time or at any time in the foreseeable future. If, in fact, a deposition needs to be obtained from him in Fond du Lac, Wisconsin, the interview should only be conducted for half an hour periods of time because of the stress on his general health."[39]

As Carl's condition worsened, he began to prepare for the inevitable. During one of his visits, Fred, along with fiance Dr. Carol Stafford, conferred with Carl about the future of Kiekhaefer Aeromarine Motors. As they were leaving for the airport, Rose asked Fred what Carl had said. "Fred said, 'Oh, Dad said he's going to sell or give me the company.' I told Fred, 'Before you quit [your job at Price Waterhouse] ... you make certain that you have it in writing.'"[40] When Rose asked Carl about the conversation, he said, 'Oh, that boy's got it all mixed up.'" Soon, though, Carl issued new orders to his trusted friend and counselor, attorney Wayne Roper.

> "These instructions are being given [in case] ... I become disabled or unable to continue. ... If something happens to me, it probably would be best if Wayne Roper were President and Tom Asher and Rose Smiljanic were both Vice Presidents. This would insulate them from family pressures, particularly from my son, Fred. It is possible that Rose and Tom could be considered as purchasers and KAM would have to loan them most of the purchase price. My son, Fred, would also like to own the business but is not aware that the Company could not support a substantial salary for him. ... Fred, of course, has the most business experience and his advice should be sought."[41]

Tom Asher, previously the controller of KAM, became executive vice president following Fred's resignation. The signature on the letter, executed in a jerky and crude manner, is in stark contrast to Carl's normally smooth, rounded and flourishing style, an indication that his pain and illness made even signing papers difficult.

In July 1983, Carl was notified by Jack Reichert that the Brunswick Foundation was establishing a $50,000 scholarship in Carl's name, in recognition of his outstanding contributions to both the industry and the community. An honor normally reserved for retiring chief executive officers of the corporation, Reichert felt that Carl's special place in Brunswick history warranted the special consideration. Also lending his support to the Kiekhaefer scholarship was the new president of Mercury Marine, Richard J. "Dick" Jordan, a career Brunswick executive who replaced Charles Alexander after five years on the job. Honored as Carl was about the scholarship, which he designated for Marian College in Fond du Lac, he refused to allow Jordan to designate a Mercury Marine employee award *The E.C. Kiekhaefer Award for Excellence.* "As much as I appreciated your considering honoring me in this way," Carl wrote to Jordan, "it is my feeling that I would not have any control on how my name would be used as well as not knowing in advance of other details regarding the issuance of this award. It would be almost like putting my name on something 'carte blanche.'"[42] In a previous, unsent draft of his letter to Jordan, he approved the use of the name for the employee award, but his real concern became evident. "I would not care to have the award placed directly on any product Brunswick makes or may make in the future since that would indicate I entirely approve of the product without even knowing what the product is, its internal construction or its performance. Being an engineer, it would not be ethical."[43]

Fumbling around in his garage, Carl had dropped a log on his left big toe. It was extremely painful, and he had to have the nail removed to reduce the pressure caused from swelling. His circulation had deteriorated in his extremities, perhaps as a latent cause of his diabetic condition. The toe refused to heal, and began to turn black. His physicians had no choice but to amputate. Rose remembers Carl's great fear of going back into surgery.

> "When he heard he had to have that toe cut off, I had to actually hold his hand. They wouldn't let me into the surgery room, but this doctor said, 'Come on Rose, he said he wants

you with him. You can be with him until they take him into the antiseptic room.'

"And there I was. Standing there. It's like six in the morning, and I'm in Mt. Sinai Hospital, you know, comforting him. And he said, 'I'm really afraid this time.' Here, I thought of this big, giant of a man reduced to this where he is, you know, afraid because they were going to cut his toe off, and I just thought that was horrible."[44]

The surgeons removed his toe, leaving a small stump. Again, it refused to heal. Less than two weeks later, on September 24, 1983, Carl returned to the hospital, and his surgeons were faced with a frightful dilemma. "They took the bandage off," Rose said, "and it was all black ... black and green. Gangrene had set in."[45] Discussions ranged from removing his entire foot, his entire leg, to implanting new blood vessels to increase the circulation to slow the progress of the horrible and rapidly worsening condition. To further complicate matters, his system had become so weakened that surgeons were uncertain if he could survive any operation at all. The decision would have to be postponed, and Carl was made as comfortable as possible in his hospital room.

Carl's oldest daughter, Helen, visited with him in the hospital as he struggled for life, and pleaded for death. In chilling testimony to the extent of the advanced cancer, his once massive form had dwindled to 110 pounds, and his wide-shouldered frame, once 5'8", had withered to 5'4". He told a nurse's aid named Marge who attended him, that "I can't die as long as my family is here."[46] To Helen he admitted his physical defeat. "He kept saying to me, 'Can't a guy call it quits?' He was asking me to help. He wanted to kill himself ... he just wanted to get on with the dying."[47] Later, the evening of October 4, 1983, he summoned his nurses, saying, "Get me up. I need to wash myself so I'm clean for the Lord."

The following morning, moments before he died at 6:27 a.m., he gently squeezed the hand of the nurse who was at his side, and the old engineer issued his last order: "Tell Charlie to forgive me." His final words acknowledged the suffering he had caused Charlie Strang, not only by selling Mercury to Brunswick in the first place, but by turning his back on his great and loyal friend when he wrongly accused him of betrayal in league with Brunswick and Jack Hanigan. Years earlier he had spelled it out in a letter to Strang. "No one need ever remind me of the great mistake I made in my lifetime. Thank Goodness you were young enough to recover, but it's too late for me to start over."[48]

A wake was held for Carl at the Zacheral Funeral Home in Fond du Lac, where throngs of mourners filed by his coffin for six hours in seemingly endless streams, dripping from the pouring rain and moistened eyes. "Scads of people," remembered Carl's dictation secretary Ruth Backus, "a lot of people from Mercury, plus the plant up on the ledge, lots of business acquaintances, and a lot of people came from out of town ... you just stood in line. ... When I left, they were still lined up to the door."[49]

The following Saturday morning at 11:00 a.m., Carl's funeral was held at St. Peter's Lutheran Church in Fond du Lac. Hundreds of Carl's friends and associates gathered from all points of the compass to pay their last respects. Jim Wynne, whose friendship with Carl had overcome his stormy resignation during the *Christmas Mutiny* of 1957, was there, awash in emotion.

> "It was a very emotional experience, because I felt that he had more influence on my life than any other man, including my father and my grandfather. Carl did more to influence my life professionally and personally than probably anybody else that I've ever been in contact with. And I felt that very strongly when I went to his funeral.
>
> "It was a great gathering of a lot of people that I hadn't seen in many, many years, like a reunion. It was almost, I shouldn't say, joyous occasion, but in a way it was, because so many of these people had been involved with Carl in his life, all came and all got together to reminisce and get acquainted. ... And I think everyone felt the same way, even the family. 'Hey, now this is over, it's nice to bring all these people together, he had so much influence on all of their lives.' Every one of us. We all felt that way. It was a very, very momentous occasion to me. And I feel that he did more toward starting me into a very successful career than any other factor in my life. ... And I wish ... I wish that somehow I would have been able to let him know that before he died."[50]

Alan Edgarton, Carl's long-time friend and legal counselor, remarked to Rose Smiljanic, "You know, Rose, I should have really told him not to sell [to Brunswick]."[51] Jack Reichert remembers how Fred Kiekhaefer was surprised that the president of Brunswick and Carl had become such good friends, and compared the Kiekhaefer's to the Ford family. "Freddie is kind of the Edsel Ford of the Kiekhaefer family," Reichert explained. "Freddie got beat up by his dad so many times that I think he got the kind of resentment that sons have for domineering fathers. And Fred is a nice guy. A *really* nice guy."[52]

Charlie Strang, along with Charlie Alexander and a group of other prominent friends of Carl, were seated together in a corner of the Church during the memorial service. "They had a young preacher up there, who obviously didn't know Carl," Strang said. "He really didn't hit the real Carl at all, and he was talking about what an incredibly religious man Mr. Kiekhaefer was. His whole sermon was on how religious Carl was. Charlie Alexander is sitting next to us and he says, 'Is he talking about the same guy we know?'" Jim Jost was among the group and remembers how the minister reflected on how Carl "would come back and sit back in one of the pews and just reminisce, and think about what he did,' and there was a whole group of us that had gotten together in the corner of the church, just about where the minister was pointing. It was really hard for the group of us not to laugh."[53]

Following the service, six pallbearers lifted Carl and carried him out into the crisp October sunshine. Freda had selected them all from the rank and file of Kiekhaefer Aeromarine Motors, rather than executives. Fitting that Carl, who had for nearly four decades given of his strength to lift countless thousands of hourly workers to lives of security and advancement, would be lifted by them for his final journey. Rose Smiljanic described the six, solemn men as "down to earth, hard working, honest people, and they were so honored to be there."[54]

Carl was buried at St. Charles Cemetery, within view of the new plant he had built on the ledge above Fond du Lac. After Carl's heart surgery, Freda had considered bringing him back to Florida when the time came, but Rose had convinced her that he would be happier at home. "I don't think he wants to be buried down there," she explained, "You know he's a native of Wisconsin, and he'd be lost down there."[55] Later, Freda would thank Stan Gores, editor of the Fond du Lac *Commonwealth Reporter* for the support of the paper and of the community. "Not having spent a great deal of time in Fond du Lac, I was overwhelmed by the support of the people during these sad days and was not able to thank them personally, most of whom I did not know."[56]

Rose realized that most places in town would be too small for the crowds expected at the luncheon following the funeral, so she made arrangements with Marian College to hold the function there. It was at Marian College that Carl had been awarded his honorary doctorate, and it was to Marian College that he had donated the $50,000 Kiekhaefer scholarship contributed by the Brunswick Foundation. She organized the luncheon just the way Carl would have done it himself. "It was just a *feast*," Rose remembered, "and I thought, 'Boy, I bet he's smiling because he loved to feed people. He never put on a small thing, always a feast for a crowd."[57]

"Wherever I am going in the future, heaven or hell, I do hope they have an engineering department," Carl had said, "because I might have some fun being reduced back to a plain, old engineer." Clearly, Carl was never a plain, old engineer. He was perhaps the last of a vanishing breed of iron-fisted entrepreneurs, an industrial Caesar who conquered a marine industry empire, a sentiment Carl shared himself. "I really believe that the men in my age bracket are the last of the breed in the rapidly changing affairs in politics, in economics, and in the business community."[58] He gave proof to the philosophy that hard work and determination can overcome virtually any obstacle, and that honesty in business can be a practical reality, even within the most extreme of competitive environments. The best ideas do win. The best men can, too.

EPILOGUE

"If you want to hit my hot button today, just accuse me of being just like my dad. There's no faster way to get my hot button. I like to think I've capitalized on the strengths that he had, and I'd like to think I've suppressed what I view as weaknesses and character flaws. But from time to time people tell me I'm an awful lot like him, and on one hand that scares me, and makes me pretty proud on the other."[1]

Frederick Carl Kiekhaefer

In 1964, on the twenty-fifth anniversary of the founding of the Kiekhaefer Corporation in Cedarburg, Wisconsin, Carl reflected on the stormy youth of his company, and gazed twenty-five years into the future. Of the past, he said, "It's not really a very long time, as the affairs of men go. Yet to me, it's been a long time. It's consumed a lot of energy, and it's produced a full share of heartaches as well as rewards."[2] Most interesting is how Carl viewed the future of the outboard motor, and the future of the devices that provide power in general.

> "Of course, we can look forward 25 years. This has one advantage -- No one will remember in 1989 the predictions we make tonight. ...
>
> "I'll tell you one thing ... the outboard motor, which has survived so many 'threats' over the years, will survive 25 more years ... at least in the sense that a portable, removable, compact engine will be far easier to attain then than it is now. And it will have much higher horsepower ratings. Whether such an engine will sit on the transom, on the deck, or under it, it will be smaller and lighter than anything we dream of tonight. ...
>
> "Of course, the gasoline engine as we know it will be obsolete. It won't take anything like 25 years for some development as the fuel cell to take over as the prime source of engine power. Nuclear power will have been harnessed to engines in a practical fashion by then. The net result will be quiet, compact engines whose appetite for fuel will be so small as to eliminate today's problems caused by limited fuel capacity."[3]

In many ways, Carl Kiekhaefer would be proud of the many and sophisticated innovations which have transformed the outboard motor into some of the most reliable, practical and efficient two-cycle engines he could have imagined. Breakthroughs in electronic performance control, fuel and oil injection and metering, metallurgical and casting advances, and the micro-computerization of virtually each step of the product life cycle, from design to manufacturing, would make him feel right at home. As an engineer, he would thrill at the progress in these techniques that have enabled the major manufacturers to shed weight, increase horsepower, prolong life and reduce maintenance far beyond his predictions in 1964. On the other hand, he would have been disappointed in the alarmingly slow progress of his fellow engineers in the fields of nuclear, solar, fusion or other power cell development.

On January 7, 1988, the National Marine Manufacturers Association (NMMA) announced the establishment of the NMMA Hall of Fame. NMMA President Jeff Napier said, "The objective of the NMMA Hall Of Fame is to recognize and honor specific individuals who have or continue to make 'substantial and lasting contributions' toward the advancement of the marine industry."[4] For their inaugural selections, the inductees were Christopher Columbus Smith (founded Chris-Craft), Garfield Arthur Wood (speedboat champion and founder of Gar Wood boats), William Edward Muncey (celebrated Gold Cup racing competitor), Charles Frederic Chapman (Long-time editor of *Motorboating* magazine and APBA official), Ole Evinrude (founder of Evinrude and ELTO), and Elmer Carl Kiekhaefer, founder of Mercury Marine. All were inducted into the marine industry's hall of fame posthumously. Carl Kiekhaefer would have felt extremely honored to have been inducted along with Ole Evinrude and the others.

In 1989, James R. Wynne, along with Ralph Evinrude and others, joined this prestigious group. Wynne was one of the first living recipients of the honor. He was honored for his extraordinary career in the industry, which included the design of both practical and high-performance boats for Carver, Chris-Craft, Cobalt, Cruisers Inc., Donzi, Grady-White, Hatteras, Larson, Stamas and Trojan. The selection committee commented that, "Wynne Marine has been acclaimed as one of the most influential independent design houses in the boating world."[5] On December 21, 1990, Jim Wynne passed away in Miami, Florida due to respiratory failure.

Following his father's death on October 5, 1983, Fred Kiekhaefer met with Carl's attorneys and estate trustees, Rose Smiljanic and KAM executive vice president Tom Asher, to forge an alliance for a smooth continuation of the business. Following spirited debate, Kiekhaefer Aeromarine Motors became Kiekhaefer Aeromarine, Inc. with Fred Kiekhaefer as president. Rose Smiljanic and Tom Asher each retained 25% of the company, Fred acquired 50%, and a few percent were placed in a voting trust to avoid shareholder deadlock on issues vital to the corporation. Though the financial condition of the company was anything but sound, Fred was able to remain philosophical about his father's last years. "It was, after all, his money - and these were his autumn years. He earned the right to do what he damned well pleased with both. But now, I have to help my mom and the company's employees."[6]

Once again, as he had done in 1977, Fred acted quickly to streamline operations at Kiekhaefer Aeromarine, and to focus the company's attention on the long-term promise of the recreational boating industry. Products, proven in seven national and world championship offshore racing seasons, were re-introduced to pleasure boat manufacturers and buyers, and carefully positioned to maintain the sterling reputation for quality and performance that had become the living legacy of the unique company. To assist in setting the new products apart in the market, Fred developed the advertising slogan, "Anything else is just a toy." He discontinued the unprofitable racing engine programs to concentrate on hydraulic trim tabs, control quadrants, propellers and exhaust silencing systems. To reduce the likelihood of periodic industry recession, he also expanded the company's subcontract work in machining and other marine accessories for major industry leaders like OMC and Mercury Marine. Within two years, Fred's effort had quadrupled sales of KAM products.

On August 31, 1985, Fred and Carol Marie Stafford were married in New Fane, Wisconsin. Dr. Carol Kiekhaefer has become a much admired trauma specialist, and for years donated her considerable services to the American Power Boat Association to assist in providing emergency medical assistance at offshore powerboat races. Dr. Kiekhaefer's unusual career includes general and trauma surgery as well as orthopaedics, and she spent a year with a specialized team of surgeons who performed reattachment of amputated digits. Remarkably, Fred and Carol met in a machine shop course while both were completing graduate courses at The University of Wisconsin at Madison, Wisconsin. She also has a masters degree in engineering, and a masters degree in physical chemistry.

In the last few weeks of 1986, fundamental change swept the modern marine industry, as both Brunswick Corporation and Outboard Marine Corporation began the aggressive acquisition of over one-third of the boat building production in America. Brunswick Chairman Jack Reichert made the first dramatic announcement: the purchase of both industry leading Bayliner and Sea Ray organizations for a total investment in excess of three-quarters of a billion dollars. Within weeks, OMC Chairman Charlie Strang would announce their own acquisitions of Four Winns, Carl A. Lowe Industries, Stratos Boat Company, Sunbird Boat Company and Bramco. For both Brunswick and OMC, the strategy was a simple one: combine boats and engines, both stern drives and outboards, and sell them as a package. In the space of forty days, the world's largest manufacturers of marine propulsion had become the world's largest producers of boats as well. The third emerging giant in the eighties was Genmar Corporation, owners of Wellcraft, Hatteras, Larson, Lund and Silverline boat companies. Chairman Irwin Jacobs welcomed the engine makers into the boat building business, and warned that "Mercury and OMC have created a window of opportunity that an elephant could walk through," and "they better hope for ten solid years of great business to get some pay back on their investment."[7] Not since Ole Evinrude and the Johnson brothers designed "detachable" engines for rowboats had the marine industry experienced such widespread metamorphosis.

How does all this fit into the life of Carl Kiekhaefer? Among the most interesting of the acquisitions which followed in the wake of the industry's realignment was the purchase of Kiekhaefer Aeromarine by Brunswick Corporation on July 20, 1990, for an undisclosed amount purportedly in excess of $4 million dollars. Fred Kiekhaefer had previously purchased shares in the company held by Tom Asher, and before the sale to Brunswick Corporation was completed, he acquired the shares owned by Rose Smiljanic. Engineered by Jack Reichert, the acquisition brought the Kiekhaefer name full circle back into the Brunswick fold. Kiekhaefer Aeromarine and Mercury Performance Products were merged into a new organization, Brunswick Marine Hi-Performance, an autonomous business unit of Brunswick Marine Power, and Fred Kiekhaefer was installed as president. Of Fred Kiekhaefer, Reichert said, "Fred is highly respected within the marine industry because of his engineering and entrepreneurial skills. ... As the son of the founder of Mercury Marine, he also brings back to Brunswick the Kiekhaefer heritage."[8]

Among the enticements that attracted Reichert and Brunswick to Kiekhaefer Aeromarine was a new generation of stern drive introduced by Fred Kiekhaefer in early 1988. The ruggedly engineered surface-piercing drives are capable of handling 1000-horsepower engine loads in competition, and up to 750-horsepower for recreational use. Soon after their introduction, the rakish silhouette of the Kiekhaefer drive was propelling the fastest boats around the offshore circuit, returning Kiekhaefer Aeromarine to the winner's circle once again. In the first three years of competition, Kiekhaefer stern drives, renamed *MerCruiser* VI following the merger, have won four National and five World Championship offshore titles. The very first time these new drives were used in competition resulted in a World Championship at the 1988 Offshore World Cup Races in Key West, Florida. Winning the Superboat class aboard offshore veteran Tom Gentry's boat, *Gentry Turbo Eagle*, was "Miami Vice" star, actor Don Johnson, who was propelled by three of Fred Kiekhaefer's new drives, coupled to turbo-charged Gentry engines, and whose throttleman was Bill Sirois.

In 1989, Peter Markey, sponsored by Little Caesars Pizza, became World Offshore Champion, while Charlie Marks won the U.S. National offshore championships, both propelled by Kiekhaefer stern drives. More victories followed in Open and Modified Class competition, as the nearly flawless performance of the new stern drive dominated the field, resulting in an additional five national and world offshore racing titles in 1990 alone.

On May 17, 1991, Jack Reichert and Mercury Marine President David Jones announced the promotion of Fred Kiekhaefer to Sr. Vice-President of Marketing and Sales for Mercury Marine outboard and stern drive products - the #2 position in the sprawling marine empire. OMC Chairman Charlie Strang was among the first to phone to extend his congratulations.

Much speculation surrounded Brunswick's acquisition of Kiekhaefer Aeromarine, as long-time Mercury Marine employees wondered if one day another Kiekhaefer would head the company, or even eventually Brunswick Corporation itself. It would be considered the most fitting irony should either possibility arise, and for the Kiekhaefer family would also be a most fitting tribute to both the exceptional abilities of the son, and the enduring legacy of the father.

FOOTNOTES TO SOURCES

FOREWORD

1. In a letter to Luther Evans of the *Miami Herald*, from E.C. Kiekhaefer of December 9, 1964.

2. In a letter to Dr. Giles A. Koelsche of the Mayo Clinic in Rochester, Minnesota, from E.C. Kiekhaefer of June 27, 1973.

3. In a letter to The Honorable William A. Steiger, member of congress from Wisconsin, from E.C. Kiekhaefer of June 10, 1976.

4. In a letter to Mrs. Maurine C. Dupuy from E.C. Kiekhaefer of May 23, 1979.

5. In a letter to Joy (Kiekhaefer) Martin, the Kiekhaefer family historian, from E.C. Kiekhaefer of March 28, 1979.

6. In a letter to Mrs. Nancy Dean, field editor of *Motorboat* magazine from E.C. Kiekhaefer of May 23, 1977.

7. In a letter to Ms Eileen Crimmin from E.C. Kiekhaefer of August 25, 1978.

8. In a letter to John Ross from E.C. Kiekhaefer of December 12, 1957.

CHAPTER ONE
Origins

1. Richard Wagner, in Deutsche Kunst und Deutsche Politik (German Art and German Politics), 1867.

2. From the Kiekhaefer Family Register, edited by Kiekhaefer family historian Joy (Kiekhaefer) Martin, 1979, DePere, Wisconsin.

3. From the deed of sale of the Kiekhaefer homestead from Carl and Friedericka Kiekhaefer to Heinrich (Henry) Kiekhaefer dated December 6, 1876, as recorded in the Kiekhaefer Family Register, edited by Kiekhaefer family historian Joy (Kiekhaefer) Martin, DePere, Wisconsin, 1979.

4. "Kiekhaefer Family Register," loc. cit.

5. In a letter to Fred M. Young, Young Radiator Company, July 22, 1982.

6. Interview of Isabelle (Kiekhaefer) Gottinger, a sister, on March 2, 1987.

7. In a draft of a letter to Beeline, column of the Chicago Daily News, October 30, 1974.

8. Interview of Palma Moerschel, a sister, of March 22, 1987.

9. In a letter to Mr. William Martin-Hurst, on February 16, 1978.

10. Ibid.

11. Interview of Almira (Kiekhaefer) Westendorf, a sister, on March 13, 1987.

12. Ibid.

13. Interview of Ruth (Kiekhaefer) Pike, a sister, on March 22, 1987, and also from the Kiekhaefer Family Reunion of July 1987 at the family homestead in Mequon, Wisconsin.

14. Interview of Palma Moerschel, loc. cit.

15. Interview of Freda Kiekhaefer, wife of E.C. Kiekhaefer, on August 16, 1987 in Fond du Lac, Wisconsin.

CHAPTER TWO
Farmboy

1. In a letter Elmer wrote to his mother, Clara Kiekhaefer, on August 10, 1955, defending the great amount of work he performed on the farm as a child.

2. Interview of Almira (Kiekhaefer) Westendorf, a sister, on April 13, 1987.

3. Ibid.

4. In a letter to William Martin-Hurst from E.C. Kiekhaefer on February 16, 1978.

5. In a letter to Mr. Gerald Taines, T.F.C. Marine Engineering, from E.C. Kiekhaefer on December 18, 1972.

6. From a telephone interview recorded and transcribed by E.C. Kiekhaefer with Bob Brown of Powerboat Magazine, July 8, 1981.

7. Evidence in support of his bout with pneumonia arresting his college aspirations is contained in the transcribed interview of July 8, 1989, given by E.C. Kiekhaefer to Bob Brown of Powerboat Magazine. However, though he says, "Six weeks in bed. That's what killed my college career," his history displays many instances of historical rationalization, applying later justification to previous events.

8. Interview of Almira (Kiekhaefer) Westendorf, loc. cit.

CHAPTER THREE
Dues

1. In a letter from Elmer Kiekhaefer to his mother, Clara Kiekhaefer, dated August 12, 1955.

2. From a speech by E.C. Kiekhaefer entitled *Getting the Job Done*, delivered at a Mercury Sales Conference on August 8-9, 1966 in Fond du Lac, Wisconsin.

3. From the Dun & Bradstreet, Inc. report of February 5, 1935, which had been requested by Elmer Kiekhaefer through his father's company, the Ozaukee Sand & Gravel Company to disguise his identity as the requester of the report.

4. Interview of Freda Greenfield Kiekhaefer on August 16, 1987, in Fond Du Lac, Wisconsin.

5. Ibid.

6. *Getting the Job Done* speech, loc. cit.

7. In a letter to a prospective employer, Z-103--Journal, from E.C. Kiekhaefer of November 11, 1935.

8. Evidence of his unsuccessful patent application for an air pump is contained in a letter from W.A. Stark, Engineer, a specialist in patent development and applications, to Elmer Kiekhaefer of August 5, 1935.

9. In a handwritten draft of a letter to Mr. R.A. Manigold of Dings Magnetic Separator Company, from Elmer Kiekhaefer of November 11, 1935.

10. In a letter addressed to Z-103--Journal, in response to a blind ad ran in the *Milwaukee Journal*, November 11, 1935.

11. In a letter to Fred M. Young, President, Young Radiator Company from E.C. Kiekhaefer of July 22, 1982.

12. Evidence is contained in letters of transactions between the McGraw-Hill Book Company and Elmer Kiekhaefer in January of 1934.

13. *Getting the Job Done* speech, loc. cit.

CHAPTER FOUR
Cedarburg

1. From the 1966 address delivered by E.C. Kiekhaefer at the Mercury Sales Conference entitled, "Getting the Job Done", August 8-9, in Fond du Lac, Wisconsin.

2. In a letter to Joy (Kiekhaefer) Martin, Kiekhaefer family historian, from E. C. Kiekhaefer of March 28, 1979.

3. From the "Getting the Job Done" speech, loc. cit.

4. From the Cedarburg Manufacturing Company's first promotional pamphlet entitled: "Presenting the Newest Outboard Motor: THOR", copyright 1934 by Thor Hansen.

5. From the "Market Analysis For The Thor Outboard Motor", written by E.C. Freshwaters, Sales Manager of the Cedarburg Manufacturing Company, as addressed to Mr. Thor Hansen, on January 21, 1935.

6. In a letter from W.J. Webb to Lloyd M. Titcombe, Jr., of September 6, 1977. Titcombe is head of the Thor outboard motor special interest group of the Antique Outboard Motor Club, a national organization dedicated to the history and preservation of classic outboard motors. In 1952, W.J. Webb became vice president and general manager of Evinrude Motors, a position which he held for eleven years, until his retirement in 1963. Webb was also, along with Robert W. Carrick, author of the popular 139 page book, The Pictorial History of Outboard Motors, which was published by Renaissance Editions, Inc., 527 Madison Avenue, New York, Y.Y. 10022 in 1967.

7. Boating Business, December, 1935, "Outboard Motor Specifications."

8. In a letter from W.C. Clausen, OMC general sales manager, to Thor Hansen of the Cedarburg Manufacturing Company, of December 18, 1935.

9. Letter from Lester H. Gunsburg, Lawyer, dated November 8, 1938, but not mailed until December 3, 1938 on behalf of the Cedarburg investors.

10. Letter from the law firm of Field & Doll, representing Royal Hansen and The Cedarburg Manufacturing Company, dated December 5, 1938, to Mr. S.E. Miller of the Montgomery Ward & Company, of Chicago, Illinois.

11. The terms and conditions of the loan to Royal Hansen by John G. Blank are contained in a letter from Royal Hansen, president of the Cedarburg Manufacturing Company, to Blank on December 3, 1938, and accepted and signed by J.G. Blank on December 5, 1938.

12. Interview of Almira (Kiekhaefer) Westendorf, a sister, of April 13, 1987.

13. In a letter of agreement between Elmer Kiekhaefer, John G. Blank and Edgar H. Roth, on the letterhead of the Cedarburg State Bank, on December 30, 1938.

CHAPTER FIVE
New Directions

1. Interview of Bob Stuth at his home in Slinger, Wisconsin on July 29, 1987.
2. In a speech delivered by E.C. Kiekhaefer to close the Kiekhaefer Mercury Atlanta dealer meeting on October 10, 1960.
3. Interview of Isabelle (Kiekhaefer) Gottinger, a sister, on March 2, 1987.
4. *MercComment*, 1979 Special Issue, on the 40th Anniversary of Mercury Marine, a publication of Mercury Marine, Fond du Lac, Wisconsin for the 1979 celebration.
5. From the first Kiekhaefer Corporation Thor Motor brochure, written and published in February, 1939.
6. The identity of Earnway Edwards is disclosed in a letter from J.A. Allan, Kiekhaefer's consulting engineer for government contracts, dated September 8, 1942.
7. From a speech entitled, "Getting the Job Done", given by E.C. Kiekhaefer for the 1967 Mercury Marine Sales Conference on August 9, 1966 in Fond du Lac, Wisconsin.
8. Ibid. 9. Ibid.
10. From the March, 1939 four page flyer to Thor dealers announcing "Red" Parkhurst representing the New Thor Line of Outboard Motors. This is the first piece of promotional literature distributed by the Kiekhaefer Corporation.
11. Interview of Bob Stuth, Slinger, Wisconsin on July 29, 1987.
12. Contained in a transcribed interview between E.C. Kiekhaefer and Bob Brown of *Powerboat* magazine, July 8, 1981.
13. "Getting the Job Done" speech, loc. cit.
14. Ibid.

CHAPTER SIX
Messenger of the Gods

1. From an address by E.C. Kiekhaefer entitled "Getting the Job Done", delivered at the yearly Mercury Sales Conference, August 9, 1966 in Fond du Lac, Wisconsin.
2. From the 1934 Silver Anniversary Outboards brochure prepared by the Elto-Evinrude divisions of Outboard Motors Corporation.
3. "Getting the Job Done" speech, loc. cit.
4. Ibid.
5. In a letter from E.C. Kiekhaefer to Guy S. Conrad, regarding his legal options of recourse against the Eisemann Magneto Corporation of March 14, 1940.
6. Interview of Bob Stuth, July 29, 1987, at his home in Slinger, Wisconsin.
7. In a letter from E.C. Kiekhaefer to Guy S. Conrad of May 16, 1940.
8. From the 1941 Mercury Outboard Motors brochure.
9. Ibid. 10. Ibid.
11. From the first Kiekhaefer Corporation dealer newsletter, the *Mercury Outboard Motor News*, published in March 1941 in Cedarburg, Wisconsin.
12. Ibid. 13. Ibid.

CHAPTER SEVEN
Suspended Dreams

1. *Mercury Outboard Motor News*, July 1941.
2. In a radio address delivered by President Franklin D. Roosevelt on September 5, 1939 over the N.B.C. and C.B.S. radio networks.
3. In a letter to Colonel Donald Armstrong of Chicago, Illinois, from E.C. Kiekhaefer of June 12, 1940.
4. In a letter to Mr. Wm. C. Baker, Jr., Captain, Corps of Engineers, War Department, Fort Belvoir, Virginia, from E.C. Kiekhaefer of August 10, 1940.
5. In a letter to Charles Hoge, Priorities Division, Washington, D.C., from E.C. Kiekhaefer of April 7, 1941.
6. In a letter to Elmer C. Kiekhaefer from Major C. Rodney Smith of the Army Corps of Engineers, Fort Belvoir, Virginia, of April 10, 1941.
7. In special edition of the Ozaukee County-Wee Wonder *Cedarburg News* dedicated entirely to the award of the Kiekhaefer Corporation of the Army-Navy "E" Performance Award, dated Wednesday, October 27, 1943.
8. In a letter to E.C. Kiekhaefer from Major C. Rodney Smith of the Army Corps of Engineers, Fort Belvoir, Virginia of May 3, 1941.

9. Ibid. 10. Ibid. 11. Ibid.

12. In a letter to Mr. C. W. Gallagher of the Reed-Prentice Corporation, Worcester, Massachusetts, from C. Rodney Smith, Major, Corps of Engineers, Assistant Executive Officer, War Department, The Engineer Board, Fort Belvoir, Virginia on June 20, 1941.

13. In a letter to Mr. F.W. McIntyre, vice president and general manager, Reed-Prentice Corporation, Worcester, Massachusetts, from E.C. Kiekhaefer of October 7, 1941.

14. Ibid.

15. In a letter to Mr. F.W. McIntyre, vice president & general manager, Reed-Prentice Corporation, Worcester, Massachusetts, from E.C. Kiekhaefer of August 5, 1941.

16. Ibid.

17. In a letter from E.C. Kiekhaefer to Mr. F.W. McIntyre, vice president and general manager of Reed-Prentice Corporation, Worcester, Massachusetts, from E.C. Kiekhaefer of August 21, 1941.

18. In a letter to Mr. C.S. Colby, Office of Production Management, Aluminum Division, Washington, D.C. from E.C. Kiekhaefer of November 22, 1941.

19. Ibid. 20. Ibid.

CHAPTER EIGHT
Machines of War

1. In a draft of a letter to the War Department by E.C. Kiekhaefer, probably 1943, entitled, "Contributions of the Kiekhaefer Corporation Toward the War Effort."

2. In a speech entitled, "Getting the Job Done", delivered by E.C. Kiekhaefer for the 1967 Sales Conference on August 9, 1966 in Fond du Lac, Wisconsin.

3. In a letter to Mr. F.R. Erbach, Facilities Staff, War Production Board, Washington, D.C., from E.C. Kiekhaefer of July 25, 1942.

4. Ibid.

5. From the Kiekhaefer Corporation brochure entitled, "It's a Great Place to Work," published and distributed between 1942-1945.

6. Ibid.

7. Contained within notes to the 1942 Kiekhaefer Corporation financial balance sheets.

8. In a letter to Colonel John S. Seyboldt, Chief Contracting Officer, U.S. Corps of Engineers, Washington, D.C. from E.C. Kiekhaefer of October 30, 1942.

9. Ibid.

10. In a letter from Robert P. Patterson, Under Secretary of War, Washington, D.C., to "The Men and Women of the Kiekhaefer Plant," dated October 2, 1943.

11. In a proclamation signed by H.A. Zeurnert, Mayor, City of Cedarburg, dated October 25, 1943.

12. Contained in a special edition of the Ozaukee County-Wee Wonder, Cedarburg News, of Wednesday, October 27, 1943.

13. Ibid.

14. The text of Colonel C. Rodney Smith's Army-Navy "E" award speech on October 30, 1943 was contained in it's entirety in a letter from E.C. Kiekhaefer to H.E. Pollock, Chief, Price Adjustment Section, War Department, Office of Division Engineer, Great Lakes Division, Chicago, Illinois, of July 15, 1944.

15. Ibid.

16. Interview of Freda, Mrs. E.C. Kiekhaefer of August 16, 1987 in Fond du Lac, WI.

17. Interview of Charles D. Strang of May 20, 1987, at the Outboard Marine Corporation salt water testing facility at Stuart, Florida.

CHAPTER NINE
Stockholder Rebellion

1. In a letter to *The Cedarburg News* by E.C. Kiekhaefer, of November 20, 1943.

2. *Ozaukee Press*, Thursday, November 18, 1943.

3. In a letter to the Kiekhaefer Corporation from the legal firm of Schanen & Schanen of Port Washington, Wisconsin, dated November 11, 1943.

4. In a letter to Schanen and Schanen, Attorneys at Law, Port Washington, Wisconsin, from Guy S. Conrad, attorney for the Kiekhaefer Corporation, of November 16, 1943.

5. In a letter to the Kiekhaefer Corporation, attention Guy S. Conrad, by Schanen & Schanen, Attorneys-At-Law, Schanen Building, Port Washington, Wisconsin, of November 17, 1943.

6. "A Statement By the Officers of the Kiekhaefer Corporation", of approximately November 19, 1943, signed by A.C. Kiekhaefer, E.C. Kiekhaefer, Willis Blank, and A.S. Horn.

7. Additional draft of "A Statement By the Officers of the Kiekhaefer Corporation, of approximately November 19, 1943.

8. In a letter to *The Cedarburg News* by E.C. Kiekhaefer of November 20, 1943.

CHAPTER TEN
Higher Sights

1. In a "Memo to all Executives and Office Employees" from E.C. Kiekhaefer dictated December 25, 1944, and dated December 26, 1944.

2. In a letter to the War Department, Office of Division Engineer, Great Lakes Division, Chicago, Illinois from E.C. Kiekhaefer of August 7, 1944.

3. In a letter to the War Production Board, Automotive Division, Washington, D.C. signed by Guy S. Conrad for E.C. Kiekhaefer of September 16, 1944.

4. *Mercury News*, July, 1945.

5. In a memo to E.C. Kiekhaefer from Fred L. Hall of July 6, 1945.

6. In the "Memorandum Report from the Army Air Forces Air Technical Service Command" concerning testing of the new "OQ-17 Radio Airplane Target" of August 11, 1945. The report was prepared for Captain J.H. Jacobsen.

7. From the "Summary of Consolidated Profit and Loss for the Year Ended September 30 , 1944 of Outboard Marine Corporation," a report to the stockholders issued on December 11, 1944.

8. *Mercury News*, March, 1945, in a column known as "The Hot Stove."

CHAPTER ELEVEN
Fond du Lac

1. Interview of Herman Steig of January 17, 1987

2. From the Souvenir Edition of the *Fond du Lac Reporter*, celebrating the sesquicentennial of Fond du Lac on June 25, 1986, in an article by Steve Sandberg.

3. Souvenir Edition, *Fond du Lac Reporter*, loc. cit., in an article by Stan Gores, managing editor, entitled, "The night Carry Nation pulled out her hatchet at E.J. Schmidt's bar."

4. Ibid. 5. Ibid.

6. Steig interview, loc. cit.

7. Ibid.

8. Fond du Lac *Commonwealth Reporter*, Saturday, July 6, 1968, on the occasion of the razing of the Corium Farms barn, after over a million outboard motors had been assembled in the site during over twenty years of operation by the Kiekhaefer Corporation. The facts in the article were based on interviews with Dirk S. Van Pelt, Clarence Sheridan and R.W. Mills, from letters and records of F.J. Rueping.

9. Ibid.

10. Steig interview, loc. cit.

11. Ibid.

12. In an transcribed telephone interview of E.C. Kiekhaefer with Bob Brown of *Powerboat* Magazine on Wednesday, July 8, 1981.

13. In a letter to Jack F. Reichert, president, Mercury Marine, division of Brunswick Corporation, Fond du Lac, Wisconsin, by E.C. Kiekhaefer of April 18, 1977.

14. Interview of Bob Stuth of July 29, 1987 at his home in Slinger, Wisconsin.

15. In a transcribed telephone conversation between E.C. Kiekhaefer and Frank Wilfred of the Western Auto Supply Company on September 25, 1947.

16. In the transcribed telephone interview of E.C. Kiekhaefer with Bob Brown of *Powerboat* Magazine on Wednesday, July 8, 1981, loc. cit.

17. *Milwaukee Journal*, Sunday, December 19, 1943, in an article entitled, "Sales Higher, Profit Lower."

18. Steig interview, loc. cit.

CHAPTER TWELVE
Shifting Threats

1. Quoted in the *Mercury Messenger*, March 1949, as a guest editorial by Captain Eddie V. Rickenbacker, president and general manager of Eastern Air Lines, Inc.

2. *Mercury News*, June 1945, in a column known as "The Hot Stove."

3. *Fortune*, April 1951, in an article entitled "Young Men From Milwaukee".

4. Ibid.

5. Interview of Charles D. Strang at the OMC saltwater testing grounds in Stuart, Florida on May 20, 1987.

6. Ibid. 7. Ibid. 8. Ibid.

9. Speech delivered to all employees by M.E. Smith, representing Kiekhaefer Corporation management on May 7, 1948.

10. Carl Kiekhaefer admitted the true horsepower of the 25-horsepower *Thunderbolt* as actually 40-horsepower in the "Background History of the Kiekhaefer Corporation" written as a pre-merger document for Brunswick Corporation analysis in 1960.

11. In a marketing summary called "Interesting Facts About The Mercury 25 With The Thunderbolt Engine", undated, written before release of the *Thunderbolt* prototype for production by E.C. Kiekhaefer.

12. Referring to the book, *The Four Men From Terre Haute*, by J.M. Van Vleet, et al, a history of the Johnson Brothers Engineering Corporation, 1918 to 1935, printed for the Thousand Islands Shipyard Museum in Clayton, NY on November 23, 1981. The other three men were Louis Johnson, Harry Johnson and Clarence Johnson.

13. Ibid.

14. Interview of Clay Conover at his home in Waukegan, Illinois on August 25, 1989.

15. Ibid.

16. Interview of Clay Conover, loc. cit.

17. Ibid.

18. From the Silver Anniversary brochure of the Outboard Motors Corporation, introducing the new 1934 lines of Evinrude and ELTO outboard motors, quoting from the introduction by Ole Evinrude.

19. Ibid.

20. Interview of Clay Conover, loc. cit.

21. Ibid. 22. Ibid.

23. In an official "History of the Outboard Marine Corporation" and corporate background brief written by Oristano Associates on behalf of OMC in 1965.

CHAPTER THIRTEEN
The Race Is On

1. In a letter to Jim Wright of *Rudder* magazine from E.C. Kiekhaefer of December 4, 1973, describing the somewhat historically unfair advantages of clandestine factory supported racing.

2. In a letter to Mr. F.E. Southard, Champion Spark Plug Company, Toledo, Ohio from E.C. Kiekhaefer of November 22, 1949.

3. Results for the race were published in *Motor Boating*, July, 1948 issue, and described by F.W. Horenburger.

4. From the June, 1948 fold-out poster distributed to all Mercury distributors and dealers.

5. In a special dealer letter signed by Evinrude Vice-President H. Biersach of June 24, 1948.

6. Ibid.

7. From the 1948 line announcement brochure entitled, "Motor Magic" by Champion Motors of Minneapolis, Minnesota.

8. Ibid. 9. Ibid. 10. Ibid. 11. Ibid.

12. Contained in the 1953 Champion line brochure, Champion Motors, Minneapolis, Minnesota.

13. Ibid.

14. Proof of the cover-up exists in the text for the 1953 Champion line brochure, loc. cit.

CHAPTER FOURTEEN
Staggered Faith

1. In a letter to Otis E. Johnson of Western Auto Supply Company, Kansas City, MO, from E.C. Kiekhaefer, of January 24, 1951.

2. In a letter to J. Paxton Hill, Utility Racing Secretary of the American Powerboat Association from E.C. Kiekhaefer of February 20, 1951.

3. In a letter to W.F. Rockwell, Jr., president of the Rockwell Manufacturing Company of Pittsburgh, PA from Fred L. Hall, vice president of sales for the Kiekhaefer Corporation of July 8, 1949.

4. In a letter from Walter F. Rockwell, Jr., to Fred Hall of July 14, 1949.

5. In a transcribed telephone conversation between Mr. Jim Ashman of Rockwell Manufacturing Company of Pittsburgh, PA and Carl Kiekhaefer of June 27, 1950.

6. In a letter to Mr. Jim Hait, vice president of engineering, Food Machinery & Chemical Company, San Jose, California from E.C. Kiekhaefer of November 18, 1950.

7. In a letter from B.C. Carter, vice president and Controller of Food Machinery and Chemical Company of San Jose, California, to E.C. Kiekhaefer of December 1, 1950.

8. In a draft of a letter never sent to Mr. B.C. Carter, vice president and controller of the Food Machinery and Chemical Company, San Jose, California from E.C. Kiekhaefer of December 6, 1950.

9. In a draft of a letter never mailed to Mr. Paul L. Davies, president, Food Machinery and Chemical Company from E.C. Kiekhaefer of February 26, 1951.

10. In a letter from B.C. Carter, vice president of Food Machinery and Chemical Company to E.C. Kiekhaefer of May 28, 1951.

11. *Chicago Journal of Commerce*, January 30, 1951, in an article entitled "Net for Food Machinery Put Up to $3 Share."

12. In a letter from B.C. Carter, vice president, Food Machinery and Chemical Company, San Jose, California, to E.C. Kiekhaefer of August 15, 1951.

13. *Ozaukee County-Wee Wonder Cedarburg News*, October 11, 1950.

14. In a letter to James C. Sargent, M.D. of Milwaukee, Wisconsin, from E.C. Kiekhaefer of March 3, 1951.

15. Contained in the "Last Will and Testament of Arnold C. Kiekhaefer", signed and witnessed in Milwaukee, Wisconsin on February 2, 1950.

16. Interview of Isabelle Kiekhaefer Gottinger of March 2, 1987.

17. Ibid.

18. Interview of Almira Kiekhaefer Westendorf of March 13, 1987.

19. Ibid. 20. Ibid.

21. Interview of Isabelle Kiekhaefer Gottinger, loc. cit.

22. Ibid. 23. Ibid.

24. In a letter dated June 13, 1951, from E.C. Kiekhaefer to his mother and six sisters.

25. Typed with the original punctuation from the hand-written letter from Clara Kiekhaefer to her son, Elmer Kiekhaefer, of August 10, 1955.

26. The original, hand-typed letter dated August 12, 1955, from Elmer Kiekhaefer to his mother, Clara Kiekhaefer.

CHAPTER FIFTEEN
Smart Move

1. In a letter to Charles D. Strang, Jr., Department of Mechanical Engineering, Massachusetts Institute of Technology, from E.C. Kiekhaefer, of December 11, 1950.

2. In a letter from Charles D. Strang, Jr., Department of Mechanical Engineering, Massachusetts Institute of Technology, to E.C. Kiekhaefer of November 20, 1950.

3. Interview of Charles D. Strang, Jr., of May 20, 1987 at OMC's Stuart, Florida Saltwater Proving Grounds. Strang was then Chairman and Chief Executive Officer of Outboard Marine Corporation.

4. In a letter to Charles D. Strang, Jr. from E.C. Kiekhaefer of December 11, 1950, loc. cit.

5. In a letter to E.C. Kiekhaefer from Charles D. Strang, Jr., Department of Mechanical Engineering, Massachusetts Institute of Technology, from E.C. Kiekhaefer of December 14, 1950

6. Ibid.

7. In a letter to the author from Charles D. Strang of July 14, 1987.

8. In a letter to Charles D. Strang, Jr., Department of Mechanical Engineering, Massachusetts Institute of Technology, from E.C. Kiekhaefer, of April 19, 1951.

9. Strang interview of May 20, 1987, loc. cit.

10. In a letter to E.C. Kiekhaefer from Charles D. Strang, Jr. from the Carlton Hotel, Nurnberg, Germany of July 30, 1951.

11. Strang interview of May 20,1987, loc. cit.

12. Ibid. 13. Ibid.

CHAPTER SIXTEEN
Operation Mexico

1. Vincent Thomas Lombardi, 1913-1970, head coach of the Green Bay Packers of the National Football League, who in nine seasons (1959-1967) led the team to five NFL championships and victory in the first two Super Bowls. Carl Kiekhaefer was known to have admired the tough, winning style of the legendary coach, and shared the same all-or-nothing winning attitude.

2. *Peterson's Circle Track*, July 1985, in an article entitled, "Little Jewel, Murrell Belanger's Indy-winning Offy".

3. Interview of Charles D. Strang of May 20, 1987 at the OMC Salt Water Proving Grounds at Stuart, Florida.

4. *Speed Age* magazine, April 1952, in an article by Charles Strang entitled "Operation Mexico". Strang's skills as a journalist were not lost when he joined the

Kiekhaefer Corporation in 1951. He was able to commandeer the cover of Speed Age for this issue with a photo of Bettenhausen's #7, along with a full-length feature article about Kiekhaefer racing efforts.

5. Ibid. 6. Ibid.

7. Interview of Charles D. Strang of February 8, 1989 in Waukegan, Illinois.

8. Ibid.

9. *Speed Age* magazine, February 1953, in an article by Vince McDonald.

10. Ibid.

11. *Speed Age*, April, 1952, loc. cit.

12. Strang interview of May 20, 1987, loc. cit.

13. In a letter to Marc R. Prass, Staff Representative of the Chrysler Sales Division of Chrysler Corporation, from Carl Kiekhaefer of July 21, 1953.

14. *Speed Age*, February 1953, loc. cit.

15. In a letter to Don O'Reilly, publisher and editor of *Speed Age* magazine, from Carl Kiekhaefer of December 15, 1952.

16. In a letter to Enrique Martin Moreno, a leading official of the Carrera Panamericana from Carl Kiekhaefer of January 30, 1953.

17. In a letter to E.C. Quinn, vice president and general manager, Chrysler Corporation from Carl Kiekhaefer of February 6, 1953.

18. Ibid.

19. Interview of Donald E. Castle in his offices in Fond du Lac, Wisconsin on August 4, 1987.

20. In a letter to Mr. Marc. R. Prass, Chrysler Sales Division, Chrysler Corporation, from Carl Kiekhaefer of July 21, 1953.

21. *Speed Age* magazine, November 1953, in an article by Don O'Reilly, editor, entitled *Viva Mexico*.

22. *New York Times*, November 20, 1953, front page sports section.

23. Ibid.

24. In a letter to Mr. O.E. Johnson, buyer, Western Auto Supply Company, from Carl Kiekhaefer of November 30, 1953.

CHAPTER SEVENTEEN
Loyal Assistance

1. In a memo sent to and countersigned by 17 members of Kiekhaefer's race car mechanic and engineering pool by Carl Kiekhaefer. Undated, but from the content was issued in August 1956.

2. Interview of Fred "Fritz" Shoenfeldt of July 31, 1987 at his home in Wautoma, WI.

3. Ibid. 4. Ibid. 5. Ibid. 6. Ibid.

7. Ibid. 8. Ibid.

9. Interview of Rose Smiljanic, August 14, 1986 at her home in Fond du Lac, Wisconsin.

10. Ibid. 11. Ibid. 12. Ibid. 13. Ibid.

14. In a letter to John F. Goetz, Jr. of Madison, Wisconsin, from E.C. Kiekhaefer of August 13, 1953.

15. Interview with Charles D. Strang of May 20, 1987 at the OMC salt water proving grounds at Stuart, Florida.

16. Ibid. 17. Ibid. 18. Ibid.

CHAPTER EIGHTEEN
A Bumber Crop of Talent

1. In a letter to Dr. Vernon S. Dick of the Lahey Clinic, Boston, Massachusetts, of November 25, 1955 by E.C. Kiekhaefer, written on behalf of James R. Wynne.

2. *True Magazine*: The Man's Magazine, in an article by Robert Lee Behme entitled "Ted Jones: Racing Renegade." Though undated, the issue was undoubtedly published in 1957.

3. Interview of Ted O. Jones of Vashon, Washington on March 13, 1987.

4. Ibid. 5. Ibid. 6. Ibid. 7. Ibid.

8. Ibid.

9. Interview of Charles D. Strang of May 20, 1987, at the OMC salt water proving grounds at Stuart, Florida.

10. Interview of Ted O. Jones, loc. cit.

11. In a letter from Ted O. Jones to Rose Smiljanic of Thursday, May 3, 1952.
12. In a letter from Ted O. Jones to Rose Smiljanic of July 4, 1952.
13. In a letter from Ted O. Jones to Rose Smiljanic of July 13, 1952.
14. In a letter from Ted O. Jones to Rose Smiljanic of May 22, 1952.
15. In a letter from Ted O. Jones to Rose Smiljanic, of May 19, 1952.
16. Interview of Ted O. Jones, loc. cit.
17. Ibid.
18. In a letter from Ted O. Jones to Rose Smiljanic of June 13, 1952.
19. Interview of Ted O. Jones, loc. cit.
20. Interview of Charles Alexander of December 1, 1986.
21. Ibid. 22. Ibid. 23. Ibid. 24. Ibid.
25. Ibid.
26. Interview of Charles D. Strang of May 20, 1987, loc. cit.
27. Ibid. 28. Ibid. 29. Ibid.
30. Interview of James R. Wynne of October 17, 1987 at his offices at Wynne Marine in Miami, Florida.
31. Ibid. 32. Ibid. 33. Ibid.
34. In the letter to Dr. Vernon S. Dick of November 25, 1955 from E.C. Kiekhaefer, loc. cit.
35. Ibid. 36. Ibid. 37. Ibid.
38. Interview of Rose Smiljanic of August 14, 1986 at her home in Fond du Lac, Wisconsin.
39. Interview of James R. Wynne, loc. cit.
40. Ibid.
41. Interview of Rose Smiljanic, loc. cit.

CHAPTER NINETEEN
Stock Car Fever

1. Bill France, NASCAR president, in an interview to Hank Schoolfield, staff reporter, *Journal and Sentinel*, Winston-Salem, North Carolina, July 8, 1956, in an article written about Carl Kiekhaefer entitled "King of the Stock Cars."
2. *Stock Car Racing*, July 1985, in an article by Jonathan Ingram entitled, "Carl Kiekhaefer: A Grand National Giant."
3. *True*, The Man's Magazine, February 1957, in an article by Sandy Grady entitled, "Kiekhaefer - The Scuderia That Walks Like a Man."
4. In an article written by Carl Kiekhaefer for *Popular Mechanics* magazine, "How to Win Stock-Car Races," appearing in the February 1956 issue.
5. *True*, February, 1957, loc. cit.
6. Ibid.
7. *True's Automobile Yearbook* #7, 1958, in an article by Sandy Grady about Bill France entitled, "Big Bill, The Stocker's Boy."
8. *Mopar* magazine, Summer of 1988, in an article by Jesse Thomas entitled, "MOPAR MERCS, The legend of Carl Kiekhaefer and his Mercury Outboards 300s."
9. Ibid. 10. Ibid. 11. Ibid.
12. *Popular Mechanics*, Feb. 1956, loc. cit.
13. Ibid. 14. Ibid.
15. *Circle Track*, November 1988, in an article by Smokey Yunick entitled, "Smokey Tells All: Thirty Years of Fun, Games, and Playing Fast and Loose With the Rules."
16. Ibid. 17. Ibid.
18. *Stock Car Racing*, August 1985 loc. cit.
19. Ibid.
20. "How to Win Stock-Car Races," by Carl Kiekhaefer, loc. cit.
21. In a draft letter to NASCAR officials by Carl Kiekhaefer of May 1, 1955. It is not clear whether or not this particular draft was converted to a letter and mailed. It serves, though, to illustrate Carl's growing suspicions about the fairness of stock car racing.
22. *Stock Car Racing*, August 1985, loc. cit.
23. *True*, February 1957, loc. cit.
24. In a letter composed but never sent to Ed Otto, vice president, NASCAR of June 29, 1955, containing a hand-written note indicating that the matter was dealt with by phone instead.
25. *Stock Car Racing*, July 1985, loc. cit.
26. Ibid.

CHAPTER TWENTY
The Fever Breaks

1. As often overheard by Bobby Allison while working for Kiekhaefer at the Kiekhaefer Mercury race car garages at Charlotte, North Carolina.

2. *True*, The Man's Magazine, February 1957 in an article by Sandy Grady entitled, "Kiekhaefer -- The Scuderia That Walks Like a Man."

3. Ibid. 4. Ibid.

5. Interview of William C. Newberg, who at this period of his dealings with Carl was president of Dodge, division of Chrysler Corporation, of June 15, 1989.

6. Ibid.

7. *Stock Car Racing*, August 1985, loc. cit.

8. The same advertisement appeared in *Hot Rod Magazine*, May 1956; *Road & Track Magazine*, May 1956 and *Speed Age Magazine*, May 1956.

9. *True*, February 1957, loc. cit.

10. *Stock Car Racing*, August 1985. In part two of the two-part article by Jonathan Ingram entitled "Carl Kiekhaefer, A Grand National Giant," July 1985, loc. cit.

11. *True*, February 1957, loc. cit.

12. *Stock Car Racing*, August 1985, loc. cit.

13. *True*, February 1957, loc. cit.

14. *Peterson's Circle Track*, July 1985, in an article by Don O'Reilly entitled, "Kiekhaefer." [Don O'Reilly was director of the NASCAR News Bureau during the years of Kiekhaefer dominance in stock car racing.]

15. In a memo sent to and countersigned by 17 members of Kiekhaefer's race car mechanic and engineering pool by Carl Kiekhaefer. Undated, but from the content was issued in August 1956.

16. Ibid.

17. *Stock Car Racing*, July 1985, loc. cit.

18. *Old Cars*, September 20, 1977, in an article written by Robert C. Ackerson entitled, "Kiekhaefer's Big, White Cars Big Winners in the Fifties."

19. *Peterson's Circle Track*, July 1985, loc. cit.

20. *Atlanta Journal 500*, 21st annual Souvenir Magazine, November 2, 1980, as told to Jonathan Ingram, in an article entitled, "Kiekhaefer, A Man Ahead of His Time."

21. In a letter to Mrs. Vicki Wood from E.C. Kiekhaefer, of October 11, 1956.

22. In a transcribed telephone conversation between William C. Newberg of Chrysler Corporation and Carl Kiekhaefer of January 14, 1957.

23. In a transcribed telephone conversation between Bernhard Kahn, Sports Editor of the Daytona Beach Evening News, Community Journal, Daytona Beach, Florida, on January 14, 1957.

24. In an uncirculated News Release from E.C. Kiekhaefer regarding the Daytona Beach Speed Weeks, dated February 12, 1957.

25. *Motor Guide*, August 1957, in an article by Kermit Moreland entitled, "Why Kiekhaefer Quit Racing".

26. Ibid.

27. In a letter never mailed, addressed to Mr. Edward T. Clapp, by E.C. Kiekhaefer, of May 22, 1957.

28. *Motor Guide*, August 1957, loc. cit.

29. In a letter to Mr. Edward T. Clapp, May 22, 1957, loc. cit.

30. Ibid.

CHAPTER TWENTY-ONE
Lake X and Operation Atlas

1. In a letter from E.C. Kiekhaefer to Mr. John Ross of December 12, 1957.

2. Interview of Charles D. Strang, of May 20, 1987 at the OMC Salt Water Proving Grounds at Stuart, Florida.

3. In a press release issued by E.C. Kiekhaefer on June 7, 1957.

4. Interview of Charles D. Strang, loc. cit.

5. Interview of James R. Wynne of October 17, 1987 at his offices at Wynne Marine in Miami, Florida.

6. Interview of Charles F. Alexander, Jr. of December 1, 1986.

7. Interview of Charles D. Strang, loc. cit.

8. Interview of Fred "Fritz" Shoenfeldt of July 31, 1987 at his home in Wautoma, Wisconsin.

9. Interview of Charles D. Strang, loc. cit.

10. Ibid.

11. Interview of James R. Wynne, loc. cit.

12. Interview of Frank Scalpone of February 11, 1987 in Miami, Florida.

13. In a memo to E.C. Kiekhaefer from C.D. Strang on Kiekhaefer Aeromarine Motors stationary of July 8, 1957 entitled, "Organization of 40 Day Run."

14. Interview of Wayne and Mildred "Millie" Meyer of September 28, 1987 at their home in Fond du Lac, Wisconsin.

15. Ibid.

16. In a letter to Jim Wynne from Duane Carter, USAC director of competition of August 21, 1957.

17. Interview of James R. Wynne, loc. cit.

18. Handwritten entry contained in Jim Wynne's Lake X endurance run log book for the night of September 21, 1957.

19. Carl's contributions to the St. Cloud Hospital began in 1961, as the hospital was financially unable to complete plans for expansion and modernization. In a letter to Bob Slough, *St. Cloud News* from E.C. Kiekhaefer of August 12, 1970.

CHAPTER TWENTY-TWO
The Enemy

1. In a transcribed telephone conversation between E.C. Kiekhaefer and Joseph G. Rayniak, president of Outboard Marine Corporation on September 28, 1957.

2. *New York Herald Tribune*, January 1950, in an article by Red Smith entitled "The Outboard Heresy" and contained in his regular column known as "Views of the Sport."

3. In an article by Everett B. Morris entitled, "Boat Men Hail C.B. Waterman For Invention." Though not identified as to source, the article reprinted by Kiekhaefer Mercury public relations staff was first published in a New York newspaper in early January 1950.

4. *New York Herald Tribune*, January 1950, in an article by Red Smith, loc. cit.

5. In a letter from Cameron B. Waterman of Detroit, Michigan to E.C. Kiekhaefer of January 27, 1955.

6. Contained in a complaint to the Federal Trade Commission, Bureau of Antimonopoly, filed by E.C. Kiekhaefer and the Kiekhaefer Corporation on March 22, 1955.

7. In a letter from The State of Wisconsin, State Historical Society, Raymond S. Siversind, Supervisor, Historic Sites, to Mr. Frederic Sammond of the legal firm of Fairchild, Foley and Sammond, representing the Kiekhaefer Corporation, of July 19, 1955.

8. In a letter to M.E. Klitsner of the legal firm of Fairchild, Foley and Sammond, from E.C. Kiekhaefer of February 18, 1957.

9. Taken from the Official Report of Proceedings Before The Federal Trade Commission, Docket No. 5882, In the Matter of Outboard, Marine & Manufacturing Company, Pages 191 to 207, of February 13, 1952.

10. From the 1951 annual report to stockholders of the Outboard Marine & Manufacturing Company.

11. In a memorandum from Guy S. Conrad to E.C. Kiekhaefer of February 11, 1952.

12. Official Report of Proceedings Before The Federal Trade Commission, loc. cit.

13. In a letter from Max H. Fruhauf, attorney with the legal firm of Butzel, Levin, Winston & Quint of Detroit, Michigan, to Guy Conrad of the Kiekhaefer Corporation of May 19, 1952.

14. From the annual report to stockholders of the Outboard Marine & Manufacturing Company of 1956.

15. In a draft complaint to the Federal Trade Commission prepared on June 14, 1957 by Thomas B. Fifield of the legal firm of Fairchild, Foley & Sammond, Milwaukee, Wisconsin on behalf of E.C. Kiekhaefer and the Kiekhaefer Corporation.

16. In a transcribed telephone conversation between Willis Blank and Tom Fifield of the legal firm of Fairchild, Foley & Sammond, Milwaukee, Wisconsin on October 2, 1958.

17. In a letter to Mrs. Nadean Brummett from E.C. Kiekhaefer of March 20, 1972.

18. The unedited version of this story appeared in *Boy's Life*, written by P. Reidy. It was reprinted by *Reader's Digest* in the March 1990 issue.

19. Interview of Ted Karls on August 6, 1987 at his home in Fond du Lac, Wisconsin.

20. In a letter to Mrs. Nadean Brummett by E.C. Kiekhaefer of March 20, 1972.

21. Interview of John C. Hull of July 31, 1988 in Fond du Lac, Wisconsin.

22. Interview of Charles D. Strang of May 20, 1987 at the OMC salt water proving grounds at Stuart, Florida.

23. Ibid.

24. In a memo entitled "Rumor Campaign by Mercury Competitors" issued to all Mercury outboard dealers on July 2, 1957, signed by E.C. Kiekhaefer.

25. In a transcribed telephone interview between E.C. Kiekhaefer and OMC President Joseph G. Rayniak on September 28, 1957.

26. Ibid. 27. Ibid. 28. Ibid.

29. In a letter from William C. Scott, executive vice president and general manager of Outboard Marine Corporation to E.C. Kiekhaefer of October 9, 1957.

30. In a letter from E.C. Kiekhaefer to William C. Scott, executive vice president and general manager of OMC of October 11, 1957.

31. *Cincinnati Post*, March 6, 1958 in an article entitled, "Mercury Easy Winner in 'Fifty' Challenge."

32. In a letter from John S. Hawley, vice president of sales for Kiekhaefer Corporation to Thomas B. Fifield of the legal firm of Fairchild, Foley & Sammond, Milwaukee, Wisconsin, of February 6, 1959.

33. Interview of John C. Hull, loc. cit.

34. Ibid. 35. Ibid.

36. In a letter to Hal Steeger of *Argosy* magazine from E.C. Kiekhaefer of June 21, 1961.

37. In a complaint to the Federal Trade Commission, Bureau of Antimonopoly from E.C. Kiekhaefer of March 22, 1955.

38. In a letter to M.E. Klitsner of the legal firm of Fairchild, Foley and Sammond, representing the Kiekhaefer Corporation, from E.C. Kiekhaefer of March 4, 1957.

39. Interview of Wilson Snyder of August 5, 1987 at his home in Fond du Lac, Wisconsin.

40. Ibid.

CHAPTER TWENTY-THREE
Merger Mania

1. In a liberally annotated letter, never mailed, to The Honorable Senator Alexander Wiley, United States Senator for the State of Wisconsin, of May 18, 1957 from E.C. Kiekhaefer.

2. Interview of Donald E. Castle at his offices in Fond du Lac, Wisconsin on August 4, 1987. Castle was controller of the Kiekhaefer Corporations from 1952 to 1957.

3. Ibid. 4. Ibid. 5. Ibid.

6. Interview of Charles D. Strang of May 20, 1987 at the OMC Salt Water Proving Grounds at Stuart, Florida.

7. Ibid.

8. In a letter to Daimler-Benz A.-G. of Stuttgart-Untertuerkheim, Germany from E.C. Kiekhaefer of May 2, 1955.

9. Interview of Rose Smiljanic of August 14, 1986 at her home in Fond du Lac, Wisconsin.

10. Interview of Charles D. Strang, loc. cit.

11. Hand written by E.C. Kiekhaefer on the back of an American Airlines post card, contained in the briefing files of the proposed Kiekhaefer Corporation/Daimler-Benz merger. Undated, but assumed written en route to the scheduled closing meeting at the Carlyle Hotel in New York City of January 1956.

12. Interview of Donald E. Castle, loc. cit.

13. In a two-page telegram to E.C. Kiekhaefer of April 16, 1957 from Carl F. Giese, president, Daimler-Benz of North America, Inc.

14. In a letter from Mr. Benjamin C. Carter, executive vice president, Food Machinery and Chemical Corporation, Machinery Divisions, of May 9, 1957 to Willis Blank.

15. Interview of Charles D. Strang, loc. cit.

16. Ibid.

17. In a memo from Willis Blank to E.C. Kiekhaefer of March 10, 1959 entitled: "Inquiries Regarding Licensing to Manufacture in Foreign Countries And Requests for Stock, Financial Reports, Mergers, etc."

18. In a letter from R.C. Ingersoll, chairman of the board of Borg-Warner Corporation from E.C. Kiekhaefer of September 20, 1957.

19. In a draft of a letter never mailed to R.C. Ingersoll, chairman of Borg-Warner Corporation from E.C. Kiekhaefer of October 9, 1957.

20. In a transcribed telephone conversation between E.C. Kiekhaefer, Willis Blank and J.J. Flannery of National Automotive Fibres Incorporated, NAFI, on November 6, 1957.

21. In a letter from Richard S. Frazer, assistant to the president for new business

interests of Motorola Corporation of Chicago, Illinois to E.C. Kiekhaefer of December 2, 1957.

22. In a transcribed telephone conversation between Willis Blank and Mr. Davis of the Chrysler Corporation on January 13, 1957.

23. Interview of Charles D. Strang, loc. cit.

CHAPTER TWENTY-FOUR
Bowling Bride

1. In a press release issued by E.C. Kiekhaefer on February 8, 1961.

2. *Time*, May 19, 1961 in a brief article entitled "Diversified Success."

3. *Brunswick, The Story of an American Company from 1845 to 1985*, Rick Kogan, published by the Brunswick Corporation in 1985.

4. Ibid. 5. Ibid.

6. *Time*, May 19, 1961, loc. cit.

7. 1960 Annual Report, American Machine and Foundry Company.

8. In a letter from Lucien L. Leeds of Ocean Issues Limited to E.C. Kiekhaefer of February 4, 1960.

9. In a handwritten diary of the proceeds of meetings with AMF maintained by Charles D. Strang on March 16-17, 1961.

10. Ibid.

11. In a letter from Brunswick President B.E. Bensinger to E.C. Kiekhaefer of September 10, 1958.

12. Ibid.

13. *Brunswick, The Story of an American Company from 1845 to 1985*, Rick Kogan, loc. cit.

14. Ibid. 15. Ibid.

16. From transcribed telephone conversations between Willis Blank and B.E. Bensinger on November 2, 1960 and November 8, 1960.

17. Interview of Robert C. Anderegg of February 23, 1987.

18. In a press release issued by E.C. Kiekhaefer on February 8, 1961, loc. cit.

19. In a letter from F.E. Troy, vice president of Brunswick Corporation to E.C. Kiekhaefer of March 13, 1961.

20. In a private work sheet developed by E.C. Kiekhaefer entitled, "Pros and Cons", though undated, was almost certainly produced in the summer of 1961.

21. Interview of Charles D. Strang of January 26, 1989 at the OMC corporate offices in Waukegan, Illinois.

22. In a letter from Arthur A. Burck of McClellan & Burck, Inc., consultants, to E.C. Kiekhaefer of August 1, 1961.

23. Interview of Alan Edgarton of July 30, 1987 at his offices in Fond du Lac, Wisconsin.

24. Interview of Robert C. Anderegg, loc. cit.

25. Interview of Rose Smiljanic of August 14, 1986 at her home in Fond du Lac, Wisconsin.

26. Interview of Alan Edgarton, loc. cit.

27. Interview of Robert C. Anderegg, loc. cit.

28. Ibid. 29. Ibid.

30. In a letter/memorandum from B.E. Bensinger to E.C. Kiekhaefer of August 30, 1961.

31. *Milwaukee Journal*, October 13, 1961, in an article by Emil A. Schneider, staff writer. Carl would confirm the accuracy of the final amount received from the Brunswick shares in a letter to Schneider of October 18, 1961 in which he writes, "It was a very accurate description of our interview and was well written."

32. In a telegram mailed to all U.S. and Canadian distributors and branch managers by Tom King on about August 15, 1961.

33. Interview of Rose Smiljanic, loc. cit.

34. Ibid. 35. Ibid.

CHAPTER TWENTY-FIVE
The Christmas Mutiny and Lessons of History

1. Contained in the 1930 Johnson Tilting Stern Drive brochure produced by the Johnson Motor Company.

2. Interview of James R. Wynne, October 17, 1987 at the offices of Wynne Marine in Miami, Florida.

3. Interview of John C. Hull of July 31, 1988 in Fond du Lac, Wisconsin.

4. Ibid.

5. Interview of John C. Hull, loc. cit.

6. Ibid.

7. Interview of James R. Wynne, loc. cit.

8. Ibid.

9. Contained in the 1930 color flyer produced by the Johnson Motor Company entitled, "Outboard Motors and Boats by Johnson," with text written and signed by Warren Ripple, president.

10. *Go Boating*, August 1973, in a James R. Wynne interview with Ann Adams entitled, "Sterndrive Charts Prosperous Course."

11. *Boating Industry*, October 1989, in an interview with James R. Wynne by Dick Porter entitled, "Elder Statesman: James R. Wynne."

12. Interview of James R. Wynne of January 2, 1989 at the offices of Wynne Marine in Miami, Florida.

CHAPTER TWENTY-SIX
The Great Stern Drive Conspiracy

1. Interview of Charles D. Strang of May 20, 1987 at OMC's salt water proving grounds in Stuart, Florida.

2. Ibid. 3. Ibid.

4. In a letter to Robert H. Soelke of Miami Outboard and Sport Sales, from E.C. Kiekhaefer of March 10, 1958.

5. Interview of Charles D. Strang, loc. cit.

6. Interview of James R. Wynne, October 17, 1987 at the offices of Wynne Marine in Miami, Florida.

7. Interview of Harald Wiklund, president of Volvo-Penta, 1949-1977, on December 13, 1990 from his home in Gothenburg, Sweden.

8. *Mirror News*, Los Angeles, California, July 24, 1958 in an article entitled "Outboard Crosses Atlantic in 10 Days".

9. Ibid.

10. In a press release issued on August 6, 1958, and sent to "All Mercury Distributors, Dealers and Area Representatives".

11. Ibid.

12. In an article clipped from a Massachusetts newspaper of approximately July 25, 1958, otherwise unidentified. Contained in a file assembled by Carl Kiekhaefer of news clippings and releases concerning the Atlantic outboard crossing by Wynne.

13. Interview of Charles D. Strang, loc. cit.

14. Interview of Harald Wiklund, loc. cit.

15. In a memo to C.D. Strang and T.B. King of January 30, 1960 from E.C. Kiekhaefer, writing from Lake X.

16. In a letter to Thomas B. Fifield of the legal firm of Foley, Sammond & Lardner of Milwaukee, Wisconsin, from Willis Blank of June 22, 1960.

17. Interview of Harald Wiklund, loc. cit.

18. Interview of James R. Wynne, loc. cit.

19. Interview of Harald Wiklund, loc. cit.

20. Evidence of the terms of the proposed Mercury-Volvo agreement are contained in notes assembled by E.C. Kiekhaefer on June 30, 1960 entitled, "Questions Re Mercury-Volvo Agreement."

21. Interview of Harald Wiklund, loc. cit.

22. Ibid.

23. Contained in a list of 12 points to be covered by E.C. Kiekhaefer at the 1960 sales and marketing conference, discussing the state of the industry.

24. In a speech delivered by E.C. Kiekhaefer to the Kiekhaefer Mercury distributors meeting held on January 13, 1961 in Fond du Lac, Wisconsin.

25. Interview of Charles D. Strang, loc. cit.

26. Contained within meeting notes from an "Outdrive Program Meeting" held Saturday, March 18, 1961, conducted by Thomas B. King, prepared in memo form and sent to E.C. Kiekhaefer on March 22, 1961. Marked "Confidential."

27. Interview of Charles D. Strang, loc. cit.

28. In a report by Ralph E. Lambrecht, OMC Stern Drive division of Outboard Marine Corporation, of July 1979.

29. Patent number 3,376,842, issued on April 9, 1968 to J.R. Wynne for "Boat Propulsion Mechanism." Patents number 3,368,516 and 3,368,517 were issued to C.E. MacDonald on February 13, 1968.

30. Interview of Harald Wiklund, loc. cit.

CHAPTER TWENTY-SEVEN
Crisis of Control

1. Contained in a letter to Jim Martenhoff, boating journalist, from E.C. Kiekhaefer of May 15, 1962.

2. In a letter from R.A. McCarroll to Carl Kiekhaefer of August 18, 1961.

3. In a memo to E.C. Kiekhaefer from Charles D. Strang of September 9, 1961.

4. Contained in the 1955 Mercury line brochure.

5. Interview of Charles D. Strang of May 20, 1987 at the OMC salt water proving grounds at Stuart, Florida.

6. Ibid.

7. In a letter to "an enthusiastic Mercury owner in Massachusetts," quoted in a letter from Tom King to "All Mercury Dealers, Distributors, Branch Managers and Area Representatives," of April 27, 1960.

8. In a memo issued to "All Department Heads, Mail Clerks and Switchboard Operators," from E.C. Kiekhaefer of January 5, 1962.

9. In a memo from B.E. "Ted" Bensinger to various Brunswick staff members of February 9, 1962.

10. In a memo to Ted Bensinger from E.C. Kiekhaefer of October 26, 1962.

11. In a letter to N.A. George, vice president of Brunswick Corporation from E.C. Kiekhaefer of March 23, 1962.

12. Contained in a memo from H.W. Wiedeman entitled "Brunswick National Agreements," to E.C. Kiekhaefer on June 4, 1962.

13. In a memo from T.B. King to E.C. Kiekhaefer entitled, "Review of Brunswick-Kiekhaefer Relationship During the 1962 Model Year," of October 26, 1962.

14. In a memo from B.S. Kubale of the legal firm of Foley, Sammond & Lardner of Milwaukee, to Frederic Sammond and Marvin E. Klitsner of the same firm, regarding "E.C. Kiekhaefer - Repurchase," of February 19, 1963.

15. In a letter to B.E. Bensinger, president of Brunswick Corporation from E.C. Kiekhaefer of January 29, 1963.

16. In a draft of a letter never mailed to Howard F. Baer, chairman of Aloe, division of Brunswick in St. Louis, from E.C. Kiekhaefer of January 30, 1963, and revised on February 6, 1963.

17. *Brunswick, The Story of an American Company from 1845 to 1985*, written by Rick Kogan under commission by the Brunswick Corporation.

18. A handwritten letter to Rose Smiljanic, mailed from Frankfurt, Germany from E.C. Kiekhaefer on August 9, 1964.

19. In a draft letter to A.S. Puelicher, Carl's close friend, fellow Brunswick director and chairman of the Marshall & Ilsley Bank of Milwaukee, from E.C. Kiekhaefer of August 14, 1964.

20. In a letter to J.L. Hanigan from E.C. Kiekhaefer of September 17, 1964.

21. Interview of Charles D. Strang, loc. cit.

22. Ibid.

23. Contained in a list of complaints against Brunswick entitled, "Staff Versus Line Management," compiled by E.C. Kiekhaefer sometime during in 1963.

24. Handwritten notes concerning the history of relationships between Brunswick Corporation and Kiekhaefer Corporation by E.C. Kiekhaefer, produced sometime in the latter half of 1963.

CHAPTER TWENTY-EIGHT
Days of Darkness

1. In a letter never mailed to fellow Brunswick Corporation director Walter M. Heymann, also a director of the First National Bank of Chicago, from E.C. Kiekhaefer of January, 22, 1969.

2. Interview of John C. Hull of July 31, 1988 at his home in Fond du Lac, Wisconsin.

3. Interview of Charles D. Strang, January 26, 1989, in the office of the chairman, Outboard Marine Corporation, Waukegan, Illinois.

4. Ibid. 5. Ibid.

6. Reconstructed from a detailed log of the conversation kept by Charles D. Strang, completed immediately after the conversation, with further editing and additions being made following subsequent conversations with Carl Kiekhaefer. Strang has kept telephone and meeting logs throughout his career, and today, as chairman of Outboard Marine Corporation, he continues this practice.

7. Interview of Charles D. Strang, loc. cit.

8. Ibid.

9. Edward A. Stephan was a general partner in the Chicago legal firm of Mayer, Friedlich, Spiess, Tierney, Brown & Platt.

10. Interview of Charles D. Strang, loc. cit.

11. Ibid. 12. Ibid. 13. Ibid.

14. From the transcribed dictabelt recording of the conversation between Charles D. Strang and Jack Hanigan of May 29, 1964.

15. Interview of Charles D. Strang, May 20, 1987, at the OMC salt water proving grounds in Stuart, Florida.

16. Ibid.

17. Interview of Charles D. Strang of January 26, 1989, loc. cit.

18. Ibid.

19. Interview of James H. Jost of May 20, 1987, at the OMC salt water proving grounds in Stuart, Florida.

20. In a letter to Simpson E. Meyers, member of the board of directors of Brunswick Corporation, from E.C. Kiekhaefer of August 6, 1965.

21. In a draft letter never mailed to Mr. M. Harned, Hughes Tool Company, from E.C. Kiekhaefer of May 1, 1965.

22. Taken from the A.O.K. Club card of James R. Wynne.

23. Interview of C.W. "Doc" Jones of December 13, 1990.

24. Interview of James H. Jost of December 14, 1990.

CHAPTER TWENTY-NINE
Fast, Faster, Fastest

1. Vincent Thomas Lombardi, 1913-1970, head coach of the Green Bay Packers of the National Football League, who in nine seasons (1959-1967) led the team to five NFL championships and victory in the first two Super Bowls. Carl Kiekhaefer was known to have admired the tough, winning style of the legendary coach, and shared the same all-or-nothing winning attitude.

2. Language contained in a contract prepared March 14, 1960 between the Kiekhaefer Corporation, represented by Thomas B. King, and by Hubert Entrop and Jack Leek.

3. *Little Rock Gazette*, Little Rock, Arkansas, July 6, 1958, in a column by Hank Bowman called "Water Line," in a piece entitled, "U.S. Outboard Title Can't Last." Bowman was an eye-witness to the attempts by Entrop at Lake X in early 1958.

4. In a letter to Bob Brister, Outdoors Editor of *The Houston Chronicle*, from E.C. Kiekhaefer of June 20, 1958.

5. Ibid.

6. Interview of Charles D. Strang of May 20, 1987 at the OMC salt water proving grounds in Stuart, Florida.

7. In a letter to Jimmy Jost of the American Power Boat Association, from E.C. Kiekhaefer of June 27, 1958.

8. In a letter to Bob Brister, editor of the *Houston Chronicle*, loc. cit.

9. Interview of C.W. "Doc" Jones of December 13, 1990 from his home in Comanche, Texas.

10. In an Evinrude press release, March 29, 1960.

11. In a press release from the Kiekhaefer Corporation dated May 5, 1960.

12. In a letter to Elwin A. Andrus, of the legal firm of Andrus and Starke, patent attorneys, from E.C. Kiekhaefer of August 15, 1960.

13. In a letter to E.C. Kiekhaefer from O.F. Christner of Quincy Welding in Quincy, Illinois of April 8, 1960.

14. In a letter never mailed to James R. "Jimmy" Jost, from E.C. Kiekhaefer of May 16, 1960.

15. Contained in a lengthy memorandum from C.D. Strang to E.C. Kiekhaefer entitled, "Items of Possible Interest" dated December 6, 1960.

16. Interview of James H. "Jimmy" Jost on December 14, 1990 from his home in Wauwatosa, Wisconsin. Jost retired from OMC in early 1990, following a remarkable 45 year career of accomplishment in the marine industry, including a quarter century at OMC alone.

17. Ibid. 18. Ibid.

19. Interview of C.W. "Doc" Jones, loc. cit.

20. Ibid.

21. Interview of James H. Jost, loc. cit.

22. *Sports Illustrated*, September 23, 1974, in an article entitled, "Grim climax to a thunderous run."

23. Interview of James H. Jost, loc. cit.

24. *Lakeland Boating*, February 1990 in an article entitled, "New Speed Record Set."

CHAPTER THIRTY
The Limits of Patience

1. In a draft of a letter never sent to Walter M. Heymann, director of Brunswick and of the First National Bank of Chicago, from E.C. Kiekhaefer of August 9, 1965.

2. From a speech entitled, "These Are Wonderful Years, Why?" delivered by Carl Kiekhaefer at the Savoy Hilton in New York City on January 18, 1964, on the occasion of the 25th anniversary of the Kiekhaefer Corporation.

3. Ibid.

4. Interview of Charles Freeman Alexander, Jr. of December 1, 1986.

5. Ibid.

6. Interview of Owen L. "Billy" Steele of August 4, 1987, at his home in Fond du Lac, Wisconsin.

7. Interview of Fred "Fritz" Shoenfeldt of July 31, 1987 at his home in Wautoma, Wisconsin.

8. Interview of Wilson Snyder of August 5, 1987 at his home in Fond du Lac, Wisconsin.

9. Interview of Fred Shoenfeldt, loc. cit.

10. Ibid.

11. Interview of Wayne Herbert Predmore of April 5, 1987.

12. Interview of Neil Freer Oleson of July 30, 1987 at his home in Fond du Lac, Wisconsin.

13. Ibid. 14. Ibid.

15. Interview of Fred "Fritz" Shoenfeldt, loc. cit.

16. Interview of Wilson Snyder, loc. cit.

17. Ibid. 18. Ibid. 19. Ibid.

20. Interview of Owen L. "Billy" Steele, loc. cit.

21. Interview of Francis J. "Frank" Scalpone on February 11, 1987, at Cy's Restaurant, during the Miami International Boat Show. Also in attendance were Joe Swift, John Crouse and Karine Rodengen.

22. Interview of Rose Smiljanic of February 15, 1987 at the author's home in Fort Lauderdale, Florida.

23. Ibid.

24. Excerpts from "A Request To All The Citizens Of Fond du Lac From The Kiekhaefer Corporation," a paid insertion advertisement prepared for the Fond du Lac *Commonwealth Reporter*, February 1964. A long series of paid ads and editorials can be found concerning this controversy from February to July, 1964.

25. *Commonwealth Reporter*, Fond du Lac, Wisconsin, April 26, 1964.

26. Contained in a letter to Peter P. Weidenbruch, president of Damrow Brothers Company in Fond du Lac, from E.C. Kiekhaefer of May 22, 1964.

27. Interview of Wilson Snyder, loc. cit.

28. Ibid.

29. Contained in remarks prepared for the Brunswick Corporation board of directors meeting, and delivered by Carl Kiekhaefer at the Kiekhaefer Corporation's new assembly building auditorium, June 1, 1965.

30. *Fortune*, July 15, 1966.

31. *Forbes*, August 15, 1965, in an article entitled, "Heavy, Heavy ... is the load of debt and uncertainty still hanging over Brunswick Corp."

CHAPTER THIRTY-ONE
The Stuff of Legends

1. Contained in a letter to Charles F. Chapman, the venerable and widely respected editor of MoToR BoatinG magazine, from E.C. Kiekhaefer of December 28, 1967.

2. In a letter to Mrs. Maurine C. Dupuy of North Miami, Florida from E.C. Kiekhaefer of July 10, 1978.

3. Interview of Joe Swift on February 11, 1987, at Cy's Restaurant in Miami Beach, during the Miami International Boat Show. Also in attendance were Frank Scalpone, Jerry Haber, John Crouse and Karine Rodengen.

4. Quoted in the speech delivered by Charles D. Strang at the *Kiekhaefer - St. Augustine Classic* offshore race testimonial dinner held for Carl Kiekhaefer in St. Augustine, Florida on September 11, 1982.

5. Interview of Wayne Herbert Predmore of April 5, 1987. Predmore is as of this writing, Aviation Manager of Brunswick Corporation, reporting directly to Brunswick Corporation Chairman of the Board Jack Reichert.

6. Interview of Frank Scalpone on February 11, 1987, at Cy's Restaurant in Miami Beach, during the Miami International Boat Show. In attendance were Joe Swift, Jerry Haber, John Crouse and Karine Rodengen.

7. Interview of Wayne Predmore, loc. cit.

8. In a letter to Senator William Proxmire from E.C. Kiekhaefer of April 21, 1980.

9. In a letter to Zack Taylor, boating editor of *Sports Afield*, from E.C. Kiekhaefer of January 10, 1972, with the "fertilizer umbrella" quote added from a letter to Senator William Proxmire from E.C. Kiekhaefer of April 21, 1980.

10. Interviews of Gene Wagner, April 8 and April 10, 1987, at Mercury Marine's salt water proving grounds at MerCabo, near Placida, Florida, and at Cabbage Key, Florida.

11. Interview of Wilson Snyder of August 5, 1987 at his home in Fond du Lac, Wisconsin.

12. Interview of Wilson, loc. cit.

13. Contained in various letters from Carl to others, including one to Lee S. Siebert, Mark Twain Marine Industries in West Frankfort, Illinois from E.C. Kiekhaefer of March 30 , 1976.

14. Interview of Fred "Fritz" Shoenfeldt at his home in Wautoma, Wisconsin, of July 31, 1987.

15. Interview of Wilson Snyder, loc. cit.

16. Interview of Fred "Fritz" Shoenfeldt, loc. cit.

17. Interview of Sylvan "Ham" Hamberger of February 13, 1987, at the Miami International Boat Show.

18. Interview of Joe Swift, loc. cit.

19. Ibid.

CHAPTER THIRTY-TWO
A Bitter Reward

1. In a letter drafted but never mailed to Edward H. Nabb of the legal firm of Harrington, Harrington & Nabb in Cambridge, Maryland, from E.C. Kiekhaefer of December 28, 1967.

2. Contained in a speech given to Mercury distributors, branch managers, field sales and service representatives and corporate executives by E.C. Kiekhaefer on the occasion of the 1968 Sales Conference held on August 14-18 in Fond du Lac, Wisconsin.

3. Contained in the 1968 Outboard Marine Corporation Annual Report, in the opening remarks signed by Ralph S. Evinrude, chairman of the board, and William C. Scott, president.

4. In a letter never mailed to fellow Brunswick director Walter M. Heymann, drafted by E.C. Kiekhaefer of January 22, 1969.

5. In a letter of dismissal to P.D. Humleker from E.C. Kiekhaefer of September 26, 1967.

6. In a letter to J.A. Puelicher from E.C. Kiekhaefer of October 7, 1967.

7. Ibid.

8. Contained in a report by Will Piette of AC Industrial Electronics of Milwaukee, Wisconsin to Mr. John Gassaway, Security Chief of the Kiekhaefer Corporation on October 30, 1967. Accompanying the report was a full description of the search conducted at the residence of E.C. Kiekhaefer.

9. In a transcribed telephone conversation between E.C. Kiekhaefer and Jack Hanigan, president of Brunswick Corporation, on July 22, 1966.

10. Ibid. 11. Ibid.

12. In a memo to R.C. Anderegg from A.A. Hauser of March 24, 1967.

13. In a letter to George Trimper from E.C. Kiekhaefer of January 17, 1967.

14. Contained in a memo from Armand A. Hauser to E.C. Kiekhaefer entitled "Competitive Activity," undated, but certainly drafted during the late summer of 1967.

15. *New York Times*, February 10, 1967, in an article by Steve Cady entitled, "Gas Turbine Boat's Performance Not Its Best, Says Its '66 Driver."

16. In a letter to Melvin Crook, editor of *Yachting* magazine from E.C. Kiekhaefer of March 11, 1967.

17. *The Boston Sunday Globe*, February 12, 1967.

18. Contained in a statement prepared by E.C. Kiekhaefer following the appearance of articles containing comments by Jim Wynne of alleged racing impropriety in the 1967 Sam Griffith Memorial Race, of approximately the first week of March 1967.

19. In a memo to Herman Stieg from Armand Hauser entitled, "Operation 100," of September 20, 1967.

20. Ibid.

21. Contained in an undated draft of a letter to Brunswick president Jack Hanigan by E.C. Kiekhaefer, written in the early months of 1966.

22. In a draft of a letter never sent to J.A. "Jack" Puelicher by E.C. Kiekhaefer of December 23, 1966.

23. In a letter from J.L. Hanigan to E.C. Kiekhaefer of October 13, 1969.

24. Contained in the "Suggested Terms To Implement Mr. Hanigan's Latest Proposal" drafted by E.C. Kiekhaefer on November 21, 1969.

25. Contained in the E.C. Kiekhaefer "Retirement Memorandum" drafted in November 1969.

26. Contained in transcripts of the November 25, 1969 management meeting held in the Kiekhaefer Mercury auditorium of plant 10 at 9:30 a.m. to announce K.B. "Brooks" Abernathy as the new president of the company.

27. *Milwaukee Sentinel*, November 26, 1969.

28. Contained in transcripts of the November 25, 1969 management meeting, loc. cit.

29. Interview of Rose Smiljanic of August 14, 1986 at her home in Fond du Lac, Wisconsin.

30. Ibid.

31. Interview of R.C. "Bob" Anderegg of February 23, 1987.

32. Interview of Charles F. Alexander, Jr. of December 1, 1986.

33. Interview of Rose Smiljanic, loc. cit.

34. Ibid. 35. Ibid

36. Interview of Fred "Fritz" Shoenfeldt of July 31, 1987 at his home in Wautoma, Wisconsin.

37. In a letter to J.L. Hanigan, president of Brunswick Corporation from E.C. Kiekhaefer of December 30, 1969, with copies indicated to "Board of Director Members" and "K.B. Abernathy." The letter was enclosed in a Christmas card mailed to Fred "Fritz" Shoenfeldt.

38. Contained in a letter to Mr. C.K. Davis of Dallas, Texas from E.C. Kiekhaefer of January 12, 1969.

39. Contained in a letter to Ray Waters of Cedarburg, Wisconsin from E.C. Kiekhaefer of October 2, 1970.

40. Contained in a letter to A.A. Hauser from E.C. Kiekhaefer of March 24, 1970.

CHAPTER THIRTY-THREE
Second Lifetime

1. Contained in a letter to Roger C. Hanks of Midland, Texas, from E.C. Kiekhaefer of March 4, 1972.

2. In a letter to Jim Roe of Wheaton, Illinois from E.C. Kiekhaefer of February 7, 1970.

3. In a letter to John B. Gunter of Gardner Advertising Company, Inc. of St. Louis, Missouri from E.C. Kiekhaefer of December 3, 1969.

4. In a letter to C.D. Strang of Outboard Marine Corporation from E.C. Kiekhaefer of January 12, 1970.

5. Interview of Owen L. "Billy" Steele at his home in Fond du Lac, Wisconsin of August 4, 1987.

6. In a letter memo from John Hull to E.C. Kiekhaefer of February 8, 1965.

7. In a memo from R.C. Anderegg and A.A. Hauser to E.C. Kiekhaefer of January 11, 1967.

8. In a memo to J.L. Hanigan from E.C. Kiekhaefer of March 8, 1967.

9. In a letter to Master William Pike of Oshkosh, Wisconsin from E.C. Kiekhaefer of January 22, 1971.

10. Contained in a letter from Thomas E. Holloran of the legal firm of Wheeler & Fredrikson of Minneapolis, Minnesota to Carl Kiekhaefer of March 27, 1967.

11. Interview of John C. Hull of July 31, 1988 at his home in Fond du Lac, Wisconsin.

12. Ibid. 13. Ibid.

14. In a letter to Brian S. Inglis of the Ford Motor Company of Australia from E.C. Kiekhaefer of September 3, 1970.

15. Interview of Jack Cotter of July 29, 1987 at his home in Fond du Lac, Wisconsin.

16. In a letter to the Honorable William A. Steiger, Member of Congress, from E.C. Kiekhaefer of February 15, 1974.

17. Contained in the program for the "Annual Midwest Championship Snowmobile Races" on January 11-12, 1975.

18. In a letter to G. Contegiacomo of Aspera Motors in Torino, Italy from E.C. Kiekhaefer of September 8, 1970.

19. In a letter to Richard Witkin of the *New York Times* from E.C. Kiekhaefer of August 3, 1970.

20. In the letter to Brian S. Inglis, loc. cit.

21. *Commonwealth Reporter*, Fond du Lac, Wisconsin, November 16, 1970 in an article entitled "New Phase in Career of E.C. Kiekhaefer."

22. *Commonwealth Reporter*, Fond du Lac, Wisconsin, Saturday, November 14, 1970 in an article by Harley Bucholz entitled, "E.C. Kiekhaefer Starts New Industrial Complex on Ledge."

23. In a letter to C.P. "Pete" Abele of Cedarburg, Wisconsin from E.C. Kiekhaefer of December 22, 1970.

24. *Commonwealth Reporter*, Fond du Lac, Wisconsin, November 16, 1970, loc. cit.

CHAPTER THIRTY-FOUR
The Invisible Family

1. Leo Nikolaveich Tolstoi in *Anna Karenina*, 1875-1877, Part 1, Chapter 1.

2. Interview of Charles D. Strang of January 26, 1989 in his offices at Outboard Marine Corporation, Waukegan, Illinois.

3. Interview of Anita Rae Kiekhaefer of August 16, 1987 at the home of Fred and Carol Kiekhaefer in Fond du Lac, Wisconsin.

4. Interview of Helen Kiekhaefer Wimberly of September 26, 1989.

5. Interview of Freda, Mrs. Carl Kiekhaefer of August 16, 1987, at the home of Fred and Carol Kiekhaefer in Fond du Lac, Wisconsin.

6. Interview of Rose Smiljanic of August 14, 1986, at her home in Fond du Lac, Wisconsin.

7. Interview of Helen Kiekhaefer Wimberly, loc. cit.

8. Ibid. 9. Ibid.

10. Interview of Anita Rae Kiekhaefer of August 16, 1987, loc. cit.

11. In a handwritten letter from Anita Kiekhaefer to Carl Kiekhaefer of January 22, 1969.

12. In a letter to C.A. Pound of the Baird Hardware Company of Gainesville, Florida from E.C. Kiekhaefer of April 25, 1955.

13. Contained in a listing of disparaging representations concerning Robert Burford, entitled "Robert Burford", prepared by E.C. Kiekhaefer on April 18, 1961.

14. In a handwritten letter from Helen Kiekhaefer Burford to Carl Kiekhaefer of September 28, 1961.

15. Ibid.

16. In a sworn deposition given under authority of the Circuit Court of the Twelfth Judicial Circuit of Florida, Helen Burford vs. Robert Burford, Case No. 18195, in a direct examination of Helen Burford by J. Hardin Peterson, Jr., Esq. on January 27, 1965 in Lakeland, Florida.

17. Ibid, in redirect examination.

18. In a letter to Carol Stafford of Madison, Wisconsin from E.C. Kiekhaefer of October 30, 1971.

19. In a letter to Richard H. Bennet of Jensen Beach, Florida from E.C. Kiekhaefer of February 2, 1973.

20. Interview of Freda, Mrs. Carl Kiekhaefer of August 16, 1987, loc. cit.

21. Ibid. 22. Ibid. 23. Ibid.

24. In a handwritten note from Freda Kiekhaefer to Carl Kiekhaefer received by Carl on July 3, 1965.

25. Interview of Fred Kiekhaefer of August 16, 1987 at his home in Fond du Lac, Wisconsin.

26. Interview of Rose Smiljanic, loc. cit.

27. Interview of Fred Kiekhaefer of August 16, 1987, loc. cit.

28. In a letter to Helen Kiekhaefer Wimberly from E.C. Kiekhaefer of November 7, 1973.

CHAPTER THIRTY-FIVE
Dashing Through the Snow

1. In a letter to the Honorable William A. Steiger, Congressman from the state of Wisconsin, from E.C. Kiekhaefer of December 13, 1971. For many years Carl maintained a regular correspondence with Congressman Steiger, to express his firm convictions on American foreign and domestic policy issues.

2. In a memo from Don Thompson to E.C. Kiekhaefer entitled "Summary of Industry Impact and Reactions - Kiekhaefer Aeromarine Motors 1971" of July 28, 1971.

3. In the Kiekhaefer Aeromarine Motors press release dated March 21, 1971, announcing the victories at Booneville, New York during the World Series U.S. Snowmobile Open Class competition.

4. In a memo from Don Thompson to E.C. Kiekhaefer entitled "Summary of Industry Impact and Reactions - Kiekhaefer Aeromarine Motors 1971" of July 28, 1971.

5. Ibid.

6. In a letter to Stan McRoberts, president of Northway Snowmobiles Limited of Pointe Claire, Quebec, Canada from E.C. Kiekhaefer of August 19, 1971.

7. In a draft of a letter never sent to Wayne J. Roper of May 1, 1973.

8. In a letter to R. Noel Penny of Solihull, Warwickshire, England from E.C. Kiekhaefer of November 29, 1971.

9. *Commonwealth Reporter*, Fond du Lac, Wisconsin, August 13, 1971, in an article entitled "Huge Machine Arrives in Area."

10. In a letter to Mr. R.H. "Dick" Bennet of Venture Out at Indian River in Jensen Beach, Florida from E.C. Kiekhaefer of February 2, 1973.

11. In a letter to David G. Findlay, president of MHV Industries, Limited of Ottawa,

Ontario, Canada from E.C. Kiekhaefer of January 10, 1972.

12. In a letter to Mel Crook, editor of *Yachting* from E.C. Kiekhaefer of March 17, 1972.

13. In the letter to R.H. Bennet, loc. cit.

14. In a letter to J.R. Krieger, Yacht Club Argentino, Buenos Aires, Argentina from E.C. Kiekhaefer of December 23, 1972.

15. From the Bombardier Limited and Kiekhaefer Aeromarine Motors joint press release of May 15, 1972.

16. *The Toronto Star*, Wednesday, May 16, 1973, in an article by Stan Davies, *Star* staff writer, entitled "Bombardier agrees to buy U.S. company."

17. *The Globe And Mail*, Wednesday, May 16, 1973 in an article by Glenn Somerville entitled "Bombardier to buy U.S. engine company."

18. In a letter to E.C. Kiekhaefer from Wayne J. Roper of the legal firm of Gibbs, Roper & Fifield in Milwaukee, Wisconsin of July 7, 1973.

19. In a draft of a letter to Laurent Beaudoin of Bombardier, Ltd from E.C. Kiekhaefer of August 1, 1973.

20. In a letter to Laurent Beaudoin, president of Bombardier Limited from E.C. Kiekhaefer of August 16, 1973.

21. In a letter from Laurent Beaudoin, President - Chief Executive Officer of Bombardier, Limited, to E.C. Kiekhaefer of September 21, 1973.

22. In a letter to Laurent Beaudoin, president of Bombardier Limited from E.C. Kiekhaefer of December 21, 1973.

23. In a letter to John Hall of Bede Aircraft Co. from Donald V. Thompson of September 16, 1971, which included copies of the Bede proposed art work.

24. In a memo from Don Thompson to E.C. Kiekhaefer of November 6, 1971 entitled, "Bede Aircraft Order."

25. In a letter from James R. Bede of Bede Aircraft, Inc. to Don Thompson of January 18, 1972.

26. Specified in Purchase Order # 5-6259 from Bede Aircraft, Inc. of February 16, 1971.

27. In a letter to Jim Bede of Bede Aircraft, Inc. from E.C. Kiekhaefer of March 3, 1971.

28. In a transcribed telephone conversation between E.C. Kiekhaefer and Ray Atchison and Mr. Suderman of Midland National Bank, Newton Kansas on April 18, 1972.

29. Excerpts from a transcribed telephone conversation between E.C. Kiekhaefer and Jim Bede of Bede Aircraft, Inc. on April 18, 1972.

30. In a letter from Dr. Alexander M. Lippisch of Cedar Rapids, Iowa to E.C. Kiekhaefer of May 5, 1972.

31. In a letter to James R. Bede of Bede Aircraft, Inc., Newton, Kansas from E.C. Kiekhaefer of June 14, 1972.

32. In a letter from James. R. Bede of Bede Aircraft, Inc., Newton, Kansas to E.C. Kiekhaefer of June 30, 1972.

33. *Air Progress*, September, 1976, in an article by Editor Keith Connes entitled, "Bede Explains Everything. Now follow this closely. ..."

34. *Popular Mechanics*, December, 1989 in an article by Fred Mackerodt entitled, "BD-5: The First Do-It-Yourself Jet", a side-bar to the principle article by William Garvey entitled, "Homebuilt Machbuster," the cover story for the issue.

35. From the 1971 *BD-5* Information Kit from Bede Aircraft, Inc. The kit also included detailed specifications and additional information on the Kiekhaefer Aeromarine engines which were to power the aircraft.

36. In a letter from Ralph P. Malloof of Calabasas Park, California to E.C. Kiekhaefer of November 4, 1974.

37. In a letter to G.E. Tuttle from E.C. Kiekhaefer of July 6, 1974.

38. In a memo from LeRoy Kirchstein to E.C. Kiekhaefer of December 2, 1976.

39. In a memo to LeRoy Kirchstein from E.C. Kiekhaefer of June 29, 1977.

40. In a memo from LeRoy E. Kirchstein, KAM Controller, to E.C. Kiekhaefer of June 29, 1977.

41. In a letter to King P. Kirchner, president of Unit Drilling Co. of Tulsa Oklahoma from E.C. Kiekhaefer of February 3, 1977.

42. In a speech delivered at the press conference to announce the Kiekhaefer Aeromarine Motors *K-Drive* at the Miami International Boat Show in February, 1974.

43. *New York Times*, Sunday, March 10, 1974 in an article by William N. Wallace, entitled "Kiekhaefer Is Weighing Next Step on Two New Products."

44. In the "Corporate Resume" section of the 1975 prospectus prepared by Carl and

Fred Kiekhaefer to assist in selling or licensing the *K-Drive* stern drive.

45. Interview of Fred Kiekhaefer of August 16, 1987 at his home in Fond du Lac, Wisconsin.

46. In a letter to Charles B. Stern of Ventnor, New Jersey from E.C. Kiekhaefer of June 6, 1973.

CHAPTER THIRTY-SIX
Return To Glory

1. 1 Corinthians 9:24. Quoted in a letter from an unnamed fan to Carl Kiekhaefer in 1971.

2. *Go Boating*, November 1970 in an article without byline entitled "Magoon Wins Key West Race."

3. In a letter from Robert C. Magoon, M.D. of Miami Beach, Florida to E.C. Kiekhaefer of January 12, 1971.

4. In a letter to Thomas Crimmins of Schieffelin & Company, New York, from E.C. Kiekhaefer of October 4, 1971.

5. In a letter to William Wishnick of Witco Chemical, New York, from E.C. Kiekhaefer of December 7, 1971. Wishnick was at the time, Offshore Racing Commission Chairman of the American Power Boat Association.

6. In a letter to Sherman F. Crise of the Bahamas Power Boat Association in Fort Lauderdale, Florida from E.C. Kiekhaefer of May 25, 1971.

7. In a memo to all "Executives and Telephone Operators" from E.C. Kiekhaefer of July 12, 1971.

8. *Saga*, October 1972, in an article by Charles Gnaegy entitled, "Deep Sea Powerboat Racing."

9. In a letter never mailed to Mel Crook, editor of *Yachting* magazine in New York, from E.C. Kiekhaefer of March 17, 1971.

10. In a letter to Henry Ford II from E.C. Kiekhaefer of June 22, 1971.

11. In a letter from R.G. Harris, Assistant General Counsel to Henry Ford II and the Ford Motor Corporation, to E.C. Kiekhaefer of July 12, 1971.

12. In a memo to "All Concerned" from K.B. Abernathy of November 23, 1971.

13. In a letter to K.B. Abernathy, president of Mercury Marine, from E.C. Kiekhaefer of March 6, 1972.

14. In a transcribed telephone conversation between Gary Garbrecht of Mercury Marine and E.C. Kiekhaefer of December 7, 1972.

15. In a letter to Bob Rankin, Outdoors Editor of the *Cincinnati Enquirer* from E.C. Kiekhaefer of October 31, 1972.

16. In a letter to Jack K. Burke of Key West, Florida from E.C. Kiekhaefer of October 31, 1972.

17. *Miami Pictorial*, November 1972, in an article by Tom Pratt entitled, "The Dawning of Another Hey-Day!"

18. In a letter to George Couzens of Bertram Yacht Company in Miami, Florida from E.C. Kiekhaefer of November 4, 1972.

19. In a letter to William Wishnick of Witco Chemical Corporation in New York, from E.C. Kiekhaefer of August 28, 1972. Wishnick was at the time, Offshore Racing Commission Chairman of the American Power Boat Association.

20. In a letter to Gerald Taines, T.F.C. Marine Engineering, from E.C. Kiekhaefer of December 18, 1972.

21. In a letter to Albert R. Said of Cincinnati, Ohio from E.C. Kiekhaefer of October 31, 1972.

22. In a letter to Dr. Carlo C. Bonomi, Subalpina Investimenti S.p.A., Milano, Italy, from E.C. Kiekhaefer of September 12, 1972.

23. In a letter to William Wishnick of Witco Chemical Corporation in New York, from E.C. Kiekhaefer of December 13, 1972. Wishnick was at the time, Offshore Racing Commission Chairman of the American Power Boat Association.

24. In a statement given by Jack Carlisle Stuteville on Carl Kiekhaefer's behalf during the discovery process of the Kiekhaefer Aeromarine Motors vs. Vincenzo Balestrieri lawsuit on September 11, 1973.

25. Contained in a proposed press release drafted by Kiekhaefer Aeromarine Motors, Inc. on October 9, 1973.

26. In a letter to Wayne J. Roper, of the legal firm of Gibbs, Roper & Fifield, Milwaukee, Wisconsin, from E.C. Kiekhaefer of November 19, 1973.

27. Excerpts of a transcribed telephone conversation between E.C. Kiekhaefer and Vincenzo Balestrieri on November 22, 1973.

28. Ibid.

29. Interview of Richie Powers at the author's home in Ft. Lauderdale, Florida on February 21, 1987.

30. Ibid. 31. Ibid. 32. Ibid. 33. Ibid.

34. Ibid. 35. Ibid. 36. Ibid.

37. In a letter to Dr. Carlo C. Bonomi, Subalpina Investimenti S.p.A., Milano, Italy from E.C. Kiekhaefer of April 7, 1973.

38. Interview of Fred Kiekhaefer at his home in Fond du Lac, Wisconsin, on August 16, 1987.

39. Ibid.

40. Interview of Richie Powers, loc. cit.

41. Ibid.

42. *Miami Herald*, Sunday, February 17, 1974, in an article by Jim Martenhoff entitled, "Kiekhaefer to Receive Hall of Fame Award."

43. Crouse, John, *Searace - A History of Offshore Powerboat Racing*, Crouse Publications, LTD., 20740 S.W. 248 Street, Homestead Florida 33031. (305) 247-7547. Copies of Crouse's exhaustive chronology (673 pp) of offshore racing are available as of this writing at $69.95.

44. Interview of Richie Powers, loc. cit.

45. Interview of Fred Kiekhaefer, loc. cit.

46. Ibid.

47. In a letter from Ralph Evinrude, Chairman of Outboard Marine Corporation, to E.C. Kiekhaefer of May 16, 1977.

48. Excerpts of a transcribed telephone conversation between Charles D. Strang, president and chief executive officer of Outboard Marine Corporation and E.C. Kiekhaefer of August 7, 1974.

49. In a letter to James H. Jost, then Public Relations Manager of Evinrude Motors, from E.C. Kiekhaefer of January 27, 1976.

50. *Dock Lines*, March 19, 1976, the newsletter of Evinrude Motors, Milwaukee, Wisconsin.

51. From the speech by Charles D. Strang delivered at the Ole Evinrude Awards ceremony, February 22, 1976 at Key Biscayne, Florida.

52. In the letter to James H. Jost, loc. cit.

53. In a letter to R.W. Page of Fond du Lac, Wisconsin from E.C. Kiekhaefer of March 2, 1976.

54. In a letter to Ralph Evinrude from E.C. Kiekhaefer of March 9, 1976.

55. In a letter to James H. Jost, Public Relations Manager of Evinrude Motors, from E.C. Kiekhaefer of October 18, 1977.

56. Ibid.

CHAPTER THIRTY-SEVEN
Changes of Heart

1. In a draft of a letter to Robert Kell of Sportsman Publications in Farmington Hills, Michigan from E.C. Kiekhaefer of April 6, 1977.

2. In a letter to Mrs. Joyce Schoenberger at Mercury Marine from E.C. Kiekhaefer of November 28, 1978.

3. Interview of Rose Smiljanic of August 14, 1986 at her home in Fond du Lac, Wisconsin.

4. In the letter to Robert Kell, loc. cit.

5. In a letter to Frank Scalpone of the National Association of Engine and Boat Manufacturers in New York City from E.C. Kiekhaefer of June 10, 1977.

6. In the letter to Robert Kell, loc. cit.

7. Interview of Ruth Olive Backus of July 29, 1987 in her home in Fond du Lac, Wisconsin.

8. Interview of General Mabry Edwards, ret. U.S.A.F. of February 28, 1987.

9. In a memorandum of his medical progress entitled, "Subject: The comedy-tragedy of E.C.K.'s medical problems," addressed to "To Whom It May Concern," by E.C. Kiekhaefer of January 26, 1977.

10. Ibid.

11. In a letter to King P. Kirchner, president of Unit Drilling Co. of Tulsa Oklahoma from E.C. Kiekhaefer of March 29, 1977.

12. Interview of Rose Smiljanic, loc. cit.

13. Ibid.

14. In a letter to Lloyl [sic] McLeod of Fond du Lac, Wisconsin from E.C. Kiekhaefer of April 6, 1977.

15. Interview of Fred Kiekhaefer at his home in Fond du Lac, Wisconsin on August 16, 1987.

16. Ibid.

17. Interview of Rose Smiljanic, loc. cit.

18. Interview of Jack F. Reichert of March 26, 1987 at his condominium in West Palm Beach, Florida.

19. Interview of Charles F. Alexander of December 1, 1986.

20. Interview of Jack F. Reichert, loc. cit.

21. Ibid. 22. Ibid.

23. In a letter to Ted Collins of St. Cloud, Florida from E.C. Kiekhaefer of March 18, 1976.

24. In a letter to Dave Craig of Miami, Florida from E.C. Kiekhaefer of March 28, 1977.

25. In a letter to Jack F. Reichert, president of Mercury Marine from E.C. Kiekhaefer of March 28, 1977.

26. Interview of Jack Reichert, loc. cit.

27. In a letter to William Martin-Hurst of Wales, U.K., from E.C. Kiekhaefer of February 16, 1978.

28. Ibid.

29. In a letter to Jack F. Reichert during his last few days as president of Mercury Marine before becoming president of Brunswick Corporation from E.C. Kiekhaefer of May 2, 1977.

30. Interview of Jack F. Reichert, loc. cit.

31. In a letter to R. T. Johnson at Mercury Marine in Oshkosh, Wisconsin from E.C. Kiekhaefer of November 5, 1977.

32. In a letter to Ralph P. Maloof of Calabassas, California from E.C. Kiekhaefer of April 8, 1977.

33. In a letter to Odell and Mona Lou Lewis from E.C. Kiekhaefer of May 9, 1977.

34. Interview of Rose Smiljanic, loc. cit.

35. In a letter to Mrs. Alexander M. Lippisch of April 8, 1977.

36. In the letter to William Martin-Hurst, loc. cit.

37. In a letter to Dr. Giuseppe Pasini of Milan, Italy from E.C. Kiekhaefer of April 18, 1977.

38. In the letter to Odell and Mona Lou Lewis, loc. cit.

39. In a letter to Giles Koelsche, M.D. from E.C. Kiekhaefer of December 12, 1977.

40. Ibid.

41. Interview of Fred Kiekhaefer, loc. cit.

42. Ibid.

43. Interview of Ruth O. Backus, loc. cit.

44. Interview of Rose Smiljanic, loc. cit.

45. In a letter to Bill Backus of Mahwah, New Jersey from E.C. Kiekhaefer of February 1, 1977.

46. Interview of Richie Powers at the author's home in Ft. Lauderdale on February 21, 1989.

47. Ibid.

48. Interview of Fred Kiekhaefer, loc. cit.

49. Ibid. 50. Ibid.

51. In a letter to Tom Gentry, chairman of Gentry Pacific, Ltd in Honolulu, Hawaii from E.C. Kiekhaefer of October 11, 1978.

52. Ibid.

53. Excerpts from a handwritten letter from Anita Rae Kiekhaefer, signed for Helen, Fred, Mom & Anita, to E.C. Kiekhaefer of October 29, 1978.

54. Interview of Fred Kiekhaefer, loc. cit.

55. In a transcribed telephone conversation between Richie Powers and E.C. Kiekhaefer on March 21, 1980.

CHAPTER THIRTY-EIGHT
Twilight

1. In a letter to Dr. Giles A. Koelsche at the Mayo Clinic of Rochester, Minnesota, from E.C. Kiekhaefer of June 27, 1973.

2. In a letter to Dr. Giles A. Koelsche, a retired physician of the Mayo Clinic, living in Sun City, Arizona from E.C. Kiekhaefer of March 6, 1980.

3. In a letter to Dr. Donald H. Schmidt, Head, Cardiovascular Disease Section, Mount Sinai Medical Center in Milwaukee, Wisconsin from E.C. Kiekhaefer of January 31, 1980.

4. In a letter to Dean Snider, director, Florida Keys Marine Institute from E.C. Kiekhaefer of January 4, 1980. The Charles F. Chapman Memorial Award came with a $1,000 honorarium, which Carl donated to the Florida Keys Marine Institute. In donating the funds, in the same above referenced letter Carl said, "I hope this contribution to the program will help the teen-agers at your rehabilitation and vocational training center find satisfying jobs in the boating industry."

5. The chronological deterioration of Carl's condition was outlined and summarized by Carl in a document he prepared in May of 1983 for the benefit of physicians reviewing his condition, entitled "E.C. Kiekhaefer Medical History."

6. Interview of Rose Smiljanic of August 14, 1986 at her home in Fond du Lac, Wisconsin.

7. Ibid.

8. In a letter to Dr. Donald H. Schmidt, Head, Cardiovascular Disease Section, Mount Sinai Medical Center, Milwaukee,

Wisconsin from E.C. Kiekhaefer of February 23, 1981.

9. In a letter to Dr. Dean A. Emanuel of the Marshfield Clinic in Marshfield, Wisconsin from E.C. Kiekhaefer of February 25, 1981.

10. From the chronological record of Carl's condition, "E.C. Kiekhaefer Medical History," loc. cit.

11. In a letter to Dr. Giles A. Koelsche of Sun City, Arizona from E.C. Kiekhaefer of June 3, 1982.

12. In a letter to Jim and Irene Jost of Wauwatosa, Wisconsin from E.C. Kiekhaefer of June 23, 1981.

13. Ibid. 14. Ibid.

15. *New York Times*, Friday, September 4, 1981, in an article by James Barron entitled, "20 Tons of Marijuana and 33 Seized in L.I. Raids."

16. Interview of Richie Powers at the author's home in Ft. Lauderdale, Florida, on February 21, 1987.

17. *Wall Street Journal*, March 3, 1982 in an article by L. Erik Calonius entitled, "'Hot, Lean, Sleek, Fast, Sexy, Ready'--This Is a Cigarette?"

18. *Miami Herald*, December 29, 1982 in an article entitled "Lauderdale speedboat racer pleads guilty to tax evasion," concerning the conviction of Edward "Fast Eddie" Trotta for tax evasion.

19. Excerpts of a four-page letter from Richie Powers to E.C. Kiekhaefer of June 3, 1982.

20. In a letter written to "To Whom It May Concern," by E.C. Kiekhaefer of April 27, 1982.

21. In a letter to Dr. Carol Stafford, the future Mrs. Fred Kiekhaefer, from E.C. Kiekhaefer of May 4, 1982.

22. Ibid.

23. In a letter to Dr. Giles A. Koelsche of Sun City, Arizona from E.C. Kiekhaefer of June 3, 1982.

24. Ibid.

25. In a letter to Dr. J. Paul Marcoux of the Mayo Clinic in Rochester, Minnesota from E.C. Kiekhaefer of September 17, 1982.

26. *The Marian*, May 1982, the newsletter of Marian College in Fond du Lac, Wisconsin.

27. In a letter to Peter E. Stone, president, National Exchange Bank and Trust of Fond du Lac, Wisconsin from E.C. Kiekhaefer of May 17, 1982.

28. *Milwaukee Journal*, Tuesday, July 20, 1982.

29. In a news clipping entitled, "Mystery Chemical Found In Schizophrenic's Blood," not identified as to source, on which Carl had made notations and comments.

30. Quoted in the speech delivered by Charles D. Strang at the *Kiekhaefer - St. Augustine Classic* offshore race testimonial dinner held for Carl Kiekhaefer in St. Augustine, Florida on September 11, 1982.

31. Ibid.

32. The chronological record of Carl's condition, "E.C. Kiekhaefer Medical History," loc. cit.

33. Ibid. 34. Ibid.

35. In a letter to Helen Kiekhaefer Wimberly from Rose Smiljanic of March 18, 1983.

36. In a letter to E.C. Kiekhaefer from Rose Smiljanic of August 27, 1974.

37. In a hand written note to Rose Smiljanic from E.C. Kiekhaefer of August 27, 1974.

38. Ibid.

39. In a letter from H. Clark Hoagland, M.D. of the Mayo Clinic in Rochester, Minnesota on behalf of E.C. Kiekhaefer of May 10, 1983.

40. Interview of Rose Smiljanic, loc. cit.

41. A letter addressed to "To Whom It May Concern," signed by E.C. Kiekhaefer on June 21, 1983.

42. In a letter to Richard J. Jordan, president of Mercury Marine in Fond du Lac, Wisconsin from E.C. Kiekhaefer of July 12, 1983.

43. In a prior draft of Carl's July 12, 1983 letter to Richard J. Jordan, president of Mercury Marine, dated July 7, 1983 from E.C. Kiekhaefer.

44. Interview of Rose Smiljanic, loc. cit.

45. Ibid.

46. Interview of Helen Kiekhaefer Wimberly of September 26, 1989.

47. Ibid.

48. In a letter to Charles D. Strang, president, Outboard Marine Corporation from E.C. Kiekhaefer of October 18, 1977.

49. Interview of Ruth Olive Backus of July 29, 1987 at her home in Fond du Lac, Wisconsin.

50. Interview of James R. Wynne of October 17, 1987 at his offices at Wynne Marine in Miami, Florida.

51. Interview of Rose Smiljanic, loc. cit.

52. Interview of Jack F. Reichert of March 26, 1987 at his condominium in West Palm Beach, Florida.

53. Interview of James H. Jost with Charles D. Strang on May 20, 1987 at the OMC salt water proving grounds in Stuart, Florida.

54. Interview of Rose Smiljanic, loc. cit.

55. Ibid.

56. In a letter to Stan Gores, editor, Fond du Lac *Commonwealth Reporter* from Freda Kiekhaefer of October 24, 1983.

57. Interview of Rose Smiljanic, loc. cit.

58. In a letter to Dr. Giles A. Koelsche at the Mayo Clinic in Rochester, Minnesota from E.C. Kiekhaefer of June 27, 1973.

EPILOGUE

1. Interview of Fred Kiekhaefer of August 16, 1987 at his home in Fond du Lac, Wisconsin.

2. From the 1964 distributor, branch manager and executives speech by Carl Kiekhaefer on the 25th anniversary of the Kiekhaefer Corporation in 1964.

3. Ibid.

4. Contained in the January 7, 1988 press release distributed by the National Marine Manufacturers Association, NMMA.

5. Contained in the 1989 National Marine Manufacturers Association Hall of Fame Awards booklet honoring inductees.

6. From the press release issued in 1990 by Brunswick Marine Hi-Performance, an autonomous business unit of Brunswick Marine Power.

7. Interview of Irwin Jacobs, Chairman of the Board of Genmar Corporation of September 29, 1987 at his offices in Minneapolis, Minnesota.

8. *Profile*, Summer 1990, the Brunswick employee news magazine.

INDEX